P9-AFO-731

Multiresolution Signal Decomposition

Transforms, Subbands, and Wavelets

Second Edition

Series in Telecommunications

Series Editor
T. Russell Hsing

Bell Communications Research
Morristown, NJ

Multiresolution Signal Decomposition: Transforms, Subbands, and Wavelets

Ali N. Akansu and Richard A. Haddad

New Jersey Institute of Technology
Newark, NJ

Other Books in the Series

Hseuh-Ming Hang and John W. Woods, ***Handbook of Visual Communications***: 1995

John J. Metzner, ***Reliable Data Communications***: 1997

Tsong-Ho Wu and Noriaki Yoshikai, ***ATM Transport and Network Integrity***: 1997

Shuo-Yen and Robert Li, ***Algebraic Switching Theory and Broadband Applications***: 1999

Winston I. Way, ***Broadband Hybrid Fiber Coax Access System Technologies***: 1999

Multiresolution Signal Decomposition

Transforms, Subbands, and Wavelets

Second Edition

Ali N. Akansu

and

Richard A. Haddad

New Jersey Institute of Technology
Newark, NJ

San Diego San Francisco New York Boston
London Sydney Tokyo

This book is printed on acid-free paper. ∞

ACADEMIC PRESS
A Harcourt Science and Technology Company
525 B Street, Suite 1900, San Diego, CA 92101-4495 USA
http://www.academicpress.com

Academic Press
Harcourt Place, 32 Jamestown Road, London NW1 7BY UK

Library of Congress Catalog Number: 99-68565

International Standard Book Number: 0-12-047141-8

Printed in the United States of America
00 01 02 03 04 EB 9 8 7 6 5 4 3 2 1

To Bilge and Elizabeth

Contents

Preface

Since the first edition of this book in 1992 we have witnessed a flood of books—texts, monographs, and edited volumes describing different aspects of block transforms, multirate filter banks, and wavelets. Some of these have been mathematically precise, designed for the rigorous theoretician, while others sought to interpret work in this arena for engineers and students.

The field is now mature, yet active. The theory is much better understood in the signal processing community, and applications of the multiresolution concept to situations in digital multimedia, communications, and others abound. In the first edition and in the early days of multirate filter banks a prime emphasis was on signal compaction and coding. Today, multiresolution decomposition and time-frequency concepts have opened up new vistas for further development and application. These ideas concerning orthogonal signal analysis and synthesis have led to applications in digital audio broadcasting, digital data hiding and watermarking, wireless and wireline communications, audio and video coding, and many others.

In this edition, we continue to treat block transforms, subband filter banks, and wavelets from a common unifying standpoint. We demonstrate the commonality among these signal analysis and synthesis techniques by showing how the block transform evolves gracefully into the more general multirate subband filter bank, and then by establishing the multiresolution decomposition features common to both the dyadic subband tree structure and the orthonormal wavelet transform. In order to achieve this unification, we have focused mainly on *orthonormal* decompositions and presented a unified and integrated treatment of multiresolution signal decomposition techniques using the property of orthonormality as the unifying theme.(A few exceptions, such as the oversampled Laplacian pyramid and biorthogonal filter banks are also presented because they provide an historical perspective and serve as foils to the critically sampled, orthonormal subband structures we emphasize.)

Our second focus in the first edition was the application of decomposition techniques to signal *compression* and coding. Accordingly, we describe *objective* performance criteria that measure this attribute and then compare the different techniques on this basis. We acknowledge that subjective evaluations of decomposition are important in applications such as image and video processing and coding, machine vision, and pattern recognition. Such aspects are treated adequately in the literature cited and are deemed beyond the scope of this book. A new focus in this edition is the time-frequency properties of signals and decomposition techniques. Accordingly, this text provides tables listing the coefficients of popular block transforms, subband and wavelet filters, and also their time-frequency features and compaction performance for both theoretical signal models and standard test images. In this respect, we have tried to make the book a reference text as well as a didactic monograph.

Our approach is to build from the fundamentals, taking simple representative cases first and then extending these to the next level of generalization. For example, we start with block transforms, extend these to lapped orthogonal transforms, and then show both to be special cases of subband filter structures. We have avoided the theorem-proof approach, preferring to give explanation and derivations emphasizing clarity of concept rather than strict rigor.

Chapter 2 on orthogonal transforms introduces block transforms from a least-squares expansion in orthogonal functions. Signal models and decorrelation and compaction performance measures are then used to evaluate and compare several proposed block and lapped transforms. The biorthogonal signal decomposition is mentioned.

Chapter 3 presents the theory of perfect reconstruction, orthonormal two-band and M-band filter banks with emphasis on the finite impulse response variety. A key contribution here is the time-domain representation of an arbitrary multirate filter bank, from which a variety of special cases emerge—paraunitary, biorthogonal, lattice, LOT, and modulated filter banks. The two-channel, dyadic tree structure then provides a multiresolution link with both the historical Laplacian pyramid and the orthonormal wavelets of Chapter 6. A new feature is the representation of the transmultiplexer as the synthesis/analysis dual of the analysis/synthesis multirate filter bank configuration.

Chapter 4 deals with specific filter banks and evaluates their objective performance. This chapter relates the theory of signal decomposition techniques presented in the text with the applications. It provides a unified performance evaluation of block transforms, subband decomposition, and wavelet filters from a signal processing and coding point of view. The topic of optimal filter banks presented in this chapter deals with solutions based on practical considerations in image coding. The chapter closes with the modeling and optimum design of quantized filter banks.

Chapter 5 on time-frequency (T-F) focuses on joint time-frequency properties of signals and the localization features of decomposition tools. There is a discussion of techniques for synthesizing signals and block transforms with desirable T-F properties and describes applications to compaction and interference excision in spread spectrum communications.

Chapter 6 presents the basic theory of the orthonormal and biorthogonal wavelet transforms and demonstrates their connection to the orthonormal dyadic subband tree of Chapter 3. Again, our interest is in the linkage to the multiresolution subband tree structure, rather than with specific applications of wavelet transforms.

Chapter 7 is a review of recent applications of these techniques to image coding, and to communications applications such as discrete multitone (DMT) modulation, and orthogonal spread spectrum user codes. This chapter links the riches of linear orthogonal transform theory to the popular and emerging transform applications. It is expected that this linkage might spark ideas for new applications that benefit from these signal processing tools in the future.

This book is intended for graduate students and R&D practitioners who have a working knowledge of linear system theory and Fourier analysis, some linear algebra, random signals and processes, and an introductory course in digital signal processing. A set of problems is included for instructional purposes.

For classroom presentation, an instructor may present the material in the text in three packets:

(1) Chapters 2 and 5 on block transforms and time-frequency methods
(2) Chapters 3 and 4 on theory and design of multirate filter banks
(3) Chapters 6 and 7 on wavelets and transform applications

As expected, a book of this kind would be impossible without the cooperation of colleagues in the field. The paper preprints, reports, and private communications they provided helped to improve significantly the quality and timeliness of the book. We acknowledge the generous help of N. Sezgin and A. Bircan for some figures. Dr. T. Russell Hsing of Bellcore was instrumental in introducing us to Academic Press. It has been a pleasure to work with Dr. Zvi Ruder during this project. Dr. Eric Viscito was very kind to review Chapter 3. The comments and suggestions of our former and current graduate students helped to improve the quality of this book. In particular, we enjoyed the stimulating discussions and interactions with H. Caglar, A. Benyassine, M. Tazebay, X. Lin, N. Uzun, K. Park, K. Kwak, and J.C. Horng. We thank them all. Lastly, we appreciate and thank our families for their understanding, support, and extraordinary patience during the preparation of this book.

Ali N. Akansu
Richard A. Haddad
April 2000

Chapter 1

Introduction

1.1 Introduction

In the first edition of this book, published in 1992, we stated our goals as threefold:
(1) To present orthonormal signal decomposition techniques—transforms, subbands, and wavelets—from a unified framework and point of view.
(2) To develop the interrelationships among decomposition methods in both time and frequency domains and to define common features.
(3) To evaluate and critique proposed decomposition strategies from a compression coding standpoint using measures appropriate to image processing.
The emphasis then was signal coding in an analysis/synthesis structure or codec. As the field matured and new insights were gained, we expanded our vistas to communications systems and other applications where objectives other than compression are vital — as for example, interference excision in CDMA spread spectrum systems. We can also represent certain communications systems such as TDMA, FDMA, and CDMA as synthesis/analysis structures, i.e., the conceptual dual of the compression codec. This duality enables one to view all these systems from one unified framework.

The Fourier transform and its extensions have historically been the prime vehicle for signal analysis and representation. Since the early 1970s, block transforms with real basis functions, particularly the discrete cosine transform (DCT), have been studied extensively for transform coding applications. The availability of simple fast transform algorithms and good signal coding performance made the DCT the standard signal decomposition technique, particularly for image and video. The international standard image-video coding algorithms, i.e., CCITT

H.261, JPEG, and MPEG, all employ DCT-based transform coding.

Since the recent research activities in signal decomposition are basically driven by visual signal processing and coding applications, the properties of the human visual system (HVS) are examined and incorporated in the signal decomposition step. It has been reported that the HVS inherently performs multiresolution signal processing. This finding triggered significant interest in multiresolution signal decomposition and its mathematical foundations in multirate signal processing theory. The multiresolution signal analysis concept also fits a wide spectrum of visual signal processing and visual communications applications. Lower, i.e., coarser, resolution versions of an image frame or video sequence are often sufficient in many instances. Progressive improvement of the signal quality in visual applications, from coarse to finer resolution, has many uses in computer vision, visual communications, and related fields.

The recognition that multiresolution signal decomposition is a by-product of multirate subband filter banks generated significant interest in the design of better-performing filter banks for visual signal processing applications.

The wavelet transform with a capability for variable time-frequency resolution has been promoted as an elegant multiresolution signal processing tool. It was shown that this decomposition technique is strongly linked to subband decomposition. This linkage stimulated additional interest in subband filter banks, since they serve as the only vehicle for fast orthonormal wavelet transform algorithms and wavelet transform basis design.

1.2 Why Signal Decomposition?

The uneven distribution of signal energy in the frequency domain has made signal decomposition an important practical problem. Rate-distortion theory shows that the uneven spectral nature of real-world signals can provide the basis for source compression techniques. The basic concept here is to divide the signal spectrum into its subspectra or subbands, and then to treat those subspectra individually for the purpose at hand. From a signal coding standpoint, it can be appreciated that subspectra with more energy content deserve higher priority or weight for further processing. For example, a slowly varying signal will have predominantly low-frequency components. Therefore, the low-pass subbands contain most of its total energy. If one discards the high-pass analysis subbands and reconstructs the signal, it is expected that very little or negligible reconstruction error occurs after this analysis-synthesis operation.

The decomposition of the signal spectrum into subbands provides the mathematical basis for two important and desirable features in signal analysis and processing. First, the monitoring of signal energy components within the subbands or subspectra is possible. The subband signals can then be ranked and processed independently. A common use of this feature is in the spectral shaping of quantization noise in signal coding applications. By bit allocation we can allow different levels of quantization error in different subbands. Second, the subband decomposition of the signal spectrum leads naturally to multiresolution signal decomposition via multirate signal processing in accordance with the Nyquist sampling theorem.

Apart from coding/compression considerations, signal decomposition into subbands permits us to investigate the subbands for contraband signals, such as band-limited or single tone interference. We have also learned to think more globally to the point of signal decomposition in a composite time-frequency domain, rather than in frequency subbands as such. This expansive way of thinking leads naturally to the concept of wavelet packets (subband trees), and to the block transform packets introduced in this text.

1.3 Decompositions: Transforms, Subbands, and Wavelets

The signal decomposition (and reconstruction) techniques developed in this book have three salient characteristics:
(1) Orthonormality. As we shall see, the block transforms will be square unitary matrices, i.e., the rows of the transformation matrix will be orthogonal to each other; the subband filter banks will be *paraunitary*, a special kind of orthonormality, and the wavelets will be orthonormal.
(2) Perfect reconstruction (PR). This means that, in the absence of encoding, quantization, and transmission errors, the reconstructed signal can be reassembled perfectly at the receiver.
(3) Critical sampling. This implies that the signal is subsampled at a minimum possible rate consistent with the applicable Nyquist theorem. From a practical standpoint, this means that if the original signal has a data rate of f_s samples or pixels per second, the sum of the transmission rates out of all the subbands is also f_s.

The aforementioned are the prime ingredients of the decomposition techniques. However, we also briefly present a few other decomposition methods for contrast or historical perspective. The oversampled Laplacian pyramid, biorthogonal filter banks, and non-PR filter banks are examples of these, which we introduce for didactic value.

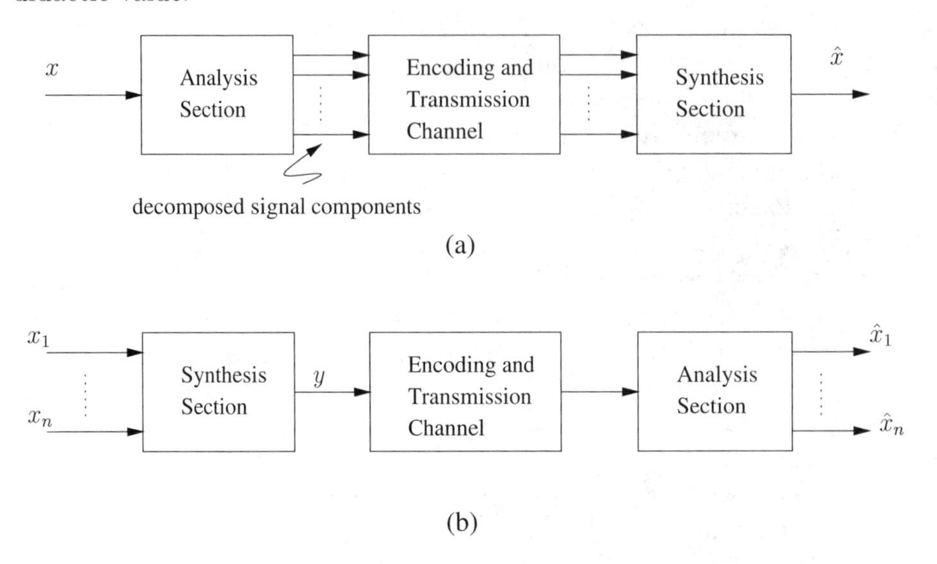

Figure 1.1: (a) Analysis-synthesis structure; (b) synthesis/analysis system.

As shown in Fig. 1.1(a), the input signal x is decomposed in the analysis section, encoded, and transmitted. At the receiver or synthesis section, it is reconstructed as $\hat{x}$. In a perfect reconstruction system $\hat{x} = x$ within an allowable delay. In a critically sampled system, the sum of the data rates of the decomposed signal components equals that of the input signal.

In Fig. 1.1(b), the dual operation is shown. Typically, the synthesis section could be a TDMA or FDMA multiplexer wherein several signals are separated in time (TDMA), frequency (FDMA), or in time-frequency (CDMA), and combined into one signal for transmission. The received signal is then separated into components in the analysis section.

1.3.1 Block Transforms and Filter Banks

In block transform notation, the analysis or decomposition operation suggested in Fig. 1.1 is done with a blockwise treatment of the signal. The input signal is first segmented into nonoverlapping blocks of samples. These signal blocks or vectors are transformed into spectral coefficient vectors by the orthogonal matrix. The spectral unevenness of the signal is manifested by unequal coefficient energies by this technique and only transform coefficients with significant energies need be considered for further processing. Block transforms, particularly the discrete cosine transforms, have been used in image-video coding. Chapter 2 introduces and discusses block transforms in detail and provides objective performance evaluations of known block transforms. The Karhunen-Loeve transform, or KLT, is the unique input-signal dependent optimal block transform. We derive its properties and use it as a standard against which all other fixed transforms can be compared.

In block transforms, the duration or length of the basis functions is equal to the size of the data block. This implies that the transform and inverse transform matrices are square. This structure has the least possible freedom in tuning its basis functions. It can meet only an orthonormality requirement and, for the optimal KLT, generate uncorrelated spectral coefficients. Limited joint time-frequency localization of basis functions is possible using the concept of block transform packets (Chapter 5).

More freedom for tuning the basis functions is possible if we extend the duration of these functions. Now this rectangular transform or decomposition has overlapping basis functions. This overlapping eliminates the "blockiness" problem inherent in block transforms. Doubling the length of the basis sequences gives the lapped orthogonal transform, or LOT, as discussed in Section 2.5.

In general, if we allow arbitrary durations for the basis sequence filters, the finite impulse response (FIR) filter bank or subband concept is reached. Therefore, block transforms and LOTs can be regarded as special filter banks. The multirate signal processing theory and its use in perfect reconstruction analysis-synthesis filter banks are discussed in depth in Chapter 3. This provides the common frame through which block transforms, LOTs, and filter banks can be viewed.

Figure 1.2 shows a hierarchical conceptual framework for viewing these ideas. At the lowest level, the block transform is a bank of M filters whose impulse responses are of length $L = M$. At the next level, the LOT is a bank of M filters, each with impulse responses (or basis sequences) of length $L = 2M$. At the top of the structure is the M-band multirate filter bank with impulse responses of any length $L \geq M$. On top of that is the M-band multirate filter bank with impulse responses of arbitrary length $L \geq M$. This subband structure is illustrated in Fig. 1.3(a), where the signal is decomposed into M equal bands by the filter bank.

The filter bank often used here has frequency responses covering the M-bands from 0 to $f_s/2$. When these frequency responses are translated versions of a low-frequency prototype, the bank is called a modulated filter bank.

The Nyquist theorem in a multiband system can now be invoked to subsample each band. The system is critically subsampled (or maximally decimated) when the decimation factor D or subsampling parameter equals the number of subbands M. When $D < M$, the system is oversampled.

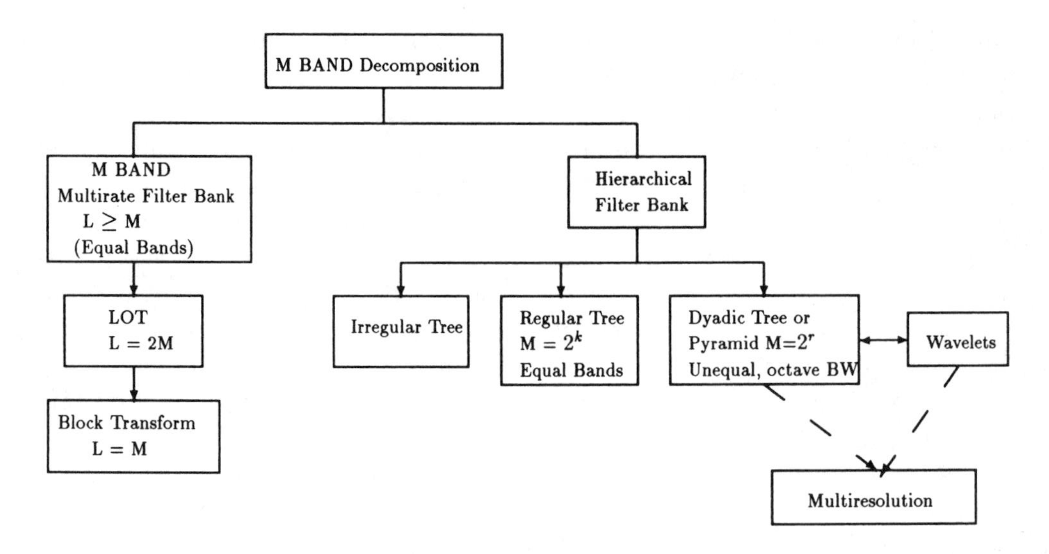

Figure 1.2: An overview of M-band signal decomposition.

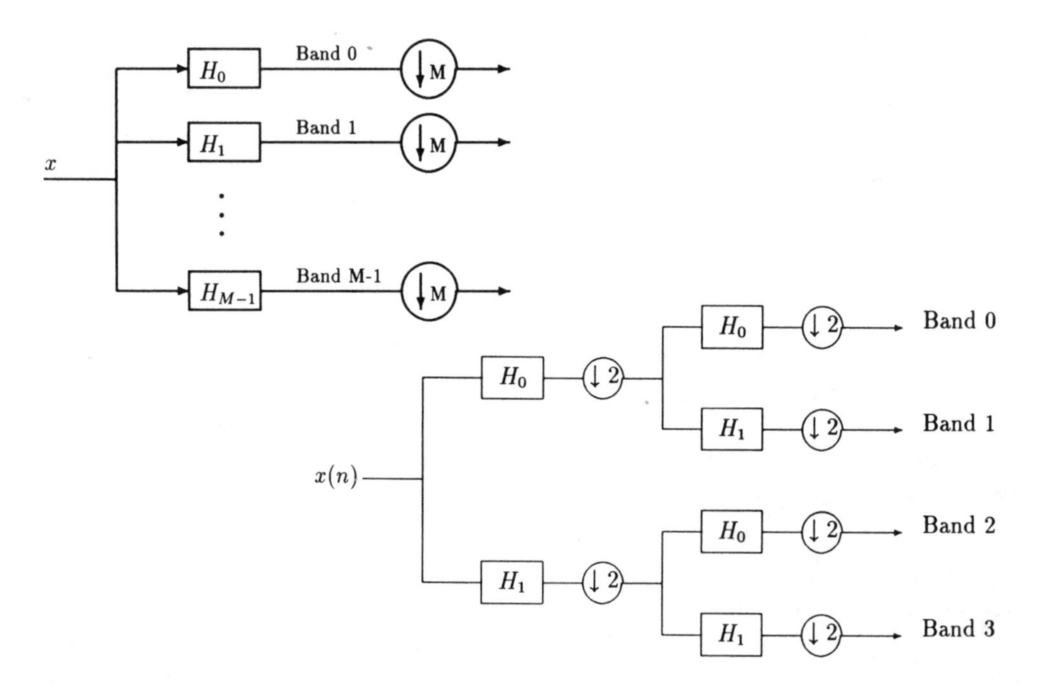

Figure 1.3: Multirate filter bank with equal bandwidths: (a) M-band; (b) four-band, realized by a two-level binary (regular) tree.

Another way of realizing the decomposition into M equal subbands is shown by the hierarchical two-band subband tree shown in Fig. 1.3(b). Each level of the tree splits the preceding subband into two equal parts, permitting a decomposition into $M = 2^k$ equal subbands. In this case the M-band structure is said to be realized by a dilation of the impulse responses of the basic two-band structure at each level of the tree, since splitting each subband in two dilates the impulse response by this factor.

1.3.2 Multiresolution Structures

Yet another possible decomposition is shown in Fig. 1.4, which represents a "dyadic tree" decomposition. The signal is first split into low- and high-frequency components in the first level. This first low-frequency subband, containing most of the energy, is subsampled and again decomposed into low- and high-frequency subbands. This process can be continued into K levels. The coarsest signal is the one labeled LLL in the figure. Moving from right to left in this diagram, we see a progression from coarser to finer signal representation as the high-frequency "detail" is added at each level. The signal can thus be approximately represented by different resolutions at each level of the tree.

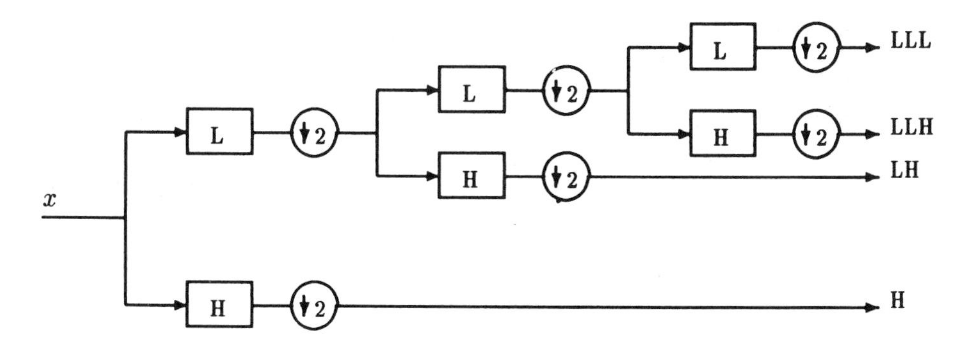

Figure 1.4: Multiresolution dyadic tree: L and H represent low-pass and high-pass filters, respectively.

An oversampled version of this tree, called the Laplacian pyramid, was first introduced for image coding by Burt and Adelson (1983). These topics are explained in detail and the reference is given in Chapter 3.

Wavelet transforms recently have been proposed as a new multiresolution decomposition tool for continuous-time signals. The kernel of the wavelet transform is obtained by dilation and translation of a prototype bandpass function. The discrete wavelet transform (DWT) employs discretized dilation and translation parameters. Simply stated, the wavelet transform permits a decomposition of a signal into the sum of a lower resolution (or coarser) signal plus a detail, much like the dyadic subband tree in the discrete-time case. Each coarse approximation in turn can be decomposed further into yet a coarser signal and a detail signal at that resolution. Eventually, the signal can be represented by a low-pass or coarse signal at a certain scale (corresponding to the level of the tree), plus a sum of detail signals at different resolutions. In fact, the subband dyadic tree structure conceptualizes the wavelet multiresolution decomposition of a signal. We show in Chapter 6 that the base or prototype function of the orthonormal wavelet transform is simply related to the two-band unitary perfect reconstruction quadrature mirror filters (PR-QMF), and that the fast wavelet transform algorithm can also be strongly linked to the dyadic tree filter bank. Hence, from our perspective, we view wavelets and dyadic subband trees as multiresolution decomposition techniques of the continuous-time and discrete-time signals, respectively, as suggested in Fig. 1.2.

1.3.3 The Synthesis/Analysis Structure

Figure 1.5 shows the dual synthesis/analysis system. As mentioned earlier and explained in Chapter 3, this structure could represent any one of several multiplexing systems depending on the choice of the synthesis and analysis filters. Interesting enough, the conditions for alias cancellation in Fig. 1.3 and for zero cross-talk in Fig. 1.5 are the same. Additionally, the conditions for "perfect reconstruction," $\hat{x}(n) = x(n-n_0)$ in Fig. 1.1(a) and $\hat{x}_i(n) = x_i(n-n_0)$ in Fig. 1.1(b), are the same!

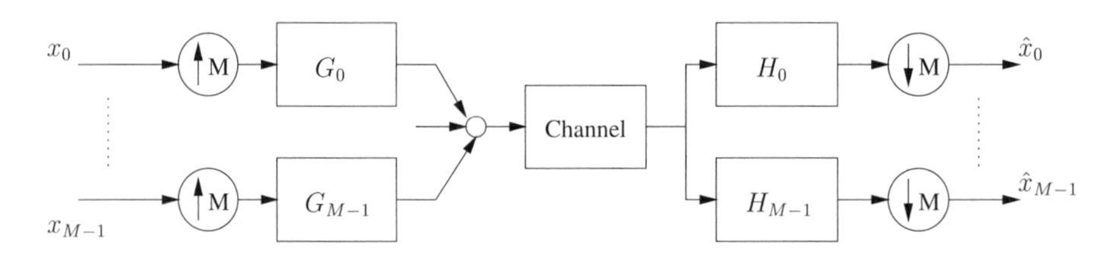

Figure 1.5: Transmultiplexer as a synthesis/analysis structure

1.3.4 The Binomial-Hermite Sequences: A Unifying Example

This book tries to provide a common framework for the interpretation and evaluation of all orthonormal signal decomposition tools: block transforms, subbands, and wavelets. The Binomial-Hermite sequences provide a family of functions with applications that touch all these categories. This elegant family will be used as a vehicle to illustrate and link together all these topics. At the simplest level they provide a set of functions for orthogonal signal expansions. Suitably modified, they generate block transform, called the modified Hermite transform (MHT). Then we linearly combine members of the Binomial family to obtain the unique, maximally flat squared magnitude, two-band paraunitary QMF. These in turn are recognized as the orthonormal wavelet filters devised by Daubechies. These functions also play a major role as kernels in discrete time-frequency analysis.

1.4 Performance Evaluation and Applications

One of the objectives of this book is a comparative evaluation of several of the more popular decomposition techniques. In Chapter 2, families of transforms are described and their compaction properties are evaluated both from a block transform and LOT realization. Chapter 4 presents the comparative evaluation of known filter families and wavelet filters. It provides criteria by which all these strategies can be compared.

Chapter 5 compares the time-frequency localization properties of block transforms and wavelet filters.

In addition to comparative evaluations, in Chapter 4 we also introduce an optimal *design* approach for filters wherein the design parameters are made part of the performance criteria that can be optimized. These optimal solutions set upper performance bounds for FIR subband decomposition in a manner conceptually similar to the performance bound that the optimal transform, the KLT, sets for all block transforms.

Chapter 7 describes a medley of applications of these techniques to the solutions of problems in communications and multimedia.

Chapter 2

Orthogonal Transforms

The purpose of transform coding is to decompose a batch of correlated signal samples into a set of uncorrelated spectral coefficients, with energy concentrated in as few coefficients as possible. This compaction of energy permits a prioritization of the spectral coefficients, with the more energetic ones receiving a greater allocation of encoding bits. For the same distortion level, the total number of bits needed to transmit encoded spectral coefficients is less than the number needed to transfer the signal samples directly. This reduction in bit rate is termed *compression*.

In this chapter we introduce the subject of orthogonal transforms—from the standpoint of function expansions in orthogonal series. For the most part, the signals and basis functions have finite support—that is, are of finite duration—as required for block transforms. For completeness of presentation, however, we also include signals of semi-infinite extent. We group the block transforms considered here in three broad categories—sinusoidal, polynomial, and rectangular—and describe leading members in each category.

The decorrelation and energy compaction properties of these transforms constitute the central issues in the applicability of these transforms for signal coding purposes. Of equal importance is the computational complexity associated with the respective transformations. In the sequel, we formulate criteria that permit comparative evaluation of the performance of different transforms in signal coding applications. These measures are also used on standard test images.

At low bit rates, transform-coded images often exhibit a "blockiness" at the borders. The lapped orthogonal transform (LOT) was introduced to counteract this effect. The spectral coefficients are calculated using data windows that overlap batch boundaries. The LOTs are described and evaluated using the criteria developed for transform coding. The block transform and LOT are interpreted as special cases of the multirate filter bank introduced in Chapter 3.

The signals and theory developed here for the most part represent functions of one variable—i.e., one dimensional (1D) signals. The extension to the multidimensional case is usually straightforward for separable transforms. In fact, in some instances, the Kronecker product expansion of a two dimensional (2D) transformation permits a factorization and grouping of terms that simply is not possible in 1D.

2.1 Signal Expansions in Orthogonal Functions

The orthogonal expansion of a continuous variable function is a subject extensively addressed in the classical literature (Sansone, 1959). Our focus is the expansion of sampled signals, i.e., sequences $\{f(k)\}$. Milne (1949) briefly treats this subject. During the decade of the 1970s, the study of the subject gained in intensity as several authors Campanella and Robinson (1977); Ahmed and Rao (1975) developed expansions for discrete-variable functions to meet the needs of transform coding. Orthogonal expansions provide the theoretical underpinnings for these applications.

2.1.1 Signal Expansions

Our task here is the representation of a sequence (or discrete-time or discrete-variable signal) $\{f(k)\}$ as a weighted sum of component sequences. These possess special properties that highlight certain features of the signal. The most familiar is

$$f(n) = \sum_{k=-\infty}^{\infty} f(k)\delta(n-k), \tag{2.1}$$

where the component sequence $\delta(n-k)$ is the Kronecker delta sequence:

$$\delta(n-k) = \begin{cases} 1, & n-k=0, \\ 0, & \text{otherwise.} \end{cases} \tag{2.2}$$

Here the weights are just the sample values themselves—not very interesting. Instead we seek to represent $\{f(k)\}$ as a superposition of component sequences which can extract identifying features of the signal in a compact way. But in any event, the component sequences should be members of an orthogonal family of functions.

In what follows, we borrow some geometric concepts from the theory of linear vector spaces. To fix ideas, consider a sequence $\{f(k)\}$ defined on the interval

$0 \leq k \leq N-1$. We can think of $\{f(k)\}$ as an N dimensional vector $\underline{f}$ and represent it by the superposition[1]

$$\begin{aligned} \underline{f} &= \begin{bmatrix} f_0 \\ f_1 \\ f_2 \\ \vdots \\ f_{N-1} \end{bmatrix} = f_0 \begin{bmatrix} 1 \\ 0 \\ 0 \\ \vdots \\ 0 \\ 0 \end{bmatrix} + f_1 \begin{bmatrix} 0 \\ 1 \\ 0 \\ \vdots \\ 0 \\ 0 \end{bmatrix} + \cdots + f_{N-1} \begin{bmatrix} 0 \\ 0 \\ 0 \\ \vdots \\ 0 \\ 1 \end{bmatrix} \\ &= f_0 \underline{e}_0 + f_1 \underline{e}_1 + \cdots + f_{N-1} \underline{e}_{N-1}. \end{aligned} \tag{2.3}$$

The form of Eq. (2.3) is a finite-dimensional version of Eq. (2.1). The norm of $\underline{f}$ is defined as

$$norm(\underline{f}) = \left[\sum_{k=0}^{N-1} |f(k)|^2 \right]^{1/2}. \tag{2.4}$$

The sequence $\{f(k)\}$ is now represented as a point in the N dimensional Euclidean space spanned by the basis vectors $\{\underline{e}_0, \underline{e}_1, \ldots, \underline{e}_{N-1}\}$. These basis vectors are linearly independent, since the linear combination

$$c_0 \underline{e}_0 + c_1 \underline{e}_1 + \cdots + c_{N-1} \underline{e}_{N-1} \tag{2.5}$$

can vanish only if $c_0 = c_1 = \cdots = c_{N-1} = 0$. Another way of expressing this is that no one basis vector can be represented as a linear combination of the others.

We say that the two sequences $\{g(k)\}$ and $\{f(k)\}$ with the same support, i.e., the interval outside of which the sequence is zero, are orthogonal if the inner product vanishes, or

$$\sum_{k=0}^{N-1} g(k) f(k) = \underline{g}^T \underline{f} = 0. \tag{2.6}$$

Clearly the finite dimensional Kronecker delta sequences are orthogonal, since

$$\sum_{k=0}^{N-1} e_i(k) e_j(k) = \underline{e}_i^T \underline{e}_j = 0, \qquad \text{for } i \neq j. \tag{2.7}$$

[1] For notational convenience, we will use f_n and $f(n)$ interchangeably.

Moreover, the norm of each basis vector is unity,

$$\sum_{k=0}^{N-1} |e_i(k)|^2 = 1, \qquad \text{for} \quad i = 0, 1, ..., N-1. \tag{2.8}$$

Again, we note that the weights in the expansion of Eq. (2.3) are just the sample values that, by themselves, convey little insight into the properties of the signal. Suppose that a set of basis vectors can be found such that the data vector $\underline{f}$ can be represented closely by just a few members of the set. In that case, each basis vector identifies a particular feature of the data vector, and the weights associated with the basis vectors characterize the features of the signal. The simplest example of this is the Fourier trigonometric expansion wherein the coefficient value at each harmonic frequency is a measure of the signal strength at that frequency. We now turn to a consideration of a broad class of orthogonal expansions with the expectation that each class can characterize certain features of the signal.

Suppose we can find $\{x_n(k), \quad 0 \leq n, k \leq N-1\}$, a family of N linearly independent sequences on the interval $[0, N-1]$. This family is orthogonal if

$$\sum_{k=0}^{N-1} x_n(k) x_s^*(k) = c_n^2 \delta_{n-s} = \begin{cases} c_n^2 & n = s, \\ 0 & \text{otherwise}, \end{cases} \tag{2.9}$$

where c_n is the norm of $\{x_n(k)\}$. (The asterisk denotes the complex conjugate.) The orthonormal family is obtained by the normalization,

$$\phi_n(k) = \frac{1}{c_n} x_n(k); \qquad 0 \leq n \leq N-1, \tag{2.10}$$

which in turn shows that

$$\sum_{k=0}^{N-1} \phi_n(k) \phi_s^*(k) = \delta_{n-s}. \tag{2.11}$$

Any nontrivial set of functions satisfying Eq. (2.11) constitutes an *orthonormal* basis for the linear vector space. Hence $\{f(k)\}$ can be uniquely represented as

$$f(k) = \sum_{n=0}^{N-1} \theta_n \phi_n(k), \qquad 0 \leq k \leq N-1, \tag{2.12}$$

where

$$\theta_s = \sum_{k=0}^{N-1} f(k)\phi_s^*(k), \qquad 0 \le s \le N-1. \tag{2.13}$$

The proof of Eq. (2.13) is established by multiplying both sides of Eq. (2.12) by $\phi_s^*(k)$ and summing over the k index. Interchanging the order of summation and invoking orthonormality shows that

$$\sum_k f(k)\phi_s^*(k) = \sum_n \theta_n \sum_k \phi_n(k)\phi_s^*(k) = \sum_n \theta_n \delta_{n-s} = \theta_s.$$

The set of coefficients $\{\theta_s, \quad 0 \le s \le N-1\}$ constitute the spectral coefficients of $\{f(k)\}$ relative to the given orthonormal family of basis functions. Classically, these are called *generalized Fourier coefficients* even when $\{\phi_n(k)\}$ are not sinusoidal.

The *energy* in a signal sequence is defined to be the square of the norm. The Parseval theorem, which asserts that

$$\sum_{k=0}^{N-1} |f(k)|^2 = \sum_{n=0}^{N-1} |\theta_n|^2 \tag{2.14}$$

can be proved by multiplying both sides of Eq. (2.12) by their conjugates and summing over k. This theorem asserts that the signal energy is preserved under an orthonormal transformation and can be measured by the square of the norm of either the signal samples or the spectral coefficients.

As we shall see, one of the prime objectives of transform coding is the redistribution of energy into a few spectral coefficients.

On a finite interval, the norm is finite if all samples are bounded on that interval. On the other hand, convergence of the norm for signals defined on $[0, \infty)$ or $(-\infty, \infty)$ requires much more stringent conditions, an obvious necessary one being $|f(k)| \to 0$, as $k \to \pm\infty$. Sequences with finite energy are said to be L^2.

The Z-transform provides an alternative signal description, which is particularly useful in a filtering context. To this end, let $\Phi_n(z)$ be the Z-transform (one-sided or two-sided, as required by the region of support) of $\{\phi_n(k)\}$. Now the orthogonality relationship of Eq. (2.11) can be restated as the contour integral

$$\sum_k \phi_n(k)\phi_s^*(k) = \frac{1}{2\pi j} \oint \Phi_n(z)\Phi_s^*\left(\frac{1}{z^*}\right)\frac{dz}{z} = \delta_{n-s}, \tag{2.15}$$

where the contour is taken on the unit circle of the Z-plane. From the Cauchy residue theorem, the sum of the residues in all the poles of the integrand within the unit circle must vanish for $n \neq s$. But on the unit circle, $z = e^{j\omega}$. With this substitution Eq. (2.15) becomes

$$\sum_k \phi_n(k)\phi_s^*(k) = \frac{1}{2\pi}\int_{-\pi}^{\pi} \Phi_n(e^{j\omega})\Phi_s^*(e^{j\omega})d\omega = \delta_{n-s}. \tag{2.16}$$

The latter form permits us to generate families of orthogonal sequences by specifying the behavior of $\Phi_n(e^{j\omega})$. (See Problem 2.4.)

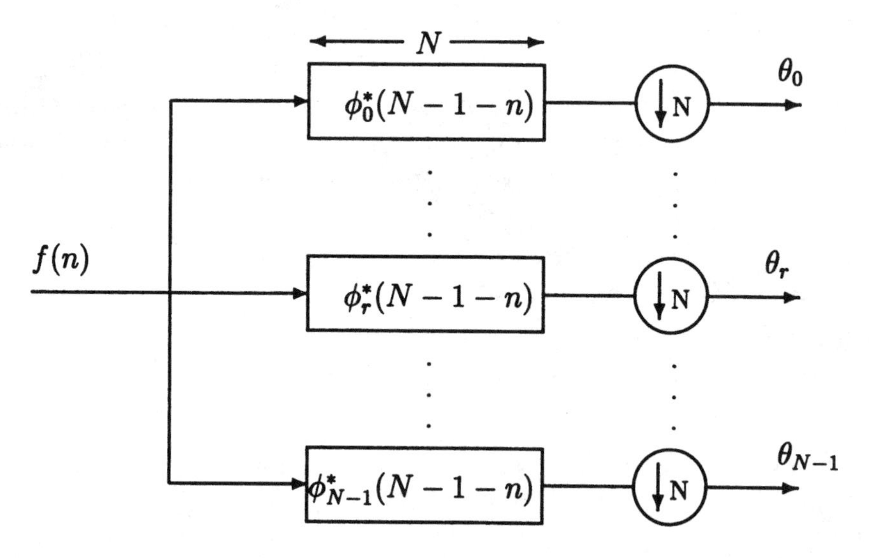

Figure 2.1: Orthonormal spectral analyzer as a multirate filter bank.

The form of Eq. (2.13) suggests that the spectral coefficients $\{\theta_n\}$ for the sequence $\{f(k)\}$ can be measured by the spectral analyzer shown as Fig. 2.1. The signal $f(k)$ is fed into a bank of FIR filters whose impulse responses are *time reversed* and translated basis sequences $\{\phi_r^*(N-1-k) = h_r(k)\}$. The output of the rth filter is the convolution

$$y_r(n) = h_r(n) * f(n) = \sum_k \phi_r^*(N-1-k)f(n-k). \tag{2.17}$$

Sampling this output at $n = N-1$ gives the coefficient $\theta_r = y_r(N-1)$. The collection of sampled outputs at this time gives the spectral coefficient vector $\underline{\theta}^T \triangleq [\theta_0, ..., \theta_{N-1}]$ for the first block of data $\underline{f}^T \triangleq [f_0, ..., f_{N-1}]$.

The circle with the downward-pointing arrow in this diagram indicates that the output sequence of each filter is subsampled, i.e., every Nth sample is retained. If the input $\{f(k)\}$ is a continuing stream of data, the subsampled outputs at times $N-1, 2N-1, \ldots$ represent successive spectral coefficient vectors corresponding to successive blocks of data. In Chapter 3, we will interpret Fig. 2.1 as a multirate filter bank that functions as the front end of a subband coder.

2.1.2 Least-Squares Interpretation

The set of coefficients $\{\theta_n\}$ in Eq. (2.13) also provides the *least-squares* approximation to $\{f(k)\}$. Suppose we want to approximate $\{f(k)\}$ by a superposition of the first L of the N basis sequences, using weighting coefficients $\{\gamma_i,\ \ i = 0, 1, \ldots, L-1\}$. Then the best least-squares choice for these coefficients is

$$\gamma_i = \theta_i, \qquad i = 0, 1, 2, \ldots, L-1.$$

The proof is as follows. Let the approximation be

$$\hat{f}(k) = \sum_{r=0}^{L-1} \gamma_r \phi_r(k). \tag{2.18}$$

and the error is then

$$\epsilon(k) = f(k) - \hat{f}(k). \tag{2.19}$$

The $\{\gamma_r\}$ are to be chosen to minimize the sum squared error

$$J_L = \sum_{k=0}^{N-1} |\epsilon(k)|^2. \tag{2.20}$$

Expanding the latter and invoking orthonormality[2] gives

$$J_L = \sum_k |f(k)|^2 - 2\left(\sum_k f(k)\right)\left(\sum_r \gamma_r \phi_r(k)\right) + \sum_r (\gamma_r)^2.$$

Next, setting the partial of J_L with respect to γ_s to zero gives

$$\frac{\partial J_L}{\partial \gamma_s} = -2\sum_k f(k)\phi_s(k) + 2\gamma_s = 0$$

with solution

$$\gamma_s = \sum_{k=0}^{N-1} f(k)\phi_s(k) \equiv \theta_s. \tag{2.21}$$

[2] For convenience, we pretend that $f(k), \gamma_i$, and $\phi_r(k)$ are real.

When $L = N$, we note that $J_N = 0$. Thus, the sum squared error in using $\hat{f}(k)$ as an approximation to $f(k)$ is minimized by selecting the weights to be the orthonormal spectral coefficients.

This choice of coefficients has the property of *finality*. This means that if we wish to reduce the error by the addition of more terms in Eq. (2.18), we need not recalculate the previously determined values of $\{\gamma_r\}$. Also, we can show

$$J_{L+1} \leq J_L, \qquad \text{for any} \quad L. \tag{2.22}$$

The resulting minimized error sequence is

$$\epsilon(k) = f(k) - \sum_{r=0}^{L-1} \theta_r \phi_r(k) = \sum_{r=L}^{N-1} \theta_r \phi_r(k). \tag{2.23}$$

Thus, the error sequence $\{\epsilon(k)\}$ lies in the space spanned by the remaining basis functions

$$\{\phi_L(k), \ldots, \phi_{N-1}(k)\} \triangleq V_2,$$

whereas the estimate $\{\hat{f}(k)\}$ lies in V_1, the space spanned by $\{\phi_s(k), \quad 0 \leq s \leq L-1\}$. The term V_2 is the *orthogonal complement* of V_1 with the property that every vector in V_2 is orthogonal to every vector in V_1. Furthermore, the space V of all basis vectors is just the direct sum[3] of V_1 and V_2. It is easy to see that $\{\epsilon(k)\}$ is orthogonal to $\{\hat{f}(k)\}$, i.e.,

$$\sum_{k=0}^{N-1} \epsilon(k)\hat{f}(k) = \underline{\epsilon}^T \underline{\hat{f}} = 0. \tag{2.24}$$

In fact, it can be shown that the orthogonality of error and approximant is necessary and sufficient to minimize the sum squared error. (Prob. 2.1)

A simple sketch depicting this relationship is shown in Fig. 2.2 for the case $N = 3, L = 2$. This sketch demonstrates that $\{\hat{f}(k)\}$, the least squares approximation to $\{f(k)\}$ is the orthogonal projection of $\{f(k)\}$ onto the two-dimensional subspace spanned by basis sequences $\{\phi_1(k)\}, \{\phi_2(k)\}$. For complex valued sequences $\{\hat{f}^*(k)\}$ is used in Eq. (2.24).

Before closing this section, we note that all of these results and theorems are valid for infinite dimensional spaces as well as finite dimensional ones, as long as the norms of the sequences are bounded, i.e., are L^2. We also note for later reference two additional theorems to be used subsequently. First, the Cauchy-Schwarz inequality asserts that

[3]In our representation $V = R^N$,the set of all real N tuples. The direct sum is $V = V_1 \oplus V_2$ if and only if $V_1 \cap V_2 = \phi$, and $V = V_1 \cup V_2$.

$$|\sum_k x(k)y^*(k)| \le \left[\sum_k |x(k)|^2\right]^{1/2} \left[\sum_k |y(k)|^2\right]^{1/2}. \tag{2.25}$$

The second relates inner products in the temporal(spatial) domain and the spectral domain. With $\{\alpha_r\}, \{\beta_r\}$ the spectral coefficients corresponding to $\{x(k)\}$, $\{y(k)\}$, respectively, we have the extended Parseval theorem,

$$\sum_{k=0}^{N-1} x(k)y^*(k) = \sum_{r=0}^{N-1} \alpha_r \beta_r^*. \tag{2.26}$$

Proof of these is left as an exercise for the reader. (Prob. 2.2 and 2.3)

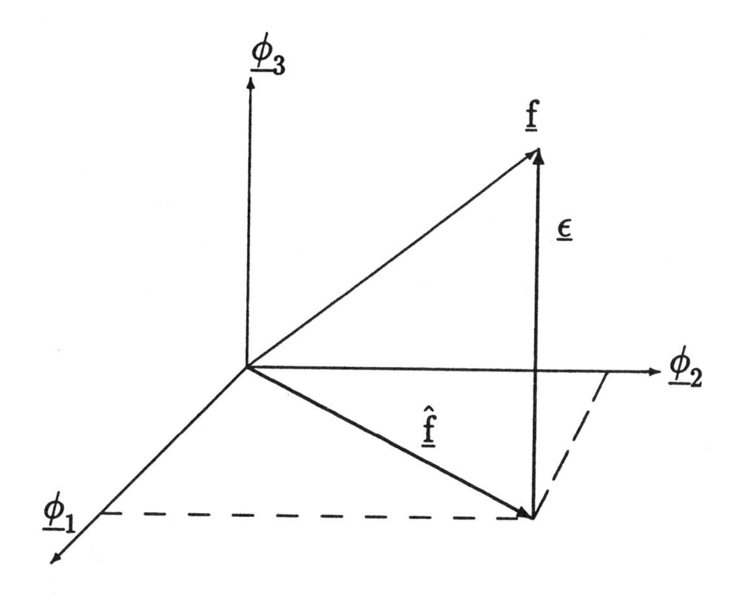

Figure 2.2(a): Orthogonality principle demonstration.

2.1.3 Block Transforms

The orthonormal expansions of the preceding section provide the foundations for signal classification and identification, particularly for speech and images.

A vector-matrix reformulation provides a succinct format for block transform manipulation and interpretation. The signal and spectral vectors are defined as

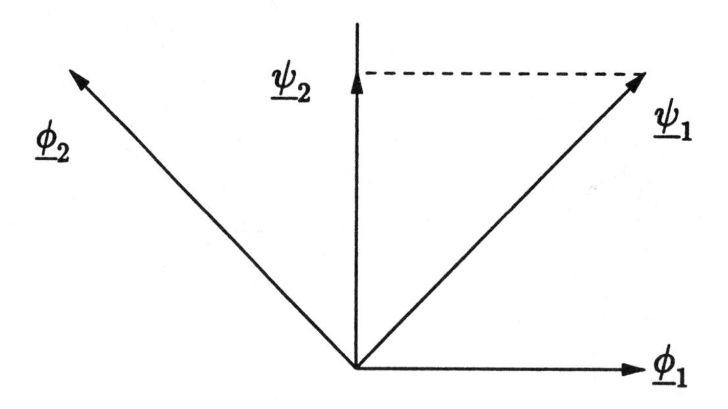

Figure 2.2(b): Biorthogonal bases.

$$\begin{aligned} \underline{f}^T &= [f_0, f_1, \ldots, f_{N-1}] \\ \underline{\theta}^T &= [\theta_0, \theta_1, \ldots, \theta_{N-1}]. \end{aligned} \tag{2.27}$$

Let the real orthonormal sequences $\phi_r(k)$ be the rows of a transformation matrix, $\phi(r,k)$,

$$\Phi = [\phi(r,k)]. \tag{2.28}$$

It is evident that

$$\underline{\theta} = \Phi \underline{f} \tag{2.29}$$

and

$$\underline{f} = \Phi^{-1}\underline{\theta} = \Phi^T \underline{\theta}; \tag{2.30}$$

thus,

$$\Phi^{-1} = \Phi^T, \tag{2.31}$$

a property that identifies Φ as an *orthogonal* matrix. Now let $\underline{\Phi}_r$ be a *column* vector representing the basis sequence $\{\phi_r(k)\}$; i.e.,

$$\underline{\Phi}_r^T = [\phi_r(0), \phi_r(1), \ldots, \phi_r(N-1)]. \tag{2.32}$$

We can write $\underline{f}$ as a weighted sum of these basis vectors,

$$\underline{f} = [\underline{\Phi}_0 \underline{\Phi}_1 \ldots \underline{\Phi}_{N-1}] \begin{bmatrix} \theta_0 \\ \theta_1 \\ \vdots \\ \theta_{N-1} \end{bmatrix} = \theta_0 \underline{\Phi}_0 + \theta_1 \underline{\Phi}_1 + \ldots + \theta_{N-1} \underline{\Phi}_{N-1}. \tag{2.33}$$

Noting that the orthonormality condition is

$$\underline{\Phi}_n^T \underline{\Phi}_s = \delta_{n-s},$$

we see that the coefficient θ_s is just the inner product

$$\underline{\Phi}_s^T \underline{f} = \sum_{n=0}^{N-1} \theta_n \underline{\Phi}_s^T \underline{\Phi}_n = \sum_{n=0}^{N-1} \theta_n \delta_{n-s} = \theta_s. \tag{2.34}$$

For complex valued signals and bases, the transformation becomes

$$\underline{\theta} = \Phi^* \underline{f} \longleftrightarrow \underline{f} = \Phi^T \underline{\theta} \tag{2.35}$$

with the property

$$\Phi^{-1} = (\Phi^*)^T. \tag{2.36}$$

This last equation, asserting that the inverse of Φ is its conjugate transpose, defines a *unitary* matrix.

The Parseval relation is given by the inner product

$$\underline{\theta}^T \underline{\theta}^* = \underline{f}^T \underline{f}^*, \tag{2.37}$$

which again demonstrates that a unitary (or orthonormal) transformation is *energy preserving*.

The transform coding application is shown in Fig. 2.3. The orthonormal spectrum of a batch of N signal samples is evaluated. These coefficients are then quantized, encoded, and transmitted. The receiver performs inverse operations to reconstruct the signal.

The purpose of the transformation is to convert the data vector $\underline{f}$ into a spectral coefficient vector $\underline{\theta}$ that can be optimally quantized. Typically, the components of $\underline{f}$ are correlated, and each component has the same variance. For example, $\{f(n)\}$ is a sequence of zero mean, correlated random variables, each with the constant variance σ^2. The orthogonal transformation tries to decorrelate the signal samples

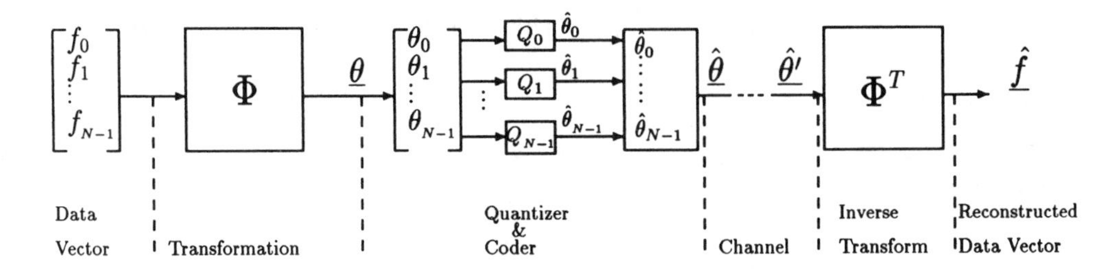

Figure 2.3: Transform *co*der *dec*oder.

(i.e., to *whiten* the sequence $\{\theta_r\}$). Moreover, the variances of the individual components of $\underline{\theta}$ will generally differ, which simply states that the sequence $\{\theta_r\}$ is nonstationary. We can exploit this fact by allocating quantization bits to each coefficient in accordance with the power (or variance) in that spectral component. Thus, some coefficients are quantized more finely than others.

Therefore, we recognize another purpose of the transformation—to repack the signal energy implied by Eq. (2.37) into a relatively small number of spectral coefficients $\{\theta_r\}$. Hence the power or worth of an orthonormal transformation from a signal coding standpoint depends on its *signal decorrelation* and *energy repacking* properties.

In the absence of channel noise, the mean square reconstruction error of the transform coder equals the mean square quantization error. From Fig. 2.3, we define

$$\begin{aligned} \underline{\tilde{f}} &= \underline{f} - \underline{\hat{f}}, \quad \text{the reconstruction error,} \\ \underline{\tilde{\theta}} &= \underline{\theta} - \underline{\hat{\theta}}, \quad \text{coefficient quantization error.} \end{aligned} \tag{2.38}$$

It can be shown that (Prob. 2.5)

$$E\{\underline{\tilde{f}}^T \underline{\tilde{f}}\} = E\{\underline{\tilde{\theta}}^T \underline{\tilde{\theta}}\}. \tag{2.39}$$

Hence we can optimally code the data stream $\underline{f}$ by using an orthogonal transformation followed by a quantizer whose characteristics depend on the probability density function (pdf) of $\underline{\theta}$. This can be optimized using the Lloyd-Max procedure described in Section 2.2.2. Moreover, the number of encoding bits assigned to each spectral coefficient is optimally allocated on the basis of the logarithm of its variance.

The codec (coder-decoder) can be optimized by a fixed transformation and quantizer based on an a priori model of the signal. At the cost of greater complexity, both transform and quantizer can be adapted, on line, to the statistics of the observed data.

Zonal sampling is a term used to indicate an approximation wherein only a subset of the N spectral coefficients is used to represent the signal vector. But this is nothing but the least squares approximation addressed in the previous subsection. The truncation error was found to be as in Eq. (2.23),

$$\epsilon(k) = f(k) - \sum_{n=0}^{L-1} \theta_n \phi_n(k) = \sum_{n=L}^{N-1} \theta_n \phi_n(k),$$

from which we conclude that

$$\underline{\epsilon}^T \underline{\epsilon} = \sum_{k=0}^{N-1} |\epsilon(k)|^2 = \sum_{n=L}^{N-1} |\theta_n|^2. \tag{2.40}$$

The best zonal sampler is therefore one that packs the maximum energy into the first L coefficients. The Karhunen-Loeve transform (KLT), a signal-dependent transform, has this property.

The discrete cosine transform is an example of a signal independent transform. Based on an a priori signal model (e.g., a low-frequency process), the optimum fixed quantizer allocates bits based on the precomputed variances in the spectral coefficients. Zonal sampling simply discards those coefficients that the signal model predicts will have small variances.

Biorthogonal Block Transforms and Dual Bases

We have defined orthonormal block transforms as a matrix whose row vectors $\{\underline{\phi}_r\}$, satisfy $\underline{\phi}_r^T \underline{\phi}_r = \delta_{r-s}$. In Chapter 3 and in the wavelet Chapter 6, we will discuss biorthogonal filter banks and wavelets. These are extensions of the biorthogonal block transforms which are represented as follows. We start with two non-orthogonal bases, called *dual* bases $\{\underline{\phi}_1, ..., \underline{\phi}_N\}$ and $\{\underline{\psi}_1, ..., \underline{\psi}_N\}$ with the property (Pei and Yeh, 1997) that the orthogonality is carried across the bases, i.e.,

$$\underline{\phi}_r^T \underline{\psi}_r = k_{rs} \delta_{r-s},$$

and when $k_{rs} = 1$, the system becomes biorthonormal. The key property is that any vector $\underline{f}$ can be projected onto $\{\underline{\phi}_r\}$, and that projection is the coefficient for the expansion in the dual basis $\{\underline{\psi}_s\}$, i.e.,

$$\underline{f} = \sum_{i=1}^{N} \alpha_i \underline{\psi}_i,$$

and

$$\alpha_i = \underline{f}^T \underline{\phi}_i.$$

In Fig. 2.3, the analysis matrix would be $\Phi = [\underline{\phi}_i]$, and the reconstruction or synthesis matrix would be $\Psi = [\underline{\psi}_j]$.

The example shown in Fig. 2.2(b) illustrates this property very nicely. In that figure the planar vectors $\underline{\phi}_1$ and $\underline{\phi}_2$ are not orthogonal, nor are $\underline{\psi}_1$ and $\underline{\psi}_2$. However, $\underline{\psi}_1$ and $\underline{\phi}_2$ are orthogonal, and so are $\underline{\phi}_1$ and $\underline{\psi}_2$. We have $\underline{\psi}_1^T = [1, 0]$, $\underline{\psi}_2^T = [-1, 1]$, $\underline{\phi}_1^T = [1, 1]$, and $\underline{\phi}_2^T = [0, 1]$. Any vector $\underline{f}^T = [a, b]$ can be expressed as a combination of $\{\underline{\phi}_1, \underline{\phi}_2\}$ or as a linear combination of $\{\underline{\psi}_1, \underline{\psi}_2\}$. It is easy to verify that

$$\underline{f} = \alpha_1 \underline{\psi}_1 + \alpha_2 \underline{\psi}_2,$$

where $\alpha_1 = \underline{f^T \phi_1}$, and $\alpha_2 = \underline{f^T \phi_2}$. Moreover, the roles of ϕ and ψ can be interchanged. The reader can verify that $\underline{f}$ can be resolved into

$$\underline{f} = (a + b)\underline{\psi}_1 + b\underline{\psi}_2 = a\underline{\phi}_1 + (b - a)\underline{\phi}_2.$$

2.1.4 The Two-Dimensional Transformation

The 2D version of transform coding is easily extrapolated from the foregoing. As shown in Fig. 2.4, the image array is divided into subblocks, each of which is separately encoded. These blocks are usually square, with 4×4, 8×8, and 16×16 being representative sizes.

Let the $N \times N$ subblock image array be denoted by

$$F = [f(m, n)], \qquad 0 \leq m, n \leq N - 1. \tag{2.41}$$

The forward transform is

$$\theta(i, j) = \sum_{m=0}^{N-1} \sum_{n=0}^{N-1} \alpha(i, j; m, n) f(m, n), \qquad 0 \leq i, j \leq N - 1, \tag{2.42}$$

and the inverse is

$$f(m, n) = \sum_{i=0}^{N-1} \sum_{j=0}^{N-1} \beta(m, n; i, j) \theta(i, j), \tag{2.43}$$

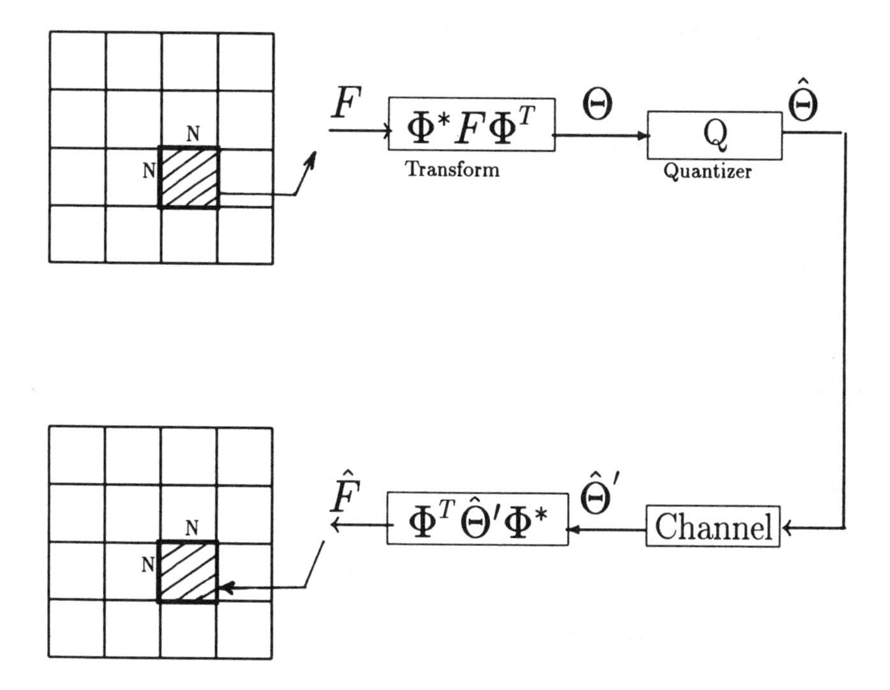

Figure 2.4: Two-dimensional transform coding.

where $\alpha(\cdot)$, and $\beta(\cdot)$ are the forward and inverse transform kernels.

In this text, as well as in all practical cases, the kernels are separable and symmetric so that the 2D kernel is simply the product of two 1D orthogonal basis functions

$$\alpha(i,j;m,n) = \phi(i,m)\phi(j,n) = \phi_i(m)\phi_j(n), \qquad 0 \le i,j,m,n \le N-1. \quad (2.44)$$

In Section 2.6, this separability is interpreted as a Kronecker product factorization.

Just as with the 1D formulation, the basis functions constitute the rows of the unitary matrix. The forward and inverse transformations have the form

$$\begin{aligned} \Theta &= \Phi^* F \Phi^T \\ F &= \Phi^T \Theta \Phi^*. \end{aligned} \quad (2.45)$$

Examination of Eq. (2.45) reveals that the image transformation can be done in two stages: First, we take the unitary transform Φ to each row of the image array to obtain an intermediate array $S = F\Phi^T$. Then we apply the transformation Φ^* to each column of S to obtain the final transformed image, $\Theta = \Phi^* S$.

The 2D Parseval relation is a simple extension of Eq. (2.14),

$$\sum_{m=0}^{N-1}\sum_{n=0}^{N-1} |f(m,n)|^2 = \sum_{i=0}^{N-1}\sum_{j=0}^{N-1} |\theta(i,j)|^2. \tag{2.46}$$

The 2D version of the basis vector is the basis array. This is a direct extrapolation of the 1D result given by Eq. (2.33). From Eq. (2.45), we can expand the image array into a superposition of basis arrays via

$$\begin{aligned} F = \Phi^T \Theta \Phi^* &= [\underline{\Phi}_0, \underline{\Phi}_1, \ldots, \underline{\Phi}_{N-1}][\Theta] \begin{bmatrix} \underline{\Phi}_0^T \\ \underline{\Phi}_1^T \\ \vdots \\ \underline{\Phi}_{N-1}^T \end{bmatrix}^* \\ &= \theta_{00}[\underline{\Phi}_0(\underline{\Phi}_0^*)^T] + \theta_{01}[\underline{\Phi}_0(\underline{\Phi}_1^*)^T] + \cdots \end{aligned}$$

or

$$F = \sum_{i=0}^{N-1}\sum_{j=0}^{N-1} \theta_{ij} B_{ij}, \tag{2.47}$$

where the basis image B_{ij} is the *outer product*

$$B_{ij} = \underline{\Phi}_i \underline{\Phi}_j^{*^T}. \tag{2.48}$$

Equation (2.47) expresses the image F as a linear combination of the N^2 basis images. Examples of commonly used basis images are shown in Fig. 2.5. Again by extrapolation of the 1D result of Eq. (2.34), we can show that the transform coefficient θ_{ij} is the inner product of B_{ij} with the input image block

$$\theta_{ij} = \sum_{m=0}^{N-1}\sum_{n=0}^{N-1} B_{ij}(m,n) F(m,n). \tag{2.49}$$

2.1.5 Singular Value Decomposition

The 2D version of a least squares fit to an image leads to an efficient image dependent decomposition known as the *singular value decomposition* (SVD). Consider an $N \times N$ image (or block) F. In Eq. (2.47), we expressed F as the

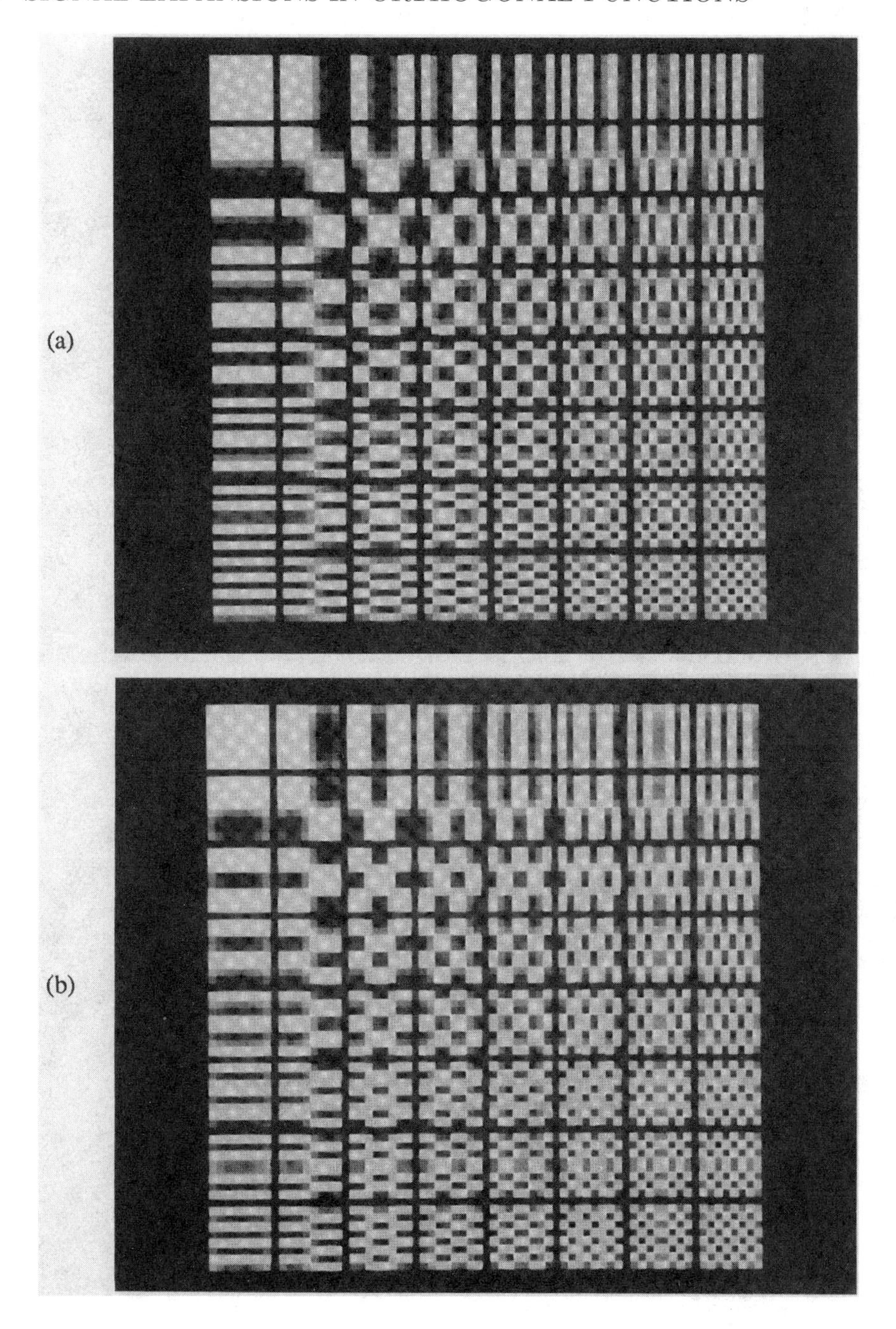

Figure 2.5: Basis images of 8×8 2D block transforms: (a) DCT; (b) DST; (c) WHT; and (d) MHT.

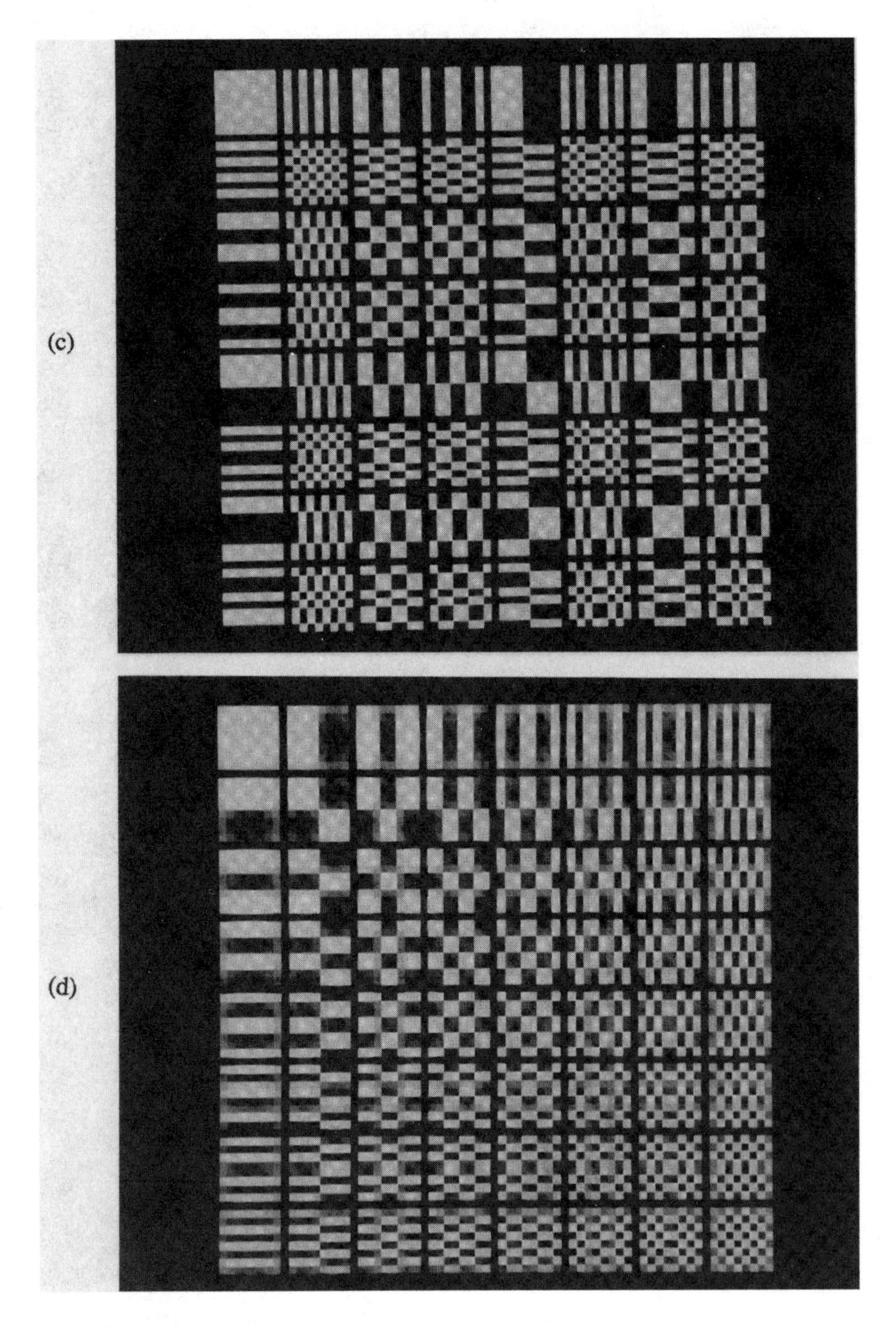

Figure 2.5: (continued)

weighted sum of the N^2 basis arrays $B_{ij} = (\underline{\Phi}_i \underline{\Phi}_j^T)$, wherein the $\{\underline{\Phi}_i\}$ are columns of a *preselected* unitary matrix. We now seek an outer product expansion similar to Eq. (2.47) but where the outer products are matched specifically to the particular image so that the double sum over N^2 basis images reduces to a single sum over r arrays, $r \leq N$. The expansion that achieves this has the form

$$\begin{aligned} F &= \sum_{i=1}^{r} \sqrt{\lambda_i} \underline{\Phi}_i \underline{\Psi}_i^T \\ r &= rank F \end{aligned} \tag{2.50}$$

and is called the *SVD*. It is constructed as follows.

We define $N \times r$ matrices Φ and Ψ such that the r columns of Φ and Ψ are the r non-zero eigenvectors of $(F^T F)$ and (FF^T), respectively. Furthermore since (FF^T) and $(F^T F)$ have the same eigenvalues, then

$$\begin{aligned} (F^T F)\underline{\Phi}_k &= \lambda_k \underline{\Phi}_k \\ (FF^T)\underline{\Psi}_k &= \lambda_k \underline{\Psi}_k, \qquad k = 1, 2, \cdots, r, \end{aligned}$$

where r is the rank of F. These non-zero eigenvalues $\{\lambda_k\}$ are the *singular values* of F.

It can be shown (Golub and Reinsch, 1970) that F can now be written as

$$F = \Psi \Lambda^{1/2} \Phi^T,$$

which, upon expansion, gives Eq. (2.50)

The form of Eq. (2.50) suggests that the SVD has excellent compaction properties for $r \ll N$. Instead of an N^2 image samples, we need encode only $2Nr$ samples(N samples each for $\underline{\Psi}_k$ and $\underline{\Phi}_k$, and there are r of these). The difficulty with SVD is that the transformation matrices Ψ and Φ are tuned to the particular image being examined. They must be recalculated for each image block. In the Karhunen-Loeve transformation, Section 2.2.1, the unitary transformation depends on the image covariance matrix that represents the ensemble of image blocks. It gives the minimum *mean*-squared error averaged over that class. The SVD gives the *least*-squares error for that particular image. For fixed transforms, the unitary matrix is preassigned. For this reason the SVD has also been called a *deterministic* least-squares expansion.

For any expansion in the form

$$\hat{F}_s = \sum_{i=1}^{m} \alpha_i \underline{u}_i \underline{v}_i^T, \qquad s \leq r,$$

the optimum choice of weights and outer product arrays that minimizes

$$\sum_{n=1}^{N}\sum_{m=1}^{N}|F(m,n)-\hat{F}_s(m,n)|^2$$

is that given explicitly by Eq. (2.50), when the eigenvalues are ordered according to decreasing value.

2.2 Transform Efficiency and Coding Performance

Signal coding tries to achieve data compression, i.e., a reduction in the number of bits needed to store or transmit the signal at a given level of distortion. Transform coding attains this objective by decorrelating the signal and repacking the energy among the spectral coefficients. Hence unitary transformations can be compared on the basis of criteria that measure these properties. The optimal transform among all unitary transforms then constitutes the ideal against which all other transforms may be compared.

But we also need to range beyond transform coding and develop measures for comparing different coding methods. A criterion suitable for this purpose is the distortion of the signal coder achievable at a given data bit rate. This performance measure allows us to compare different coding schemes—e.g., transform coding, and subband coding—against each other. For convenience, we take as the base of this measure the distortion induced in the most primitive coding method, that of pulse code modulation (PCM), and compare the coding gain of other methods to this base.

2.2.1 Decorrelation, Energy Compaction, and the KLT

Transform efficiency is measured by the decorrelation and energy compaction provided by the given transformation. To develop these measures, we need to model the data source in a statistical way, in particular, by the variances and covariances of the signal source. For this purpose we need to represent only the wide-sense properties of the signal—i.e., means and autocorrelation function—which we assume are wide-sense stationary (WSS). This simplification is not only for mathematical simplicity, but also for the very real and practical realization that the stationary assumption is reasonable over a short segment of a 1D signal or over a small block in a 2D array. It is noted, however, that in the design of the optimum quantizer with the attendant optimum allocation of bits the wide-sense properties are not enough. We also need to know or model the probability density function of the individual data samples.

The wide-sense properties of the signal vector $\{f(n), 0 \leq n \leq N-1\}$ are denoted by

$$\begin{aligned} E\{f(n)\} &= \mu(n) \\ E\{f(n)f(n+k)\} &= R(n+k,n). \end{aligned}$$

(For ease of notation, we assume that the data samples are real.) Wide-sense stationarity implies that the mean μ is a constant, independent of n, and that the autocorrelation simplifies to $R(n+k,n) = R(k)$, a function of only the time difference between the signal samples. Thus,

$$E\{f(n)\} = \mu, \text{ or } \qquad E\{\underline{f}\} = \mu \begin{bmatrix} 1 \\ 1 \\ \vdots \\ 1 \end{bmatrix},$$

and

$$E\{f(n)f(n+k)\} = R(k).$$

The autocovariance is then

$$\begin{aligned} C(k) &\triangleq E\{(f(n)-\mu)(f(n+k)-\mu)\} \\ &= R(k) - \mu^2. \end{aligned} \tag{2.51}$$

The correlation and covariance matrices are

$$\begin{aligned} R &= E\{\underline{f}\ \underline{f}^T\} \\ C &= cov\{\underline{f}\} = E\{(\underline{f}-\underline{\mu})(\underline{f}-\underline{\mu})^T\} \\ &= R - \underline{\mu}\ \underline{\mu}^T. \end{aligned} \tag{2.52}$$

For the WSS case, we see that

$$R = \begin{bmatrix} R(0) & R(1) & \cdots & R(N-1) \\ R(1) & R(0) & \cdots & R(N-2) \\ \cdots & \cdots & \cdots & \cdots \\ R(N-1) & R(N-2) & \cdots & R(0) \end{bmatrix} \tag{2.53}$$

$$\underline{\mu}\ \underline{\mu}^T = \mu^2 \begin{bmatrix} 1 & 1 & \cdots & 1 \\ 1 & 1 & \cdots & 1 \\ \cdots & & & \\ 1 & 1 & \cdots & 1 \end{bmatrix}. \tag{2.54}$$

It is common practice to simplify the notation (without loss of generality) by assuming that the constant mean either is zero or has been removed from the data. Thus, for zero mean, $R = C$, and using

$$R(k) = \sigma^2 \rho(k), \qquad \rho(0) = 1, \tag{2.55}$$

we get

$$R = cov\{\underline{f}\} = \sigma^2 \begin{bmatrix} 1 & \rho(1) & \cdots & \rho(N-1) \\ \rho(1) & 1 & \cdots & \rho(N-2) \\ \vdots & \vdots & \vdots & \vdots \\ \rho(N-1) & \cdots & \cdots & 1 \end{bmatrix}. \tag{2.56}$$

This symmetric matrix with equal entries along the main diagonal and along lines parallel to the main diagonal is called a *Toeplitz* matrix. It is also known as an *autocorrelation* matrix in the speech processing community (Jayant and Noll, 1984). In particular $R(0) = E\{|f(n)|^2\} = \sigma^2$ represents the variance or "power" in the signal samples.

The simplest example of an autocorrelation is that of a stationary, zero-mean, white sequence. In this instance, $\rho(k) = 0, \quad k \neq 0, \quad$ and R is diagonal

$$R = \sigma^2 I. \tag{2.57}$$

Another typical signal representation is that of the first-order autoregressive AR(1) sequence, modeled by

$$f(n) = \rho f(n-1) + \xi(n), \tag{2.58}$$

where

$$\begin{aligned} E\{\xi(n)\xi(n+k)\} &= A\delta(k) \\ A &= (1-\rho^2)\sigma^2. \end{aligned} \tag{2.59}$$

The autocorrelation is simply

$$R(k) = \sigma^2 \rho^{|k|}, \tag{2.60}$$

and the covariance reduces to

$$R = \sigma^2 \begin{bmatrix} 1 & \rho & \rho^2 & \cdots & \rho^{N-1} \\ \rho & 1 & \rho & \cdots & \cdots \\ \vdots & \vdots & \vdots & \vdots & \vdots \\ \rho^{N-1} & \cdots & \cdots & \cdots & 1 \end{bmatrix}. \tag{2.61}$$

For ease of notation, we assume real signals, and zero means. The transformation $\underline{\theta} = \Phi \underline{f}$ leads to

$$\begin{aligned} cov\{\underline{\theta}\} &= R_\theta \stackrel{\triangle}{=} E\{\underline{\theta}\ \underline{\theta}^T\} \\ &= E\{\Phi \underline{f}\ \underline{f}^T \Phi^T\} = \Phi E\{\underline{f}\ \underline{f}^T\}\Phi^T \\ &= \Phi R_f \Phi^T, \end{aligned} \tag{2.62}$$

where the subscripts on the covariance matrices denote the variables in question. The energy preserving properties of a unitary transformation have already been developed in Eq. (2.37). From a statistical standpoint, we now have

$$\begin{aligned} E\{\underline{\theta}^T \underline{\theta}\} &= \sum_{r=0}^{N-1} E\{\theta_r^2\} = \sum_{r=0}^{N-1} \sigma_r^2 \\ &= E\{\underline{f}^T(\Phi^T \Phi)\underline{f}\} = E\{\underline{f}^T \underline{f}\} \\ &= \sum_{k=0}^{N-1} E\{|f(k)|^2 = N\sigma^2. \end{aligned}$$

Therefore, the energy preservation property emerges as

$$\sum_{r=0}^{N-1} \sigma_r^2 = N\sigma^2. \tag{2.63}$$

Note that the transformation results in a *nonstationary* sequence of spectral coefficients. The σ_r^2 are not constant *by design*. In fact, we would like to whiten the $\{\theta_r\}$ sequence by making the off-diagonal terms in R_θ zero, while making it nonstationary by compacting the energy into as few coefficients as possible. Viewed in this light, the purpose of the transformation is to generate a diagonal covariance matrix R_θ whose elements are unevenly distributed.

The foregoing considerations lead to two measures of transform efficiency (Clarke, 1985). The decorrelation efficiency η_c compares the sum of the off-diagonal terms in R_θ, and R_f. We define

$$\lambda_f = \sum_{i=1}^{N-1} \sum_{\substack{j=1 \\ i \neq j}}^{N-1} |R_f(i,j)|, \quad \lambda_\theta = \sum_{i=1}^{N-1} \sum_{\substack{j=1 \\ i \neq j}}^{N-1} |R_\theta(i,j)|. \tag{2.64}$$

Then the decorrelation efficiency is

$$\eta_c = 1 - \frac{\lambda_\theta}{\lambda_f}. \tag{2.65}$$

For completely decorrelated spectral coefficients, $\eta_c = 1$.

The second parameter η_E measures the energy compaction property of the transform. Defining J'_L as the expected value of the summed squared error J_L of Eq. (2.20)

$$\lambda_E = \frac{J'_L}{J'_0} = \frac{E\{J_L\}}{E\{J_0\}} = \frac{\sum_{r=L}^{N-1} \sigma_r^2}{\sum_{r=0}^{N-1} \sigma_r^2} = \frac{\sum_{r=L}^{N-1} \sigma_r^2}{N\sigma^2}. \tag{2.66}$$

This J'_L has also been called the *basis restriction error* by Jain (1989). Then the compaction efficiency is

$$\eta_E = 1 - \lambda_E = \frac{\left(\sum_{r=0}^{L-1} \sigma_r^2\right)}{N\sigma^2}. \tag{2.67}$$

Thus η_E is the fraction of the total energy in the first L components of $\underline{\theta}$, where $\{\sigma_r^2\}$ are indexed according to decreasing value.

The unitary transformation that makes $\eta_c = 0$ and minimizes J'_L is the Karhunen-Loeve transform (Karhunen, 1947; Hotelling, 1933). Our derivation for real signals and transforms follows.

Consider a unitary transformation Φ such that

$$\begin{aligned} \underline{\theta} &= \Phi \underline{f}, \quad \underline{f} = \Phi^T \underline{\theta} \\ \underline{f} &= \sum_{r=0}^{N-1} \theta_r \underline{\Phi}_r. \end{aligned} \tag{2.68}$$

The approximation $\underline{f}_L$ and approximation error $\underline{e}_L$ are

$$\begin{aligned} \underline{f}_L &= \sum_{r=0}^{L-1} \theta_r \underline{\Phi}_r \\ \underline{e}_L &= \underline{f} - \underline{f}_L = \sum_{r=L}^{N-1} \theta_r \underline{\Phi}_r. \end{aligned} \tag{2.69}$$

By orthonormality, it easily follows that

$$\underline{e}_L^T \underline{e}_L = \sum_{r=L}^{N-1} \theta_r^2$$

and

$$J'_L = E\{\underline{e}_L^T \underline{e}_L\} = \sum_{r=L}^{N-1} E\{\theta_r^2\}. \tag{2.70}$$

From Eq. (2.34),

$$\theta_r = \underline{\Phi}_r^T \underline{f}.$$

Therefore,

$$\begin{aligned} E\{\theta_r^2\} &= E\{\underline{\Phi}_r^T(\underline{f}\underline{f}^T)\underline{\Phi}_r\} \\ &= \underline{\Phi}_r^T R_f \underline{\Phi}_r, \end{aligned} \tag{2.71}$$

so that the error measure becomes

$$J_L' = \sum_{r=L}^{N-1} \underline{\Phi}_r^T R_f \underline{\Phi}_r. \tag{2.72}$$

To obtain the optimum transform, we want to find the $\underline{\Phi}_r$ that minimizes J_L' for a given L, subject to the orthonormality constraint, $\underline{\Phi}_r^T \underline{\Phi}_s = \delta_{r-s}$. Using Lagrangian multipliers, we minimize

$$J = \sum_{r=L}^{N-1} [\underline{\Phi}_r^T R_f \underline{\Phi}_r - \lambda_r(\underline{\Phi}_r^T \underline{\Phi}_r - 1)]. \tag{2.73}$$

Each term in the sum is of the form

$$J = \underline{x}^T R \underline{x} - \lambda(\underline{x}^T \underline{x} - 1). \tag{2.74}$$

Taking the gradient[4] of this with respect to $\underline{x}$ (Prob. 2.6),

$$\nabla_x J = \frac{\partial J}{\partial \underline{x}} = 2R\underline{x} - 2\lambda \underline{x} = 0,$$

or

$$R\underline{x} = \lambda \underline{x}.$$

Doing this for each term in Eq. (2.74) gives

$$R_f \underline{\Phi}_r = \lambda_r \underline{\Phi}_r, \tag{2.75}$$

which implies

$$R_f \Phi^T = \Phi^T \Lambda$$

where

$$\Lambda = diag(\lambda_0, \ldots, \lambda_{N-1}).$$

[4]The gradient is a vector defined as $\nabla f = \frac{\partial f}{\partial \underline{x}} = \left[\frac{\partial f}{\partial x_1} \cdots \frac{\partial f}{\partial x_N}\right]^T$

(The reason for the transpose is that we had defined $\underline{\Phi}_r$ as the rth column of Φ.) Hence $\underline{\Phi}_r$ is an eigenvector of the signal covariance matrix R_f, and λ_r, the associated eigenvalue, is a root of the characteristic polynomial, $det(\lambda I - R_f)$. Since R_f is a real, symmetric matrix, all $\{\lambda_i\}$ are real, distinct, and nonnegative. The value of the minimized J'_L is then

$$(J'_L)_{min} = \sum_{r=L}^{N-1} \underline{\Phi}_r^T(\lambda_r \underline{\Phi}_r) = \sum_{r=L}^{N-1} \lambda_r. \tag{2.76}$$

The covariance matrix for the spectral coefficient vector is diagonal, as can be seen from

$$R_\theta = \Phi R_f \Phi^T = \Phi\Phi^T \Lambda = \Lambda. \tag{2.77}$$

Thus Φ is the unitary matrix that does the following:
(1) generates a diagonal R_θ and thus completely decorrelates the spectral coefficients resulting in $\eta_c = 1$,
(2) repacks the total signal energy among the first L coefficients, maximizing η_E.

It should be noted, however, that while many matrices can decorrelate the input signal, the KLT both decorrelates the input perfectly and optimizes the repacking of signal energy. Furthermore, it is unique. The difficulty with this transformation is that it is input signal specific—i.e., the matrix Φ^T consists of the eigenvectors of the input covariance matrix R_f. It does provide a theoretical limit against which signal-independent transforms (DFT, DCT, etc.) can be compared. In fact, it is well known that for an AR(1) signal source Eq. (2.61) with ρ large, on the order of 0.9, the DCT performance is very close to that of the KLT. A frequently quoted result for the AR(1) signal and N even (Ray and Driver, 1970) is

$$\lambda_k = \frac{1-\rho^2}{1-2\rho\, cos(\omega_k) + \rho^2} \tag{2.78}$$

where $\{\omega_k\}$ are the positive roots of

$$\begin{aligned} tan(N\omega_k) &= -\frac{(1-\rho^2) sin(\omega_k)}{cos(\omega_k) - 2\rho + \rho cos(\omega_k)} \\ \Phi(r,k) &= \frac{2}{(N+\lambda_r)} sin\{\omega_r(k - \frac{N-1}{2}) + (r+1)\frac{\pi}{2}\}. \end{aligned} \tag{2.79}$$

This result simply underscores the difficulty in computing the KLT even when applied to the simplest, nontrivial signal model. In the next section, we describe other fixed transforms and compare them with the KLT. (See also Prob. 2.19)

For $\rho = 0.91$ in the AR(1) model and $N = 8$, Clarke (1985) has calculated the packing and decorrelation efficiencies of the KLT and the DCT:

	L	1	2	3	4	5	6	7	8
η_E	KLT	79.5	91.1	94.8	96.7	97.9	98.7	99.4	100
η_E	DCT	79.3	90.9	94.8	96.7	97.9	98.7	99.4	100

These numbers speak for themselves. Also for this example, $\eta_c = 0.985$ for the DCT compared with 1.0 for the KLT.

2.2.2 Comparative Performance Measures

The efficiency measures η_c, η_E in Section 2.2.1 provide the bases for comparing unitary transforms against each other. We need, however, a performance measure that ranges not only over the class of transforms, but also over different coding techniques. The measure introduced here serves that purpose.

In all coding techniques, whether they be pulse code modulation (PCM), differential pulse code modulation (DPCM), transform coding (TC), or subband coding (SBC), the basic performance measure is the reconstruction error (or distortion) at a specified information bit rate for storage or transmission.

We take as the basis for all comparisons, the simplest coding scheme, namely the PCM, and compare all others to it. With respect to Fig. 2.3, we see that PCM can be regarded as a special case of TC wherein the transformation matrix Φ is the identity matrix I, in which case we have simply $\underline{\theta} = \underline{f}$. The reconstruction error is $\underline{\tilde{f}}$ as defined by Eq. (2.38), and the mean square reconstruction error is

$$\sigma_\epsilon^2 = \frac{1}{N} E\{\underline{\tilde{f}}^T \underline{\tilde{f}}\}. \tag{2.80}$$

The TC performance measure compares σ_ϵ^2 for TC to that for PCM. This measure is called the *gain* of transform coding over PCM and defined (Jayant and Noll, 1984) as

$$G_{TC} = \frac{\sigma^2_{\epsilon(PCM)}}{\sigma^2_{\epsilon(TC)}}. \tag{2.81}$$

In the next chapter on subband coding, we will similarly define

$$G_{SBC} = \frac{\sigma^2_{\epsilon(PCM)}}{\sigma^2_{\epsilon(SBC)}}. \tag{2.82}$$

In Eq. (2.39) we asserted that for a unitary transform, the mean square reconstruction error equals the mean square quantization error. The proof is easy. Since

$$\underline{\tilde{f}} = \underline{f} - \underline{\hat{f}} = \Phi^T(\underline{\theta} - \underline{\hat{\theta}}) = \Phi^T \underline{\tilde{\theta}},$$

then

$$\underline{\tilde{f}}^T \underline{\tilde{f}} = \underline{\tilde{\theta}}^T (\Phi\Phi^T)\underline{\tilde{\theta}} = \underline{\tilde{\theta}}^T \underline{\tilde{\theta}}, \tag{2.83}$$

where $\underline{\tilde{\theta}}$ is the quantization error vector.

The average mean square (m.s.) error (or distortion) is

$$\sigma_\epsilon^2 = \frac{1}{N} E\{\underline{\tilde{f}}^T \underline{\tilde{f}}\} = \frac{1}{N} E\{\underline{\tilde{\theta}}^T \underline{\tilde{\theta}}\} = \frac{1}{N} \sum_{k=0}^{N-1} \sigma_{q_k}^2, \tag{2.84}$$

where $\sigma_{q_k}^2$ is the variance of the quantization error in the Kth spectral coefficient,

$$\sigma_{q_k}^2 = E\{|\tilde{\theta}_k|^2\}, \tag{2.85}$$

as depicted in Fig. 2.6.

Suppose that R_k bits are allocated to quantizer Q_k. Then we can choose the quantizer to minimize $\sigma_{q_k}^2$ for this value of R_k and the given probability density function for θ_k. This minimum mean square error quantizer is called the Lloyd-Max quantizer, (Lloyd, 1957; Max, 1960). It minimizes separately each $\sigma_{q_k}^2$, and hence the sum $\sum_k \sigma_{q_k}^2$. The structure of Fig. 2.6 suggests that the quantizer can be thought of an estimator, particularly so since a mean square error is being minimized. For the optimal quantizer it can be shown that the quantization error is unbiased, and that the error is orthogonal to the quantizer output (just as in the case for optimal linear estimator), (Prob. 2.9)

$$E\{\tilde{\theta}_k\} = 0, \qquad E\{\tilde{\theta}_k \hat{\theta}_k\} = 0 \tag{2.86}$$

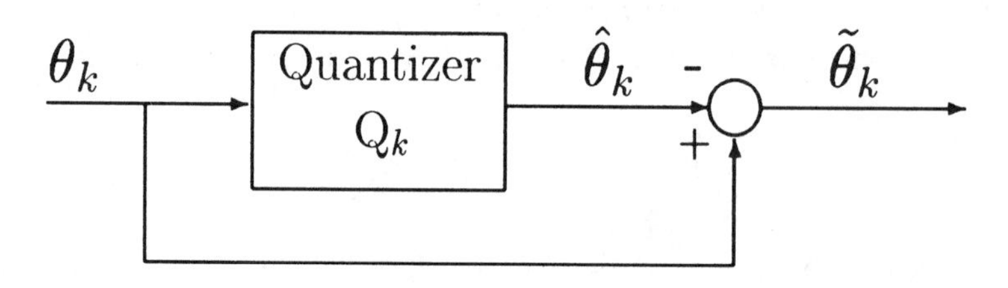

Figure 2.6: The coefficient quantization error.

The resulting mean square error or distortion, depends on the spectral coefficient variance σ_k^2, the pdf, the quantizer (in this case, Lloyd-Max), and the number of bits R_k allocated to the kth coefficient. From rate-distortion theory (Berger, 1971), the error variance can be expressed as

$$\sigma_{q_k}^2 = f(R_k)\sigma_k^2, \tag{2.87}$$

where $f(R_k)$ is the quantizer distortion function for a unity variance input. Typically,

$$f(R_k) = \gamma_k 2^{-2R_k}, \tag{2.88}$$

where γ_k depends on the pdf for θ_k and on the specific quantizer. Jayant and Noll (1984) report values of $\gamma = 1.0$, 2.7, 4.5, and 5.7 for uniform, Gaussian, Laplacian, and Gamma pdfs, respectively. The average mean square reconstruction error is then

$$\sigma_\epsilon^2 = \frac{1}{N}\sum_{k=0}^{N-1}\sigma_{q_k}^2 = \frac{1}{N}\sum_{k=0}^{N-1}\gamma_k 2^{-2R_k}\sigma_k^2. \tag{2.89}$$

Next, there is the question of bit allocation to each coefficient, constrained by NR, the total number of bits available to encode the coefficient vector $\underline{\theta}$

$$NR = \sum_{k=0}^{N-1} R_k, \tag{2.90}$$

and R is the average number of bits per coefficient. To minimize Eq. (2.89) subject to the constraint of Eq. (2.90), we again resort to Lagrangian multipliers. First we assume γ_k to be the *same for each coefficient*, and then solve

$$\frac{\partial}{\partial R_k}\{\frac{1}{N}\sum_{k=0}^{N-1}\gamma\sigma_k^2 2^{-2R_k} - \lambda(R - \frac{1}{N}\sum_{k=0}^{N-1} R_k)\} = 0, \tag{2.91}$$

to obtain (Prob. 2.7)

$$R_k = R + \frac{1}{2}log_2\frac{\sigma_k^2}{(\prod_{j=0}^{N-1}\sigma_j^2)^{1/N}}. \tag{2.92}$$

This result is due to Huang and Schultheiss (1963) and Segall (1976). The number of bits is proportional to the logarithm of the coefficient variance, or to the power in that band, an intuitively expected result.

It can also be shown that the bit allocation of Eq. (2.92) results in equal quantization error for each coefficient, and thus the distortion is spread out evenly among all the coefficients, (Prob. 2.8)

$$\sigma_{q_k}^2 = \gamma 2^{-2R} (\prod_{j=1}^{N-1} \sigma_j^2)^{1/N} = constant. \tag{2.93}$$

The latter also equals the average distortion, since

$$\sigma_\epsilon^2 = \frac{1}{N} \sum_{k=0}^{N-1} \sigma_{q_k}^2 = \sigma_{q_k}^2. \tag{2.94}$$

The preceding result is the pdf and R_k optimized distortion for any unitary transform. For the PCM case, $\Phi = I$, and σ_ϵ^2 reduces to

$$\sigma_\epsilon^2 = \frac{1}{N} \sum_k \sigma_{q_k}^2 = \frac{1}{N} \sum_k \gamma 2^{-2R} \sigma^2 = \gamma 2^{-2R} \sigma^2. \tag{2.95}$$

There is a tacit assumption here that the γ in the PCM case of Eq. (2.95) is the same as that for TC in Eq. (2.93). This may not be the case when, for example, the transformation changes the pdf of the input signal. We will neglect this effect.

Recall from Eq. (2.63) that, for a unitary transform,

$$\sum_{j=0}^{N-1} \sigma_j^2 = N\sigma^2. \tag{2.96}$$

The ratio of distortions in Eqs. (2.95) and (2.93) gives

$$max\ {}^N G_{TC} = \frac{\sigma_{\epsilon(PCM)}^2}{min\, \sigma_{\epsilon(TC)}^2} = \frac{\frac{1}{N} \sum_{j=0}^{N-1} \sigma_j^2}{(\prod_{j=0}^{N-1} \sigma_j^2)^{1/N}}. \tag{2.97}$$

The maximized G_{TC} is the ratio of the arithmetic mean of the coefficient variances to the geometric mean.

Among all unitary matrices, the KLT minimizes the geometric mean of the coefficient variances. To appreciate this, recall that from Eq. (2.77) the KLT produced a diagonal R_θ, so that

$$|R_f| = |R_\theta| = |\Lambda| = \prod_{k=0}^{N-1} \lambda_k = \prod_{k=0}^{N-1} \sigma_k^2. \tag{2.98}$$

Hence,

$$max_{\Phi}\{^N G_{TC}\} =^N G_{KLT} = \frac{\sigma^2}{(\prod_{k=0}^{N-1} \lambda_k)^{1/N}} = \frac{\sigma^2}{|R_f|^{1/N}}. \tag{2.99}$$

The limiting value of G_{KLT} for $N \to \infty$ gives an upper bound on transform coding performance. The denominator in Eq. (2.99) can be expressed as

$$(\prod_{i=0}^{N-1} \lambda_i)^{1/N} = e^{\frac{1}{N}\sum_i \ln(\lambda_i)}. \tag{2.100}$$

Jayant and Noll (1984) show that

$$\lim_{N\to\infty} \frac{1}{N} \sum_{i=0}^{N-1} \ln \lambda_i = \frac{1}{2\pi} \int_{-\pi}^{\pi} \ln\{S_f(e^{j\omega})\} d\omega, \tag{2.101}$$

where S_f is the power spectral density of the signal

$$S_f(e^{j\omega}) = \mathcal{F}\{R_f(k)\}.$$

Hence,

$$\lim_{N\to\infty} |R_f|^{1/N} = exp\{\frac{1}{2\pi} \int_{-\pi}^{\pi} \ln\{S_f(e^{j\omega})\} d\omega\}, \tag{2.102}$$

and the numerator in Eq. (2.99) is recognized as

$$\sigma^2 = \frac{1}{2\pi} \int_{-\pi}^{\pi} S_f(e^{j\omega}) d\omega = R_f(0). \tag{2.103}$$

Hence,

$$^{\infty}G_{TC} = \frac{\frac{1}{2\pi}\int_{-\pi}^{\pi} S_f(e^{j\omega}) d\omega}{exp\{\frac{1}{2\pi}\int_{-\pi}^{\pi} \ln[S_f(e^{j\omega})] d\omega\}}, \tag{2.104}$$

is the reciprocal of the *spectral flatness measure* introduced by Makhoul and Wolf (1972). It is a measure of the predictability of a signal. For white noise, $^{\infty}G_{TC} = 1$ and there is no coding gain. This measure increases with the degree of correlation and hence predictability. Accordingly, coding gain increases as the redundancy in the signal is removed by the unitary transformation.

2.3 Fixed Transforms

The KLT described in Section 2.2 is the optimal unitary transform for signal coding purposes. But the DCT is a strong competitor to the KLT for highly correlated signal sources. The important practical features of the DCT are that it is signal independent (that is, a fixed transform), and there exist fast computational algorithms for the calculation of the spectral coefficient vector. In this section we define, list, and describe the salient features of the most popular fixed transforms. These are grouped into three categories: sinusoidal, polynomial, and rectangular transforms.

2.3.1 Sinusoidal Transforms

The discrete Fourier transform (DFT) and its linear derivatives the discrete cosine transform (DCT) and the discrete sine transform (DST) are the main members of the class described here.

2.3.1.1 The Discrete Fourier Transform

The DFT is the most important orthogonal transformation in signal analysis with vast implication in every field of signal processing. The fast Fourier transform (FFT) is a fast algorithm for the evaluation of the DFT.

The set of orthogonal (but not normalized) complex sinusoids is the family

$$\begin{aligned} x_r(n) &= e^{j2\pi rn/N} = W^{rn}, \qquad r = 0, 1, \cdots, N-1 \\ W &\triangleq e^{j2\pi/N}, \end{aligned} \tag{2.105}$$

with the property

$$\sum_{n=0}^{N-1} x_r(n) x_s^*(n) = N\delta_{r-s}. \tag{2.106}$$

Most authors define the forward and inverse DFTs as

$$X(k) = DFT\{x(n)\} = \sum_{n=0}^{N-1} x(n) e^{-j2\pi nk/N} \tag{2.107}$$

$$x(n) = DFT^{-1}\{X(k)\} = \frac{1}{N} \sum_{k=0}^{N-1} X(k) e^{j2\pi nk/N}. \tag{2.108}$$

The corresponding matrices are

$$\Phi = [W^{-nk}], \qquad \Phi^{-1} = \frac{1}{N}[W^{nk}] = \frac{1}{N}\Phi^*. \tag{2.109}$$

This definition is consistent with the interpretation that the DFT is the Z-transfrom of $\{x(n)\}$ evaluated at N equally-spaced points on the unit circle. The set of coefficients $\{X(k)\}$ constitutes the frequency spectrum of the samples. From Eqs. (2.107) and (2.108) we see that both $X(k)$ and $x(n)$ are periodic in their arguments with period N. Hence Eq. (2.108) is recognized as the discrete Fourier series expansion of the periodic sequence $\{x(n)\}$, and $\{X(k)\}$ are just the discrete Fourier series coefficients scaled by N. Conventional frequency domain interpretation permits an identification of $X(0)/N$ as the "DC" value of the signal. The fundamental $\{x_1(n) = e^{j2\pi n/N}\}$ is a unit vector in the complex plane that rotates with the time index n. The first harmonic $\{x_2(n)\}$ rotates at twice the rate of fundamental and so on for the higher harmonics. The properties of this transform are summarized in Table 2.1. For more details, the reader can consult the wealth of literature on this subject, e.g., Papoulis (1991), Oppenheim and Schafer (1975), Haddad and Parsons (1991).

The unitary DFT is simply a normalized DFT wherein the scale factor N appearing in Eqs. (2.106)–(2.109) is reapportioned according to

$$X'(k) = \frac{1}{\sqrt{N}} X(k) = \frac{1}{\sqrt{N}} \sum_{n=0}^{N-1} x(n) W^{-nk}.$$

This makes

$$x(n) = \frac{1}{\sqrt{N}} \sum_{k=0}^{N-1} X'(k) W^{nk}, \tag{2.110}$$

and the unitary transformation matrix is

$$\begin{aligned} \Phi &= \frac{1}{\sqrt{N}} [W^{-nk}], \\ \Phi^{-1} &= \frac{1}{\sqrt{N}} [W^{nk}] = (\Phi^*)^T = \Phi^*. \end{aligned} \tag{2.111}$$

From a coding standpoint, a key property of this transformation is that the basis vectors of the unitary DFT (the columns of Φ^*) are the eigenvectors of a *circulant* matrix. That is, with the kth column of Φ^* denoted by

$$(\underline{\Phi}_k^*)^T = [W^0, W^{1k}, \cdots, W^{(N-1)k}], \tag{2.112}$$

we will show that $\underline{\Phi}_k^*$ are the eigenvectors in

$$\mathcal{H} \underline{\Phi}_k^* = \lambda_k \underline{\Phi}_k^*, \tag{2.113}$$

where $\mathcal{H}$ is any circulant matrix

$$\mathcal{H} = \begin{bmatrix} h_0 & h_{N-1} & \cdots & h_1 \\ h_1 & h_0 & \cdots & h_2 \\ \vdots & \vdots & \vdots & \vdots \\ h_{N-1} & h_{N-2} & \cdots & h_0 \end{bmatrix}. \tag{2.114}$$

Each column (or row) is a circular shift of the previous column (or row). The eigenvalue λ_k is the DFT of the first column of $\mathcal{H}$

$$\lambda_k = \sum_{n=0}^{N-1} h(n) W^{-kn} = H(k). \tag{2.115}$$

$$X(k) = \sum_{n=0}^{N-1} x(n) W^{-nk} \leftrightarrow x(n) = \frac{1}{N} \sum_{k=0}^{N-1} X(k) W^{nk}$$

Property	*Operation*
(1) *Orthogonality*	$\sum_{k=0}^{N-1} W^{mk} W^{-nk} = N\delta_{m-n}$
(2) *Periodicity*	$x(n+rN) = x(n)$ $X(k+lN) = X(k)$
(3) *Symmetry*	$Nx(-n) \leftrightarrow X(k)$
(4) *Circular Convolution*	$x(n) * y(n) \leftrightarrow X(k)Y(k)$
(5) *Shifting*	$x(n-n_o) \leftrightarrow W^{n_o k} X(k)$
(6) *Time Reversal*	$x(N-n) \leftrightarrow X(N-k)$
(7) *Conjugation*	$x^*(n) \leftrightarrow X^*(N-k)$
(8) *Correlation*	$\rho(n) = x(n) * x^*(-n) \leftrightarrow R(k) = \|X(k)\|^2$
(9) *Parseval*	$\sum_{n=0}^{N-1} \|x(n)\|^2 = \frac{1}{N} \sum_{k=0}^{N-1} \|X(k)\|^2$
(10) *Real Signals/ Conjugate Symmetry*	$X^*(N-k) = X(k)$

Table 2.1: Properties of the discrete Fourier transform.

We can write Eq. (2.113) as

$$\mathcal{H}[\underline{\Phi}_0^* \cdots \underline{\Phi}_{N-1}^*] = [\underline{\Phi}_0^* \cdots \underline{\Phi}_{N-1}^*]\Lambda, \tag{2.116}$$

where

$$\Lambda = diag(\lambda_0 \cdots \lambda_{N-1}),$$

which results in

$$\mathcal{H}\Phi^* = \Phi^*\Lambda,$$

or

$$\Lambda = \Phi\mathcal{H}\Phi^*.$$

The proof is straightforward. Consider a linear, time-invariant system with finite impulse response $\{h(n), 0 \leq n \leq N-1\}$, excited by the periodic input W^{kn}. The output is also periodic and given by

$$y(n) = h(n) * W^{kn} = \left[\sum_{m=0}^{N-1} h(m)W^{-mk}\right] W^{kn} = H(k)W^{kn} = \lambda_k W^{kn}. \tag{2.117}$$

Let the output vector be

$$\underline{y}^T = [y(0), y(1), \cdots, y(N-1)].$$

Then Eq. (2.117) can be stacked,

$$\underline{y} = \begin{bmatrix} W^0 \\ W^k \\ \vdots \\ W^{(N-1)k} \end{bmatrix} \lambda_k = \lambda_k \underline{\Phi}_k^*.$$

Since $y(n) = y(n + lN)$ is periodic, it can also be calculated by the *circular* convolution of W^{kn} and a periodically repeated

$$\tilde{h}(n) = \begin{cases} h(n), & 0 \leq n \leq N-1 \\ \tilde{h}(n+lN), & \text{otherwise.} \end{cases}$$

Hence,

$$\begin{aligned} y(n) &= \tilde{y}(n) = \sum_{i=0}^{N-1} \tilde{h}(n-i)W^{ki} \\ &= [\tilde{h}(n)\tilde{h}(n-1)\cdots\tilde{h}(0)\cdots\tilde{h}(n-N+1)] \begin{bmatrix} W^0 \\ W^k \\ \vdots \\ W^{(N-1)k} \end{bmatrix}. \end{aligned} \tag{2.118}$$

Stacking the output in Eq. (2.118) and recognizing the periodicity of terms such as $\tilde{h}(-1) = \tilde{h}(N-1) = h(N-1)$ gives us

$$\underline{y} = \begin{bmatrix} h_0 & h_{N-1} & \cdots & h_1 \\ h_1 & h_0 & \cdots & h_2 \\ \vdots & \vdots & \vdots & \vdots \\ h_{N-1} & h_{N-2} & \cdots & h_0 \end{bmatrix} \begin{bmatrix} W^0 \\ W^k \\ \vdots \\ W^{(N-1)k} \end{bmatrix}.$$

Equating the two stacked versions of $\underline{y}$ gives us our starting point, Eq. (2.113).

In summary, the DFT transformation diagonalizes any circulant matrix, and therefore completely decorrelates any signal whose covariance matrix has the circulant properties of $\mathcal{H}$.

2.3.1.2 The Discrete Cosine Transform

This transform is virtually the industry standard in image and speech transform coding because it closely approximates the KLT especially for highly correlated signals, and because there exist fast algorithms for its evaluation. The orthogonal set is (Prob. 2.10)

$$\begin{aligned} \Phi(r,n) = \Phi_r(n) &= (\frac{1}{c_r}) cos \frac{(2n+1)r\pi}{2N}, 0 \leq n, r \leq N-1 \qquad (2.119) \\ c_r &= \begin{cases} \sqrt{N}, & r = 0 \\ \sqrt{N/2}, & r \neq 0 \end{cases} \end{aligned}$$

and

$$\Phi = [\Phi(r,n)], \qquad \Phi^{-1} = \Phi^T. \qquad (2.120)$$

Jain (1976) argues that the basis vectors of the DCT approach the eigenvectors of the AR(1) process (Eq. 2.58) as the correlation coefficient $\rho \to 1$. The DCT is therefore near optimal (close to the KLT) for many correlated signals encountered in practice, as we have shown in the example given in Section 2.2.1. Some other characteristics of the DCT are as follows:

(1) The DCT has excellent compaction properties for highly correlated signals.

(2) The basis vectors of the DCT are eigenvectors of a symmetric tridiagonal matrix

$$Q = \begin{bmatrix} (1-\alpha) & -\alpha & 0 & \cdots & \cdots & 0 \\ -\alpha & 1 & -\alpha & \cdots & \cdots & 0 \\ \cdots & \cdots & \cdots & \cdots & \cdots & \cdots \\ \cdots & \cdots & \cdots & -\alpha & 1 & -\alpha \\ \cdots & \cdots & \cdots & 0 & -\alpha & (1-\alpha) \end{bmatrix}, \qquad (2.121)$$

whereas the covariance matrix of the AR(1) process has the form

$$R^{-1} = \frac{1}{\beta^2} \begin{bmatrix} (1-\rho\alpha) & -\alpha & 0 & \cdots & \cdots & \cdots \\ -\alpha & 1 & -\alpha & \cdots & \cdots & \cdots \\ 0 & \cdots & \cdots & \cdots & \cdots & \cdots \\ \cdots & \cdots & \cdots & -\alpha & 1 & -\alpha \\ \cdots & \cdots & \cdots & 0 & -\alpha & (1-\rho\alpha) \end{bmatrix}, \tag{2.122}$$

where

$$\beta^2 = \frac{1-\rho^2}{1+\rho^2}, \qquad \alpha = \frac{\rho}{1+\rho^2}.$$

As $\rho \to 1$, we see that $\beta^2 R^{-1} \cong Q$, confirming the decorrelation property. This is understood if we recognize that a diagonalizing unitary transformation implies

$$\Phi^T Q \Phi = \Lambda = diagonal$$

and consequently

$$\Lambda^{-1} = \Phi^T Q^{-1} \Phi.$$

Hence the matrix that diagonalizes Q also diagonalizes Q^{-1}.

Sketches of the DCT and other transform bases are displayed in Fig. 2.7. We must add one caveat, however. For a low or negative correlation the DCT performance is poor. However, for low ρ, transform coding itself does not work very well. Finally, there exist fast transforms using real operations for calculation of the DCT.

2.3.1.3 The Discrete Sine Transform

This transform is appropriate for coding signals with low or negative correlation coefficient. The orthogonal sine family is

$$x_r(n) = sin\frac{(n+1)(r+1)\pi}{(N+1)}, 0 \le r, n \le N-1,$$

with norm

$$\|x_r\|^2 = c_r^2 = \frac{1}{2}(N+1).$$

Normalization gives the unitary basis sequences as

$$\Phi(r,k) = \Phi_r(k) = \sqrt{\frac{2}{N+1}}\, sin\frac{(n+1)(r+1)\pi}{(N+1)}. \tag{2.123}$$

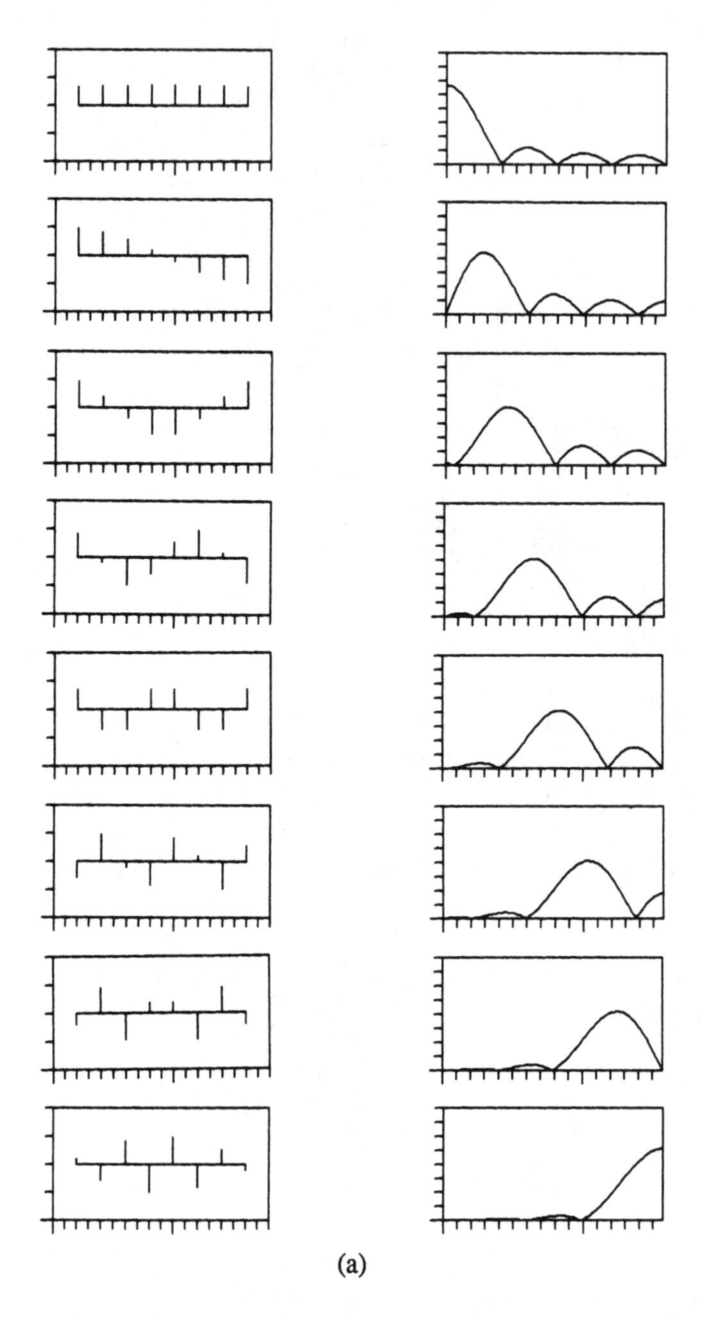

Figure 2.7: Transform bases in time and frequency domains for $N = 8$: (a) KLT ($\rho = 0.95$); (b) DCT; (c) DLT; (d) DST; (e) WHT; and (f) MHT.

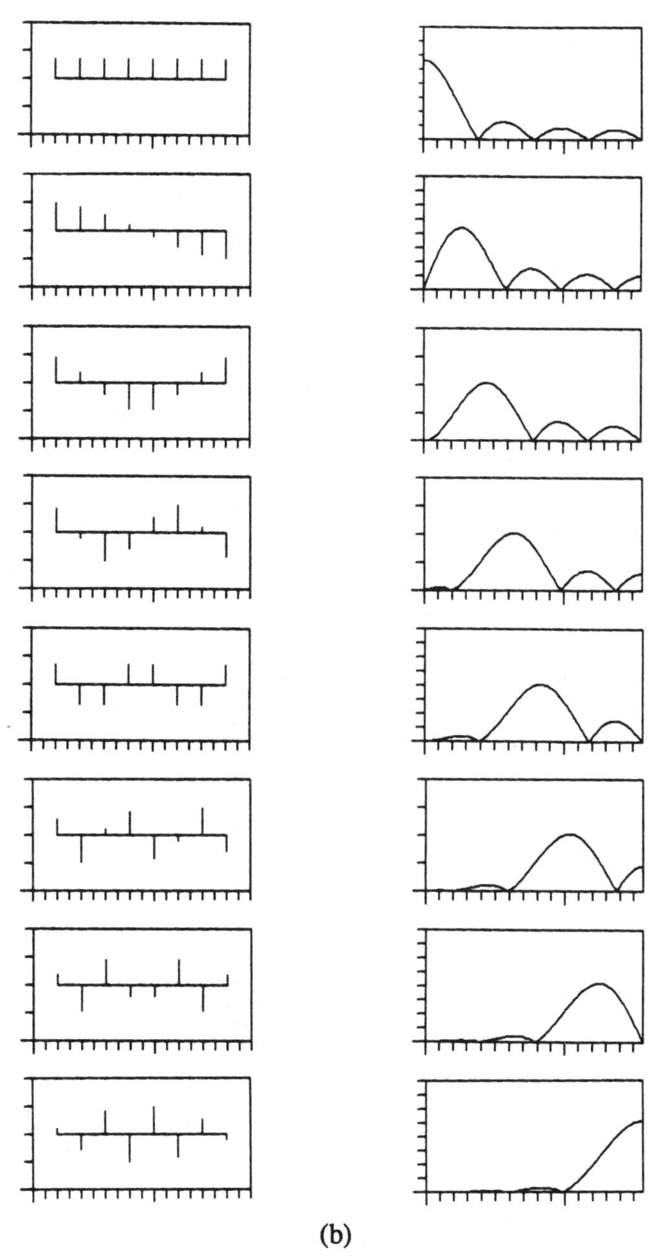

(b)

Figure 2.7 (*continued*)

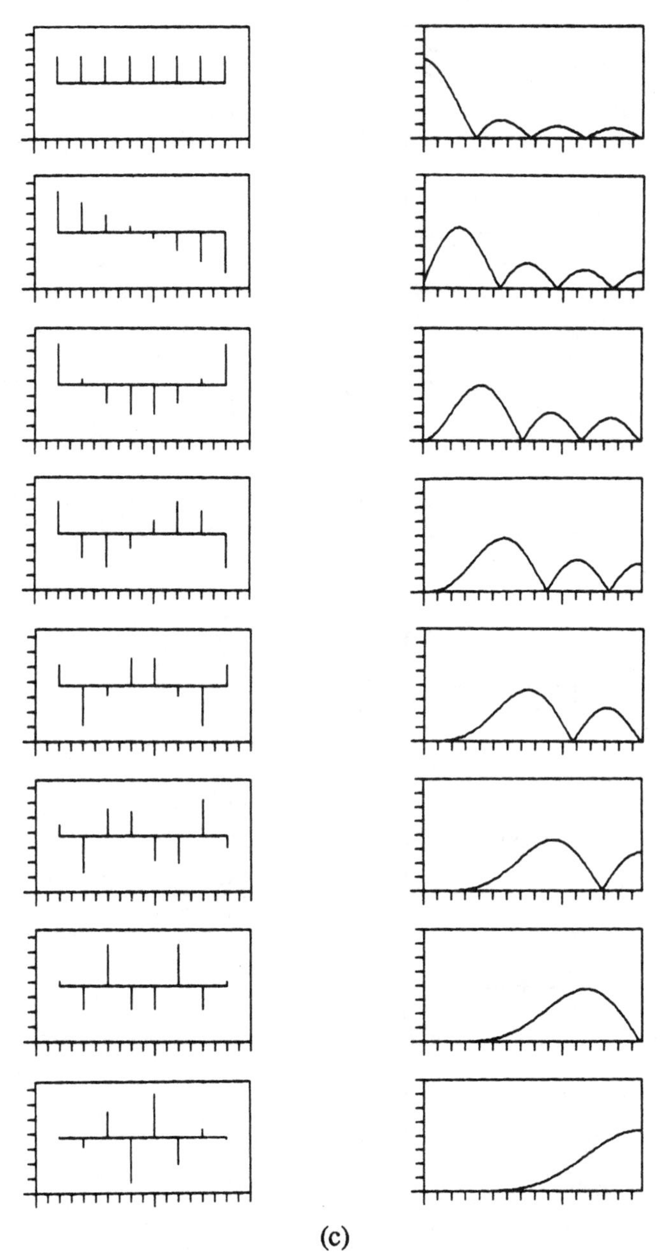

Figure 2.7 (*continued*)

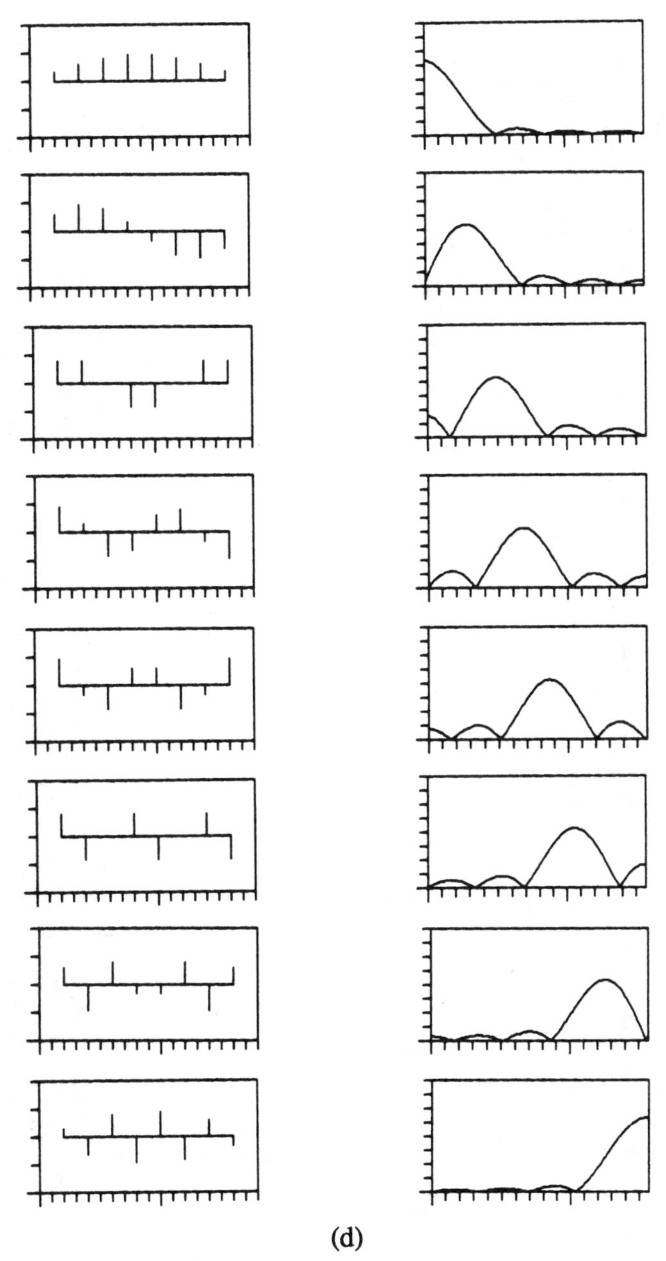

(d)

Figure 2.7 (*continued*)

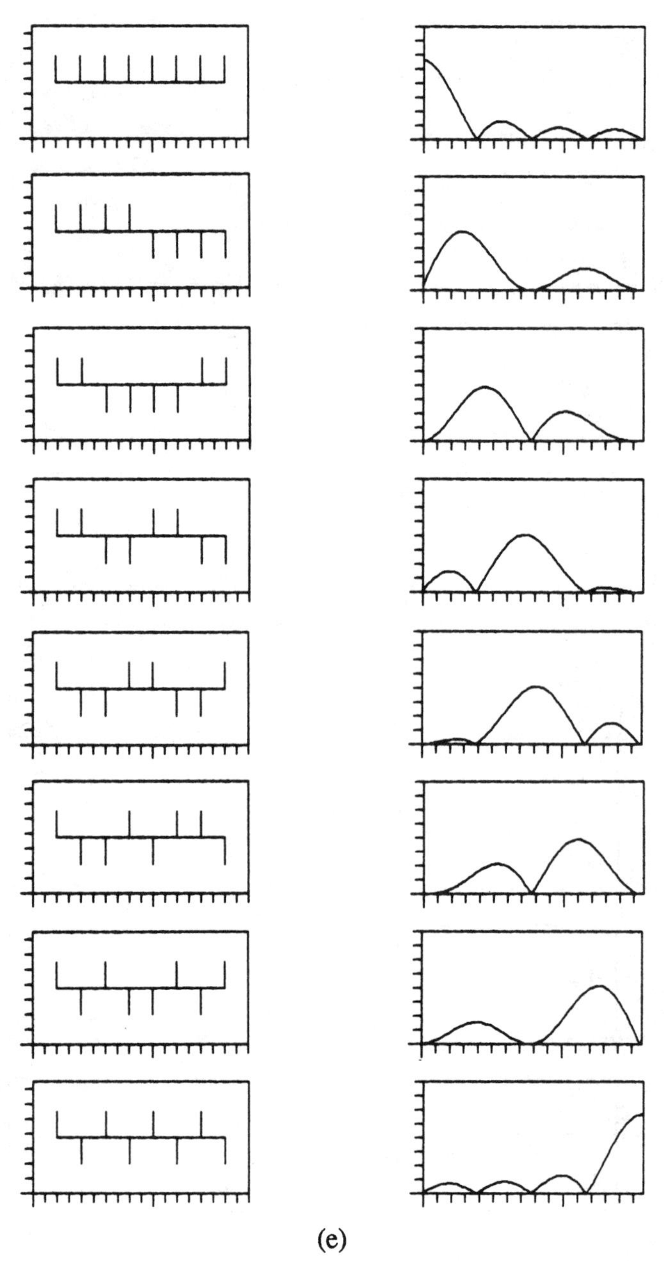

(e)

Figure 2.7 (*continued*)

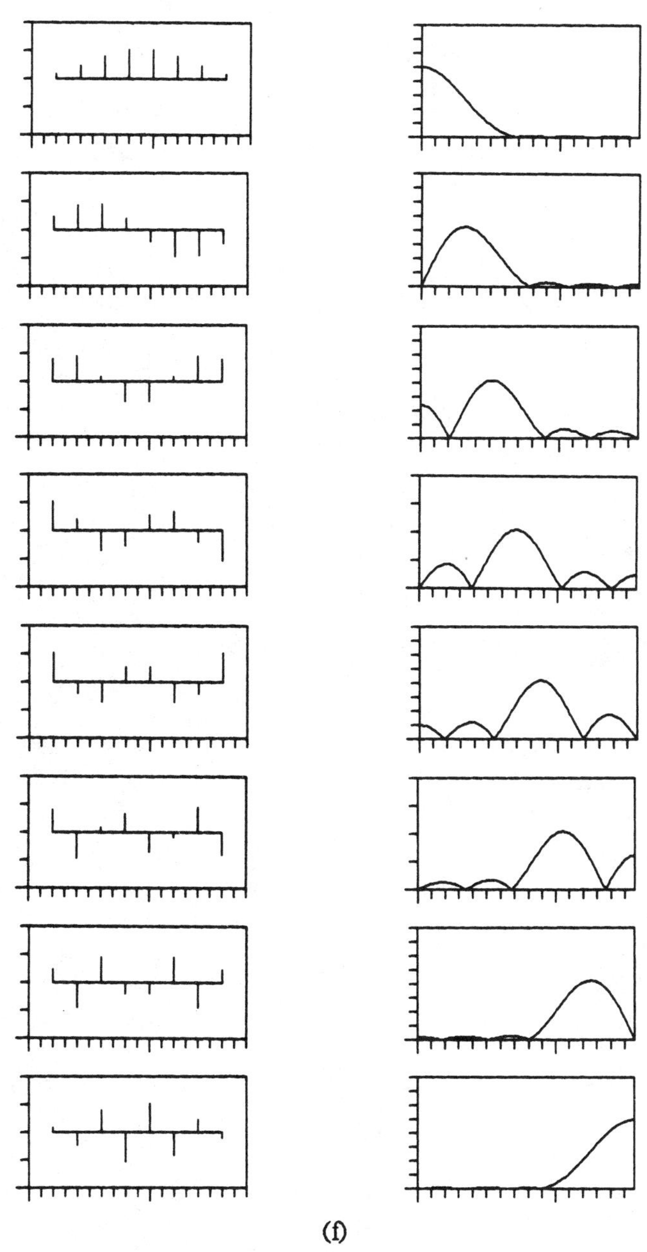

(f)

Figure 2.7 (*continued*)

It turns out the basis vectors of the DST are eigenvectors of the symmetric tridiagonal Toeplitz matrix

$$T = \begin{bmatrix} 1 & -\alpha & 0 & \cdots & 0 \\ -\alpha & 1 & -\alpha & \cdots & \cdots \\ 0 & \cdots & \cdots & \cdots & \cdots \\ \cdots & \cdots & -\alpha & 1 & -\alpha \\ \cdots & \cdots & 0 & -\alpha & 1 \end{bmatrix}. \tag{2.124}$$

The covariance matrix for the AR(1) process, Eq. (2.124), resembles this matrix for low correlated values of ρ, typically, $|\rho| < 0.5$. Of course, for $\rho = 0$, there is no benefit from transform coding since the signal is already white.

Some additional insight into the properties of the DCT, the DST, and relationship to the tridiagonal matrices Q and T in Eqs. (2.121)–(2.124) can be gleaned from the following observations (Ur, 1999):

(1) The matrices Q and T are part of a family of matrices with general structure

$$S = \begin{bmatrix} (1-k_1\alpha) & -\alpha & 0 & \cdots & 0 \\ -\alpha & 1 & -\alpha & \cdots & \cdots \\ 0 & \cdots & \cdots & \cdots & \cdots \\ \cdots & \cdots & -\alpha & 1 & -\alpha \\ k_3\alpha & \cdots & 0 & -\alpha & (1-k_2\alpha) \end{bmatrix}.$$

Jain (1979) showed that the set of eigenvectors generated from this parametric family of matrices define a family of sinusoidal transforms. Thus $k_1 = 1$, $k_2 = 1$, $k_3 = 0$ defines the matrix Q and $k_1 = k_2 = k_3 = 0$ specifies T.

(2) Clearly the DCT basis functions in Eq. (2.119) the eigenvectors of Q, must be independent of α. (But the eigenvalues of Q depend on α.) To see this, we can define a matrix $Q' = Q - (1-2\alpha)I$. Dividing by α, we obtain

$$Q'' = (1/\alpha)Q' = \begin{bmatrix} 1 & -1 & 0 & \cdots & 0 \\ -1 & 2 & -1 & \cdots & \cdots \\ 0 & \cdots & \cdots & \cdots & \cdots \\ \cdots & \cdots & -1 & 2 & -1 \\ 0 & \cdots & 0 & -1 & 1 \end{bmatrix},$$

$(1/\alpha)Q'$ is independent of α, but has the same eigenvectors as Q. (Problem 2.21)

(3) Except for the first and last rows, the rows of Q'' are $-1, 2, -1$, a second difference operator which implies sinusoidal solutions for the eigenvectors depending on initial conditions which are supplied by the first and last rows of the tridiagonal matrix S. Modifying these leads to 8 DCT forms.

(4) These comments also apply to the DST.

2.3.2 Discrete Polynomial Transforms

The class of discrete polynomial transforms are descendants, albeit not always in an obvious way, of their analog progenitors. (This was particularly true for the sinusoidal transforms.) The polynomial transforms are uniquely determined by the interval of definition or support, weighting function, and normalization. Three transforms are described here. The Binomial-Hermite family and the Legendre polynomials have finite support and are realizable in finite impulse response (FIR) form. The Laguerre family, defined on the semi-infinite interval $[0, \infty)$, can be realized as an infinite impulse response (IIR) structure.

2.3.2.1 The Binomial-Hermite Transform

This family of discrete weighted orthogonal functions was developed in the seminal paper by Haddad (1971), and subsequently orthonormalized (Haddad and Akansu, 1988).

The Binomial-Hermite family are discrete counterparts to the continuous-time orthogonal Hermite family familiar in probability theory. Before delving into the discrete realm, we briefly review the analog family to demonstrate the parental linkage to their discrete progeny.

The analog family (Sansone, 1959; Szego, 1959) is obtained by successive differentiation of the Gaussian $e^{-t^2/2}$,

$$x_n(t) = \frac{d^n}{dt^n}\{e^{-t^2/2}\} = H_n(t)e^{-t^2/2}. \tag{2.125}$$

The polynomials $H_n(t)$ in Eq. (2.125) are the Hermite polynomials. These can be generated by a two-term recursive formula

$$H_{n+1}(t) + tH_n(t) + nH_{n-1}(t) = 0 \tag{2.126}$$

$$H_0(t) = 1, H_1(t) = -t.$$

The polynomials also satisfy a linear, second-order differential equation

$$\ddot{H}_n(t) - t\dot{H}_n(t) + nH_n(t) = 0. \tag{2.127}$$

The Hermite family $\{x_n(t)\}$, and the Hermite polynomials $\{H_n(t)\}$, are orthogonal on the interval $(-\infty, \infty)$ with respect to weighting functions $e^{t^2/2}$ and $e^{-t^2/2}$, respectively:

$$\int_{-\infty}^{\infty} x_m(t)x_n(t)e^{t^2/2}dt = \int_{-\infty}^{\infty} H_m(t)H_n(t)e^{-t^2/2}dt = \sqrt{2\pi}n!\delta_{n-m}. \tag{2.128}$$

From a signal analysis standpoint, the key property of the Gaussian function is the isomorphism with the Fourier transform. We know that

$$F\{e^{-t^2/2}\} = \sqrt{2\pi}e^{-\omega^2/2}. \tag{2.129}$$

Furthermore, from Fourier transform theory, if $f(t) \leftrightarrow F(\omega)$ are a transform pair, then

$$\frac{d^n f}{dt^n} \leftrightarrow (j\omega)^n F(\omega). \tag{2.130}$$

These lead immediately to the transform pair

$$H_r(t)e^{-t^2/2} \leftrightarrow \sqrt{2\pi}(j\omega)^r e^{-\omega^2/2}. \tag{2.131}$$

In the discrete realm, we know that the Binomial sequence

$$x_0(k) = \begin{cases} \begin{pmatrix} N \\ k \end{pmatrix}, & 0 \leq k \leq N \\ 0, & \text{otherwise} \end{cases} \tag{2.132}$$

resembles a truncated Gaussian. Indeed, for large N (Papoulis, 1984),

$$\begin{pmatrix} N \\ k \end{pmatrix} \sim \frac{2^N}{\sqrt{N\pi/2}} exp\{-\frac{(k-N/2)^2}{N/2}\}.$$

We also know that the first difference is a discrete approximation to the derivative operator. Fortuitously, the members of the discrete Binomial-Hermite family are generated by successive differences of the Binomial sequence

$$x_r(k) = \nabla^r \begin{pmatrix} N-r \\ k \end{pmatrix}, \qquad r = 0, 1, \cdots, N, \tag{2.133}$$

where

$$\nabla f(n) = f(n) - f(n-1).$$

Taking successive differences gives

$$x_r(k) = \begin{pmatrix} N \\ k \end{pmatrix} \sum_{\nu=0}^{r} (-2)^\nu \begin{pmatrix} r \\ \nu \end{pmatrix} \frac{k^{(\nu)}}{N^{(\nu)}} = \begin{pmatrix} N \\ k \end{pmatrix} H_r(k), \tag{2.134}$$

where $k^{(\nu)}$, the *forward factorial function*, is a polynomial in k of degree ν

$$k^{(\nu)} = \begin{cases} k(k-1)\cdots(k-\nu+1), & \nu \geq 1 \\ 1, & \nu = 0. \end{cases} \tag{2.135}$$

The polynomials appearing in Eq. (2.134) are the discrete Hermite polynomials. They are symmetric with respect to index and argument,

$$H_r(k) = H_k(r), \qquad 0 \leq r, k \leq N, \tag{2.136}$$

which implies the symmetry

$$\frac{x_r(k)}{\begin{pmatrix} N \\ k \end{pmatrix}} = \frac{x_k(r)}{\begin{pmatrix} N \\ r \end{pmatrix}}. \tag{2.137}$$

The other members of the Binomial-Hermite family are generated by the two-term recurrence relation (Prob. 2.11)

$$x_{r+1}(k) = -x_{r+1}(k-1) + x_r(k) - x_r(k-1), \qquad 0 \leq k, r \leq N, \tag{2.138}$$

with initial values $x_r(-1) = 0$ for $0 \leq r \leq N$, and initial sequence $x_0(k) = \begin{pmatrix} N \\ k \end{pmatrix}$. In the Z-transform domain, the recursion becomes

$$\begin{aligned} X_r(z) &= (\frac{1-z^{-1}}{1+z^{-1}})X_{r-1}(z) = (\frac{1-z^{-1}}{1+z^{-1}})^r X_0(z) \\ &= (1-z^{-1})^r(1+z^{-1})^{N-r}, \end{aligned} \tag{2.139}$$

where

$$X_0(z) = Z\{\begin{pmatrix} N \\ k \end{pmatrix}\} = \sum_{k=0}^{N} \begin{pmatrix} N \\ k \end{pmatrix} z^{-k} = (1+z^{-1})^N.$$

Note that there are no multiplications in the recurrence relation, Eq. (2.138). The digital filter structure shown in Fig. 2.8 generates the entire Binomial-Hermite family.

The Hermite polynomials and the Binomial-Hermite sequences are orthogonal on $[0, N]$ with respect to weighting sequences $\begin{pmatrix} N \\ k \end{pmatrix}$ and $\begin{pmatrix} N \\ k \end{pmatrix}^{-1}$, respectively (Prob. 2.12):

$$\sum_{k=0}^{N} H_r(k)H_s(k) \begin{pmatrix} N \\ k \end{pmatrix} = \sum_{k=0}^{N} \frac{x_r(k)x_s(k)}{\begin{pmatrix} N \\ k \end{pmatrix}} = \frac{2^N}{\begin{pmatrix} N \\ k \end{pmatrix}} \delta_{r-s}. \tag{2.140}$$

This last equation is the discrete counterpart to the analog Hermite orthogonality of Eq. (2.128).

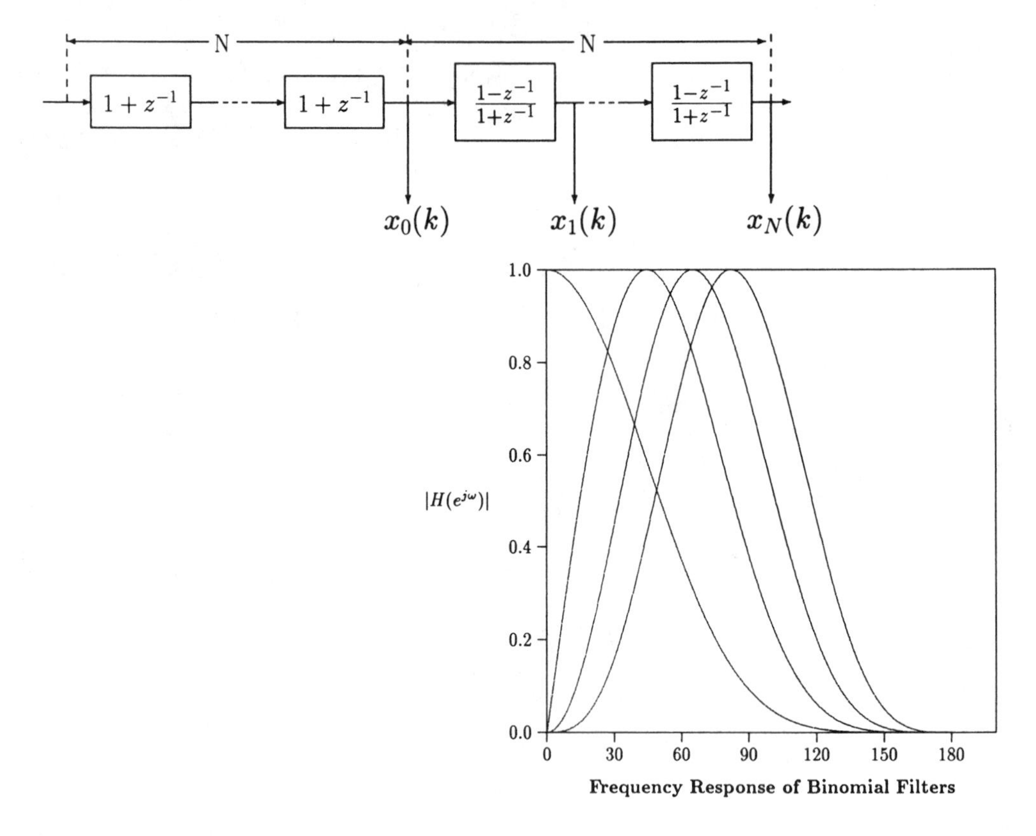

Figure 2.8: (a) Binomial filter bank structure; (b) magnitude responses of duration 8 Binomial sequences (first half of the basis).

The associated Hermite and Binomial transformation matrices are

$$\begin{aligned} H &= [H_{rk}] \\ X &= [X_{rk}], \end{aligned}$$

where we are using the notation $H_{rk} = H_r(k)$, and $X_{rk} = X_r(k)$. The matrix H is real and symmetric; the rows and columns of X are orthogonal (Prob. 2.13)

$$\begin{aligned} H &= H^T \\ X^2 &= 2^N I. \end{aligned} \tag{2.141}$$

These Binomial-Hermite filters are linear-phase quadrature mirror filters. From Eq. (2.139) we can derive

$$X_r(-z) = X_{N-r}(z), \tag{2.142}$$

which implies

$$(-1)^k x_r(k) = x_{N-r}(k) \qquad r = 0, 1, ...N. \tag{2.143}$$

Also,

$$z^{-N} X_r(z^{-1}) = (-1)^r X_r(z) \tag{2.144}$$

implies

$$x_r(N-k) = (-1)^r x_r(k). \tag{2.145}$$

These equations demonstrate the symmetry and anti-symmetry of the Binomial matrix X. Equation (2.143), for example, asserts that the filters represented by the bottom half of the Binomial matrix are mirror images of the filters in the top half. These last two equations can also be used to prove the orthogonality of rows and columns asserted by Eq. (2.141). Finally, from Eq. (2.142), we can infer that the complementary filters $X_r(z)$ and $X_{N-r}(z)$ have magnitude responses that are mirror images about $\omega = \pi/2$,

$$|X_r(e^{j(\frac{\pi}{2}-\omega)})| = |X_{N-r}(e^{j(\frac{\pi}{2}+\omega)})|. \tag{2.146}$$

Hence, the complementary rows and columns of X possess the mirror filter property (Section 3.3).

From Eq. (2.139), it is clear that $X_r(e^{j\omega}) \triangleq A_r(\omega)e^{j\theta_r(\omega)}$, has magnitude and (linear) phase responses given by

$$\begin{aligned} A_r(\omega) &= 2^N \left(\sin\frac{\omega}{2}\right)^r \left(\cos\frac{\omega}{2}\right)^{N-r} \\ \theta_r(\omega) &= \frac{r\pi}{2} - \frac{N\omega}{2}. \end{aligned}$$

The first half of the set, $r = 0, 1, \ldots, (N-1)/2$, have significant energy in the half band $(0, \pi/2)$, while the second half, $(r = (N+1)/2, \ldots, N$, span the upper half-band. These properties will be exploited in Chapter 4 in developing Binomial quadrature mirror filters, and in Chapter 5 as basis sequences for wavelets. The 8×8 Binomial matrix X follows:

$$X = \begin{bmatrix} 1 & 7 & 21 & 35 & 35 & 21 & 7 & 1 \\ 1 & 5 & 9 & 5 & -5 & -9 & -5 & -1 \\ 1 & 3 & 1 & -5 & -5 & 1 & 3 & 1 \\ 1 & 1 & -3 & -3 & 3 & 3 & -1 & -1 \\ 1 & -1 & -3 & 3 & 3 & -3 & -1 & 1 \\ 1 & -3 & 1 & 5 & -5 & -1 & 3 & -1 \\ 1 & -5 & 9 & -5 & -5 & 9 & -5 & 1 \\ 1 & -7 & 21 & -35 & 35 & -21 & 7 & -1 \end{bmatrix}.$$

The corresponding magnitude frequency responses shown in Fig. 2.8(b) have almost Gaussian-shaped low-pass and band-pass characteristics.

These weighted orthogonality properties suggest that by proper normalization the Hermite transform can provide a unitary matrix suitable for signal coding. This modified Hermite transform (MHT) is defined as

$$\Phi(r,k) = \frac{\binom{N}{r}^{1/2}\binom{N}{k}^{1/2}}{2^{N/2}} H_r(k) = \Phi(k,r), \tag{2.147}$$

or

$$\begin{aligned} \Phi &= DHD \\ D &= \frac{1}{\sqrt{2^{N/2}}} diag\{\cdots \binom{N}{k}^{1/2} \cdots\} \end{aligned} \tag{2.148}$$

with the unitary property $\Phi\Phi^T = I$.

Plots of the MHT basis functions and their Fourier transforms for size 8 are shown in Fig. 2.7(f), along with the DCT, DST, DLT, WHT, and KLT(0.95). Note that the MHT basis has no *DC* term. Signals with a Gaussian-like envelope could be represented very accurately by a few terms in the MHT expansion, whereas the DCT requires more terms. On the other hand, a constant signal is represented by one term in the DCT expansion, but requires all even indexed terms in the MHT decomposition.

We can compute the MHT spectrum in a three-step process.

(1) Multiply the signal $f(k)$ by a prewindow function, $w_1(k) = \binom{N}{k}^{-1/2}$, to form $g(k)$.

(2) Apply the time reversed signal $g(-k)$ to the Binomial network of Fig. 2.8. The output at the rth tap at $n = 0$ is the intermediate coefficient θ'_r,

$$\theta'_r = \sum_{k=0}^{N} X(r,k)g(k), \quad 0 \leq r \leq N.$$

No multiplication is needed in this stage.

(3) Multiply θ'_r by the post-window function $w_2(r)$ to obtain

$$\theta_r = w_2(r)\theta'_r$$

$$w_2(r) = \frac{\begin{pmatrix} N \\ r \end{pmatrix}^{1/2}}{2^{N/2}}.$$

This MHT algorithm can be implemented using $2N$ real multiplications, as compared with the fast DCT, which requires $(N \log_2 N - N + 2)$ multiplications.

In Section 2.4.6, we compare the coding performance (compaction) of various transforms. The MHT is clearly inferior to the DCT for positively correlated signals, but superior to it for small or negative values of ρ.

2.3.2.2 The Discrete Laguerre Polynomials

This set of functions are useful in representing signals on the semi-infinite interval $[0, \infty)$. Because of this support interval, this family can be generated by an IIR filter structure. Although this represents a departure from the FIR and block transforms discussed thus far, nevertheless we introduce it at this point as representative of a class of the infinite-dimensional polynomial-type transform. We also hold out the possibility of using a finite number of these as approximation vehicles.

The set, defined on $0 \le k < \infty$, is

$$\Phi_r(k) = A_r p_r(k) \lambda^k = L_r(k) \lambda^k. \tag{2.149}$$

In this last equation, λ is a constant, $0 < \lambda < 1$, A_r is a normalizing factor

$$A_r = \lambda^r \sqrt{1 - \lambda^2},$$

and $p_r(k)$ is a polynomial of degree r (Prob. 2.14),

$$\begin{aligned} p_r(k) &= \sum_{m=0}^{r} (-1)^m \begin{pmatrix} r \\ m \end{pmatrix} \alpha^m \frac{k^{(m)}}{m!} \\ p_0(k) &= 1, \end{aligned} \tag{2.150}$$

where $\begin{pmatrix} r \\ m \end{pmatrix}$ is the binomial coefficient, $k^{(m)}$ is the forward factorial function Eq. (2.135), and $\alpha = (1 - \lambda^2)/\lambda^2$.

Using Z-transforms, we can establish the orthonormality (Haddad and Parsons, 1991):

$$\sum_{k=0}^{\infty} \Phi_r(k) \Phi_s(k) = \delta_{r-s}. \tag{2.151}$$

For our purposes, we outline the steps in the proof. First we calculate $P_r(z)$ by induction and obtain

$$P_r(z) = (1 - \frac{\alpha}{z-1})^r (\frac{z}{z-1}). \tag{2.152}$$

Then

$$\Phi_r(z) = A_r P_r(z/\lambda). \tag{2.153}$$

By contour integration, we can evaluate

$$\sum_k \Phi_r(k)\Phi_s(k) = \frac{1}{2\pi j} \oint \frac{\Phi_r(z)\Phi_s(z^{-1})dz}{z}. \tag{2.154}$$

For $m = r - s \geq 1$, the integrand is of the form

$$\begin{aligned} G(z) &= A\frac{(1-\lambda z)^{m-1}}{(z-\lambda)^{m+1}} \\ A &= (-1)^{r+s} A_r A_s / \lambda^{r+s} \end{aligned}$$

and has an $(m+1)$th order pole inside the unit circle at $z = \lambda$, and an $(m-1)$th order zero at $z = \lambda^{-1}$. Then, for $m \geq 1$,

$$\frac{1}{2\pi j} \oint G(z)dz = \frac{1}{m!} \frac{d^m}{dz^m} (1 - \lambda z)^{m-1}|_{z=\lambda} = 0.$$

For $m \leq -1$, the integrand is

$$G(z) = \frac{K(z-\lambda)^{|m|-1}}{(1-\lambda z)^{|m|+1}}$$

with an $(m+1)$th order pole at $z = \lambda^{-1}$ outside the unit circle, and only zeros inside at $z = \lambda$. This integrand is analytic on and inside the unit circle , so that the contour integral vanishes. Finally, for $r = s$, we can obtain the normalization factor

$$\frac{A_r^2}{\lambda^{2r}} \frac{1}{2\pi j} \oint \frac{dz}{(z-\lambda)(1-\lambda z)} = \frac{A_r^2}{\lambda^{2r}(1-\lambda^2)} = 1.$$

The polynomials $L_r(k)$ in Eq. (2.149) are the discrete Laguerre polynomials

$$L_r(k) = A_r p_r(k).$$

These are orthonormal with respect to the exponential weighting factor $\theta^k = \lambda^{2k}$

$$\sum_{k=0}^{\infty} L_r(k) L_s(k) \lambda^{2k} = \delta_{r-s}. \tag{2.155}$$

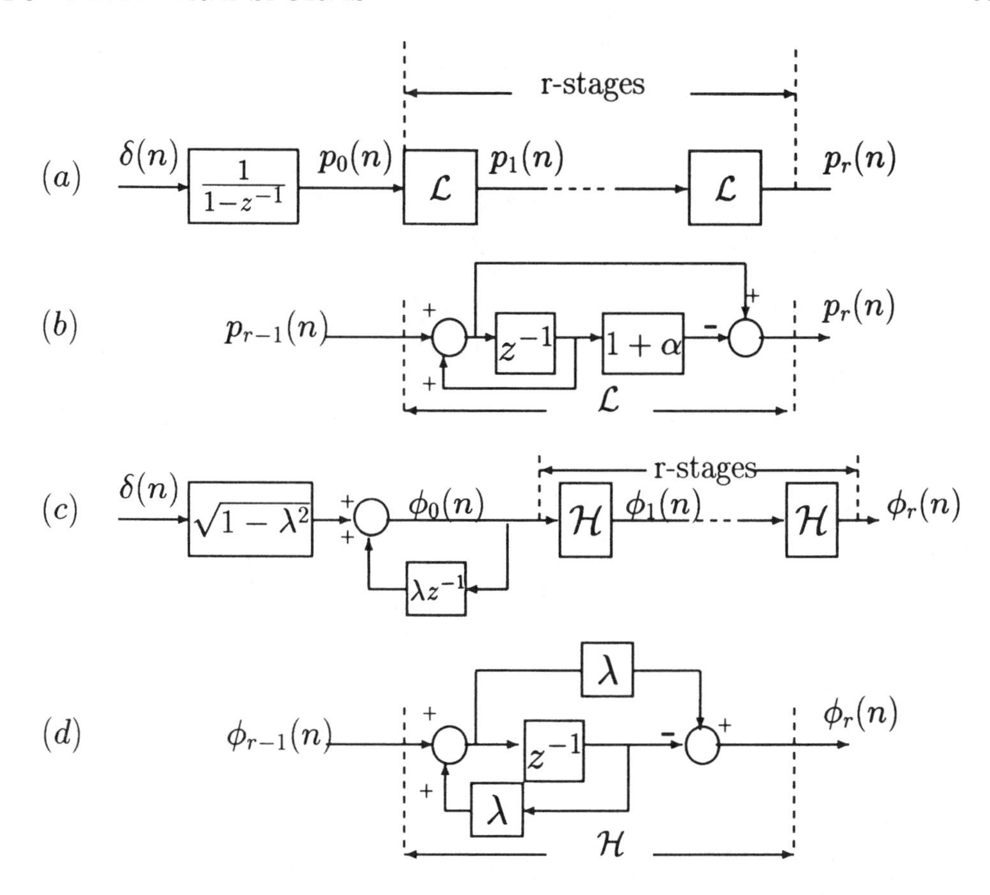

Figure 2.9: Generation of Laguerre polynomials (a,b) and family (c,d).

From the Z-transform Eq. (2.152), we can easily obtain the recurrence relation

$$\begin{aligned} p_r(k) &= p_r(k-1) + p_{r-1}(k) - (1-\alpha)p_{r-1}(k-1) \\ p_0(k) &= 1. \end{aligned} \tag{2.156}$$

Similarly, we find that the recurrence relation for the $\{\Phi_r(k)\}$ is

$$\begin{aligned} \phi_r(n) &= \lambda\phi_r(n-1) + \lambda\phi_{r-1}(n) - \phi_{r-1}(n-1) \\ \phi_0(n) &= \sqrt{1-\lambda^2}\lambda^n. \end{aligned} \tag{2.157}$$

Digital networks for the generation of these families are shown in Fig. 2.9. (See also Prob. 2.15)

2.3.2.3 The Discrete Legendre Polynomials

The discrete Hermite polynomials weighted by the Binomial sequence are suitable for representing signals with Gaussian-like features on a finite interval. Such sequences fall off rapidly near the end points of the interval $[0, N-1]$. The Laguerre functions provide a signal decomposition on the semi-infinite interval $[0, \infty)$. The discrete Legendre polynomials are uniformly weighted on a finite interval. Morrison (1969) has used these to construct finite-memory polynomial filters. Here, we outline the steps in the derivation of this family.

Let $L_r(k)$ be a polynomial of degree r on $[0, N-1]$,

$$L_r(k) = 1 + \alpha_{r1}k + \alpha_{r2}k^{(2)} + \cdots + \alpha_{rr}k^{(r)}, r = 0, 1, \cdots, N-1. \tag{2.158}$$

We choose α_{rs} to satisfy orthogonality

$$\sum_{k=0}^{N-1} L_r(k)L_s(k) = c_r^2\delta_{r-s}.$$

Morrison shows that the result is

$$\alpha_{rs} = \begin{cases} (-1)^s \binom{r}{s}\binom{r+s}{s}\frac{1}{(N-1)^{(s)}}, & 0 \le s \le r \\ 0, & s > r \end{cases} \tag{2.159}$$

and the associated norms are

$$c_r^2 = \frac{(N+r)^{(r+1)}}{(2r+1)(N-1)^{(r)}}. \tag{2.160}$$

The orthonormalized discrete Legendre transform(DLT) is therefore

$$\Phi(r,k) = \Phi_r(k) = \frac{L_r(k)}{c_r}.$$

The rows of the DLT matrix for $N = 8$ are shown in Fig. 2.7(c). The even and odd indexed rows are, respectively, symmetric and skew symmetric about $N/2$. These plots show that the DLT waveforms are similar to the DCT, and in Section 2.4.6, we see that the DLT performance is slightly inferior to that of the DCT for signals tested—both theoretical and experimental. The main drawback to the DLT is that a fast algorithm has not yet been developed.

2.3.3 Rectangular Transforms

We use the term *rectangular* transform to denote orthonormal basis sequences obtained by sampling analog (i.e, continuous-time) functions that are switched pulses in time. In the Walsh family, the pulse amplitudes are ± 1 for every member of the set. For the Haar functions, the values of the switched amplitudes can vary from row to row.

The Walsh-based transform is by far the more important of the two because of its simplicity, fast transform, and compaction properties. Accordingly, we allot the majority of this subsection to this very appealing transform.

2.3.3.1 The Discrete Walsh-Hadamard Transform

Certain continuous-time orthogonal functions, when sampled, produce orthogonal discrete-time sequences. Sampling the sinusoidal family $\{e^{jk\omega_0 t},\ \omega_0 = 2\pi/T\}$ orthogonal on $[0,T]$, at a spacing of T/N generates the finite set $\{e^{j2\pi kn/N}\}$, discrete orthogonal on $[0, N-1]$. The Walsh function (Walsh, 1923) and sequences also preserve orthogonality under sampling (as do the Haar functions described in the next subsection).

The continuous-time Walsh functions are a complete orthonormal set on the unit interval $[0,1)$. Their salient feature is that they are binary valued, ± 1, and thus consist of sequences of positive and negative pulses of varying widths.

The first two Walsh functions are

$$\begin{aligned} w(0,t) &= 1, \quad 0 \le t < 1 \\ w(1,t) &= \begin{cases} 1, & 0 \le t < 1/2 \\ -1, & 1/2 \le t < 1. \end{cases} \end{aligned} \tag{2.161}$$

The other members of the denumerably infinite set are generated by a multiplicative iteration

$$w(r,t) = w([\frac{r}{2}], 2t)w(r - 2[\frac{r}{2}], t), \tag{2.162}$$

where $[\frac{r}{2}]$ is the integer part of $r/2$. These are orthonormal,

$$\int_0^1 w(m,t)w(n,t)dt = \delta_{m-n}. \tag{2.163}$$

The first eight Walsh functions are shown in Fig. 2.10 in sequence order, which is the number of zero crossings or sign changes in $[0,1)$. In this sense they resemble the frequency of the sinusoidal functions, but differ since the spacing between zeros

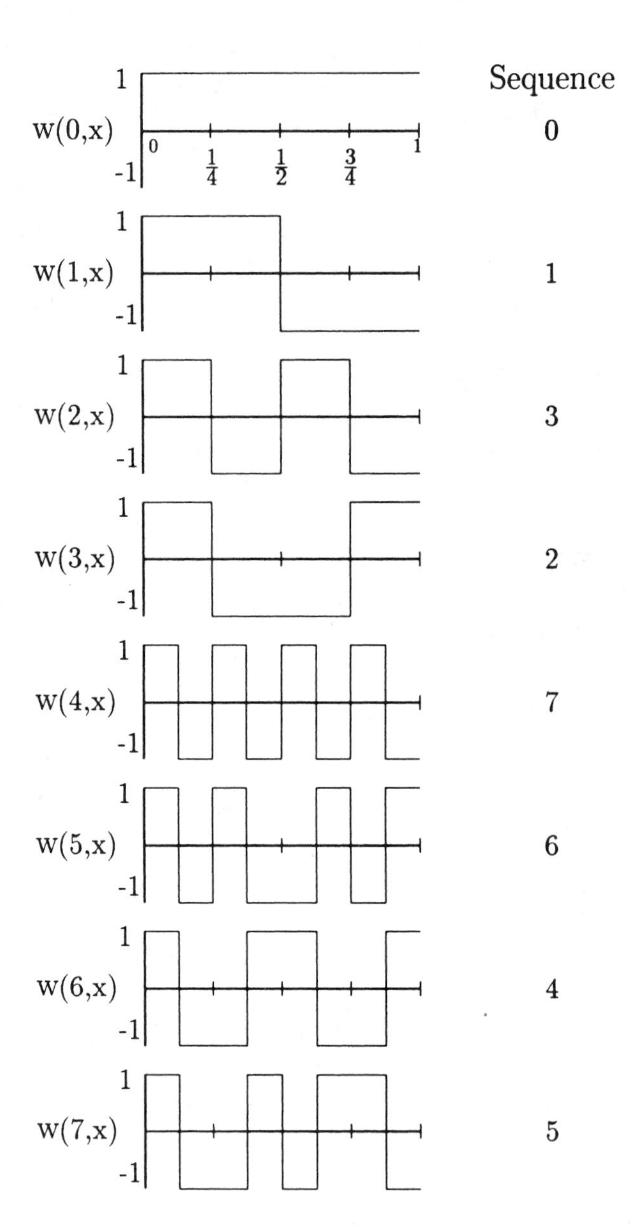

Figure 2.10: Walsh functions, $N = 8$.

is not necessarily constant. Also, the index of a Walsh function can differ from its sequence.

The discrete time Walsh functions are a finite set of sequences obtained by sampling the first N analog functions at a spacing of $\Delta T = 2^{-N}, N = 2^p$, and then relabeling the ordinate so that there is unit spacing between samples. The Walsh functions are continuous from the right, Eqs. (2.161) and (2.162). The sampled value at a discontinuity t_0 is the value at t_0^+, just to the right of t_0. Therefore, the Walsh sequences are a complete set of N orthogonal sequences on $[0, N-1]$ consisting of $+1$ and -1 values, defined by initial sequences

$$\begin{aligned} w(0,n) &= 1, \quad 0 \leq n \leq N-1 \\ w(1,n) &= \begin{cases} 1, & 0 \leq n \leq N/2-1 \\ -1, & N/2 \leq n \leq N-1 \end{cases} \end{aligned} \tag{2.164}$$

and by the iterations

$$\begin{aligned} w(r,n) &= w([r/2], 2n)w(r-2[r/2], n) \\ r &= 2, 3, \cdots, N-1. \end{aligned} \tag{2.165}$$

In order to prove the orthogonality of the Walsh sequences

$$J(r,s) = \sum_{n=0}^{N-1} w(r,n)w(s,n) = N\delta_{r-s} \quad 0 \leq r, s \leq (N-1), \tag{2.166}$$

we must introduce a binary coded notation for the integer variables. Let

$$r = i_{p-i}(2)^{p-1} + \cdots i_0(2)^0 \rightarrow i_{p-1}i_{p-2}\cdots i_0.$$

Similarly,

$$s \rightarrow j_{p-1}j_{p-2}\cdots j_0$$

$$n \rightarrow k_{p-1}k_{p-2}\cdots k_0.$$

By iterating the defining equations (2.164) and (2.165) we obtain a binary factorization of the Walsh sequences in the form

$$w(r,n) = w(i_{p-1}i_{p-2}\cdots i_0, k_{p-1}k_{p-2}\cdots k_0) = \prod_{\nu=0}^{p-1}(-1)^{k_\nu i_{p-\nu-1}}. \tag{2.167}$$

In this representation, the product term in Eq. (2.166) is

$$\begin{aligned} w(r,n)w(s,n) &= \prod_{\nu=0}^{p-1}(-1)^{k_\nu i_{p-\nu-1}}\prod_{\nu=0}^{p-1}(-1)^{k_\nu j_{p-\nu-1}} \\ &= \prod_{\nu=0}^{p-1}(-1)^{k_\nu[i_{p-\nu-1}+j_{p-\nu-1}]}. \end{aligned}$$

Next we note that the decimal indexed sum can be written as a repeated binary sum, i.e.,

$$\sum_{n=0}^{N-1} = \sum_{k_{p-1}=0}^{1}\cdots\sum_{k_1=0}^{1}\sum_{k_0=0}^{1}.$$

The inner product $J(r,s)$ now takes the form

$$J(r,s) = \sum_{k_{p-1}=0}^{1}\cdots\sum_{k_0=0}^{1}\prod_{\nu=0}^{p-1}(-1)^{k_\nu[i_{p-\nu-1}+j_{p-\nu-1}]}.$$

Interchanging the product and sum operations,

$$\begin{aligned} J(r,s) &= \prod_{\nu=0}^{p-1}\sum_{k_\nu=0}^{1}(-1)^{k_\nu[i_{p-\nu-1}+j_{p-\nu-1}]} \\ &= \prod_{\nu=0}^{p-1}[1+(-1)^{[i_{p-\nu-1}+j_{p-\nu-1}]}]. \end{aligned}$$

Suppose $r = s$. Then $i_\nu = j_\nu$ for all ν so that $(i_\nu + j_\nu) = 0$, or 2. This in turn implies that $(-1)^{i_\nu + j_\nu} = 1$, thereby rendering

$$J(r,r) = \prod_{\nu=0}^{p-1}(2) = 2^p = N.$$

Next, if $r \neq s$, then their binary representations differ in at least one bit. That is, $i_\nu \neq j_\nu$ for at least one ν in $[0, p-1]$. Hence $(i_\nu + j_\nu) = 1$, which means that there is at least one term in the product $[1 + (-1)^{i_\nu + j_\nu}] = 0$. Hence $J(r,s) = 0$, for $r \neq s$, and we have established orthogonality.

The matrix obtained by ordering the rows by their sequency is the discrete Walsh transform. These are shown in Fig. 2.7(e). There are other representations as well; the most notable is the Hadamard form. These matrices of order $N = 2^p$ are defined recursively,

$$\begin{aligned} H_1 &= \frac{1}{\sqrt{2}} \begin{bmatrix} 1 & 1 \\ 1 & -1 \end{bmatrix} \\ H_{2N} &= \frac{1}{\sqrt{2}} \begin{bmatrix} H_N & H_N \\ H_N & -H_N \end{bmatrix} = H_1 \otimes H_N, \end{aligned} \tag{2.168}$$

where the $\otimes$ indicates the Kronecker product, and the $\sqrt{2}$ term is a normalizing factor. The Hadamard matrix of size $N = 2^p$ is the same as the discrete Walsh matrix with shuffled rows. There is an algorithm for the shuffling of the row indices. The rows of the Hadamard matrix ("natural" order) correspond to the bit-reversed gray code representation of its sequence. For example, sequences 4 and 5 are gray coded as 110 and 111, respectively; bit reversal gives 011 and 111, which are the binary representations of rows 3 and 7 in the natural-ordered Hadamard matrix. Row 3 has sequence 4 and row 7 has sequence 3 5. This normalized Hadamard form is called the discrete Walsh-Hadamard transform, or WHT. (See Prob. 2.16)

The WHT has good compaction properties (see Section 2.4.6). There is a fast transform similar in structure to the FFT based on the ability to express H_N as a product of p sparse matrices, $p = \log_2 N$,

$$H_N = \frac{1}{2^{p/2}} S^p,$$

where

$$S = \left[\begin{array}{ccccccc} 1 & 1 & 0 & 0 & \cdots & 0 & 0 \\ 0 & 0 & 1 & 1 & \cdots & \cdots & \\ \cdots & \cdots & & & & & \\ \cdots & \cdots & & & & 1 & 1 \\ -- & -- & -- & -- & -- & -- & -- \\ 1 & -1 & 0 & 0 & \cdots & & \\ 0 & 0 & 1 & -1 & \cdots & & \\ \cdots & \cdots & & & & & \\ \cdots & \cdots & & & & 1 & -1 \end{array} \right] \begin{array}{c} \uparrow \\ \frac{N}{2} \\ \downarrow \\ --- \\ \uparrow \\ \frac{N}{2} \\ \downarrow \end{array}. \tag{2.169}$$

There are just two entries in each row of S and these are ± 1. Hence each pass of the S matrix is achieved in $N/2$ additions and $N/2$ subtractions. For p stages, the total number of additions and subtractions is $Np = N \log_2 N$.

2.3.3.2 The Haar Transform

The Haar functions are an orthogonal family of switched rectangular waveforms where the amplitudes can differ from one function to another. They are defined on the interval $[0, 1)$ by

$$\begin{aligned} h_r(t) &= \frac{1}{\sqrt{N}} \begin{cases} 2^{m/2}, & \frac{k-1}{2^m} \leq t < \frac{k-(1/2)}{2^m} \\ -2^{m/2}, & \frac{k-(1/2)}{2^m} \leq t < \frac{k}{2^m} \\ 0, & \text{otherwise} \quad in \quad [0,1) \end{cases} \\ h_0(t) &= \frac{1}{\sqrt{N}}. \end{aligned} \tag{2.170}$$

The index $r = 0, 1, \cdots, N-1$, and $N = 2^p$. Also, m and k represent the integer decomposition of the index r

$$r = 2^m + k - 1. \tag{2.171}$$

These are rectangular functions that can be zero in subintervals of $[0, 1)$.

Just as with the Walsh functions, sampling these functions at a spacing $\Delta T = 1/N$ gives a discrete family that retains its orthogonality. Hence with $t = n/N$, $n = 0, 1, \cdots, N-1$ we obtain the discrete Haar transform

$$H = [h_r(n)], \qquad 0 \leq r, n \leq N-1. \tag{2.172}$$

The Haar matrix is unitary (and real) and its rows are sequence ordered. Although a fast transform exists, this transform has not found practical applications in coding because of its poor energy compaction. For additional details see Shore (1973) or Ahmed and Rao (1975).

We shall see in Chapter 5 that the Haar functions serve as the simplest wavelet family in multiresolution signal decomposition. Even in that context, the time-frequency resolution is poor, so these functions are primarily of academic interest. (Prob. 2.17)

2.3.4 Block Transform Packets

The block transforms with time and frequency responses shown in Fig. 2.7 may be regarded as basically frequency selective. In Chapter 5, we revisit the block transforms from a time-frequency standpoint and show how block transform packets can be designed to have desirable time-frequency localization properties.

2.4 Parametric Modeling of Signal Sources

It is desirable to define the behavior of any signal source by a set of parameters or features. The challenge of signal source modeling spans diverse fields from economics to weather forecasting; it is an essential tool for simulation and prediction purposes. One widely used application of source modeling is in speech coding. This is called linear predictive coding (LPC) and provides the best coding performance known for speech. Since most natural signal sources are not globally stationary, the modeling operation is repeated for each segment of source output over which the stationarity assumption holds. Although the modeling of speech is useful and works well, the same is not yet true for images. Therefore, model-based image processing and coding is still an active research area and some new modeling approaches, rather than the classical waveform modeling, are being studied extensively. There are several outstanding books and tutorial papers on this subject in the literature. A brief summary will be presented here for the later use in this book.

Modeling a discrete-time signal as the output of a linear, time-invariant (LTI) system driven by a white Gaussian noise source provides a useful representation over short intervals. These systems in general have a rational Z-transform function; therefore the term *pole-zero modeling* is also widely used for this. This pole-zero modeling is also directly related to the approximation of the unit-sample response of a discrete-time system by linear combination of complex exponentials.

A wide-sense stationary, zero-mean white noise process has an autocorrelation function

$$R_{NN}(k) = E\{\eta(n)\eta(n+k)\} = \sigma_N^2\delta(k), \tag{2.173}$$

where

$$\delta(k) = \begin{cases} 1 & \text{for } k = 0 \\ 0 & \text{otherwise.} \end{cases} \tag{2.174}$$

Its power spectral density (PSD) is a constant,

$$S_{NN}(e^{j\omega}) = \sigma_N^2, \tag{2.175}$$

where σ_N^2 is the variance of the noise signal.

The rational transfer function of a linear, time-invariant system is

$$H(z) = \frac{A(z)}{B(z)} = \frac{\sum_{k=0}^{L} a_k z^{-k}}{1 - \sum_{k=1}^{P} b_k z^{-k}}. \tag{2.176}$$

The numerator polynomial $A(z)$ has L roots (zeros of the system) and the denominator polynomial $B(z)$ has P roots (poles of the system). The defining difference equation of this system with input $\eta(n)$ and output $x(n)$ is

$$x(n) = \sum_{k=1}^{P} b_k x(n-k) + \sum_{k=0}^{L} a_k \eta(n-k). \tag{2.177}$$

If this system is stable and if $\eta(n)$ is stationary white, the output signal $\{x(n)\}$ is a wide-sense stationary process with the autocorrelation function

$$R_{xx}(k) = h(k) * h(-k) * R_{NN}(k) = \sigma_N^2 h(k) * h(-k) \tag{2.178}$$

and the corresponding power spectral density function

$$S_{xx}(z) = H(z)H(z^{-1})\sigma_N^2, \tag{2.179}$$

and on the unit circle

$$S_{xx}(e^{j\omega}) = |H(e^{j\omega})|^2 \sigma_N^2. \tag{2.180}$$

Several well-known approaches in the literature deal with pole-zero modeling of sources. The details of these techniques are beyond the scope of this book. The interested readers are advised to go to the references, for example, Gardner (1988). Our interest is to present those aspects of modeling that are subsequently needed in the comparative evaluation of signal decomposition schemes.

2.4.1 Autoregressive Signal Source Models

Two special cases of pole-zero modeling that have found extensive application in the literature are the moving average (MA), and the autoregressive (AR) processes. In the first instance (MA), the denominator $B(z)$ of $H(z)$ in Eq. (2.176) is a constant, and the process is said to be "all-zero,"

$$x(n) = \sum_{k=0}^{L} a_k \eta(n-k) \leftrightarrow H(z) = \sum_{k=0}^{L} a_k z^{-k}.$$

The filter in this case is FIR, and the autocorrelation is of finite duration,

$$R_{xx}(k) = \sigma_N^2 (a_k * a_{-k}),$$

where the asterisk ($*$) implies a convolution operation. The MA process model is used extensively in adaptive equalizers and inverse system modeling (Haykin, 1986; Haddad and Parsons, 1991).

The representation used in coding performance evaluation is the AR process or "all-pole" model, wherein the numerator $A(z)$ is constant. The all-pole model with white noise input is also referred to as a Markov source. [The system of Eq. (2.176) with both poles and zeros is called an autoregressive, moving average (ARMA).] Thus, the autoregressive signal is generated by passing white noise $\{\eta(n)\}$ through an all-pole discrete-time system

$$H(z) = \frac{1}{1 - \sum_{k=1}^{P} \rho_k z^{-k}} = \frac{z^P}{z^P - \sum_{k=1}^{P} \rho_k z^{P-k}}. \tag{2.181}$$

The corresponding AR(P) signal evolves as

$$x(n) = \sum_{k=1}^{P} \rho_k x(n-k) + \eta(n). \tag{2.182}$$

Here P is called the *order of the prediction* and $\{\rho_k\}$ are called the *prediction coefficients*. The recursive relation of the autocorrelation function of an AR(P) source can be easily derived as (Prob. 2.18)

$$R_{xx}(k) = \sum_{j=1}^{P} \rho_j R_{xx}(k-j) \qquad k > 0 \tag{2.183}$$

with signal power

$$\sigma_x^2 = E\{|x(n)|^2\} = R_{xx}(0). \tag{2.184}$$

The problem in AR(P) source modeling is the estimation of the model parameters $\{\rho_j\}$ from the observed data. It turns out that the all-pole model leads to a set of P linear equations in the P unknowns, which can be solved efficiently by the Levinson algorithm or the Cholesky decomposition (Kay, 1988).

The AR(P) modeling of sources has been very efficiently used especially for speech. Natural voiced speech is well approximated by the all-pole model for a period of the glottal pulse. The stationarity assumption of the source holds during this time interval. Therefore, the predictor coefficients are calculated for approximately every 10 ms. The AR(P) sources are good models for a wide variety of stationary sources. The AR(P) is a standard model for speech sources implemented in many vocoders for low bit rate coding and transmission applications. Today, it is possible to transmit intelligible speech below 1 Kbits/sec by LPC.

2.4.2 AR(1) Source Model

The AR(1) signal source is defined by the first-order difference equation

$$x(n) = \rho x(n-1) + \eta(n), \tag{2.185}$$

where ρ is the prediction or correlation coefficient and $\{\eta(n)\}$ is the white noise sequence of Eq. (2.173). The corresponding first-order system function is

$$H(z) = \frac{1}{1 - \rho z^{-1}} \tag{2.186}$$

with the unit sample response

$$h(n) = \rho^n \qquad n = 0, 1, ... \quad . \tag{2.187}$$

The autocorrelation function of the AR(1) signal is

$$R_{xx}(k) = \sigma_x^2 \rho^{|k|} \qquad\qquad k = 0, \mp 1, \mp 2... \tag{2.188}$$

with

$$\sigma_x^2 = \frac{\sigma_N^2}{1 - \rho^2} \tag{2.189}$$

and the corresponding power spectral density function of the AR(1) source

$$S_{xx}(e^{j\omega}) = \frac{1 - \rho^2}{1 + \rho^2 - 2\rho cos\omega} \sigma_x^2. \tag{2.190}$$

The AR(1) source model is a crude, first approximation to real-world sources such as speech and images. Therefore it is a commonly used artificial source model for analytical performance studies of many signal processing techniques.

2.4.3 Correlation Models for Images

The two-dimensional extension of the 1D random sequence is the *random field*, a 2D grid of random variables. Many properties of the 2D random sequences are extrapolations of the 1D progenitor (Haddad and Parsons, 1991).

Each pixel $x(m, n)$ is a random variable with some probability density function. This collection $\{x(m, n)\}$ and the statistical relations among them constitute the random field. We are concerned primarily with wide-sense properties—means and correlations—that over a small enough region may be considered stationary. In this case, the mean and correlation are

$$E\{x(m, n)\} = \mu(m, n) = \text{constant}$$
$$E\{x(j, k)x(j + m, k + n)\} = R_{xx}(m, n).$$

The 2D power spectral density is the 2D Fourier transform of $R_{xx}(m, n)$,

$$S_{xx}(e^{j\omega_1}, e^{j\omega_2}) = \sum_m \sum_n R_{xx}(m, n) e^{-j(m\omega_1 + n\omega_2)}.$$

A white noise source has zero mean, uncorrelated pixels, and a flat PSD

$$\begin{aligned} R(m,n) &= \sigma^2\delta(m,n) = \begin{cases} \sigma^2, & m=n=0 \\ 0, & \text{otherwise} \end{cases} \\ S(e^{j\omega_1}, e^{j\omega_2}) &= \sigma^2. \end{aligned}$$

Experimental studies on real-world images have shown that the autocorrelation functions of natural scenes are better represented by nonseparable autocorrelation models. The discussion here starts with the definition of a 2D separable autocorrelation function followed by two nonseparable correlation models.

2.4.3.1 2D Separable Correlation Model

The simplest source model is generated by passing white noise through a 2D AR(1) process of the form

$$x(m,n) = \rho_h x(m-1,n) + \rho_v x(m,n-1) - \rho_h\rho_v x(m-1,n-1) + \eta(m,n), \quad (2.191)$$

where $\{\eta(m,n)\}$ is the zero-mean, white noise source with unit variance, and ρ_h, ρ_v denote the first-order horizontal and vertical prediction or correlation coefficients, respectively. Its autocorrelation function is separable and can be expressed as the product of two 1D autocorrelations

$$R_{xx}(m,n) = \sigma_x^2 \rho_h^{|m|} \rho_v^{|n|} \qquad m,n = 0, \mp 1, \mp 2, ... \quad (2.192)$$

where

$$\sigma_x^2 = \frac{1}{(1-\rho_h^2)(1-\rho_v^2)}. \quad (2.193)$$

Likewise, the 2D PSD can be expressed as the product of two 1D PSDs, one in the horizontal direction and one along the vertical.

2.4.3.2 Generalized-Isotropic Correlation Model

This is a nonseparable 2D autocorrelation model that fits real image data better than the separable correlation model (Natarajan and Ahmed, 1978). The generalized-isotropic correlation model is defined as

$$R_{xx}(m,n) = exp[-(\alpha^2 m^2 + \beta^2 n^2)^{1/2}] \quad (2.194)$$

where

$$\alpha = -ln\rho_h \quad (2.195)$$

and

$$\beta = -ln\rho_\nu. \quad (2.196)$$

2.4.3.3 Generalized Correlation Model

This correlation model is a combination of separable and generalized-isotropic correlation models of images and is defined as

$$R_{xx}(m,n) = exp\{-[(\alpha m^{r_1})^h + (\beta n^{r_2})^h]^{1/h}\}. \tag{2.197}$$

The parameter values $r_1 = 1.137$, $r_2 = 1.09$, $h = \sqrt{2}$, $\alpha = 0.025$, $\beta = 0.019$ were found optimal for many test images (Clarke, 1985). This model fits the statistical behavior of real images.

2.4.4 Coefficient Variances in Orthogonal Transforms

We have emphasized that the prime objective in transform coding is the repacking of the signal energy into as few spectral coefficients as possible. Performance assessment of a particular transform depends not only on the particular transform used but also on the statistical properties of the source signal itself.

In this section we obtain a representation of the coefficient variances in a form that effectively separates the orthogonal transform from the correlation model for that signal source. We can then compare packing efficiency for various transforms in terms of the parameterized source models of the previous sections.

The 1D Case

We showed in Section 2.1.1 that the multirate filter bank in Fig. 2.1 is a realization of the orthonormal transformation $\underline{\theta} = \Phi \underline{x}$, and that $y_i(N-1) = \theta_i$.

It is now an easy matter to calculate $\sigma_i^2 = E\{\theta_i^2\}$. From Fig. 2.1, the correlation function for the output of the ith filter bank is

$$\begin{aligned} R_{y_i}(n) &= [\phi_i^*(-n) * \phi_i(n)] * R(n) \\ &= \rho_i(n) * R(n) \\ &= \sum_k \rho_i(k) R(n-k), \end{aligned} \tag{2.198}$$

where $\rho_i(n) = \phi_i(n) * \phi_i^*(-n)$ is the time autocorrelation function for the ith basis sequence $\phi_i(n)$, and $R(n)$ represents the statistical autocorrelation function for the input signal $x(n)$, which we assume to be zero mean, stationary:

$$\begin{aligned} E\{x(n+m)x(m)\} = R_{xx}(n) &\triangleq R(n) \\ E\{x(n)\} &= 0. \end{aligned} \tag{2.199}$$

Evaluating Eq. (2.198) at $n = 0$ gives σ_i^2, the variance of the ith spectral coefficient

$$\sigma_i^2 = R_{y_i}(0) = \sum \rho_i(k)R(-k) = \sum_{k=-(N-1)}^{(N-1)} \rho_i(k)R(k). \tag{2.200}$$

Both $\rho_i(k)$ and $R(k)$ are even functions of k, so that Eq. (2.200) becomes

$$\sigma_i^2 = \rho_i(0)R(0) + 2\rho_i(1)R(1) + \cdots + 2\rho_i(N-1)R(N-1).$$

Stacking these variances to form a vector of variances gives us

$$\begin{bmatrix} \sigma_0^2 \\ \sigma_1^2 \\ \vdots \\ \sigma_{N-1}^2 \end{bmatrix} = \begin{bmatrix} \rho_0(0) & 2\rho_0(1) & \cdots & 2\rho_0(N-1) \\ \rho_1(0) & 2\rho_1(1) & \cdots & 2\rho_1(N-1) \\ \vdots & \vdots & \vdots & \vdots \\ \rho_{N-1}(0) & 2\rho_{N-1}(1) & \cdots & 2\rho_{N-1}(N-1) \end{bmatrix} \begin{bmatrix} R(0) \\ R(1) \\ \vdots \\ R(N-1) \end{bmatrix} \tag{2.201}$$

or

$$\underline{\sigma}^2 = W\underline{R}, \tag{2.202}$$

where $\underline{R}^T = [R(0), R(1), \cdots, R(N-1)]$, and W is the indicated matrix of basis function autocorrelations.

The W matrix for the discrete cosine transform, $N = 8$, is found to be $\mathrm{W}_{DCT} =$

$$\begin{bmatrix} 1 & 1.750 & 1.500 & 1.250 & 1.000 & 0.750 & 0.500 & 0.250 \\ 1 & 1.367 & 0.599 & -0.125 & -0.653 & -0.890 & -0.816 & -0.480 \\ 1 & 0.987 & -0.353 & -1.133 & -1.000 & -0.280 & 0.353 & 0.426 \\ 1 & 0.419 & -1.252 & -1.051 & 0.270 & 0.769 & 0.162 & -0.345 \\ 1 & -0.250 & -1.500 & 0.250 & 1.000 & -0.250 & -0.500 & 0.250 \\ 1 & -0.919 & -0.869 & 1.258 & -0.270 & -0.589 & 0.544 & -0.154 \\ 1 & -1.487 & -1.487 & 0.353 & 0.633 & -1.000 & 0.780 & 0.073 \\ 1 & -1.867 & 1.522 & -1.081 & 0.653 & -0.316 & 0.108 & -0.019 \end{bmatrix}. \tag{2.203}$$

The W matrix provides the link between the signal's autocorrelation function and the distribution of signal energy among the transform coefficients. This W matrix is unique and fixed for any orthonormal transformation of a given size.

This expression can explain the unique properties of a given transformation. It has been observed that the DCT behaves differently for negatively and positively correlated signals. The W_{DCT} matrix clearly predicts this behavior (Akansu and Haddad, 1990).

This variance or energy calculation can be done in the frequency domain. With $\Phi_i(e^{j\omega}) = \mathcal{F}\{\phi_i(n)\}$, Fig. 2.1 or Eq. (2.198) shows that the PSD is

$$S_{y_i}(e^{j\omega}) = |\Phi_i(e^{j\omega})|^2 S_x(e^{j\omega}) \tag{2.204}$$

and the variance is then

$$\sigma_i^2 = \frac{1}{2\pi}\int_{-\pi}^{\pi} S_{y_i}(e^{j\omega})d\omega. \tag{2.205}$$

2D Case

a. Separable Correlation:

The 2D image $[x(m,n)]$ is transformed via Eq. (2.45)

$$\Theta = \Phi^* X \Phi^T, \tag{2.206}$$

where

$$X = [x(m,n)] \quad \text{and} \quad \Theta = [\theta(i,j)]. \tag{2.207}$$

In the present case, we will calculate each coefficient element directly from

$$\theta(i,j) = \sum_m \sum_n x(m,n)\phi(i,m)\phi(j,n) \qquad 0 \leq i,j \leq N-1. \tag{2.208}$$

The variance of each coefficient is then

$$E\{\theta^2(i,j)\} = \sum_m \sum_n \sum_r \sum_s \phi(i,m)\phi(j,n)\phi(i,r)\phi(j,s)R_{xx}(m-r,n-s). \tag{2.209}$$

For the 2D AR(1) source of Eq. (2.191), this last equation separates into

$$\begin{aligned}\sigma^2(i,j) &= \sigma_x^2\left[\sum_m \sum_r \phi(i,m)\phi(i,r)\rho_h^{|m-r|}\right]\left[\sum_n \sum_s \phi(j,n)\phi(j,s)\rho_v^{|n-s|}\right] \\ &= \sigma_x^2\sigma_h^2(i)\sigma_v^2(j).\end{aligned} \tag{2.210}$$

Thus the variances for a 2D transform reduce to the product of two variances, one in each dimension. That is, $\sigma_h^2(i)$ and $\sigma_v^2(j)$ in Eq. (2.210) can be calculated using Eq. (2.202) for the 1D case. Next, we define a vector of horizontal and vertical variances

$$\underline{\sigma}_v^2 = \begin{bmatrix} \sigma_v^2(0) \\ \sigma_v^2(1) \\ \vdots \\ \sigma_v^2(N-1) \end{bmatrix} = W \begin{bmatrix} 1 \\ \rho_v \\ \rho_v^2 \\ \vdots \\ \rho_v^{N-1} \end{bmatrix} = W \underline{\rho}_v \quad (2.211)$$

Similarly,

$$\underline{\sigma}_h^2 = W \underline{\rho}_h.$$

Then the matrix of variances $V = [\sigma^2(i,j)]$ can be expressed as

$$V = \sigma_x^2 \underline{\sigma}_h^2 (\underline{\sigma}_v^2)^T = \sigma_x^2 W (\underline{\rho}_h \underline{\rho}_v^T) W^T. \quad (2.212)$$

Equation (2.212) provides the transform coefficient variances for the separable correlation case. It depends on the two correlation coefficients of the signal source and the orthogonal transformation employed.

b. Nonseparable Correlation

For the nonseparable 2D correlation function $R_{xx}(m,n)$, Eq. (2.209) becomes

$$\sigma^2(i,j) = \sum_k \sum_l w(i,l) R_{xx}(l,k) w(j,k). \quad (2.213)$$

The matrix of variances becomes

$$V = WRW^T, \quad (2.214)$$

where

$$R = [R_{xx}(m,n)], \qquad 0 \leq m,n \leq N-1. \quad (2.215)$$

and W is given in Eq. (2.201).

Equation (2.214) is a matrix representation of the variances of transform coefficients and represents a closed form time-domain expression that effectively separates the transformation from the source statistics. Thus, for a given correlation model, one can study the effects of different transformations, and conversely. The W matrix can be precalculated along with the given transformation base matrix Φ.

Adaptive transform coding techniques require the on-line computation of the coefficient variances. Equation (2.214) provides the theoretical basis for several adaptive coding scenarios.

2.4.5 Goodness of 2D Correlation Models for Images

Here, we use the results developed in the previous section to evaluate and compare the merits of the different 2D source correlation models. The 2D G_{TC} and optimum bit allocations based on these image correlation models are calculated and then compared with statistically measured results on actual image sources.

a. 2D Correlation Models

The three source correlation models of Section 2.4.3 and three transformations—the DCT, WHT, and MHT—are tested with two standard real images. Equation (2.214) provides the coefficient variances substituted into the transform gain equation,
Eq. (2.97), modified for the 2D case. We assume globally stationary source models for the two test images. The model match would be even more pronounced had we decomposed the images into smaller blocks wherein the stationarity is more realistic.

The two monochrome images tested are the standard pictures LENA and BRAIN. Each picture has 256×256 resolution with 8 bits per pixel. For each picture, we calculated the first-order correlation coefficients, ρ_h and ρ_v, using the autocovariance method over the entire frame with the assumption of spatial stationarity.

This pair of parameters for each image determines the 2D autocorrelation function. The chosen transform determines the W matrix. The W matrix and the correlation parameters are then combined to yield coefficient variances. Finally, these are employed to calculate 2D G_{TC} for the given image, and 8×8 transform. This figure of merit for the three transforms considered here is converted to decibels and displayed in Table 2.2. Also shown in this table is the statistical measured performance, which is described in the next section.

b. Statistical Test Performance

In order to check the analytical source correlation models, we used a nonparametric calculation of the variance of each coefficient. The 256×256 frame was divided into 1024 (32×32), 8×8 blocks. Each block is transformed. Then the variance of the (i, j) coefficient is calculated by averaging over all 1024 blocks.

$$\text{var}\{\theta(i,j)\} = \frac{1}{ML}\sum_{m=1}^{ML}(\theta_{ij} - \overline{\theta}_{ij})_m^2 \tag{2.216}$$

where m is the block index, and

$$\overline{\theta}_{ij} = \frac{1}{ML}\sum_{m=1}^{ML}(\theta_{ij})_m. \tag{2.217}$$

In the present instance, $M = L = 32$.

These coefficient variances are used to calculate the statistical(measured) test results tabulated in Table 2.2.

Table 2.2 clearly indicates that the DCT performs best for the two test images. It also indicates that the generalized correlation model provides the best representation with results that are very close to the measured ones. The separable correlation model, on the other hand predicts a performance that is totally inconsistent with the measured results.

It is also observed that the G_{TC} measure decreases as the image correlation decreases. As expected, the superiority of the DCT over the other transforms for highly correlated sources diminishes for low-correlation sources.

c. Optimum Bit Allocation

The 2D version of optimum bit allocation in Eq. (2.92),

$$B_{i,j} = B + \frac{1}{2} log_2 \frac{\sigma^2(i,j)}{\left[\prod_{i,j} \sigma^2(i,j)\right]^{1/N^2}}, \tag{2.218}$$

was used to encode the two test images, using the correlation models as the basis for the calculation of $\sigma^2(i,j)$. We also experimentally determined the coefficient variances via Eq. (2.216).

These tests confirmed that the generalized-correlation model was the best in generating a bit allocation matrix close to that obtained by statistical means. These bit allocation results for B=1 bit/pixel are shown in Tables 2.3 and 2.4. (Clearly scalar quantization requires integer bit allocation.) These tests suggest that the model-based prediction of bit allocation is accurate, especially at low bit rates, and could provide the basis for an a priori mask for transform operation and coding. This provides a basis for totally discarding some of the coefficients a priori. These coefficients therefore need not be calculated. These tests exercised all of the theory presented here: ranging over transforms, source models, bit allocations, and figure of merit.

2.4.6 Performance Comparison of Block Transforms

As mentioned earlier, G_{TC} is a commonly used performance criterion for orthonormal transforms. Its connection to rate-distortion theory makes it meaningful also from a source coding point of view. The only assumption made in this criterion is that of the same pdf type for all the coefficient bands as well as for the input signal. The error introduced from this assumption is acceptable for comparison purposes. Therefore, energy compaction powers of several different transforms for

Test Image	ρ_h	ρ_v	Transform	Separable correlation model	Generalized isotropic model	Generalized correlation model	Measured
LENA	0.945	0.972	DCT	89.26	19.93	22.45	21.98
			MHT	25.43	11.97	13.07	13.81
			WHT	59.01	15.66	17.59	14.05
BRAIN	0.811	0.778	DCT	5.63	3.62	3.83	3.78
			MHT	4.26	3.16	3.37	3.17
			WHT	4.32	3.02	3.15	3.66

Table 2.2: The 2D G_{TC} of several transforms with $N = 8$ for different source correlation models compared with statistical measurements.

AR(1) signal sources and standard still images are presented in this section. The results here also include different transform sizes.

Table 2.5 displays the compaction performance of discrete cosine transform (DCT), discrete sine transform (DST), modified Hermite transform (MHT), Walsh-Hadamard transform (WHT), Slant transform (ST), Haar transform (HT), and Karhunen-Loeve transform (KLT) for several different AR(1) sources and with the transform size $N = 8$. The KLT matrix was generated for the AR(1) source with $\rho = 0.95$ and held fixed for all tests. Table 2.6 assumes $N = 16$.

These tables demonstrate that the DCT performs very close to the KLT for AR(1) sources. It was theoretically shown that DCT is the best fixed transform for AR(1) sources (Jain, 1976).

It is also interesting that the performance of the discrete Legendre transform is only marginally inferior to that of the DCT, and second best to the KLT. The energy compaction of all transforms decrease for less correlated signal sources. Figure 2.11 displays the variation of energy compaction of these transforms as a function of transform size for an AR(1) source with $\rho = 0.95$. It is seen that the energy compaction increases as the transform size increases. It can be easily shown that for AR(1) sources, when $N \to \infty$ (Prob. 2.20)

$$G_{TC}^{gb} = \frac{1}{1-\rho^2}, \tag{2.219}$$

the global upper bound of energy compaction is obtained. Figure 2.11 shows that even for $N = 16$, the *DCT* performs close to G_{TC}^{gb}. One clearly prefers the smaller transform size because of practical considerations.

Tables 2.7 and Tables 2.8 provide the energy compaction performance of separable 2D DCT, DLT, DST, MHT, and WHT for the standard monochrome

(a)

i,j	1	2	3	4	5	6	7	8
	$B_{i,j}$ (bits/coefficient)							
1	6.1	4.5	3.4	2.6	2.0	1.5	1.1	0.6
2	3.6	3.4	2.7	2.1	1.8	1.2	0.9	0.6
3	2.4	2.4	2.2	1.9	1.3	1.1	0.8	0.4
4	1.5	1.6	1.8	1.4	1.3	0.9	0.5	0.2
5	0.9	1.0	1.0	1.0	0.9	0.6	0.4	0.1
6	0.3	0.4	0.4	0.5	0.5	0.2	0.0	-0.2
7	-0.2	0.1	-0.1	0.0	-0.0	0.0	-0.3	-0.4
8	-0.5	-0.4	-0.3	-0.3	-0.3	-0.4	-0.4	-0.4

(b)

i,j	1	2	3	4	5	6	7	8
	$B_{i,j}$ (bits/coefficient)							
1	6.08	4.45	3.41	2.68	2.2	1.83	1.59	1.44
2	3.76	2.8	2.22	1.74	1.39	1.13	0.94	0.83
3	2.56	1.9	1.53	1.21	0.96	0.77	0.64	0.56
4	1.83	1.27	0.99	0.77	0.60	0.47	0.38	0.32
5	1.35	0.86	0.63	0.46	0.34	0.24	0.17	0.13
6	1.02	0.57	0.38	0.24	0.14	0.07	0.02	-0.0
7	0.79	0.38	0.21	0.09	0.01	-0.03	-0.07	-0.1
8	0.66	0.27	0.11	0.01	-0.05	-0.1	-0.14	-0.15

Table 2.3: Bit allocation of 8×8 DCT coefficients, using Eq. (2.92) for LENA image at 1 bit per pixel (bpp) (a) with statistical measurement; (b) generalized correlation model.

(a)

$B_{i,j}$ (bits/coefficient)

i,j	1	2	3	4	5	6	7	8
1	4.6	2.5	2.1	1.6	1.2	1.1	0.9	1.1
2	2.9	2.0	1.4	1.2	1.2	1.0	0.5	1.1
3	2.6	1.5	1.2	1.0	1.0	0.8	0.7	0.8
4	2.3	1.1	1.1	0.9	0.8	0.5	0.5	0.6
5	1.2	0.9	0.8	0.6	0.8	0.5	0.5	0.7
6	1.4	0.7	0.7	0.6	0.6	0.5	0.4	0.7
7	1.3	0.5	0.4	0.5	0.5	0.3	0.4	0.8
8	0.9	0.5	0.5	0.3	0.5	0.5	0.5	0.5

(b)

$B_{i,j}$ (bits/coefficient)

i,j	1	2	3	4	5	6	7	8
1	4.23	3.40	2.65	1.96	1.48	1.12	0.88	0.74
2	3.47	2.79	2.17	1.58	1.15	0.84	0.63	0.50
3	2.80	2.24	1.74	1.26	0.90	0.63	0.44	0.33
4	2.16	1.71	1.33	0.95	0.66	0.44	0.28	0.19
5	1.72	1.33	1.01	0.70	0.47	0.29	0.16	0.09
6	1.37	1.04	0.77	0.52	0.33	0.18	0.07	0.01
7	1.16	0.84	0.61	0.39	0.23	0.10	0.01	-0.03
8	1.02	0.73	0.51	0.32	0.17	0.05	-0.02	-0.07

Table 2.4: Bit allocation of 8×8 DCT coefficients, using Eq. (2.92) for BRAIN image at 1 bpp (a) with statistical measurements (b) generalized correlation model.

test images; LENA, BUILDING, CAMERAMAN, and BRAIN, with $N = 8$ and $N = 16$, respectively.

All the performance results presented in this section prove the DCT superior to the other known fixed transforms. The DLT performs very closely to the DCT but the difficulty of implementation renders it impractical.

We may observe that the performance of nonsymmetrical fixed transforms is not the same for positive and negative values of ρ. This indicates that the low and high frequency basis functions of these transforms are not mirror images. The filter bank interpretation of block transforms demonstrates this very clearly. The asymmetrical performance of DCT is easily explained (Akansu and Haddad, 1990).

Discrete block transforms have been proposed as signal decomposition tech-

AR(1)	8 × 8 Transforms							
ρ	DCT	DLT	DST	MHT	WHT	ST	HT	KLT
0.95	7.631	7.372	4.877	4.412	6.232	7.314	6.226	7.666
0.85	3.039	2.935	2.642	2.444	2.601	2.915	2.589	3.070
0.75	2.036	1.971	1.938	1.849	1.847	1.960	1.799	2.061
0.65	1.597	1.553	1.574	1.534	1.471	1.546	1.456	1.616
0.50	1.273	1.248	1.277	1.265	1.217	1.246	1.206	1.286
-0.95	3.230	2.393	4.877	4.412	6.232	3.203	3.204	7.666
-0.85	2.067	1.629	2.642	2.444	2.601	1.859	1.903	3.070
-0.75	1.673	1.410	1.938	1.849	1.847	1.506	1.518	2.061
-0.65	1.440	1.284	1.574	1.534	1.471	1.327	1.323	1.616
-0.50	1.226	1.158	1.277	1.265	1.217	1.172	1.161	1.286

Table 2.5: Energy compaction performance, G_{TC}, of transforms for AR(1) sources with $N = 8$.

AR(1)	16 × 16 Transforms							
ρ	DCT	DLT	DST	MHT	WHT	ST	HT	KLT
0.95	8.822	8.097	6.000	4.718	6.598	8.034	6.580	8.867
0.85	3.294	3.058	2.984	2.579	2.656	3.019	2.627	3.326
0.75	2.148	2.013	2.082	1.923	1.836	1.993	1.809	2.170
0.65	1.657	1.570	1.644	1.575	1.481	1.559	1.459	1.673
0.50	1.301	1.253	1.303	1.283	1.221	1.250	1.207	1.309
-0.95	3.999	2.065	6.000	4.718	6.598	3.536	3.205	8.867
-0.85	2.475	1.573	2.984	2.579	2.656	1.931	1.905	3.326
-0.75	1.881	1.400	2.082	1.923	1.836	1.537	1.520	2.170
-0.65	1.551	1.283	1.644	1.575	1.481	1.343	1.324	1.673
-0.50	1.271	1.159	1.303	1.283	1.221	1.178	1.161	1.309

Table 2.6: Energy compaction performance, G_{TC}, of transforms for AR(1) sources with $N = 16$.

	8 × 8 Transforms				
Image	DCT	DLT	DST	MHT	WHT
LENA	21.988	19.497	14.880	13.817	14.056
CAMERAMAN	19.099	17.343	13.818	12.584	13.907
BUILDING	20.083	18.564	14.116	12.650	14.116
BRAIN	3.789	3.686	3.389	3.172	3.663

Table 2.7: Energy compaction performance, G_{TC}, of 2D transforms for test images with $N = 8$.

	16 × 16 Transforms			
Image	DCT	DST	MHT	WHT
LENA	25.655	19.106	16.435	14.744
CAMERAMAN	22.315	17.579	14.412	14.654
BUILDING	23.755	18.097	14.150	15.158
BRAIN	4.188	3.856	3.393	3.923

Table 2.8: Energy compaction performance, G_{TC}, of 2D transforms for test images with $N = 16$.

niques for almost two decades. They have found a wide spectrum of applications. Their good performance, especially for highly correlated sources, made them almost the only candidate for still image coding applications. DCT has become the standard transformation for image decomposition. The JPEG, MPEG, and the other standards include DCT in their adaptive image transform coding algorithms. Section 2.6.4 will briefly discuss the currently available hardware for real-time image-video coding.

The discontinuities of block transform operations become a problem especially at the low bit rate still image coding applications. There have been several attempts in the literature to circumvent this "blockiness" problem. One of these, the lapped orthogonal transform (LOT), is explained next in Section 2.5.

2.5 Lapped Orthogonal Transforms

2.5.1 Introduction

The block transform, particularly the DCT, is now an established technique for image and speech coding. However, the performance of block transforms is known to degrade significantly at low bit rates. The "blocking" effect results from the

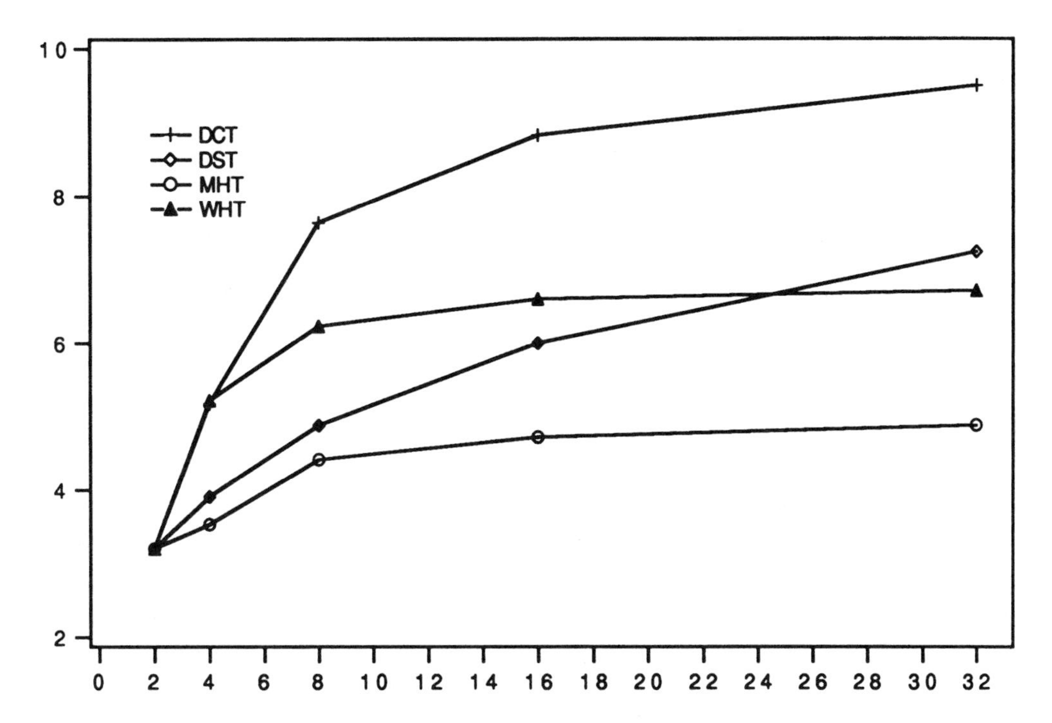

Figure 2.11: Energy compaction performance, G_{TC}, of DCT, DST, MHT, and WHT as a function of transform size N for AR(1) source with $\rho = 0.95$.

independent coding of each subimage and manifests itself as discontinuities at the subimage boundaries. Several researchers proposed techniques to overcome this problem (Reeve, and Lim, 1984).

Cassereau, Staelin, and Jager (1989) proposed an overlapping block transform called the lapped orthogonal transform (LOT), which uses pixels in adjacent blocks to smooth out the discontinuities at the subimage borders. Malvar and Staelin (1989) proposed a new LOT structure that utilizes the basis functions of the discrete cosine transform. More recently, the LOT was extended for other block transforms (Akansu and Wadas, 1992). The LOT has been used as a signal decomposition vehicle for image coding applications (Haskell, Tzou, and Hsing, 1989).

In this section, we review the properties of the lapped orthogonal transform and derive an optimal LOT. Malvar's fast LOT algorithm is extended to other block transforms. The energy compaction performance of several LOTs are compared for 1D AR(1) signal source models as well as for several test images. The effectiveness of the LOT in reducing the blocking artifacts is discussed and the 1D LOT basis functions for several transforms will be displayed in Fig. 2.14. We will show that the LOT is a special case of the more general subband decomposition. In a sense, the LOT is a precursor to the multirate filter bank.

2.5.2 Properties of the LOT

In conventional transform coding each segmented block of N data samples is multiplied by an $N \times N$ orthonormal matrix Φ to yield the block of N spectral coefficients. If the vector data sequence is labeled $\underline{x}_0, \underline{x}_1, ..., \underline{x}_i ...$, where each $\underline{x}_i$ represents a block of N contiguous signal samples, the transform operation produces $\underline{\theta}_i = \Phi \underline{x}_i$. We have shown in Fig. 2.1 that such a transform coder is equivalent to a multirate filter bank where each FIR filter has N taps corresponding to the size of the coefficient vector.

But, as mentioned earlier, this can lead to "blockiness" at the border region between data segments. To ameliorate this effect the lapped orthogonal transform calculates the coefficient vector $\underline{\theta}_i$ by using all N sample values in $\underline{x}_i$ and crosses over to accept some samples from $\underline{x}_{i-1}$ and $\underline{x}_{i+1}$. We can represent this operation by the multirate filter bank shown in Fig. 2.12. In this case, each FIR filter has L taps. Typically, $L = 2N$; the coefficient $\underline{\theta}_i$ uses N data samples in $\underline{x}_i$, $N/2$ samples from the previous block $\underline{x}_{i-1}$, and $N/2$ samples from the next block $\underline{x}_{i+1}$. We can represent this operation by the noncausal filter bank of Fig. 2.12 where the support of each filter is the interval $[-\frac{N}{2}, N-1+\frac{N}{2}]$. The time-reversed impulse responses are the basis functions of the LOT.

The matrix representation of the LOT is

$$
\begin{bmatrix} \underline{\theta}_0 \\ \underline{\theta}_1 \\ \cdot \\ \cdot \\ \cdot \\ \underline{\theta}_{M-1} \end{bmatrix}
=
\begin{bmatrix} \boxed{P_1'} & & & & \\ & \boxed{P_0'} & & & \\ & & \boxed{P_0'} & & \\ & & & \ddots & \\ & & & \boxed{P_0'} & \\ & & & & \boxed{P_2'} \end{bmatrix}
\begin{bmatrix} \underline{x}_0 \\ \underline{x}_1 \\ \cdot \\ \cdot \\ \cdot \\ \underline{x}_{M-1} \end{bmatrix}
\tag{2.220}
$$

The $N \times L$ matrix P_0' is positioned so that it overlaps neighboring blocks[5], typically by $N/2$ samples on each side. The matrices P_1', P_2' account for the fact that the first and last data blocks have only one neighboring block. The N rows of

[5] In this section, we indicate a transpose by P', for convenience.

P_0' correspond to the time-reversed impulse responses of the N filters in Fig. 2.12. Hence, there is a one-to-one correspondence between the filter bank and the LOT matrix P_0'.

We want the $MN \times MN$ matrix in Eq. (2.220) to be orthogonal. This can be met if the rows of P_0' are orthogonal,

$$P_0' P_0 = I_{N \times N}, \tag{2.221}$$

and if the *overlapping basis* functions of neighboring blocks are also orthogonal, or

$$P_0' W P_0 = P_0' W' P_0 = 0, \tag{2.222}$$

where W is an $L \times L$ shift matrix,

$$W = \begin{bmatrix} 0 & I_{(L-N)\times(L-N)} \\ 0 & 0 \end{bmatrix}. \tag{2.223}$$

A *feasible* LOT matrix P_0 satisfies Eqs. (2.221) and (2.222). The orthogonal block transforms Φ considered earlier are a subset of feasible LOTs. In addition to the required orthogonality conditions, a good LOT matrix P_0 should exhibit good energy compaction. Its basis functions should have properties similar to those of the good block transforms, such as the KLT, DCT, DST, DLT, and MHT,[6] and possess a variance preserving feature, i.e., the average of the coefficient variances equals the signal variance:

$$\sigma_x^2 = \frac{1}{N} \sum_{i=1}^{N} \sigma_i^2. \tag{2.224}$$

Our familiarity with the properties of these orthonormal transforms suggest that a good LOT matrix P_0 should be constructed so that half of the basis functions have even symmetry and the other half odd symmetry. We can interpret this requirement as a linear-phase property of the impulse response of the multirate filter bank in Fig. 2.12. The lower-indexed basis sequences correspond to the low frequency bands where most of the signal energy is concentrated. These sequences should gracefully decay at both ends so as to smooth out the blockiness at the borders. In fact, the orthogonality of the overlapping basis sequences tends to force this condition.

[6]The basis functions of the Walsh-Hadamard transform are stepwise discontinuous. The associated P matrix of Eq. (2.227) is ill-conditioned for the LOT.

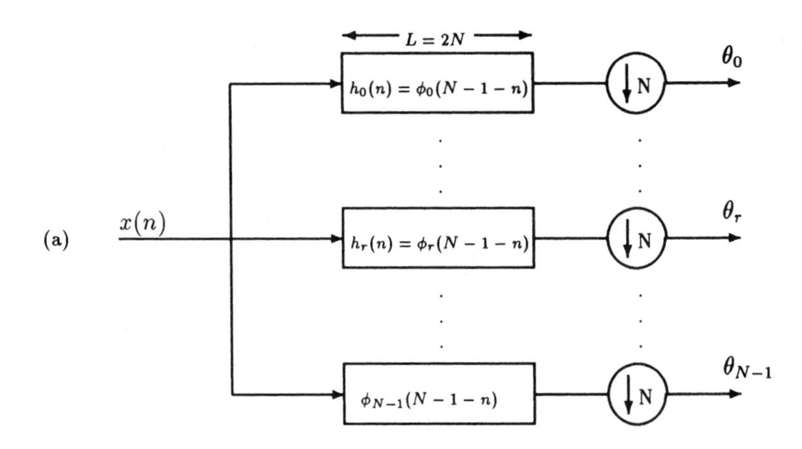

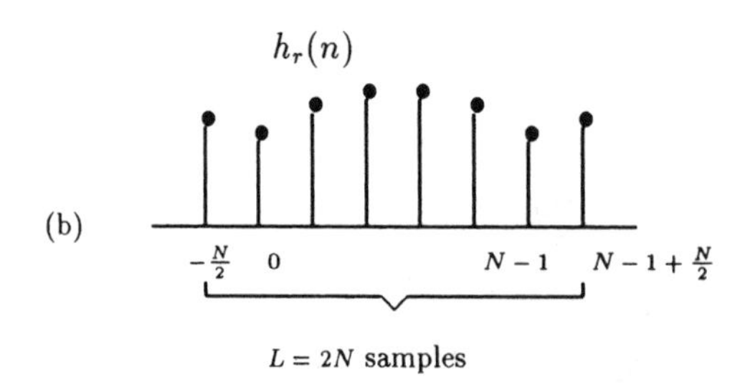

Figure 2.12: (a) The LOT as a multirate filter bank; (b) Noncausal filter impulse response.

2.5.3 An Optimized LOT

The LOT computes

$$\underline{\theta} = P_0^{'}\underline{x},$$

where $\underline{x}$ is the L dimensional data vector, $P_0^{'}$ the $N \times L$ LOT matrix, and $\underline{\theta}$ the N dimensional coefficient vector. The stated objective in transform coding is the maximization of the energy compaction measure G_{TC}, Eq. (2.97), repeated here as

$$G_{TC} = \frac{\sigma_x^2}{\left(\prod_{i=1}^{N}\sigma_i^2\right)^{1/N}}, \tag{2.225}$$

where $\sigma_i^2 = E\{\theta_i^2\}$ is the variance in the ith transform coefficient and also the ith diagonal entry in the coefficient covariance matrix

$$R_{\theta\theta} = E\{\underline{\theta}\underline{\theta}'\} = P_0' R_{xx} P_0. \tag{2.226}$$

From Eq. (2.225), the globally optimal P_0' is the matrix that minimizes the denominator of G_{TC}, that is, the geometric mean of the variances $\{\sigma_i^2\}$. Cassereau (1989) used an iterative optimization technique to maximize G_{TC}. The reported difficulty with their approach is the numerical sensitivity of iterations. Furthermore, a fast algorithm may not exist.

Malvar approached this problem from a different perspective. The first requirement is a fast transform. In order to ensure this, he grafted a perturbation on a standard orthonormal transform (the DCT). Rather than tackle the global optimum implied by Eq. (2.226), he formulated a suboptimal or locally optimum solution. He started with a feasible LOT matrix P preselected from the class of orthonormal transforms with fast transform capability and good compaction property. The matrix P is chosen as

$$P = \frac{1}{2}\begin{bmatrix} D_e - D_o & D_e - D_o \\ J(D_e - D_o) & -J(D_e - D_o) \end{bmatrix}, \tag{2.227}$$

where D_e' and D_o' are the $\frac{N}{2} \times N$ matrices consisting of the even and odd basis functions (rows) of the chosen $N \times N$ orthonormal matrix and J is the $N \times N$ counter identity matrix

$$J = \begin{bmatrix} 0 & \cdots & & 0 & 1 \\ 0 & \cdots & 0 & 1 & 0 \\ \vdots & & & & \vdots \\ 1 & 0 & & \cdots & 0 \end{bmatrix}. \tag{2.228}$$

This selection of P satisfies the feasibility requirements of Eqs. (2.221) and (2.222). In this first stage, we have

$$\underline{y} = P'\underline{x} \tag{2.229}$$

with associated covariance

$$R_{yy} = P' R_{xx} P. \tag{2.230}$$

So much is fixed a priori, with the expectation that a good transform, e.g., DCT, would result in compaction for the intermediate coefficient vector $\underline{y}$.

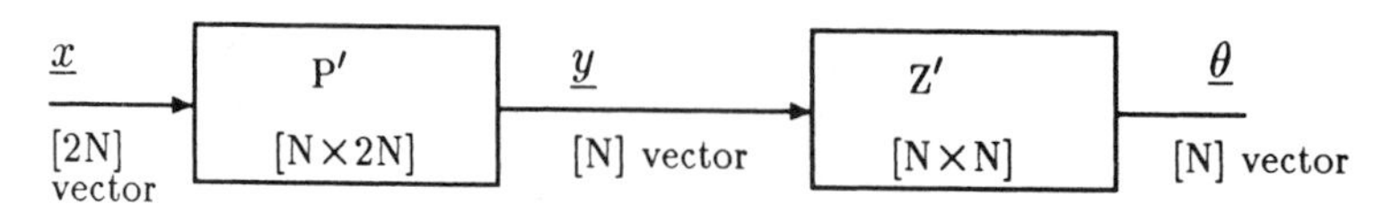

Figure 2.13: The LOT optimization configuration.

In the next stage, as depicted in Fig. 2.13, we introduce an orthogonal matrix Z, such that

$$\underline{\theta} = Z^{'}\underline{y} \tag{2.231}$$

and

$$R_{\theta\theta} = Z^{'}R_{yy}Z = Z^{'}(P^{'}R_{xx}P)Z. \tag{2.232}$$

The composite matrix is now

$$P_0^{'} = Z^{'}P^{'}, \tag{2.233}$$

which is also feasible, since

$$\begin{aligned} P_0^{'}P_0 &= Z^{'}P^{'}PZ = Z^{'}Z = I \\ P_0^{'}WP_0 &= Z^{'}P^{'}WPZ = 0. \end{aligned}$$

The next step is the selection of the orthogonal matrix Z, which diagonalizes $R_{\theta\theta}$. The columns of Z are then the eigenvectors $\{\underline{\xi}_i\}$ of R_{yy}, and

$$Z = [\underline{\xi}_1\underline{\xi}_2 \cdots \underline{\xi}_N]. \tag{2.234}$$

Since R_{yy} is symmetric and Toeplitz, half of these eigenvectors are symmetric and half are antisymmetric, i.e.

$$R_{yy}\underline{\xi} = \lambda\underline{\xi} \rightarrow J\underline{\xi} = \underline{\xi}, \quad \text{or} \quad J\underline{\xi} = -\underline{\xi}. \tag{2.235}$$

The next step is the factorization of Z into simple products so that coupled with a fast P such as the DCT, we can obtain a fast LOT. This approach is clearly locally rather than globally optimal since it depends on the a priori selection of the initial matrix P.

The matrices P_1 and P_2 associated with the data at the beginning and end of the input sequence need to be handled separately. The $N/2$ points at these boundaries can be reflected over. This is equivalent to splitting D_e into

$$D_e = \begin{bmatrix} H_e \\ J_*H_e \end{bmatrix}, \tag{2.236}$$

where H_e is the $N/2 \times N/2$ matrix containing half of the samples of the even orthonormal transform sequences and J_* is $N/2 \times N/2$. This H_e is then used in the following $(N + \frac{N}{2}) \times N$ end segment matrices

$$P_1 = \frac{1}{2} \begin{bmatrix} 2H_e & 2H_e \\ J_*(D_e - D_o) & -J_*(D_e - D_o) \end{bmatrix} \tag{2.237}$$

$$P_2 = \frac{1}{2} \begin{bmatrix} D_e - D_o & D_e - D_o \\ 2J_*H_e & -2J_*H_e \end{bmatrix}. \tag{2.238}$$

Malvar used the DCT as the prototype matrix for the initial matrix P. Any orthonormal matrix with fast algorithms such as DST or MHT could also be used. The next step is the approximate factorization of the Z matrix.

2.5.4 The Fast LOT

A fast LOT algorithm depends on the factorization of each of the matrices P and Z. The first is achieved by a standard fast transform, such as a fast DCT. The second matrix Z must be factored into a product of butterflies. For a DCT-based P and an AR(1) source model for R_{xx} with correlation coefficient ρ close to 1, Malvar shows that Z can be expressed as

$$Z \simeq \begin{bmatrix} I & 0 \\ 0 & Z_2 \end{bmatrix}, \tag{2.239}$$

where Z_2 and I are each $\frac{N}{2} \times \frac{N}{2}$, and Z_2 is a cascade of plane rotations

$$Z_2 = T_1 T_2 \cdots T_{N/2-1}, \tag{2.240}$$

where each plane rotation is

$$T_i = \begin{bmatrix} I_{i-1} & 0 & 0 \\ 0 & Y(\theta_i) & 0 \\ 0 & 0 & I \end{bmatrix}. \tag{2.241}$$

The term I_{i-1} is the identity matrix of order $i-1$, and $Y(\theta_i)$ is a 2×2 rotation matrix

$$Y(\theta_i) = \begin{pmatrix} \cos\theta_i & \sin\theta_i \\ -\sin\theta_i & \cos\theta_i \end{pmatrix}. \tag{2.242}$$

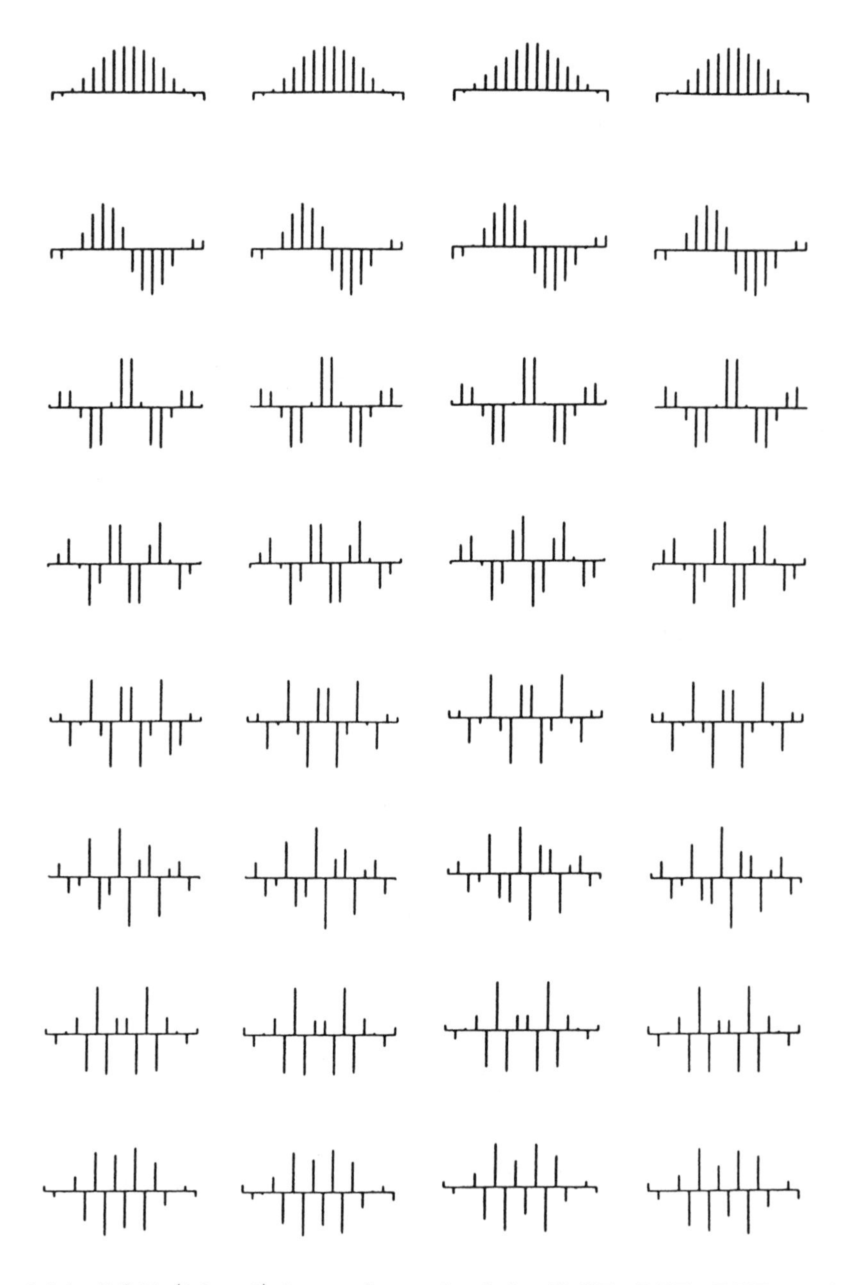

Figure 2.14: LOT (16×8) bases from the left: DCT, DST, DLT, and MHT, respectively. Their derivation assumes AR(1) source with $\rho = 0.95$.

	LOT, Markov Model, $\rho = 0.95$					
	$\tilde{Z}_1$			$\tilde{Z}_2$		
	θ_1	θ_2	θ_3	$theta_1$	θ_2	θ_3
DCT		I		$0.130\ \pi$	$0.160\ \pi$	$0.130\ \pi$
DLT	$0.005\ \pi$	$0.079\ \pi$	$0.105\ \pi$	$0.117\ \pi$	$0.169\ \pi$	$0.1\ 56\ \pi$
DST	$0.104\ \pi$	$0.149\ \pi$	$0.123\ \pi$	$0.0177\ \pi$	$0.0529\ \pi$	$0\ .0375\ \pi$
MHT	$0.152\ \pi$	$0.121\ \pi$	$0.063\ \pi$	0.0000	$0.0265\ \pi$	$0.0457\ \pi$

Table 2.9: Angles that best approximate the optimal LOT, $N = 8$.

For the other orthonormal transforms considered here, namely DST, DLT, and MHT, and an AR(1) source model

$$Z \simeq \begin{bmatrix} Z_1 & 0 \\ 0 & Z_2 \end{bmatrix} \tag{2.243}$$

$$Z_1 = T_{N/2-1} \cdots T_2 T_1 \tag{2.244}$$

and Z_2 as in Eq. (2.240).

Finally, the resulting P_0 for the general case can be written as

$$P_0 = \frac{1}{2} \begin{pmatrix} D_e & D_o & 0 & 0 \\ 0 & 0 & D_e & D_o \end{pmatrix} \begin{pmatrix} I & I & 0 & 0 \\ I & -I & 0 & 0 \\ 0 & 0 & I & I \\ 0 & 0 & I & -I \end{pmatrix} \begin{pmatrix} 0 & 0 \\ I & I \\ I & -I \\ 0 & 0 \end{pmatrix} \begin{pmatrix} \tilde{Z}_1 & 0 \\ 0 & \tilde{Z}_2 \end{pmatrix}. \tag{2.245}$$

This approximate factorization of Z into $log_2(N-1)$ butterflies is found to be satisfactory for small $N < 32$. The rotation angles that best approximate LOTs of size 16×8 for the DCT, DST, DLT, and MHT are listed in Table 2.9.

2.5.5 Energy Compaction Performance of the LOTs

Several test scenarios were developed to assess the comparative performance of LOTs against each other, and versus conventional block transforms for two signal covariance models: Markov, AR(1) with $\rho = 0.95$, and the generalized correlation model, Eq. (2.197) with $\rho = 0.9753$ and $r = 1.137$. The DCT, DST, DLT, and MHT transform bases were used for 8×8 block transforms and 16×8 LOTs.

The testing scenario for the LOT was developed as follows:

(1) An initial 16×8 matrix P was selected corresponding to the block transform being tested, e.g., MHT.

AR(1)*Input*	8 × 8 Transform s			
ρ	DCT	DST	DLT	MHT
0.95	7.6310	4.8773	7.3716	4.4120
0.85	3.0385	2.6423	2.9354	2.4439
0.75	2.0357	1.9379	1.9714	1.8491
0.65	1.5967	1.5742	1.5526	1.5338
0.50	1.2734	1.2771	1.2481	1.2649

Table 2.10(a): Energy compaction G_{TC} in 1D transforms for AR(1) signal source models.

Markov Model, $\rho = 0.95$				
AR(1) Input	LOT (16 × 8)			
ρ	DCT	DST	DLT	MHT
0.95	8.3885	8.3820	8.1964	8.2926
0.85	3.2927	3.2911	3.2408	3.2673
0.75	2.1714	2.1708	2.1459	2.1591
0.65	1.6781	1.6778	1.6633	1.6710
0.50	1.3132	1.3131	1.3060	1.3097

Table 2.10(b): Energy compaction G_{TC} in 1D transforms for AR(1) signal source models.

Generalized Correlation Model				
AR(1) Input	LOT (16 × 8)			
ρ	DCT	DST	DLT	MHT
0.95	8.3841	8.3771	8.1856	8.2849
0.85	3.2871	3.2853	3.2279	3.2580
0.75	2.1673	2.1665	2.1364	2.1523
0.65	1.6753	1.6749	1.6565	1.6663
0.50	1.3117	1.3115	1.3023	1.3071

Table 2.10(c): Energy compaction G_{TC} in 1D transforms for AR(1) signal source models.

(2) Independently of (1), a source covariance R_{xx} was selected, either AR(1), $\rho = 0.95$, or the generalized correlation model.
(3) The Z matrix is calculated for P in (1) and R_{xx} in (2).
(4) The LOT of steps (1), (2), and (3) was tested against a succession of test inputs, both matched and mismatched with the nominal R_{xx}. This was done to ascertain the sensitivity and robustness of the LOT and for comparative evaluation of LOTs and block transforms.

Table 2.10 compares compaction performance for AR(1) sources when filtered by 8×8 transforms, 16×8 LOTs optimized for Markov model, $\rho = 0.95$, and 16×8 LOTs optimized for the generalized-correlation model. In the 8×8 transforms we notice the expected superiority of DCT over other block transforms for large ρ input signals. Table 2.10 reveals that the 16×8 LOTs are superior to the 8×8 block transforms, as would be expected. But we also see that all LOTs exhibit essentially the same compaction. This property is further verified by inspection of the generalized-correlation model. Hence, from a compaction standpoint all LOTs of the same size are the same independent of the base block transform used.

Table 2.11 repeats these tests, but this time for standard test images. These results are almost a replay of Table 2.10 and only corroborate the tentative conclusion reached for the artificial data of Table 2.10.

The visual tests showed that the LOT reduced the blockiness observed with block transforms. But it was also noticed that the LOT becomes vulnerable to ringing at very low bit rates.

Our broad conclusion is that the 16×8 LOT outperformed the 8×8 block transforms in all instances and that the compaction performance of an LOT of a given size is relatively independent of the base block matrix used. Hence the selection of an LOT should be based on the simplicity and speed of the algorithm itself. Finally, we conclude that the LOT is insensitive to the source model assumed and to the initial basis function set. The LOT is a better alternative to conventional block transforms for signal coding applications. The price paid is the increase in computational complexity.

2.6 2D Transform Implementation

2.6.1 Matrix Kronecker Product and Its Properties

Kronecker products provide a factorization method for matrices that is the key to fast transform algorithms. We define the matrix Kronecker product and give a few of its properties in this section.

Block Transforms				
	8×8			
Images	DCT	DST	DLT	MHT
Lena	21.98	14.88	19.50	13.82
Brain	3.78	3.38	3.68	3.17
Building	20.08	14.11	18.56	12.65
Cameraman	19.10	13.81	17.34	12.58

Table 2.11(a): 2D energy compaction G_{TC} for the test images.

Markov Model, $\rho = 0.95$				
	LOT (16×8)			
Images	DCT	DST	DLT	MHT
Lena	25.18	24.98	23.85	24.17
Brain	3.89	3.87	3.85	3.84
Building	22.85	22.81	21.92	22.34
Cameraman	21.91	21.82	21.09	21.35

Table 2.11(b): 2D energy compaction G_{TC} for the test images.

Generalized Correlation Model				
	LOT (16×8)			
Images	DCT	DST	DLT	MHT
Lena	25.09	24.85	23.66	23.98
Brain	3.88	3.86	3.83	3.83
Building	22.70	22.65	21.65	22.11
Cameraman	21.78	21.67	20.83	21.13

Table 2.11(c): 2D energy compaction G_{TC} for the test images.

Markov Model, $\rho = 0.95$				
	LOT (16×8)			
Images	DCT	DST	DLT	MHT
Lena	24.45	24.02	23.78	23.62
Brain	3.88	3.83	3.85	3.83
Building	22.47	22.13	21.86	22.18
Cameraman	21.48	21.19	21.04	21.12

Table 2.12: Energy compaction G_{TC} of LOTs that employ the estimated Z-matrices.

The Kronecker product of an $(N_1 \times N_2)$ matrix A and $(M_1 \times M_2)$ matrix B is an $(N_1M_1 \times N_2M_2)$ matrix C defined as

$$C = A \otimes B \triangleq \begin{bmatrix} a(1,1)B & a(1,2)B & ... & a(1,N_2)B \\ a(2,1)B & a(2,2)B & ... & a(2,N_2)B \\ ... & & & \\ a(N_1,1)B & a(N_1,2)B & ... & a(N_1,N_2) \end{bmatrix} \tag{2.246}$$

where

$$A = a(i,j) \qquad i = 1,2,...,N_1 \quad j = 1,2,...,N_2.$$

The Kronecker products $A \otimes B$ and $B \otimes A$ are not necessarily equal. Several important properties of matrix Kronecker products are given as (Jain, 1989)

$$\begin{aligned} (A+B) \otimes C &= A \otimes C + B \otimes C & (2.247) \\ (A \otimes B) \otimes C &= A \otimes (B \otimes C) \\ (A \otimes B)(C \otimes D) &= (AC) \otimes (BD) \\ (A \otimes B)^T &= A^T \otimes B^T & (2.248) \\ (A \otimes B)^{-1} &= A^{-1} \otimes B^{-1}. \end{aligned}$$

2.6.2 Separability of 2D Transforms

A general 2D orthonormal transformation of an $N \times N$ image array F is defined by Eq. (2.42) and repeated here as

$$\theta(i,j) = \sum_{k=0}^{N-1} \sum_{l=0}^{N-1} f(k,l)\Phi(i,j;k,l). \tag{2.249}$$

This 2D transform operation requires $O(N^4)$ multiplications and additions for a real signal F and real transform kernel $\Phi(i,j;k,l)$.

Let us now map the image array F and the coefficient array Θ into vectors $\underline{f}$ and $\underline{\theta}$ of size N^2 each by row ordering as

$$\underline{f} \triangleq \begin{bmatrix} f_{11} \\ f_{12} \\ \vdots \\ f_{1N} \\ \cdots \\ f_{21} \\ \vdots \\ f_{2N} \\ \cdots \\ \vdots \\ f_{N1} \\ \vdots \\ f_{NN} \end{bmatrix}. \tag{2.250}$$

Let us also create an $N^2 \times N^2$ matrix T from the 2D transform kernel $\Phi(i,j;k,l)$. Now, we can rewrite the 2D transform of size N in Eq. (2.249) as a 1D transform of size N^2

$$\underline{\theta} = T\underline{f}. \tag{2.251}$$

The relations in Eqs. (2.249) and (2.251) are identical and both require the same number of multiplications and summations.

The 1D transformation in Eq. (2.251) is called *separable* if the basis matrix T can be expressed as a Kronecker product

$$T = \Phi_1 \otimes \Phi_2. \tag{2.252}$$

In this case the 1D transform of Eq. (2.251) is expressed as the separable 2D transform

$$\Theta = \Phi_1 F \Phi_2^T, \tag{2.253}$$

where F and Θ are square matrices obtained by row ordering of vectors $\underline{f}$ and $\underline{\theta}$, respectively, as

$$F = \begin{bmatrix} f_{11} & f_{12} & \cdots & f_{1N} \\ f_{21} & f_{22} & \cdots & f_{2N} \\ \cdots & & & \\ f_{N1} & f_{N2} & \cdots & f_{NN} \end{bmatrix}$$

$$\Theta = \begin{bmatrix} \theta_{11} & \theta_{12} & \dots & \theta_{1N} \\ \theta_{21} & \theta_{22} & \dots & \theta_{2N} \\ \dots & & & \\ \theta_{N1} & \theta_{N2} & \dots & \theta_{NN} \end{bmatrix}. \tag{2.254}$$

If

$$\Phi = \Phi_1 = \Phi_2, \tag{2.255}$$

Eq. (2.253) becomes

$$\Theta = \Phi F \Phi^T. \tag{2.256}$$

Equation (2.255) is the definition of a 2D separable unitary transform that was previously encountered as Eq. (2.45).

The 1D transform given in Eq. (2.251) requires $O(N^4)$ multiplications and additions for real $\underline{f}$ and T. The separable 2D unitary transform in Eq. (2.256) implies $O(N^3)$ multiplications and additions. This reduction of computational complexity is significant in practice.

2.6.3 Fast 2D Transforms

The separability of the 2D unitary transform kernel provides the foundation for a reduction of computations. This feature allows us to perform row and column transform operations in sequence, and the separable 2D transform is now given as

$$\Theta = \Phi(F\Phi^T) \triangleq \Phi S, \tag{2.257}$$

where S is an $N \times N$ intermediary matrix.

Now, the separability of the unitary matrix Φ is examined for further computational savings.

In Eq. (2.257) let the vector $\underline{s}_j$ be the jth column of S with transform

$$\underline{\theta}_j = \Phi \underline{s}_j, \tag{2.258}$$

where $\underline{\theta}_j$ is the jth column of Θ. This product requires $O(N^2)$ multiplications and summations. If the matrix Φ can be factored as a Kronecker product, then

$$\Phi = \Phi^{(1)} \otimes \Phi^{(2)}, \tag{2.259}$$

where matrices $\Phi^{(1)}$ and $\Phi^{(2)}$ are of size $(\sqrt{N} \times \sqrt{N})$. The vector $\underline{s}_j$ can now be row ordered into the matrix $S^{(j)}$ of size $(\sqrt{N} \times \sqrt{N})$ and the 1D transform of Eq. (2.258) is now expressed as a separable 2D transform of size $(\sqrt{N} \times \sqrt{N})$ as

$$\Theta^{(j)} = \Phi^{(1)} S^{(j)} (\Phi^{(2)})^T. \tag{2.260}$$

The matrix product in this last equation requires $O(2N\sqrt{N})$ multiplications and summations compared to $O(N^2)$, which was the case in Eq. (2.258). All row-column transform operations of the separable 2D transform in Eq. (2.257) can be factored into smaller-sized separable transforms similar to the case considered here.

Changing a 1D transform into a 2D or higher dimensional transform is one of the most efficient methods of reducing the computational complexity. This is also called multidimensional index mapping and in fact, this is the main idea behind the popular Cooley-Tukey and Winograd algorithms for DFT (Burrus and Parks, 1985).

The index mapping breaks larger size 1D transforms into smaller size 2D or higher dimensional transforms. It is clear that this mapping requires additional structural features from the transform basis or matrix Φ. The DFT, DCT, DST, WHT, and few other transforms have this property that provides efficient transform algorithms.

The readers with more interest in fast transform algorithms are referred to Burrus and Parks (1985), Blahut (1984), Rao and Yip (1990). and *IEEE Signal Processing Magazine* (January 1992 issue) for detailed treatments of the subject.

2.6.4 Transform Applications

The good coding performance of the DCT makes that block transform the prime signal decomposition tool of the first-generation still image and video codecs. The Joint Photographic Experts Group (JPEG) is a joint committee of the International Telegraph and Telephone Consultive Committee (CCITT) and the International Standards Organization (ISO) which was charged with defining an image compression standard for still frames with continuous tones (gray scale or color). This standard is intended for general purpose use within application-oriented standards created by ISO, CCITT, and other organizations. These applications include facsimile, video-tex, photo-telegraphy and compound office documents, and a number of others. On the other hand, CCITT has standardized a coding algorithm for video telephony and video conferencing at the bit-rate range of 64 to 1,920 kb/s, H.261. Similar to this, ISO's Moving Picture Experts Group (MPEG) has studied a possible coding standard for video storage applications below 1.5 Mb/s . This capacity allows a broad range of digital storage applications based on CD-ROM, digital audio tape (DAT), and Winchester technologies. Image and video codecs are now a reality for certain bit rates and will be feasible within 2 to 3 years for a wide range of channel capacities or storage mediums. The advances of computing power and digital storage technologies along with new digital signal processing techniques will provide engineering solutions to image and video coding at various rates.

All of the present image and video coding standards, e.g., MPEG II, employ 2D DCT as their signal decomposition technique. In principle, all of them perform 8×8 forward 2D DCT on 8×8 image or motion compensated frame difference (MCFD) blocks and obtain the corresponding 8×8 transform or spectral coefficients. These are quantized at the desired rate or distortion level. The quantization procedures incorporate the response of the human visual system to each spectral coefficient. The quantizer outputs are entropy encoded (Huffman or arithmetic encoding) and sent to the receiver. The decoder inverses the operations of the encoder to reconstruct the image or video frames at the receiver. The coding problem is discussed later in Chapter 7.

October 1991 and March 1992 issues of *IEEE Spectrum* give a very nice overview of visual communications products and coding techniques. Interested readers are referred to these journals for further information.

Although coding is one of the most popular transform applications, there are many emerging transform applications in multimedia and communications. Some of these applications are presented in Chapter 7. More detailed treatment of transform applications can be found in Akansu and Smith (1996) and Akansu and Medley (1999) for further studies.

2.7 Summary

The concept of the unitary block transform was developed from classical discrete-time signal expansions in orthogonal functions. These expansions provided spectral coefficients with energies distributed nonuniformly among the coefficients. This compaction provided the basis for signal compression.

The input-signal dependent KLT was shown to be the optimal block transform from a compaction standpoint. The reason for the popularity of the DCT as a compressive block transform was established by showing it to be very close to the KLT for highly correlated AR(1) sources.

Several block transforms—the DCT, MHT, WHT, etc.—were derived and their compaction performance evaluated both theoretically and for standard test images. The performance tables reinforce the superiority of the DCT over all other fixed transforms.

The LOT, or lapped orthogonal transform, was proposed as a structure that would reduce the blockiness observed for block transforms (including the DCT) at low bit rates. Analysis and tests demonstrated the perceptible improvement of the

DCT-based LOT over the DCT block transform. But it was also found that an LOT derived from other unitary transformations performed as well as the DCT-based LOT. The choice of LOT therefore could be based on other considerations, such as fast algorithms, parallel processing, and the like.

Both the block transform and the LOT were shown to be realizable as an M-band filter bank, which is the topic of the next chapter.

References

N. Ahmed, T. Natarajan, and K. R. Rao, "Discrete Cosine Transform," IEEE Trans. Comput. C-23, pp. 90–93, 1974.

N. Ahmed and K. R. Rao, *Orthogonal Transforms for Digital Signal Processing.* Springer-Verlag, New York, 1975.

A. N. Akansu and R. A. Haddad, "On Asymmetrical Performance of Discrete Cosine Transform," IEEE Trans. ASSP, ASSP-38, pp. 154–156, Jan. 1990.

A. N. Akansu and Y. Liu, "On Signal Decomposition Techniques," Optical Engineering, Special Issue on Visual Communication and Image Processing, Vol. 30, pp. 912–920, July 1991.

A. N. Akansu and M. J. Medley, Eds., *Wavelet, Subband and Block Transforms in Communications and Multimedia.* Kluwer Academic Publishers, 1999.

A. N. Akansu and M. J. T. Smith, Eds., *Subband and Wavelet Transforms: Design and Applications.* Kluwer Academic Publishers, 1996.

A. N. Akansu and F. E. Wadas, "On Lapped Orthogonal Transforms," IEEE Trans. Signal Processing, Vol. 40, No. 2, pp. 439–443, Feb. 1992.

H. C. Andrews, "Two Dimensional Transforms," chapter in *Picture Processing and Digital Filtering,* T. S. Huang (Ed.). Springer-Verlag, 1975.

K. G. Beauchamp, *Applications of Walsh and Related Functions.* Academic Press, 1984.

T. Berger, *Rate Distortion Theory.* Prentice Hall, 1971.

R. E. Blahut, *Fast Algorithms for Digital Signal Processing.* Addison-Wesley, 1984.

E. O. Brigham, *The Fast Fourier Transform.* Prentice-Hall, 1974.

C. S. Burrus and T. W. Parks, *DFT/FFT and Convolution Algorithms.* Wiley-Interscience, 1985.

R. N. Bracewell, "The Fast Hartley Transform," Proc. IEEE, Vol. 72, pp. 1010–1018, Aug. 1984.

S. J. Campanella and G. S. Robinson, "A Comparison of Orthogonal Transformations for Digital Signal Processing," IEEE Trans. Communications, pp. 1004–1009, Sept. 1977.

A. Cantoni and P. Butler, "Eigenvalues and Eigenvectors of Symmetric Centrosymmetric Matrices," Linear Algebra Applications, Vol. 13, pp. 275–288, 1976.

P. M. Cassereau, D. H. Staelin, and G. de Jager, "Encoding of Images Based on a Lapped Orthogonal Transform," IEEE Transactions on Communications, Vol. 37, No. 2, pp. 189–193, February 1989.

W.-H. Chen and C. H. Smith, "Adaptive Coding of Monochrome and Color Images," IEEE Transactions on Communications, Vol. COM-25, No. 11, pp. 1285–1292, Nov. 1977.

R. J. Clarke, "Relation Between Karhunen-Loeve and Cosine Transforms," IEE Proc., Part F, Vol. 128, pp. 359–360,Nov. 1981.

R. J. Clarke, "Application of Image Covariance Models to Transform Coding," Int. J. Electronics, Vol. 56 No. 2, pp. 245–260, 1984.

R. J. Clarke, *Transform Coding of Images.* Academic Press, 1985.

J. W. Cooley, and J. W. Tukey, "An Algorithm for the Machine Calculation of Complex Fourier Series," Math. Comput., Vol. 19, pp. 297–301, 1965.

J. W. Cooley, P. A. W. Lewis, and P. D. Welch, "Historical Notes on the Fast Fourier Transform," IEEE Trans. Audio. Electroacoust., Vol. AU-15, pp. 76–79, 1967.

C-CUBE Microsystems, CL550 JPEG Image Compression Processor, Product Brief, March 1990.

Draft Revision of Recommendation H.261, Document 572, CCITT SGXV, Working Party XV/1, Special Group on Coding for Visual Telephony.

D. F. Elliot and K. R. Rao, *Fast Transforms: Algorithms, Analyses, and Applications.* Academic Press, 1982.

O. Ersoy, "On Relating Discrete Fourier, Sine and Symmetric Cosine Transforms," IEEE Trans. ASSP, Vol. ASSP-33, pp. 219–222, Feb. 1985.

O. Ersoy and N. C. Hu, "A Unified Approach to the Fast Computation of All Discrete Trigonometric Transforms," Proc. ICASSP, pp. 1843–1846, 1987.

B. Fino and V. B. Algazi, "A Unified Treatment of Discrete Fast Unitary Transforms," SIAM J. Comput., Vol. 6, pp. 700–717, 1977.

W. A. Gardner, *Statistical Spectral Analysis.* Prentice-Hall, 1988.

G. H. Golub and C. Reinsch, "Singular Value Decomposition and Least Squares Solutions," Numer. Math, pp. 403–420, 14, 1970.

A. Habibi, "Survey of Adaptive Image Coding Techniques," IEEE Trans. Communications, Vol. COM-25, pp. 1275–1284, Nov. 1977.

R. A. Haddad, "A Class of Orthogonal Nonrecursive Binomial Filters," IEEE Trans. on Audio and Electroacoustics, Vol. AU-19, No. 4, pp. 296–304, Dec. 1971.

R. A. Haddad and A. N. Akansu, "A New Orthogonal Transform for Signal Coding," IEEE Trans. ASSP, pp. 1404–1411, Sep. 1988.

R. A. Haddad and T. W. Parsons, *Digital Signal Processing: Theory, Applications, and Hardware.* Computer Science Press, 1991.

M. Hamidi and J. Pearl, "Comparison of Cosine and Fourier Transforms of Markov-I Signals," IEEE Trans. ASSP, Vol. ASSP-24, pp. 428–429, Oct. 1976.

M. L. Haque, "A Two-dimensional Fast Cosine Transform," IEEE Trans. ASSP, Vol. ASSP-33, pp. 1532–1538, Dec. 1985.

H. F. Harmuth, *Transmission of Information by Orthogonal Functions*, 2nd ed. Springer-Verlag, 1972.

R. V. L. Hartley, "A More Symmetrical Fourier Analysis Applied to Transmission Problems," Proc. IRE, Vol. 30, pp. 144–150, 1942.

P. Haskell, K.-H. Tzou, and T. R. Hsing, "A Lapped Orthogonal Transform Based Variable Bit-Rate Video Coder for Packet Networks," IEEE Proc. of ICASSP, pp. 1905–1908, 1989.

S. Haykin, *Adaptive Filter Theory.* Prentice-Hall, 1986.

H. Hotelling, "Analysis of a Complex of Statistical Variables into Principal Components," J. Educ. Psychol., Vol. 24, pp. 417–441 and 498–520, 1933.

Y. Huang and P. M. Schultheiss, "Block Quantization of correlated Gaussian Random Variables," IEEE Trans. on Comm., pp. 289–296, Sept. 1963.

IEEE Signal Processing Magazine, January 1992. Special issue on DFT and FFT.

IEEE Spectrum, October 1991 issue, Video Compression, New Standards, New Chips.

IEEE Spectrum, March 1992 issue, Digital Video.

Image Communication, August 1990 issue.

A. K. Jain, "A Fast Karhunen-Loeve Transform for a Class of Random Processes," IEEE Trans. on Communications, pp. 1023–1029, Sept. 1976.

A. K. Jain, "A Sinusoidal Family of Unitary Transforms," IEEE Trans. Pattern Anal. Mach. Intelligence, PAMI, No. 8, pp. 358–385, Oct. 1979.

A. K. Jain, "Image Data Compression: A Review," Proc. IEEE, Vol. 69, pp. 349–389, March 1981.

A. K. Jain, "Advances in Mathematical Models for Image Processing," Proc. IEEE, Vol. 69, pp. 502–528, 1981.

A. K. Jain, *Fundamentals of Digital Image Processing.* Prentice-Hall, 1989.

N. S. Jayant and P. Noll, *Digital Coding of Waveforms.* Prentice-Hall, 1984.

JPEG Technical Specification, Revision 5, JPEG-8-R5, Jan. 1990.

K. Karhunen, "Ueber lineare methoden in der Wahrscheinlichkeitsrechnung," Ann. Acad. Sci. Fenn. Ser A.I. Math. Phys., vol.37, 1947.

S. M. Kay, *Modern Spectral Estimation: Theory and Application.* Prentice-Hall, 1988.

H. B. Kekre and J. K. Solanki, "Comparative Performance of Various Trigonometric Unitary Transforms for Transform Image Coding," Intl. J. Electronics, Vol. 44, pp. 305–315, 1978.

J. C. Lee and C. K. Un, "Block Realization of Multirate Adaptive Digital Filters," IEEE Trans. ASSP, Vol. ASSP-34, pp. 105–117, Feb. 1986.

J. S. Lim, *Two-Dimensional Signal and Image Processing.* Prentice-Hall, 1989.

S. P. Lloyd, "Least Squares Quantization in PCM," Inst. of Mathematical Sciences Meeting, Atlantic City, NJ, Sept. 1957; also IEEE Trans. on Information Theory, pp. 129–136, March 1982.

LSI Logic Corporation, Advance Information, Jan. 1990, Rev.A.

J. Makhoul,"On the Eigenvectors of Symmetric Toeplitz Matrices," *IEEE Trans. ASSP,* Vol. ASSP-29, pp. 868–872, Aug. 1981.

J. I. Makhoul and J. J. Wolf, "Linear Prediction and the Spectral Analysis of Speech," Bolt, Beranek, and Newman, Inc., Tech. Report, 1972.

H. S. Malvar, "Optimal Pre- and Post-filters in Noisy Sampled-data Systems," Ph.D. dissertation, Dept. Elec. Eng., Mass. Inst. Technology, Aug. 1986.(Also as Tech. Rep. 519, Res. Lab Electron., Mass. Inst. Technology, Aug. 1986.)

H. S. Malvar, *Signal Processing with Lapped Transforms.* Artech House, 1991.

H. S. Malvar, "The LOT: A Link Between Transform Coding and Multirate Filter Banks," Proc. Int. Symp. Circuits and Syst., pp. 835–838, 1988.

H. S. Malvar and D. H. Staelin, "Reduction of Blocking Effects in Image Coding with a Lapped Orthogonal Transform," IEEE Proc. of ICASSP, pp. 781–784, 1988.

H. S. Malvar and D. H. Staelin, "The LOT: Transform Coding Without Blocking Effects," IEEE Trans. on ASSP, Vol. 37, No.4, pp. 553–559, April 1989.

W. Mauersberger, "Generalized Correlation Model for Designing 2-dimensional Image Coders," Electronics Letters, Vol. 15, No. 20, pp. 664–665, 1979.

J. Max, "Quantizing for Minimum Distortion," IRE Trans. on Information Theory, pp. 7–12, March 1960.

W. E. Milne, *Numerical Calculus.* Princeton Univ. Press, 1949.

M. Miyahara and K. Kotani, "Block Distortion in Orthogonal Transform Coding-Analysis, Minimization and Distortion Measure," IEEE Trans. Communications, Vol. COM-33, pp. 90–96, Jan. 1985.

N. Morrison, *Introduction to Sequential Smoothing and Prediction.* McGraw-Hill, 1969.

H. G. Musmann, P. Pirsch, and H.-J. Grallert, "Advances in Picture Coding," Proc. IEEE, Vol. 73, No.4, pp. 523–548, April 1985.

S. Narayan, A. M. Peterson, and M. J. Narasimha, "Transform Domain LMS Algorithm," IEEE Trans. ASSP, Vol. ASSP-31, pp. 609–615, June 1983.

T. Natarajan and N. Ahmed, "Performance Evaluation for Transform Coding Using a Nonseparable Covariance Model," IEEE Trans. Comm., COM-26, pp. 310–312, 1978.

A. N. Netravali and B.G. Haskell, *Digital Pictures, Representation and Compression.* Plenum Press, 1988.

A. N. Netravali, and J.O. Limb, "Picture Coding: A Review," Proc. IEEE, Vol. 68, pp. 366–406, March 1980.

H. J. Nussbaumer, *Fast Fourier Transform and Convolution Algorithms.* Springer Verlag (Germany), 1981.

H. J. Nussbaumer, "Polynomial Transform Implementation of Digital Filter Banks," IEEE Trans. ASSP, Vol. ASSP-31, pp. 616–622, June 1983.

A. V. Oppenheim and R. W. Schafer, *Digital Signal Processing.* Prentice-Hall, 1975.

R. E. A. C. Paley, "On Orthogonal Matrices," J. Math. Phys., vol. 12, pp. 311–320, 1933.

A. Papoulis, *Signal Analysis.* McGraw Hill, 1977.

A. Papoulis, *Probability, Random Variables, and Stochastic Processes.* McGraw-Hill, 3rd Edition, 1991.

S. C. Pei and M. H. Yeh, "An Introduction to Discrete Finite Frames," IEEE Signal Processing Magazine, Vol. 14, No. 6, pp. 84–96, Nov. 1997.

W. B. Pennebaker and J. L. Mitchell, *JPEG Still Image Data Compression Standard.* Van Nostrand Reinhold, 1993.

M. G. Perkins, "A Comparison of the Hartley, Cas-Cas, Fourier, and Discrete Cosine Transforms for Image Coding," IEEE Trans. Communications, Vol. COM-36, pp. 758–761, June 1988.

W. K. Pratt, *Digital Image Processing.* Wiley-Interscience, 1978.

Programs for Digital Signal Processing. IEEE Press, 1979.

L. R. Rabiner and B. Gold, *Theory and Application of Digital Signal Processing.* Prentice-Hall, 1975.

L. R. Rabiner and R. W. Schafer, *Digital Processing of Speech Signals.* Prentice-Hall, 1978.

K. R. Rao (Ed.), *Discrete Transforms and Their Applications.* Academic Press, 1985.

K. R. Rao and P. Yip, *Discrete Cosine Transform.* Academic Press, 1990.

W. Ray, and R. M. Driver, "Further Decomposition of the K-L Series Representation of a Stationary Random Process," IEEE Trans. Information Theory, IT-16, pp. 663–668, 1970.

H. C. Reeve III, and J. S. Lim, "Reduction of Blocking Effect in Image Coding," Optical Engineering, Vol. 23, No. 1, pp. 34–37, Jan./Feb. 1984.

A. Rosenfeld and A. C. Kak, *Digital Picture Processing.* Academic Press, 1982.

G. Sansone, *Orthogonal Functions.* Wiley-Interscience, 1959.

H. Schiller, "Overlapping Block Transform for Image Coding Preserving Equal Number of Samples and Coefficients," Proc. SPIE Visual Communications and Image Processing, Vol. 1001, pp. 834–839, 1988.

A. Segall, "Bit Allocation and Encoding for Vector Sources," IEEE Trans. on Information Theory, pp. 162–169, March 1976.

J. Shore, "On the Application of Haar Functions," IEEE Trans. Comm., vol. COM-21, pp. 209–216, March 1973.

G. Szego, *Orthogonal Polynomials.* New York, AMS, 1959.

H. Ur, Tel-Aviv University, private communications, 1999.

M. Vetterli, P. Duhamel, and C. Guillemot, "Trade-offs in the Computation of Mono-and Multi-dimensional DCTs," Proc. ICASSP, pp. 999–1002, 1989.

G. K. Wallace, "Overview of the JPEG ISO/CCITT Still Frame Compression Standard," presented at SPIE Visual Communication and Image Processing, 1989.

J. L. Walsh, "A Closed Set of Orthogonal Functions," Am. J. Math., Vol. 45, pp. 5–24, 1923.

P. A. Wintz, "Transform Picture Coding," Proc. IEEE, Vol. 60, pp. 809–820, July 1972.

E. Wong, "Two-dimensional Random Fields and Representation of Images," SIAM J. Appl. Math., Vol. 16, pp. 756–770, 1968.

J. W. Woods, "Two-dimensional Discrete Markov Fields," IEEE Trans. Info. Theo., Vol. IT-18, pp. 232–240, 1972.

Y. Yemeni and J. Pearl, "Asymptotic Properties of Discrete Unitary Transforms," IEEE Trans. PAMI-1, pp. 366–371, 1979.

R. Zelinski and P. Noll, "Adaptive Transform Coding of Speech Signal," IEEE Trans. ASSP, Vol. ASSP-25, pp. 299–309, Aug. 1977.

R. Zelinski and P. Noll, "Approaches to Adaptive Transform Speech Coding at Low-bit Rates," IEEE Trans. ASSP, Vol. ASSP-27, pp. 89–95, Feb. 1979.

Chapter 3

Theory of Subband Decomposition

The second method of mutiresolution signal decomposition developed in this text is that of subband decomposition. In this chapter we define the concept, discuss realizations, and demonstrate that the transform coding of Chapter 2 can be viewed as a special case of a multirate filter bank configuration. We had alluded to this in the previous chapter by representing a unitary transform and the lapped orthogonal transform by a bank of orthonormal filters whose outputs are subsampled. The subband filter bank is a generalization of that concept.

Again, data compression is the driving motivation for subband signal coding. The basic objective is to divide the signal frequency band into a set of uncorrelated frequency bands by filtering and then to encode each of these subbands using a bit allocation rationale matched to the signal energy in that subband. The actual coding of the subband signal can be done using waveform encoding techniques such as PCM, DPCM, or vector quantization.

The subband coder achieves energy compaction by filtering serial data whereas transform coding utilizes block transformation. If the subbands have little spillover from adjacent bands (as would be the case if the subband filters have sharp cutoffs), the quantization noise in a given band is confined largely to that band. This permits separate, band-by-band allocation of bits, and control of this noise in each subband.

In Fig. 1.2, we described various structural realizations of the subband configuration. Starting with the two-channel filter bank, we first derive the conditions the filters must satisfy for zero aliasing and then the more stringent requirements for perfect reconstruction with emphasis on the orthonormal (or paraunitary) solution. Expanding this two-band structure recursively in a hierarchical subband tree generates a variety of multiband PR realizations with equal or unequal band splits, as desired.

Following this, we pursue a direct attack on the single level M-band filter bank and derive PR conditions using an alias-component (AC) matrix approach and the polyphase matrix route. From the latter, we construct a general time-domain representation of the analysis-synthesis system. This representation permits the most general PR conditions to be formulated, from which various special cases can be drawn, e.g., paraunitary constraints, modulated filter banks, and orthonormal LOTs.

This formulation is extended to two dimensions for the decidedly nontrivial case of nonseparable filters with a nonrectangular subsampling lattice. As an illustration of the freedom of the design in 2D filter banks, we describe how a filter bank with wedge-shaped (fan filter) sidebands can be synthesized in terms of appropriate 2D filters and decimation lattice.

In this second edition, we have expanded our scope to include a section on transmultiplexers. These systems, which find wide application in telecommunications, can be represented as synthesis/analysis multirate filter banks. These are shown to be the conceptual dual of the analysis/synthesis subband codecs whose major focus is data compression.

3.1 Multirate Signal Processing

In a multirate system, the signal samples are processed and manipulated at different clock rates at various points in the configuration. Typically, the band-limited analog signal is sampled at the Nyquist rate to generate what we call the full band signal $\{x(n)\}$, with a spectral content from zero to half the sampling frequency. These signal samples can then be manipulated either at higher or lower clock rates by a process called *interpolation* or *decimation.* The signal must be properly conditioned by filters prior to or after sampling rate alteration. These operations provide the framework for the subband signal decomposition of this chapter.

3.1.1 Decimation and Interpolation

The decimation and interpolation operators are represented as shown in Figs. 3.1 and 3.3, respectively, along with the sample sequences. Decimation is the process of reducing the sampling rate of a signal by an integer factor M. This process is achieved by passing the full-band signal $\{x(n)\}$ through a (typically low-pass)

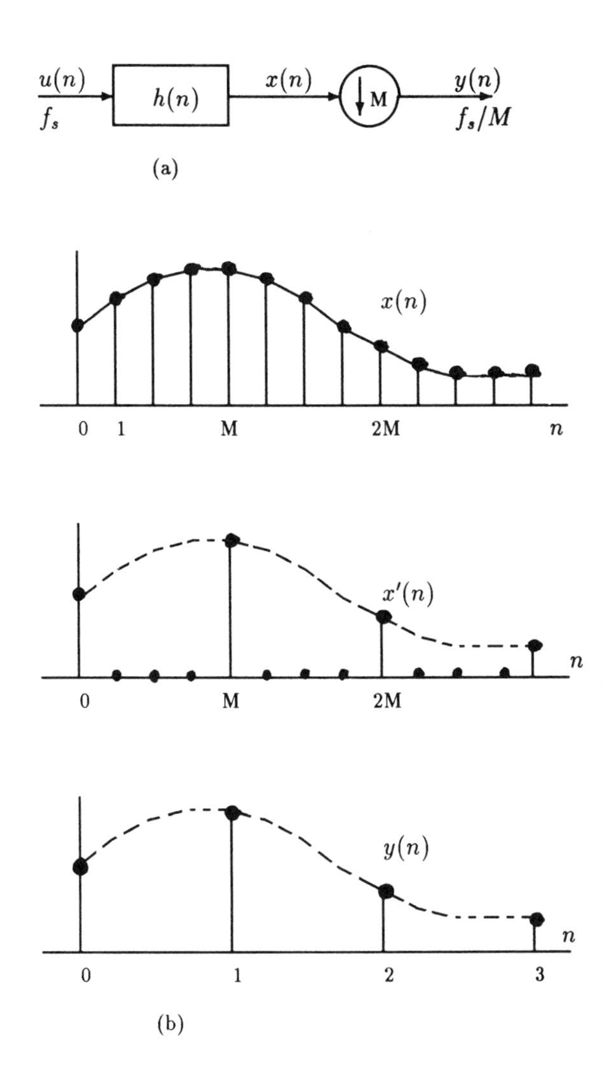

Figure 3.1: The decimation operation: (a) composite filter and down-sampler, (b) typical sequences.

antialiasing filter $h(n)$, and then subsampling the filtered signal, as illustrated in Fig. 3.1(a).

The *subsampler*, or *down-sampler* as it is also called in Fig. 3.1(a), is represented by a circle enclosing a downward arrow and the decimation factor M. The subsampling process consists of retaining every Mth sample of $x(n)$, and relabeling the index axis as shown in Fig. 3.1(b).

Figure 3.1(b) shows an intermediate signal $x'(n)$, from which the subsampled signal $y(n)$ is obtained:

$$x'(n) = \begin{cases} x(n) & n = 0, \pm M, \pm 2M, ... \\ 0 & \text{otherwise} \end{cases} \tag{3.1}$$

$$y(n) = x'(Mn) = x(Mn). \tag{3.2}$$

The intermediate signal $x'(n)$ operating at the same clock rate as $x(n)$ can be expressed as the product of $x(n)$ and a sampling function, the periodic impulse train $i(n)$,

$$x'(n) = i(n)x(n) = \left[\sum_{r=-\infty}^{\infty} \delta(n - rM) \right] x(n). \tag{3.3}$$

But $i(n)$ can be expanded in a discrete Fourier series (Haddad and Parsons, 1991):

$$i(n) = \sum_r \delta(n - rM) = \frac{1}{M} \sum_{k=0}^{M-1} e^{j\frac{2\pi}{M}nk}. \tag{3.4}$$

Hence

$$x'(n) = \frac{1}{M} \sum_{k=0}^{M-1} x(n)e^{j\frac{2\pi}{M}nk}.$$

Therefore the transform is simply

$$X'(z) = \frac{1}{M} \sum_k Z\{x(n)(e^{j\frac{2\pi}{M}k})^n\} = \frac{1}{M} \sum_{k=0}^{M-1} X(ze^{-j\frac{2\pi}{M}k}).$$

Using $W = e^{-j2\pi/M}$, this becomes

$$X'(z) = \frac{1}{M} \sum_{k=0}^{M-1} X(zW^k). \tag{3.5}$$

On the unit circle $z = e^{j\omega}$, the frequency response is just

$$X'(e^{j\omega}) = \frac{1}{M} \sum_{k=0}^{M-1} X(e^{j(\omega - \frac{2\pi k}{M})}). \tag{3.6}$$

This latter form shows that the discrete-time Fourier transform is simply the sum of M replicas of the original signal frequency response spaced $2\pi/M$ apart. [This may be compared with the sampling of an analog signal wherein the spectrum of the sampled signal $x(n) = x_a(nT_s)$ is the periodic repetition of the analog spectrum at a spacing of $2\pi/T_s$.]

Next we relabel the time axis via Eq. (3.2), which compresses the time scale by M. It easily follows that

$$Y(z) = \sum_{n=-\infty}^{\infty} x'(Mn)z^{-n} = \sum_{k=-\infty}^{\infty} x'(k)(z^{\frac{1}{M}})^{-k}$$

or

$$Y(z) = X'(z^{\frac{1}{M}}) \tag{3.7}$$

and

$$Y(e^{j\omega}) = X'(e^{\frac{j\omega}{M}}). \tag{3.8}$$

Using Eq. (3.5), the transform of the M subsampler is

$$Y(z) = \frac{1}{M}\sum_{k=0}^{M-1} X(z^{1/M}W^k) \tag{3.9}$$

or

$$Y(e^{j\omega}) = \frac{1}{M}\sum_{k=0}^{M-1} X(e^{j(\frac{\omega-2\pi k}{M})}). \tag{3.10}$$

Thus the time compression implicit in Eq. (3.2) is accompanied by a stretching in the frequency-domain so that the interval from 0 to π/M now covers the band from 0 to π. It should be evident that the process of discarding samples can lead to a loss in information. In the frequency-domain this is the aliasing effect as indicated by Eq. (3.6). To avoid aliasing, the bandwidth of the full band signal should be reduced to $\pm\pi/M$ prior to down-sampling by a factor of M. This is the function of the antialiasing filter $h(n)$. Figure 3.2 shows spectra of the signals involved in subsampling. These correspond to the signals of Fig. 3.1(b). [In integer-band sampling as used in filter banks the signal bandwidth is reduced to $\pm[\frac{k\pi}{M}, \frac{(k+1)\pi}{M}]$ prior to down-sampling. See Section 3.2.1.]

Interpolation is the process of increasing the sampling rate of a signal by the integer factor M. As shown in Fig. 3.3(a), this process is achieved by the combination of *up-sampler* and low-pass filter $g(n)$. The up-sampler is shown symbolically in Fig. 3.3(a) by an upward-pointing arrow within a circle. It is

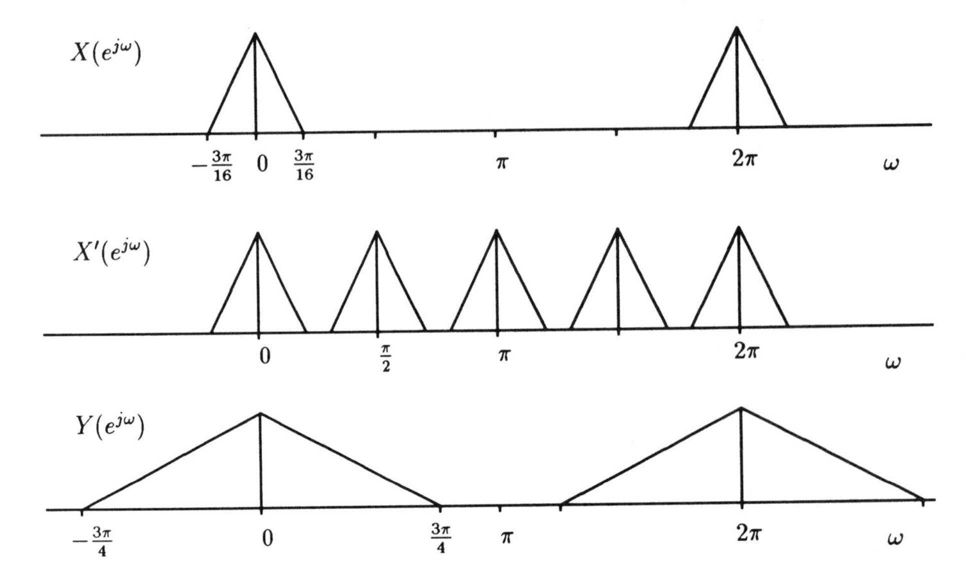

Figure 3.2: Frequency spectra of signals in down-sampling drawn for $M = 4$.

defined by

$$y(n) = \begin{cases} x(n/M) & n = 0, \pm M, \pm 2M, ... \\ 0 & \text{otherwise} \end{cases} \tag{3.11}$$

This operator inserts $(M-1)$ zeros between sample values and reindexes the time scale as shown in Fig. 3.3(b). Effectively, the clock rate increases by a factor of M.

Up-sampling has two effects. First, stretching the time axis induces a compression in frequency; second, forcing the "interpolated" signal to pass through zero between samples of $x(n)$ generates high-frequency signals or images. These effects are readily demonstrated in the transform domain by

$$Y(z) = \sum_{-\infty}^{\infty} y(n) z^{-n} = \sum_{-\infty}^{\infty} x(\frac{n}{M}) z^{-n} = \sum_{k=-\infty}^{\infty} x(k)(z^M)^{-k}$$

or

$$Y(z) = X(z^M), \qquad Y(e^{j\omega}) = X(e^{j\omega M}). \tag{3.12}$$

Figure 3.4 illustrates this frequency compression and image generation for $M = 4$. Observe that the frequency axis from 0 to 2π is scale changed to 0

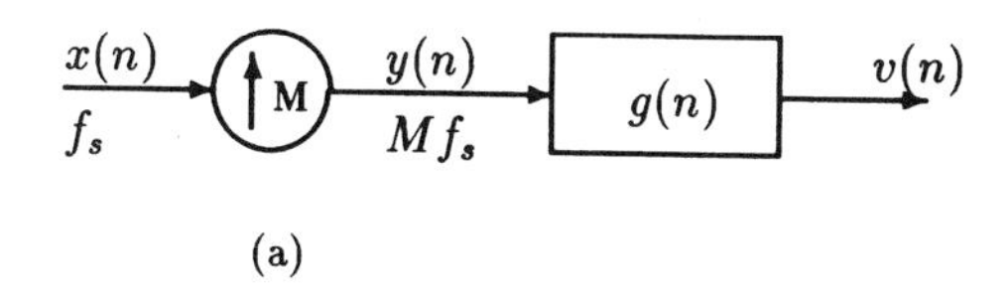

(a)

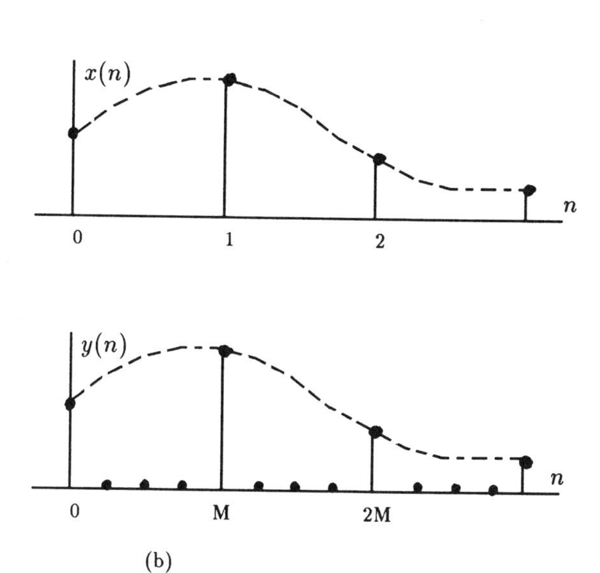

(b)

Figure 3.3: (a) Up-sampling operation, (b) input and output waveforms for $M = 4$.

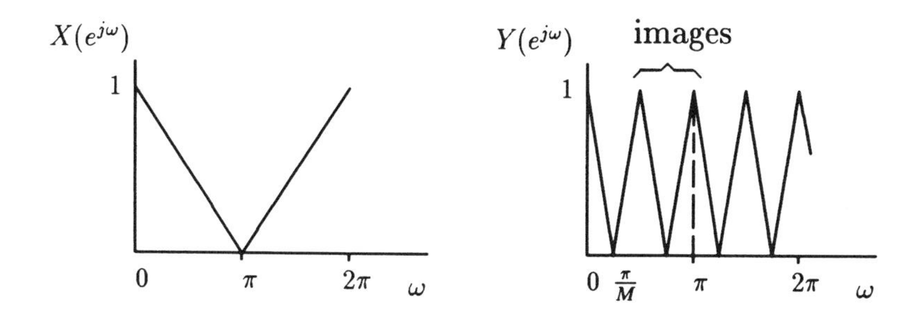

Figure 3.4: Frequency axis compression due to up-sampling for $M = 4$.

to $2\pi/M$ and periodically repeated.[1] The purpose of the low-pass filter $g(n)$ is to eliminate these images by smoothing the up-sampled signal.

It is easy to show (Prob. 3.5) that the time-domain representations of the decimator and interpolator of Figs. 3.1(a) and 3.3(b) are, respectively,

$$\begin{aligned} y(n) &= \sum_k h(Mn-k)u(k) \\ v(n) &= \sum_k g(n-Mk)x(k) \end{aligned} \tag{3.13}$$

The up- and down-sampling operations have made these systems time-varying. Shift invariance requires that $u(n) \to y(n)$, which implies that $u(n-n_0) \to y(n-n_0)$. We see from Eq. (3.13) that this latter condition is satisfied in the decimator only for n_0 a multiple of M. Similar statements can be argued for the interpolator.

Now consider what happens if we position the down-sampler and up-sampler back-to-back, as in Fig. 3.5. We can recognize the interpolator output $v(n)$ as the intermediate signal $x^{'}(n)$ in Fig. 3.1. Hence $V(z)$ in this case reduces to $X^{'}(z)$ in Eqs. (3.5) and (3.6); i.e., $V(z) = \frac{1}{M}\sum_{k=0}^{M-1} X(zW^k)$.

Sketches of the spectra of these signals are shown in Fig. 3.6. The input signal spectrum in Fig. 3.6(a) has a bandwidth greater than π/M for $M = 4$, so aliasing is expected. Three of the four terms in Eq. (3.10) are displayed as (b), (c), and (d) in that diagram; when all four are added together we obtain the decimated spectrum of (e), which shows the aliasing due to overlap of the frequency bands. Up-sampling compresses the frequency axis, as in Fig. 3.6(f), and induces the images. Therefore the spectrum of the signal following down-sampling and up-sampling exhibits both aliasing (the original bandwidth is too large for the decimation parameter used) and the images that are always the consequence of up-sampling. (See Problems 3.1, 3.2)

For the polyphase signal decomposition in the next section, we need to manipulate transfer functions across up- and down-samplers. The basic results are illustrated in Fig. 3.7. We can establish these equivalences by straightforward use of the defining equations. For example, the output of the down-sampler (filter) in Fig. 3.7(a) is

$$Y(z) = G(z)U(z) = G(z)\frac{1}{M}\sum_{k=0}^{M-1} X(z^{1/M}W^k).$$

[1] Classically, interpolation is the process of fitting a smooth continuous curve between sample values. The multirate DSP community, however, uses the term to force zeros between samples followed by a smoothing filter. Some authors use the term "expander" to indicate specifically the $\uparrow M$ operator. The "interpolator" is the composite of up-sampler followed by a filter. Similarly the $\downarrow M$ operation is sometimes called a "compressor." A filter followed by $\downarrow M$ can then be termed as a "decimator."

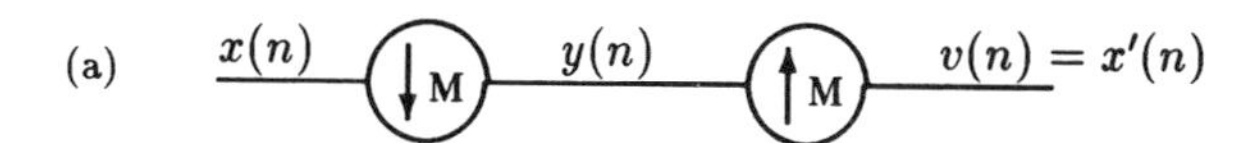

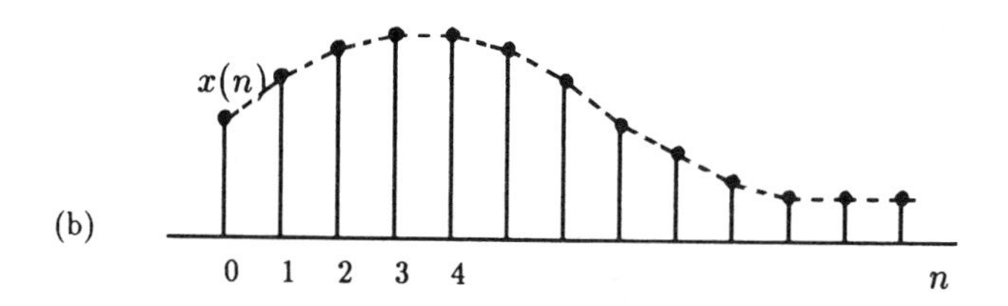

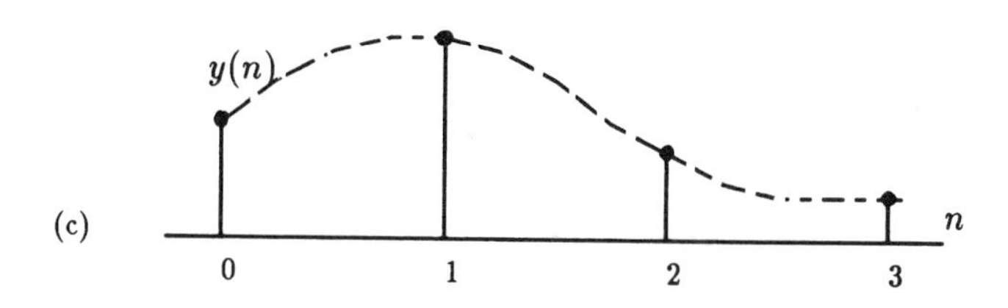

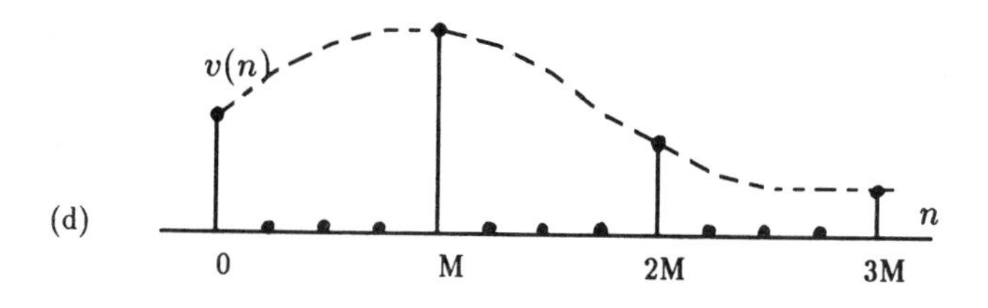

Figure 3.5: Typical signals in a down-sampler up-sampler for $M = 4$.

For the structure just to the right, we see that

$$V(z) = G(z^M)X(z).$$

Passing through the down-sampler gives the output transform:

$$\begin{aligned}
\tfrac{1}{M}\textstyle\sum_k V(z^{1/M}W^k) &= \tfrac{1}{M}\textstyle\sum_k G[(z^{1/M}W^k)^M)]X(z^{1/M}W^k) \\
&= \tfrac{1}{M}\textstyle\sum_k G(zW^{kM})X(z^{1/M}W^k) \\
&= \tfrac{1}{M}\textstyle\sum_k G(z)X(z^{1/M}W^k).
\end{aligned}$$

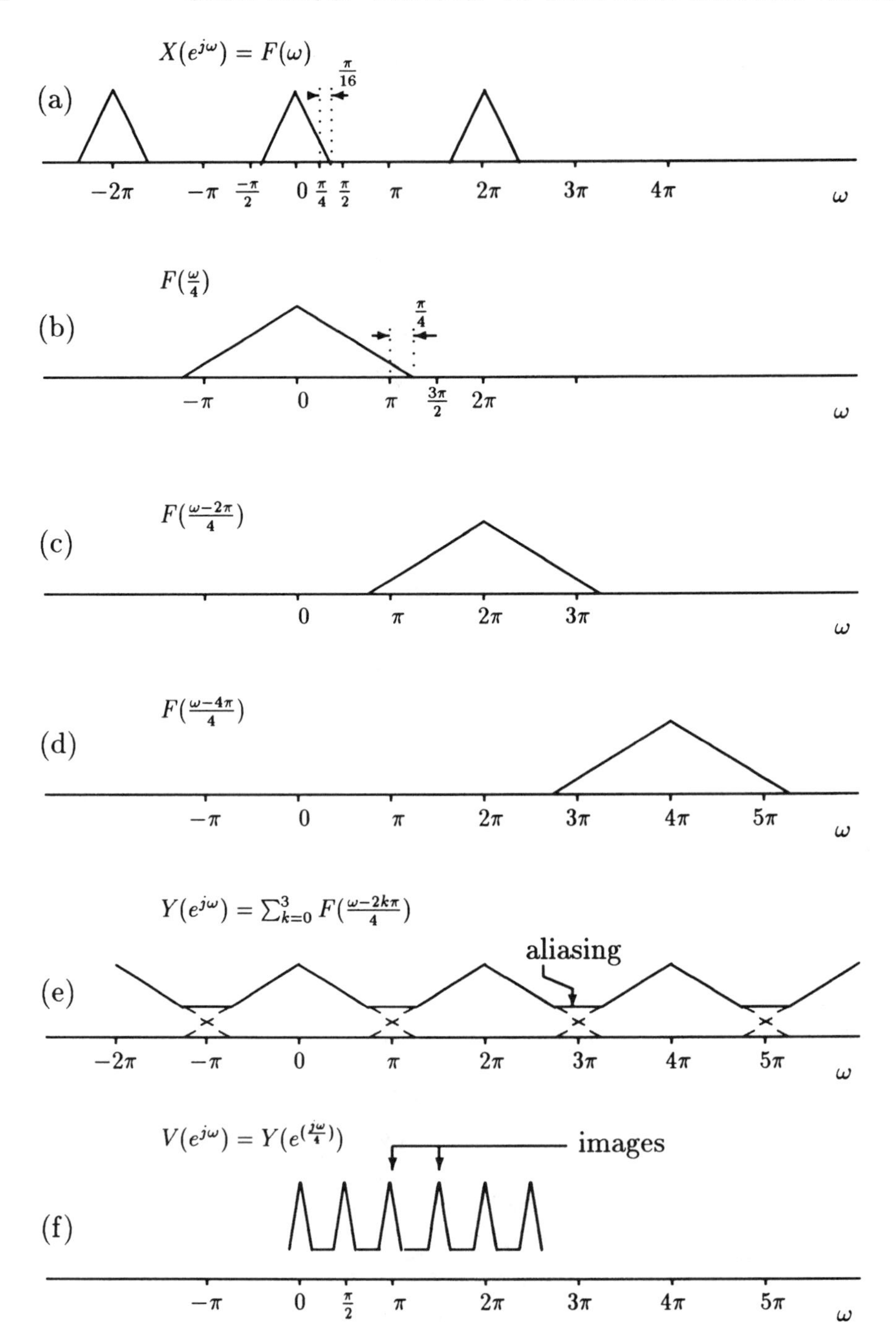

Figure 3.6: The spectra of signals shown in Fig. 3.5.

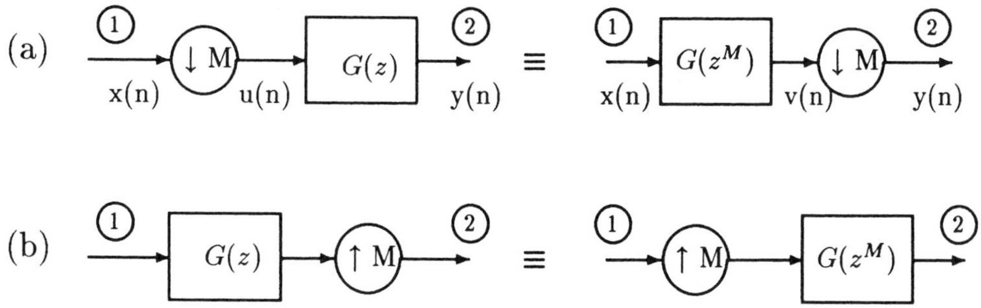

Figure 3.7: Equivalent structures.

This is the same as $Y(z)$ obtained previously. In a similar way, we can show the equivalence of the representations in Fig. 3.7(b). (See Problem 3.8; see also Prob. 3.5 for time-domain)

3.1.2 Polyphase Decomposition

To prevent or reduce the aliasing inherent in the subsampling operator, an antialiasing filter—typically low-pass— is usually placed in front of the down-sampler as in Fig. 3.8(a). We will show that this combination can be represented by the polyphase decomposition shown in Fig. 3.8(b), and given explicitly by

$$H(z) = \sum_{k=0}^{M-1} z^{-k} G_k(z^M), \tag{3.14}$$

where

$$G_k(z) = h(k) + h(k+M)z^{-1} + h(k+2M)z^{-2} + ... \tag{3.15}$$

The impulse response of the kth polyphase filter is simply a subsampling of $h(n+k)$,

$$g_k(n) = h(Mn+k), \qquad k = 0, 1, ..., M-1. \tag{3.16}$$

The proof is straightforward. We simply expand $H(z)$ and group terms

$$\begin{aligned} H(z) &= h_0 + h_1 z^{-1} + h_2 z^{-2} + ... \\ &= (h_0 + h_M z^{-M} + h_{2M} z^{-2M} + ...) + (h_1 z^{-1} + h_{M+1} z^{-(M+1)} + ...) + \\ &\quad ... + (h_{M-1} z^{-(M-1)} + h_{2M-1} z^{-(2M-1)} + ...) \end{aligned}$$

From the latter expansion, we recognize that $H(z)$ can be written as

$$H(z) = G_0(z^M) + z^{-1} G_1(z^M) + ... + z^{-(M-1)} G_{M-1}(z^M)$$

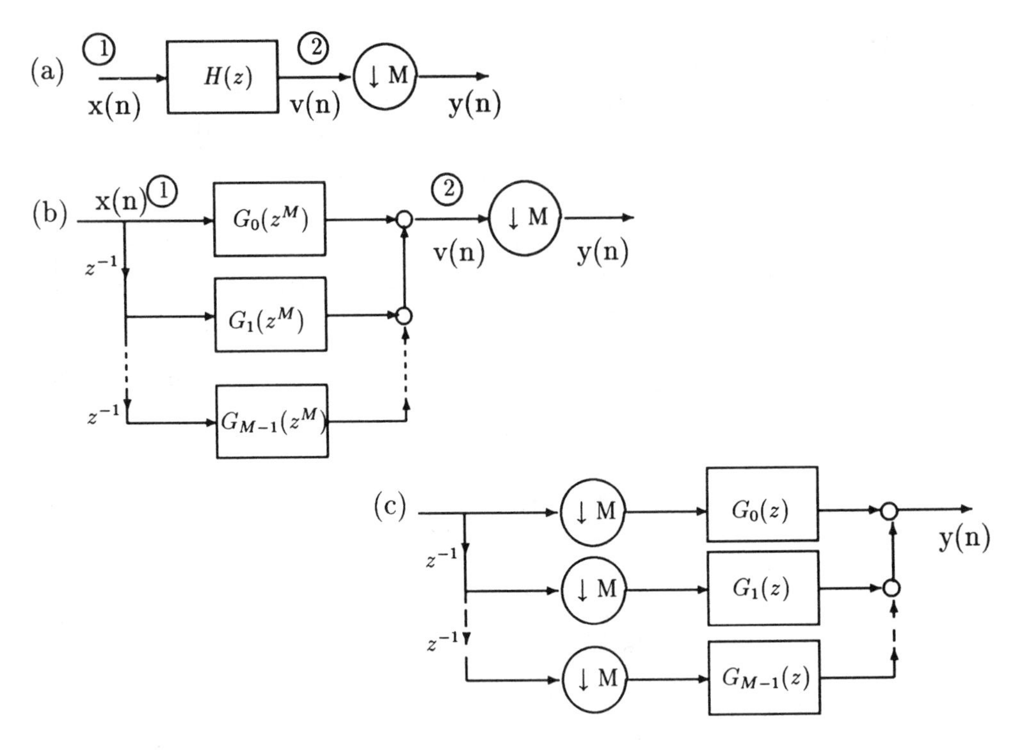

Figure 3.8: (a) Filter followed by down-sampler; (b) polyphase decomposition; (c) alternative polyphase network representation.

with $G_k(z)$ given by Eq. (3.15). More simply put, $g_k(n)$ is just a down-sampling of $h(n)$ shifted by k.

Similarly, we can show that the polyphase decomposition of the up-sampler and filter combination is as illustrated in Fig. 3.9. First $H(z)$ in Fig. 3.9(a) is replaced by the polyphase bank from point (1) to point (2) in Fig. 3.9(c). Shifting the up-sampler to the right using the equivalence suggested by Fig. 3.7(b) then yields the composite structure of Fig. 3.9(b).

Equation (3.15) represents the polyphase components $G_k(z)$ in terms of the decimated samples $\{h(k + lM)\}$. To cast this in the transform-domain, we note that $g_r(n)$ is an M-fold decimation of $h(n + r)$. Therefore with

$$h(n+r) \leftrightarrow z^r H(z) \stackrel{\Delta}{=} F_r(z)$$

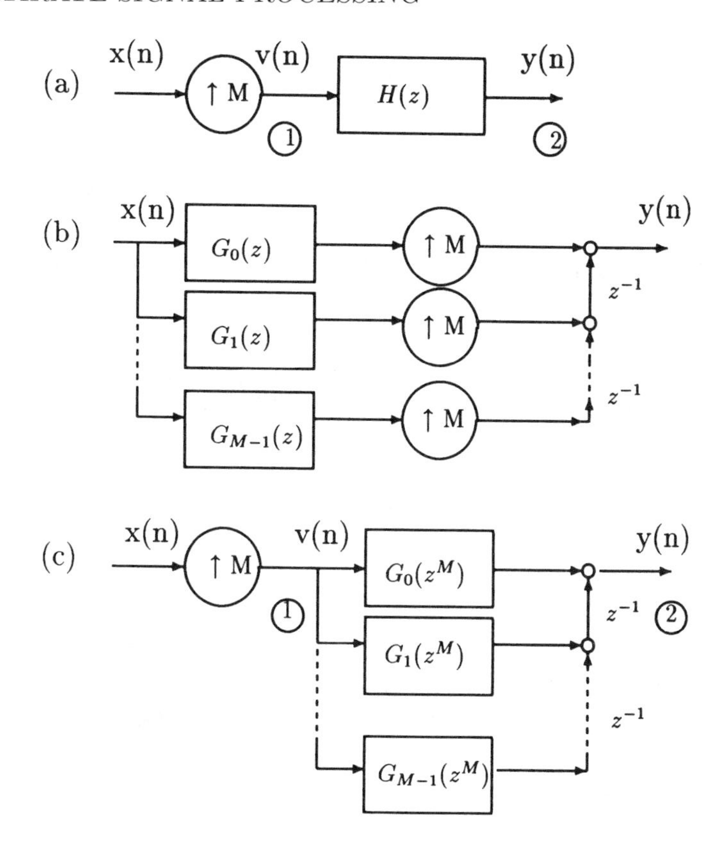

Figure 3.9: Polyphase decomposition of interpolator.

and using Eq. (3.9) for the transform of a decimated signal, we get

$$\begin{aligned} G_r(z) &= \frac{1}{M}\sum_{k=0}^{M-1} F_r(z^{1/M}W^k) \\ &= \frac{z^{r/M}}{M}\sum_{k=0}^{M-1} W^{kr}H(z^{1/M}W^k) \end{aligned} \tag{3.17}$$

and, on the unit circle

$$G_r(e^{j\omega}) = \frac{e^{jr\omega/M}}{M}\sum_{k=0}^{M-1} W^{kr}H(e^{j(\omega-2k\pi)/M}). \tag{3.18}$$

The five Eqs. (3.14)–(3.18) completely define the polyphase analysis and synthesis equations. (See Problems 3.3, 3.4)

Two simple applications of polyphase decomposition provide a first look at the role of this powerful representation for filter synthesis. In the first case, we consider the realization of an ideal low-pass filter that functions as an ideal antialiasing filter, as shown in Fig 3.10 by a polyphase filter bank.

If $H(e^{j\omega})$ is bandlimited to $\pm\frac{\pi}{M}$, then $H(e^{j\omega/M})$ occupies the full band $[-\pi,\pi]$. Moreover, the other terms $\{H(e^{j(\omega-2\pi k)/M}), k>0\}$ do not overlap onto $[-\pi,\pi]$. Hence, there is no aliasing, and only the $k=0$ term in Eq. (3.18) contributes to $G_r(e^{j\omega})$, so that

$$G_r(z) = \frac{z^{r/M}}{M} H(z^{1/M})$$

or

$$G_r(e^{j\omega}) = \frac{1}{M} e^{jr\omega/M} H(e^{j\omega/M}).$$

For the case that $H(e^{j\omega})$ is the ideal low-pass filter of Fig. 3.10, the polyphase components in Eq. (3.18) reduce to

$$G_r(e^{j\omega}) = \frac{1}{M} e^{jr\omega/M}, \quad r=0,1,...,M-1 \quad 0<\omega<2\pi.$$

Thus, the polyphase representation is just a bank of all-pass filters with leading linear-phase, as in Fig. 3.11. We have effectively replaced the requirement of the step discontinuity in the ideal $H(e^{j\omega})$ by a bank of all-pass networks. The linear-phase characteristic of the all-pass now becomes the approximation problem.

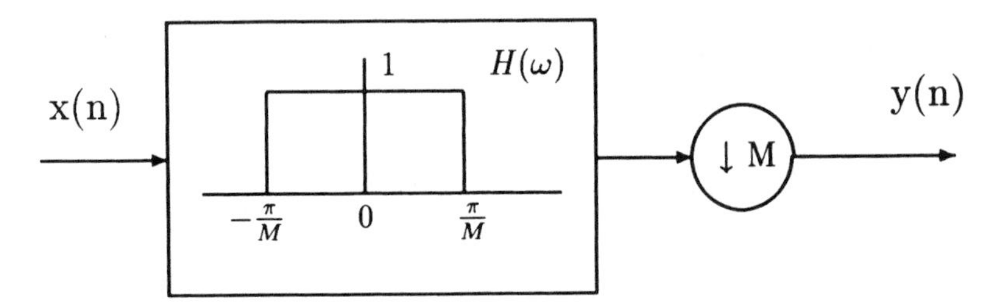

Figure 3.10: Ideal antialiasing filter.

In passing we note that the presence of the bank of decimators is no problem. In fact, they permit us to operate the all-pass filters at the reduced clock rate.

The second application is the uniform filter bank of Fig. 3.12. The frequency response of each filter is just a shifted version of the low-pass prototype $H_0(e^{j\omega})$

$$\begin{aligned} H_m(z) &= H_0(ze^{-j2\pi m/M}) \\ H_m(e^{j\omega}) &= H_0(e^{j(\omega-2\pi m/M)}), \quad m=0,1,...,M-1. \end{aligned} \tag{3.19}$$

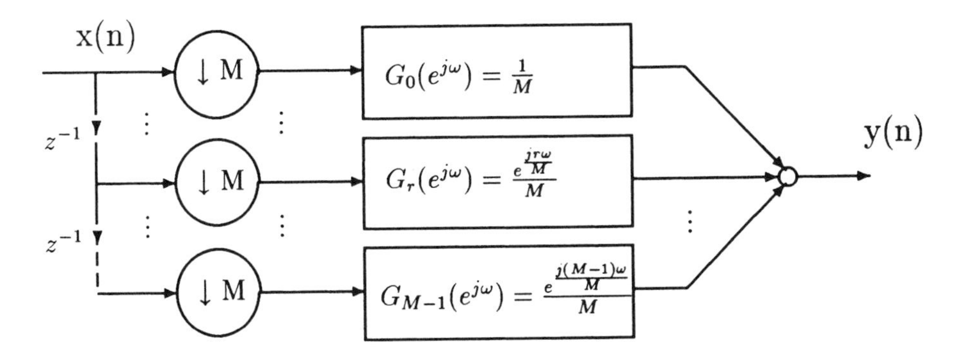

Figure 3.11: Polyphase realization of decimated ideal low-pass filter.

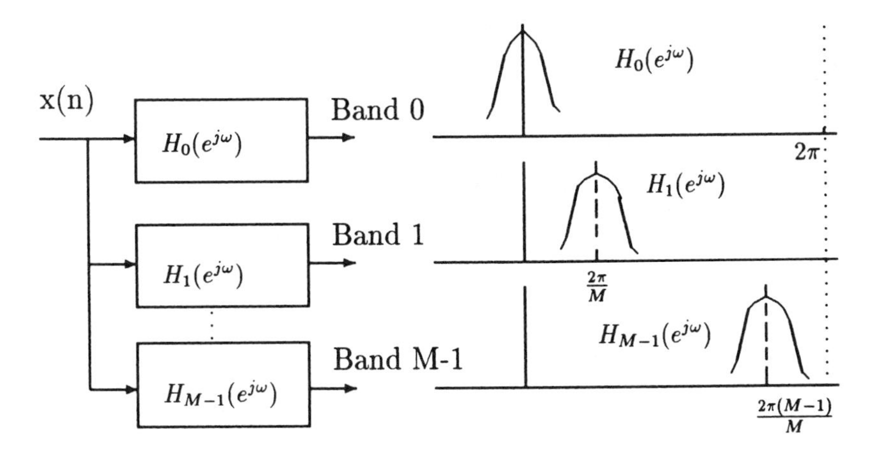

Figure 3.12: Uniform filter bank.

Applying the polyphase decomposition to $H_0(z)$ gives

$$H_0(z) = \sum_{k=0}^{M-1} z^{-k} G_k(z^M).$$

Then

$$H_m(z) \quad = \quad \sum_{k=0}^{M-1} (ze^{-j2\pi m/M})^{-k} G_k[(ze^{-j2\pi m/M})^M]$$

$$= \sum_{k=0}^{M-1} z^{-k} W^{-mk} G_k(z^M). \tag{3.20}$$

We note that G_k is independent of m.

Stacking the equations in Eq. (3.20)

$$\begin{bmatrix} H_0(z) \\ H_1(z) \\ \vdots \\ H_{M-1}(z) \end{bmatrix} = \begin{bmatrix} W^0 & W^0 & \cdots & W^0 \\ W^0 & W^{-1} & \cdots & W^{-(M-1)} \\ \vdots & \vdots & & \vdots \\ W^0 & W^{-(M-1)} & \cdots & W^{-(M-1)^2} \end{bmatrix} \begin{bmatrix} G_0(z^M) \\ z^{-1}G_1(z^M) \\ \vdots \\ z^{-(M-1)}G_{M-1}(z^M) \end{bmatrix}. \tag{3.21}$$

The matrix in Eq. (3.21) is just the DFT matrix; therefore the uniform filter bank of Fig. 3.12 can be realized by the polyphase decomposition of $H_0(z)$ followed by the DFT as in Fig. 3.13.

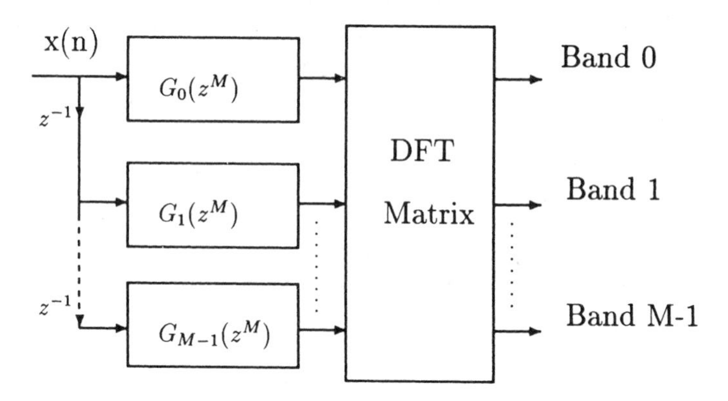

Figure 3.13: Polyphase-DFT realization of a uniform filter bank.

3.2 Bandpass and Modulated Signals

In this section we look at bandpass signals and examine how modulation and demodulation can be achieved by decimation and interpolation. The first case considered is that of integer-band sampling. This is followed by quadrature modulation of bandpass signals.

3.2.1 Integer-Band Sampling

Suppose we have a signal with spectrum $X(e^{j\omega})$ as shown in Fig. 3.14(b). Let the frequency band from 0 to π be split into M equal bands ($M = 4$ in the illustration). As shown in Fig. 3.14(a), the signal is filtered by the ideal band-pass filter H_2 to isolate Band 2, and then down-sampled by $M = 4$. This down-sampling effectively heterodynes Band 2 down to DC as shown in Fig. 3.14(e) and stretches it to occupy the entire frequency axis $[0, \pi]$. The down-sampler performs the dual tasks of modulation (heterodyning) and sampling rate reduction.

The process of reconstructing the information in Band 2 is simply the inverse of the analysis section. The low-frequency signal $Y_2(e^{j\omega})$ is up-sampled by $M = 4$ to give $Y_2^{'}(e^{j\omega}) = Y_2(e^{jM\omega})$ as shown in Fig. 3.14(d). Note that a replica of Band 2 occupies the original band $[\pi/2, 3\pi/4]$. The second band-pass filter removes the images and retains only Band 2 in its original frequency location. The signal at point 5 is therefore equal to that at point 2.

A parallel bank of band-pass filters can be used to separate the ($M = 4$) bands. The down-sampler heterodynes these into the low-frequency region. These signals can then be quantized, transmitted, and reconstructed at the receiver by up-sampling and band-pass filtering.

There is one small caveat, however. It can be shown that the odd-indexed bands are inverted by the heterodyning operation. To obtain the noninverted version, we can simply multiply $v_j(n)$ by $(-1)^n$ for $j = 1, 3$. However, there is no need for this, since the odd-indexed interpolators can handle the inverted bands.

3.2.2 Quadrature Modulation

An alternative to the integer-band sampling is the more conventional approach to heterodyning, namely sinusoidal modulation. Suppose the real signal in question $x(n)$ is a band-pass signal with the spectrum illustrated in Fig. 3.15. This is deliberately drawn to suggest an asymmetric pattern about the center frequency ω_0. We can shift the frequency spectrum down to DC and then subsample without aliasing by employing the configuration shown in Fig. 3.16.

The spectra of the signals at various points in the modulator-demodulator are shown in Fig. 3.17. For real $x(n)$ we have

$$\begin{aligned} x(n) &\leftrightarrow X(e^{j\omega}) \\ x(n)e^{j\omega_0 n} &\leftrightarrow X(e^{j(\omega-\omega_0)}) \\ X^*(e^{-j\omega}) &= X(e^{j\omega}). \end{aligned}$$

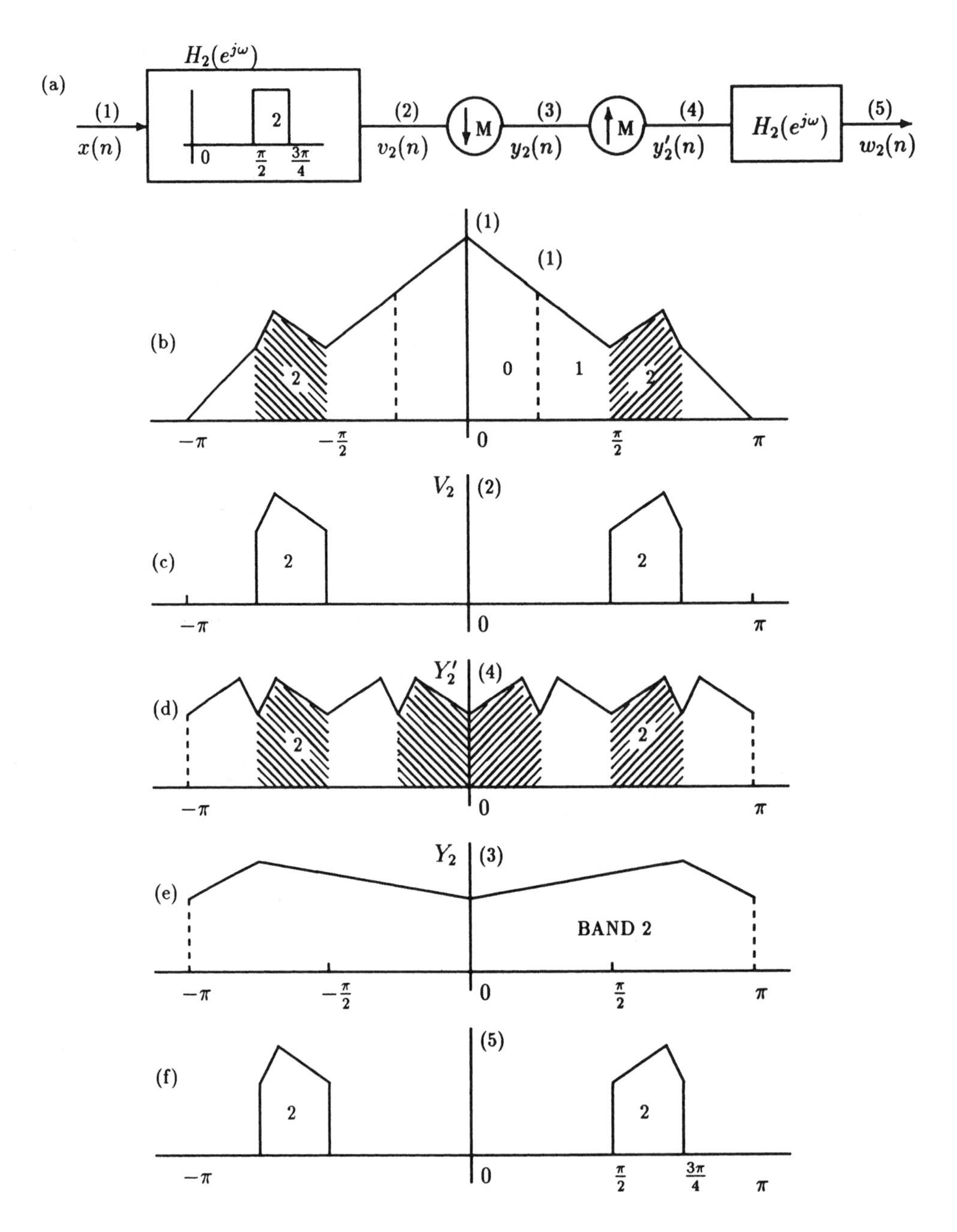

Figure 3.14: Integer-band sampling configuration.

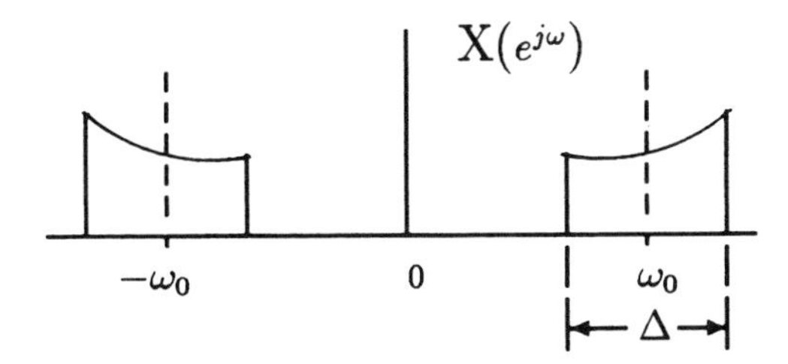

Figure 3.15: The spectrum of a band-pass signal.

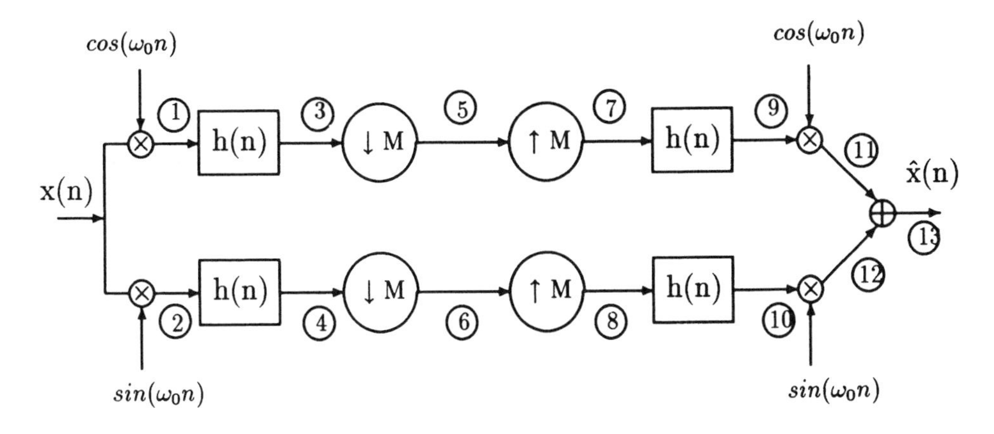

Figure 3.16: Quadrature modulation and demodulation, and frequency reduction.

At points (1) and (2) the quadrature modulation generates the spectra

$$x(n)\cos\omega_0 n \quad \leftrightarrow \quad \tfrac{1}{2}[X(e^{j(\omega-\omega_0)}) + X(e^{j(\omega+\omega_0)})]$$

$$x(n)\sin\omega_0 n \quad \leftrightarrow \quad \tfrac{1}{2j}[X(e^{j(\omega-\omega_0)}) - X(e^{j(\omega+\omega_0)})].$$

The low-pass filter in the analysis stage removes the images at $\pm 2\omega_0$, leaving us with the sum and difference spectra at points (3) and (4), respectively, with bandwidths $\pm\Delta/2$. Down-sampling stretches these spectra by a factor of M. If the original bandwidth is $(\Delta/2) = (\pi/M)$, we obtain the spectra shown at points (5) and (6). The up-sampler compresses the frequency axis as at points (7) and (8). Then the synthesis low-pass filter removes all lobes except the lobe around DC.

Quadrature sinusoidal modulation again shifts these spectra to $\pm\omega_0$. Finally, addition of the two signal components gives the reconstructed signal at point (13).

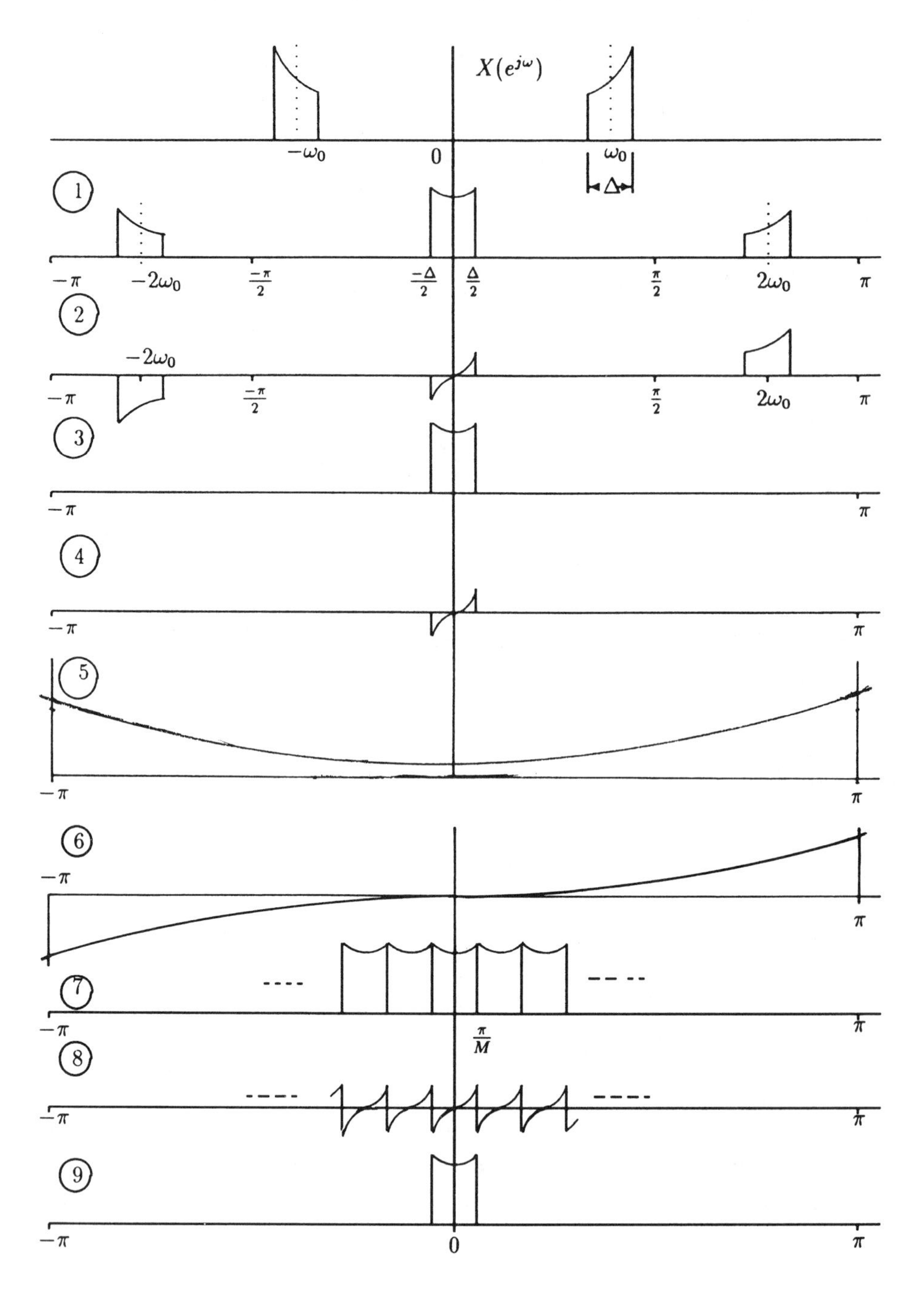

Figure 3.17: Frequency spectra of signals in the quadrature modulator-demodulator.

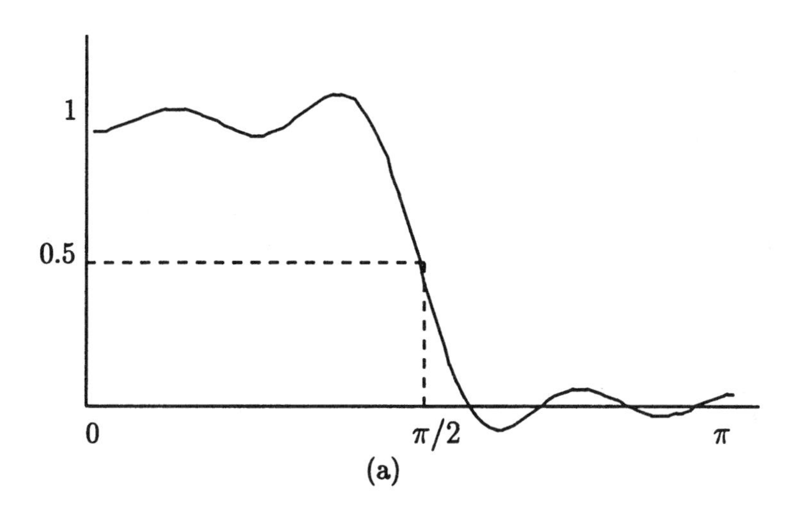

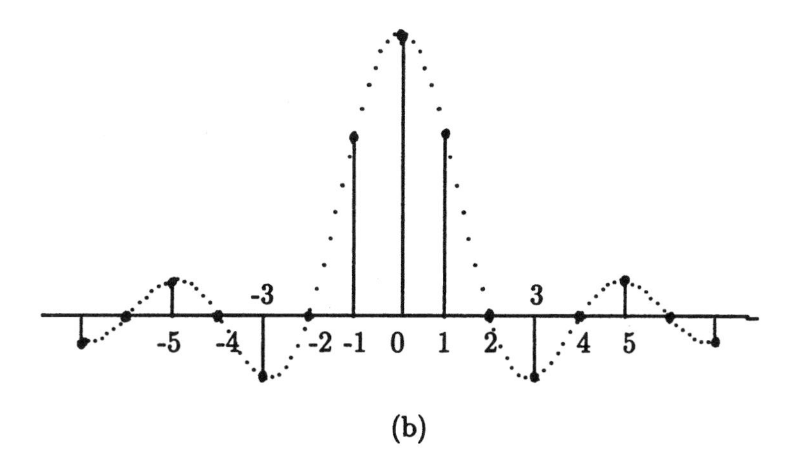

Figure 3.18: (a) Frequency response of half-band filter; (b) impulse response of half-band filter.

This second method of modulation and down-sampling is an alternative to the band-pass filter associated with the integer-band approach. Since we can choose the center frequency ω_0, this method tends to be more robust and less constrained.

3.3 Mth Band, Mirror, and Power Complementary Filters

Certain classes of filters will appear frequently in the subband filter structures of the rest of this chapter. In this section, we define and explore the special properties of these filters.

3.3.1 Mth Band Filters

A half-band filter (Mintzer, 1982) is an FIR filter with the following frequency response features:

$$\begin{aligned} H(e^{j\omega}) &= H(e^{-j\omega}) \\ H(e^{j\omega}) + H(e^{j(\pi-\omega)}) &= H(1) = 1. \end{aligned} \tag{3.22}$$

That is, $H(e^{j\omega})$ is a real even function of ω with odd symmetry about $[\pi/2, \frac{1}{2}H(1)]$, as indicated in Fig. 3.18. The first condition implies $h(n) = h(-n)$, i.e., a real, even sequence. Furthermore, since $H(e^{j\omega}) \leftrightarrow h(n) = h(-n)$, then

$$H(e^{-j(\pi-\omega)}) \leftrightarrow (-1)^n h(n).$$

The second condition now implies

$$h(n) + (-1)^n h(n) = [1 + (-1)^n]h(n) = \delta(n),$$

with the solution

$$h(n) = \left\{ \begin{array}{ll} 1/2, & \text{n=0} \\ 0, & \text{n=even, n} \neq 0. \end{array} \right\} \tag{3.23}$$

or more succinctly $h(2n) = \frac{1}{2}\delta(n)$. Hence, the even-indexed samples of $h(n)$, except $n = 0$, are zero.(Prob. 3.7)

The zeros in the impulse response reduce the number of multiplications required by almost one-half, while the symmetry about $\pi/2$ implies an equal ripple in passband and stopband.

The polyphase expansion of this half-band filter with $M = 2$ reduces to

$$H(z) = h(0) + z^{-1}G_1(z^2), \qquad h(0) = 1/2. \tag{3.24}$$

The Mth band filter is an extension of the half-band. In the time-domain, it is defined as a zero-phase FIR filter with every M^{th} sample equal to zero, except $n = 0$,

$$h(Mn) = \begin{cases} \frac{1}{M}, & \text{n=0} \\ 0, & \text{n} \neq 0 \end{cases} \tag{3.25}$$

We can expand such an $H(z)$ in polyphase form and obtain

$$H(z) = G_0(z^M) + \sum_{k=1}^{M-1} z^{-1}G_k(z^M) = \frac{1}{M} + \sum_{k=1}^{M-1} z^{-1}G_k(z^M).$$

Evaluating $G_0(z^M)$ from Eq. (3.17) gives

$$MG_0(z^M) = \sum_{k=0}^{M-1} H(zW^k) = 1 \tag{3.26}$$

or

$$\sum_{k=0}^{M-1} H(e^{j(\omega - \frac{k2\pi}{M})}) = 1 \tag{3.27}$$

which represents a generalization of Eq. (3.22). (Prob. 3.15)

3.3.2 Mirror Image Filters

Let $h_0(n)$ be some FIR low-pass filter with real coefficients. The *mirror* filter is defined as

$$h_1(n) = (-1)^n h_0(n) \tag{3.28}$$

or, equivalently, in the transform-domain,

$$\begin{aligned} H_1(z) &= H_0(-z) \\ H_1(e^{j\omega}) &= H_0(e^{j(\omega-\pi)}). \end{aligned} \tag{3.29}$$

Using the substitution $\omega \rightarrow \frac{\pi}{2} - \omega$, and noting that the magnitude is an even function of ω, leads to

$$|H_1(e^{j(\frac{\pi}{2}-\omega)})| = |H_0(e^{j(\frac{\pi}{2}+\omega)})|. \tag{3.30}$$

This last form demonstrates the mirror image property of H_0 and H_1 about $\omega = \pi/2$, and is illustrated in Fig. 3.19(a); hence, the appellation *quadrature mirror filters*, or *QMF*. The pole-zero patterns are also reflected about the imaginary axis of the Z-plane, as required by Eq. (3.29), as shown in Fig. 3.19(b). These QMFs were used in the elimination of aliasing in two-channel subband coders in the seminal paper by Esteban and Galand (1977).

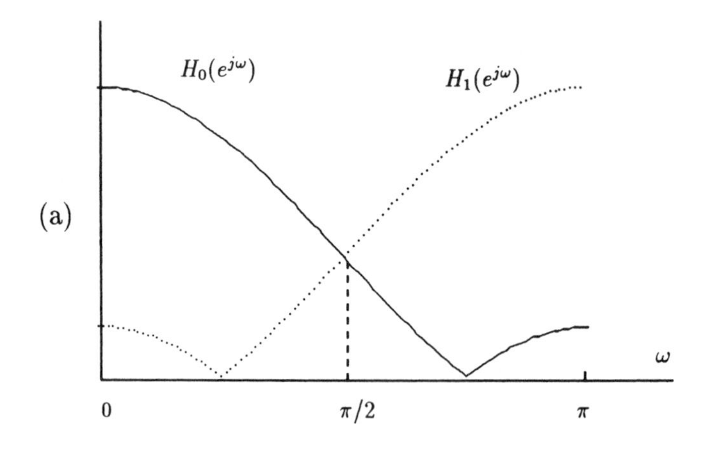

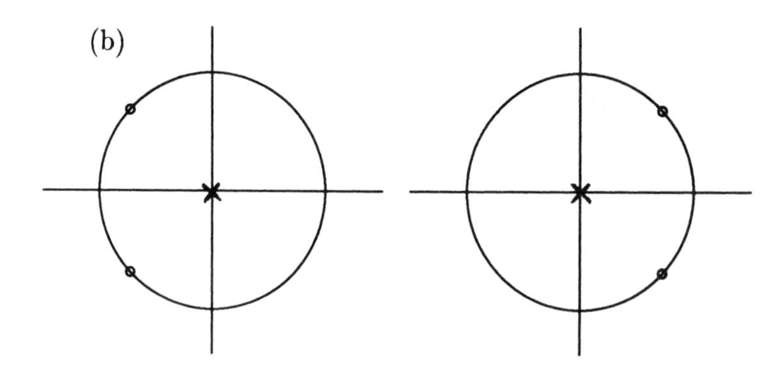

Figure 3.19: (a) Frequency responses of quadrature mirror filters; (b) pole-zero patterns.

3.3.3 Power Complementary Filters

The filter pair $\{H_0(z), H_1(z)\}$ are said to be power complementary if

$$|H_0(e^{j\omega})|^2 + |H_1(e^{j\omega})|^2 = 1. \tag{3.31}$$

The M-band extension is evidently

$$\sum_{k=0}^{M-1} |H_k(e^{j\omega})|^2 = 1. \tag{3.32}$$

As shown in Vaidyanathan (Jan. 1990), the M polyphase components $\{G_k(z), 0 \leq k \leq M-1\}$ of some filter $H(z)$ are power complementary if and only if $Q(z) = H(z)H(z^{-1})$ is an Mth band filter.(Prob. 3.9) These power complementary filters will play a role in IIR subband filter banks in Section 3.5.

3.4 Two-Channel Filter Banks

The two-channel filter bank provides the starting point for the study of subband coding systems. The purpose of a subband filter system is to separate the signal into frequency bands and then to allocate encoding bits to each subband based on the energy in that subband.

In this section we derive the requirements and properties of a perfect reconstruction, two-channel subband system. This two-band case is extended into an *M-band* structure by a binary, hierarchical, subband tree expansion. We then show how another tree structure, the dyadic tree, relates to the multiresolution pyramid decomposition of a signal. The results of this two-band case thus provide us with a springboard for the more general M-band filter bank of Section 3.5.

The two-channel filter bank is shown in Fig. 3.20. The input spectrum $X(e^{j\omega})$, $0 \leq \omega \leq \pi$ is divided into two equal subbands. The analysis filters $H_0(z)$ and $H_1(z)$ function as antialiasing filters, splitting the spectrum into two equal bands. Then, according to the Nyquist theorem $\theta_0(n)$ and $\theta_1(n)$ are each down sampled by 2 to provide the subband signals $v_0(n)$ and $v_1(n)$ as the outputs of the analysis stage. In a subband coder, these signals are quantized, encoded, and transmitted to the receiver. We assume ideal operation here, with no coding and transmission errors, so that we can focus on the analysis and synthesis filters. Therefore, we pretend that $v_0(n)$ and $v_1(n)$ are received at the synthesis stage. Each is up-sampled by 2 to give the zero-interlaced signals $f_0(n)$ and $f_1(n)$. These signals in turn are processed by the interpolation filters $G_0(z)$ and $G_1(z)$, and then summed at the output to yield the reconstructed signal $\hat{x}(n)$.

The focus of the book is on perfect reconstruction (PR) signal decomposition. Therefore, we will consider only FIR PR filter banks here. (Subsequently, in Section 3.6.2, IIR 2-band structures using all-pass filters are shown to provide causal approximations to perfect reconstruction.)

3.4.1 Two-Channel PR-QMF Bank

The conditions for perfect reconstruction in the prototype two-channel FIR QMF bank were obtained first by Smith and Barnwell (1986) and thoroughly treated by Vaidyanathan (July 1987). We will derive these conditions in both frequency- and time-domains. The time-domain PR conditions will prove to be very useful especially in the Mth band filter bank design.

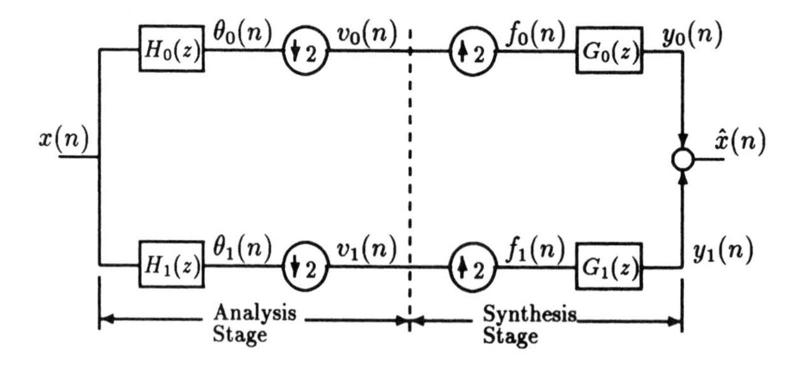

Figure 3.20: Two-channel subband filter bank.

Tracing the signals through the top branch in Fig. 3.20 gives

$$\begin{aligned} \theta_0(z) &= H_0(z)X(z) \\ Y_0(z) &= G_0(z)F_0(z) \end{aligned} \tag{3.33}$$

as the outputs of the decimation and interpolation filters, while the down-sampler and the up-sampler impose, respectively,

$$\begin{aligned} V_0(z) &= \frac{1}{2}[\theta_0(z^{1/2}) + \theta_0(-z^{1/2})]. \\ F_0(z) &= V_0(z^2). \end{aligned}$$

Combining all these gives

$$Y_0(z) = \frac{1}{2}G_0(z)[H_0(z)X(z) + H_0(-z)X(-z)]. \tag{3.34}$$

Similarly,

$$Y_1(z) = \frac{1}{2} G_1(z)[H_1(z)X(z) + H_1(-z)X(-z)].$$

The Z-transform of the reconstructed signal, is then

$$\begin{aligned} \hat{\mathrm{X}}(\mathrm{z}) &= \frac{1}{2}[H_0(z)G_0(z) + H_1(z)G_1(z)]X(z) \\ &\quad + \frac{1}{2}[H_0(-z)G_0(z) + H_1(-z)G_1(z)]X(-z) \\ &= T(z)X(z) + S(z)X(-z). \end{aligned} \tag{3.35}$$

Perfect reconstruction requires the following:

$$\begin{aligned} &(1) \quad S(z) = \; 0, \text{for all } \; z \\ &(2) \quad T(z) = \; cz^{-n_0}, \text{ where } c \text{ is a constant.} \end{aligned}$$

To eliminate aliasing and force $S(z) = 0$, we require

$$\frac{G_0(z)}{G_1(z)} = -\frac{H_1(-z)}{H_0(-z)}.$$

This can be achieved by the selection

$$\begin{aligned} G_0(z) &= -H_1(-z) \\ G_1(z) &= H_0(-z), \end{aligned} \tag{3.36}$$

leaving us with

$$T(z) = \frac{1}{2}[H_0(-z)H_1(z) - H_0(z)H_1(-z)]. \tag{3.37}$$

Several choices can be made here to force $T(z) = cz^{-n_0}$. The FIR paraunitary solution derived in Section 3.5.4 is as follows: let $H_0(z)$, $H_1(z)$ be N-tap FIR, where N is even, and let

$$H_1(z) = z^{-(N-1)} H_0(-z^{-1}). \tag{3.38}$$

This choice forces

$$H_1(-z) = -G_0(z)$$

so that

$$T(z) = \frac{1}{2} z^{-(N-1)}[H_0(z)H_0(z^{-1}) + H_0(-z)H_0(-z^{-1})]. \tag{3.39}$$

Therefore, the perfect reconstruction requirement reduces to finding an $H(z) = H_0(z)$ such that

$$\begin{aligned} Q(z) &= H(z)H(z^{-1}) + H(-z)H(-z^{-1}) = \text{constant} = 2 \\ &= R(z) + R(-z) = 2. \end{aligned} \tag{3.40}$$

This selection implies that all four filters are causal whenever $H_0(z)$ is causal.

The PR requirement, Eq. (3.40), can be readily recast in an alternate, time-domain form. First, one notes that $R(z)$ is a spectral density function and hence is representable by a finite series of the form (for an FIR $H(z)$)

$$R(z) = \gamma_{N-1}z^{N-1} + \gamma_{N-2}z^{N-2} + \ldots + \gamma_0 z^0 + \ldots + \gamma_{N-1}z^{-(N-1)}. \tag{3.41}$$

Then

$$R(-z) = -\gamma_{N-1}z^{(N-1)} + \gamma_{N-2}z^{N-2} - \ldots + \gamma_0 z^0 - \gamma_1 z^{-1} \ldots - \gamma_{N-1}z^{-(N-1)}. \tag{3.42}$$

Therefore $Q(z)$ consists only of even powers of z. To force $Q(z) =$ constant, it suffices to make all even-indexed coefficients in $R(z)$ equal to zero except γ_0. However, the γ_n coefficients in $R(z)$ are simply the samples of the autocorrelation $\rho(n)$ given by

$$\begin{aligned} \rho(n) &= \sum_{k=0}^{N-1} h(k)h(k+n) = \rho(-n) \\ &\triangleq h(n) \odot h(n), \end{aligned} \tag{3.43}$$

where $\odot$ indicates a correlation operation. This follows from the Z-transform relationships

$$R(z) = H(z)H(z^{-1}) \longleftrightarrow h(n) * h(-n) = \rho(n), \tag{3.44}$$

where $\rho(n)$ is the convolution of $h(n)$ with $h(-n)$, or equivalently, the time autocorrelation, Eq. (3.43). Hence, we need to set $\rho(n) = 0$ for n even, and $n \neq 0$. Therefore,

$$\rho(2n) = \sum_{k=0}^{N-1} h(k)h(k+2n) = 0, \qquad n \neq 0. \tag{3.45}$$

If the normalization

$$\sum_{k=0}^{N-1} |h(k)|^2 = 1 \tag{3.46}$$

is imposed, one obtains the PR requirement in the time-domain as

$$\rho(2n) = \sum_{k=0}^{N-1} h(k)h(k+2n) = \delta_n. \tag{3.47}$$

This last requirement is recognized as the same as that for a half-band filter in Eq. (3.23). Hence, $R(z) = H(z)H(z^{-1})$ should satisfy the half-band requirement of Eq. (3.22),

$$R(e^{j\omega}) + R(e^{j(\omega-\pi)}) = |H(e^{j\omega})|^2 + |H(e^{j(\omega-\pi)})|^2 = 1. \tag{3.48}$$

Finally, Eq. (3.38) permits us to convert Eq. (3.48) into

$$|H_0(e^{j\omega})|^2 + |H_1(e^{j\omega})|^2 = 1 \tag{3.49}$$

which asserts that $H_0(z)$ and $H_1(z)$ are also power complementary. In Section 3.5, we will see that this PR solution is called the *paraunitary solution.*

In summary, the two-band paraunitary PR FIR structure with N even and $H_0(z) = H(z)$ satisfies:

$$\begin{aligned}
R(z) = H(z)H(z^{-1}) \quad &\leftrightarrow \quad \rho(n) = h(n) * h(-n) \\
R(z) + R(-z) = 2 \quad &\leftrightarrow \quad \rho(2n) = \delta(n) \\
&\textit{then} \\
H_1(z) \quad &= \quad z^{-(N-1)}H_0(-z^{-1}) \\
G_0(z) \quad &= \quad z^{-(N-1)}H_0(z^{-1}) \\
G_1(z) \quad &= \quad z^{-(N-1)}H_1(z^{-1}).
\end{aligned} \tag{3.50}$$

These results will be rederived from more general principles in Section 3.5. At this point, they suffice to enable us to construct multiband PR filter banks using hierarchical subband tree structures. However, it can be shown that these filters for the 2-band case cannot be linear-phase. (Prob. 3.11; see also Prob. 3.12)

3.4.2 Regular Binary Subband Tree Structure

Multirate techniques provide the basic tool for multiresolution spectral analysis and the PR QMF bank is the most efficient signal decomposition block for this purpose. As shown in the previous section, these filter banks divide the input spectrum into two equal subbands, yielding the low (L) and high (H) bands. This two-band PR QMF split can again be applied to these (L) and (H) half bands to generate the quarter bands: (LL), (LH), (HL), and (HH).

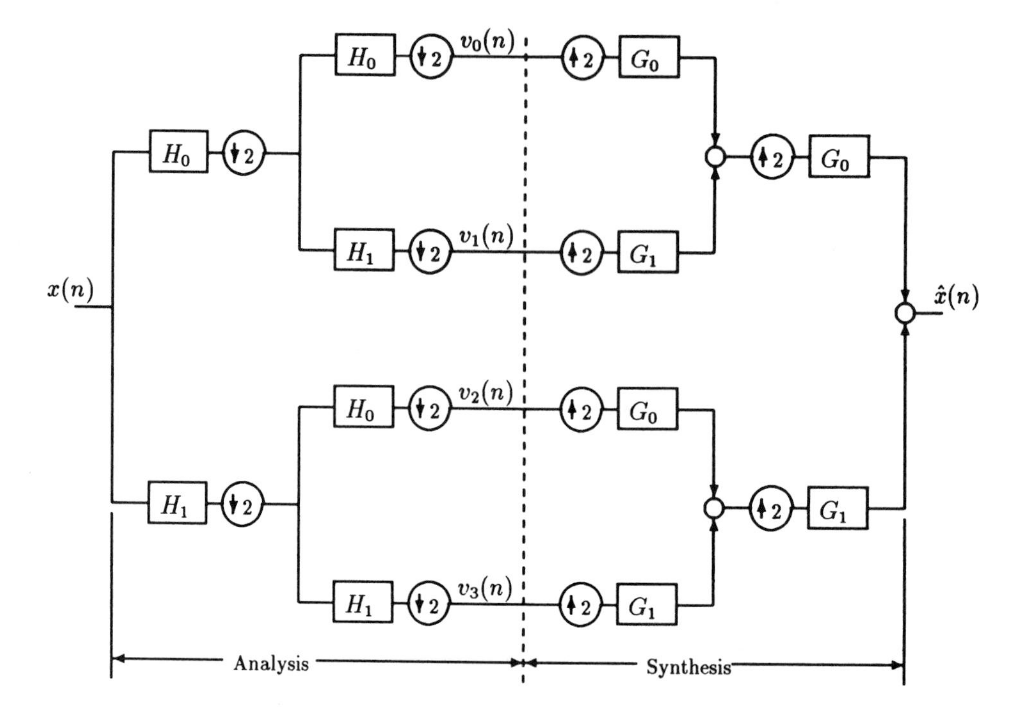

Figure 3.21: Four-band, analysis-synthesis tree structure.

Two levels of this decomposition are shown in Fig. 3.21, where the original signal at a data rate f_s is decomposed into the four subband signals $v_0(n), ..., v_3(n)$, each operating at a rate of $f_s/4$. Therefore, the net data rate at the output of the analysis section equals that of the input signal. This conservation of data rates is called *critical sampling.* Now Smith and Barnwell have shown that such an analysis-synthesis tree is perfect reconstruction if the progenitor two-band analysis-synthesis structure is PR, with $\{H_0, H_1, G_0, G_1\}$ satisfying Eq. (3.50). (Prob. 3.10) Consequently, this structure can be iterated many times with the assurance that perfect reconstruction is attained in the *absence of all error sources.* But, as we shall see, some structures are more sensitive to encoding errors than others.

We can represent the two-level, hierarchical, analysis section by the equivalent four-band analysis bank of Fig. 3.22. Consider the cascade structure of Fig. 3.23 showing three filters separated by down-samplers of rates M_1 and M_2. Using the equivalent structures of Fig. 3.7, we can successively interchange filter and subsampler with the substitution $z \to z^{M_1}$. Filter $G_3(z)$ can be commuted with

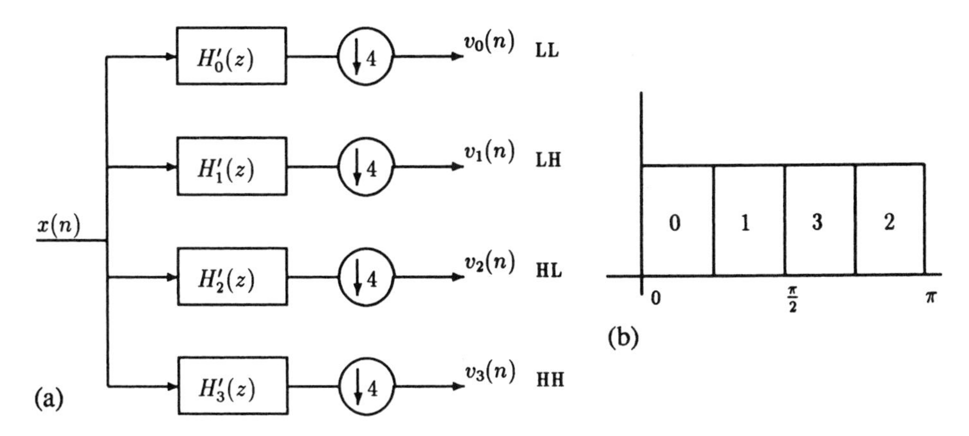

Figure 3.22: (a) Four-band equivalent to two-level regular binary tree; (b) frequency bands corresponding to the four-bands with ideal filters; (c) (see page 144) frequency bands of two-band 8-tap Binomial-QMF (Section 4.1) and a four-band hierarchical subband structure using the same filters; (d) (see page 145) frequency responses of a typical four-band paraunitary filter bank with a duration of 8-tap each (the filter coefficients are given in Table 4.15).

subsampler M_2 as in Fig. 3.23(b). Repetition of this step for subsampler M_1 gives the equivalent structure shown as Fig. 3.23(c). Therefore, the two-band and the two-level, two-band based hierarchical decompositions are equivalent if

$$\begin{aligned} H_0^{'}(z) &= H_0(z)H_0(z^2) \\ H_1^{'}(z) &= H_0(z)H_1(z^2) \\ H_2^{'}(z) &= H_1(z)H_0(z^2) \\ H_3^{'}(z) &= H_1(z)H_1(z^2). \end{aligned} \tag{3.51}$$

The four-band frequency split of the spectrum is shown in Fig. 3.22(b) for ideal band-pass filters. Figure 3.22(c) displays the imperfect frequency behavior of a finite duration, eight-tap Binomial-QMF filter employed in the two-level regular subband tree. Note that Band 3, the (HH) band, is actually centered in $[\pi/2, 3\pi/4]$, rather than $[3\pi/4, \pi]$ as might be expected. Although the two-level hierarchical analysis tree is equivalent to the four-band, one-level filter bank, the former is far more constrained than a freely chosen four-band filter bank. For example, as we shall see, a nontrivial two-band paraunitary bank cannot have a linear-phase, whereas an unconstrained four-band filter bank can have both paraunitary PR and linear-phase. For comparison purposes Figure 3.22(d) displays the frequency

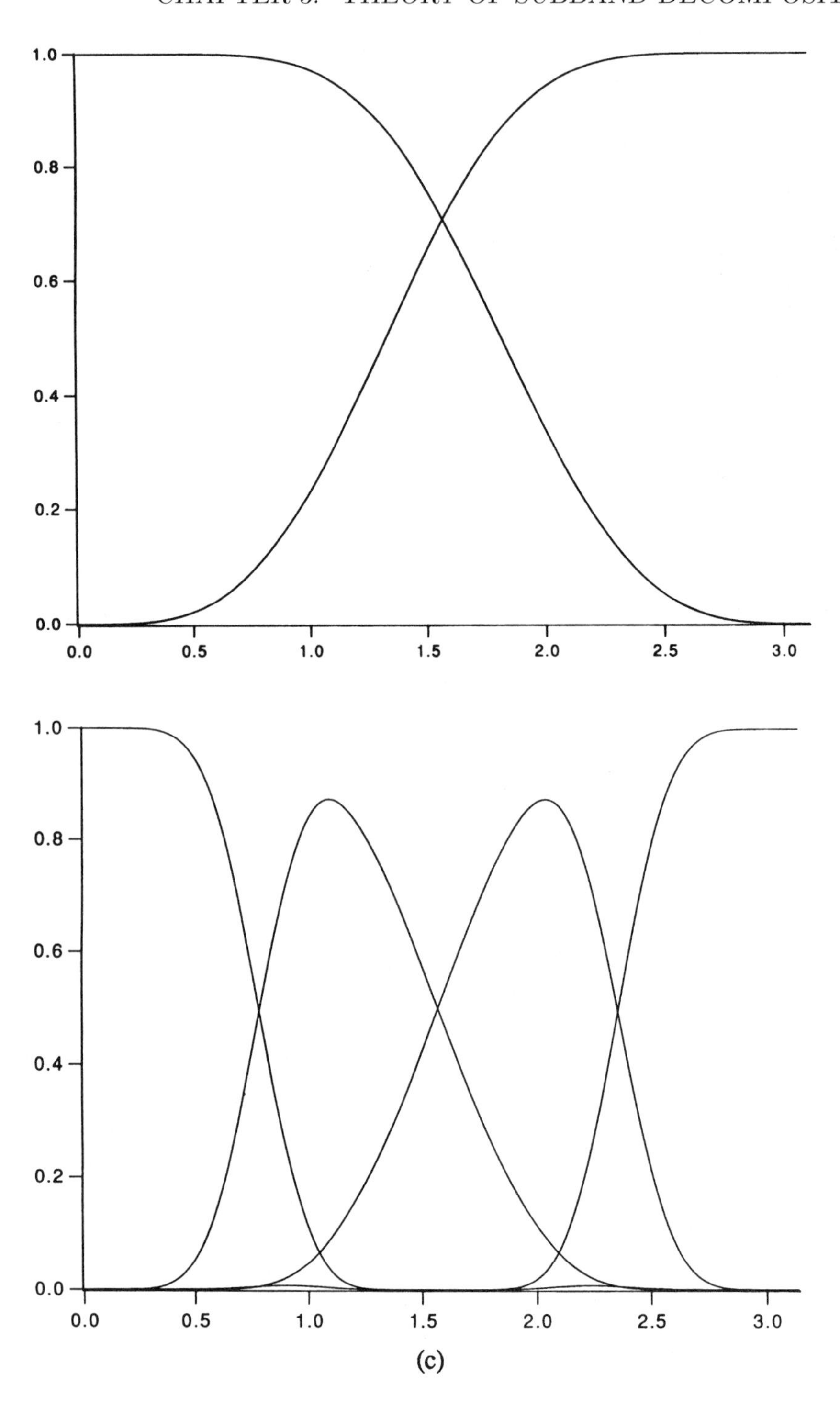

Figure 3.22 (continued)

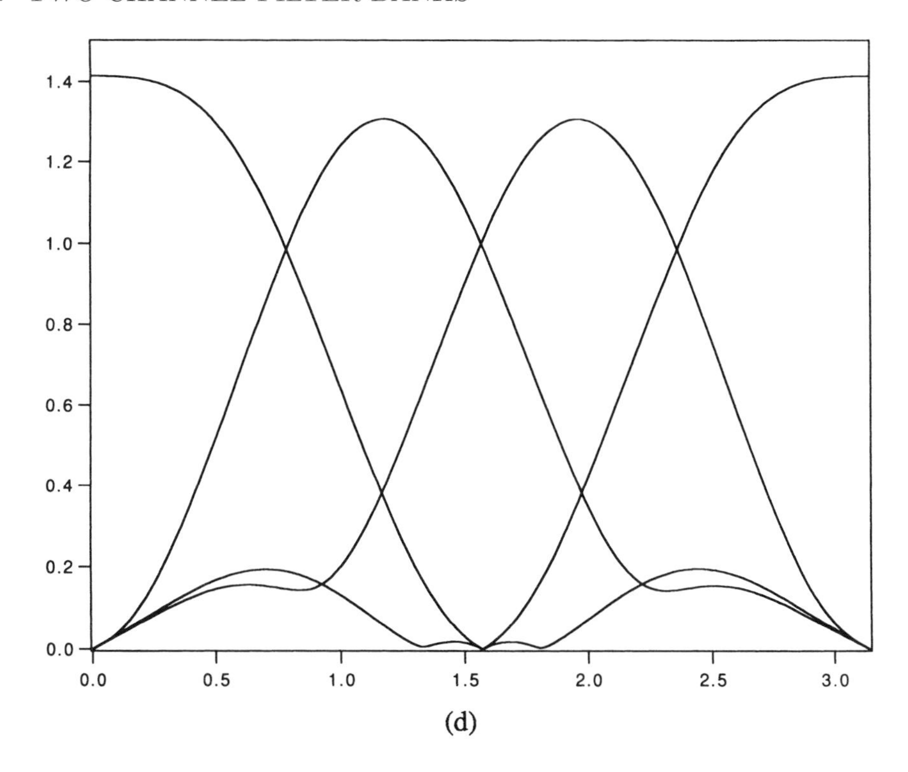

Figure 3.22 (continued)

responses of filters for length 8 each. The coefficients of these filters are given in Table 4.15. It is observed from these figures that the hierarchical tree structure degrades the filter characteristics, in both time- and frequency-domains. This is somewhat expected since these filters were originally designed for a two-band decomposition.

When this procedure is repeated L times, 2^L equal-width bands are obtained. This hierarchical subband tree approach provides the maximum possible frequency resolution of $\pi/2^L$ within L levels. This spectral analysis structure is called an *L-level regular binary* or *full subband tree.* For $L = 3$ the regular binary tree structure and the corresponding frequency band split are shown in Fig. 3.24. This figure assumes that ideal filters are employed. The imperfectness of the frequency responses increases when the level of hierarchical subband tree increases for finite duration filters.

In practice, finite length filter PR QMFs replace the ideal filters. Therefore inter-band aliasing or leakage exists. This is also the reason for interband correlations. In the presence of encoding errors in a multilevel tree structure, this

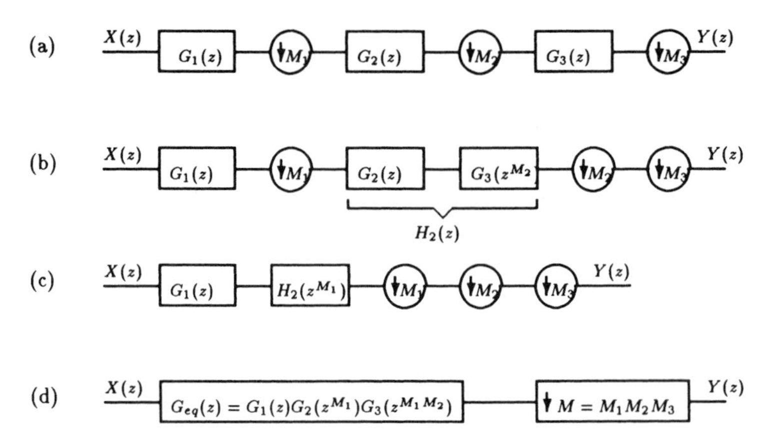

Figure 3.23: Equivalent structures for filters separated by down-samplers.

frequency leakage can cause some degradation in the frequency bands of the finer frequency resolutions. This is a disadvantage of the regular binary subband tree over a direct M-band (equal) frequency split, since the M-band approach monitors the frequency behavior of those four bands during filter bank design.

On the other hand, the multilevel hierarchical analysis-synthesis subband trees are much simpler to implement and provide a coarse-to-fine (multiresolution) signal decomposition as a by-product.

3.4.3 Irregular Binary Subband Tree Structure

Almost all real signal sources concentrate significant portions of their energies in subregions of their spectrums. This indicates that some intervals of the overall signal spectrum are more significant or important than the others. Therefore all the subbands of the regular binary tree may not be needed. Since we also aim to minimize the computational complexities of the spectral analysis-synthesis operations, some of the fine frequency resolution subbands can be combined to yield larger bandwidth frequency bands. This implies the irregular termination of the tree branches. Hence it is expected that the frequency bands of the irregular tree will have unequal bandwidths. Figure 3.25 displays an arbitrary irregular binary subband tree with the maximum tree level $L = 3$ and its corresponding frequency band split. The band split shown assumes that ideal filters are employed.

The number of bands in this irregular spectral decomposition structure is less than that in the regular tree case, $M \leq 2^L$. The regular tree provides the best possible frequency resolution for a fixed L. The regular tree has equal width frequency

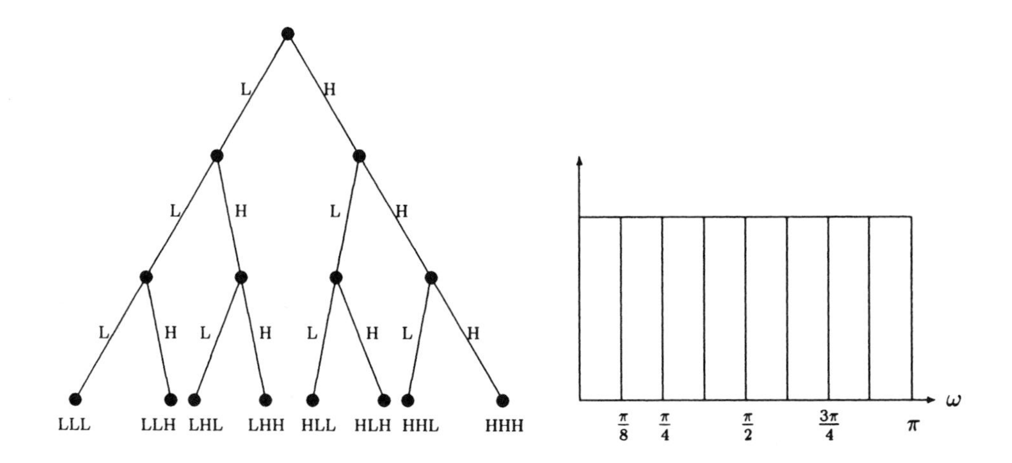

Figure 3.24: A regular tree structure for $L = 3$ and its frequency band split, assuming ideal two-band PR-QMFs are employed.

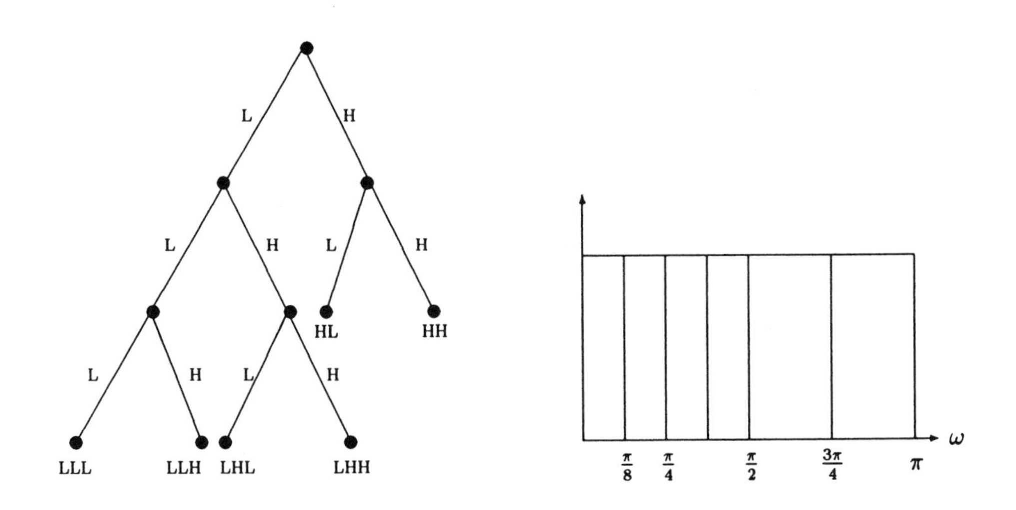

Figure 3.25: An irregular tree structure and its frequency band split, assuming ideal two-band PR-QMFs are employed.

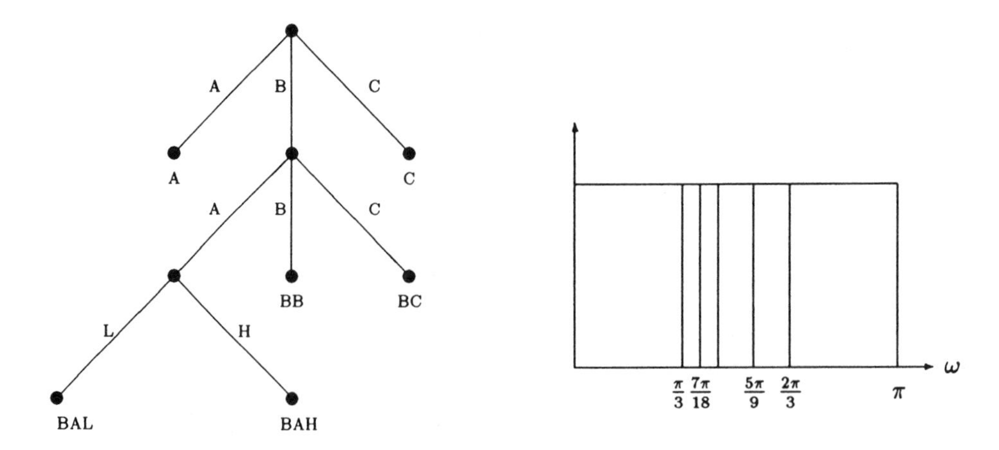

Figure 3.26: An irregular tree structure and its frequency band split, assuming ideal equal-bandwidth filter banks employed at each stage.

bands while the irregular tree provides unequal bands. Both of these structures split the spectrum as a power of 2 since they employ a two-band frequency split algorithm repeatedly. Figure 3.26 shows a non-binary irregular subband tree structure that might be used if the input signal energy is concentrated in the band-pass region $[\pi/3, 2\pi/3]$.

3.4.4 Dyadic or Octave Band Subband Tree Structure

The *dyadic* or *octave* band tree is a special irregular tree structure. It splits only the lower half of the spectrum into two equal bands at any level of the tree. Therefore the detail or higher half-band component of the signal at any level of the tree is decomposed no further. The dyadic tree configuration and its corresponding frequency resolution for $L = 3$ are given in Fig. 3.27.

An examination of this analysis-synthesis structure shows that a half-resolution frequency step is used at each level. Therefore it is also called the *octave-band or constant-Q* (see Section 6.1.2) subband tree structure. First, low (L) and high (H) signal bands are obtained here. While the band (L) provides a coarser version of the original signal, band (H) contains the detail information. If the low spectral component or band (L) is interpolated by 2, the detail information or the interpolation error is compensated by the interpolated version of band (H). Hence the original is perfectly recovered in this one-step dyadic tree structure. The approach

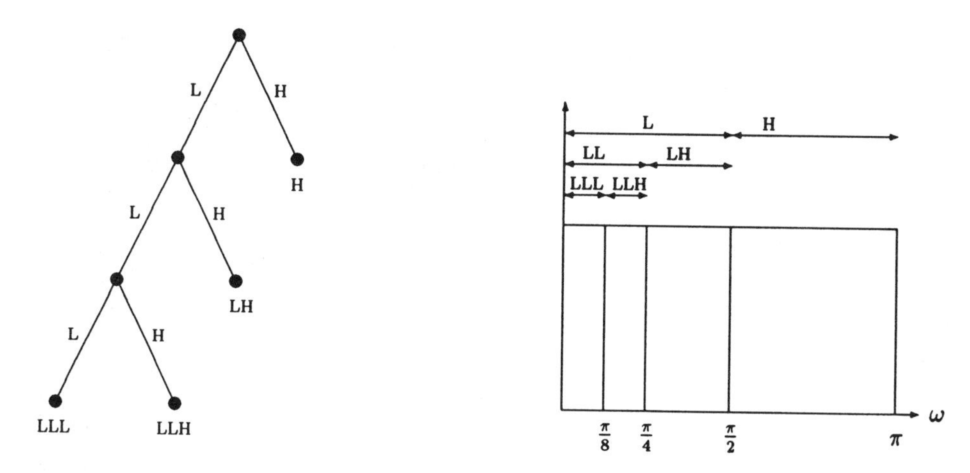

Figure 3.27: A dyadic (octave band) tree structure and its frequency band split, assuming ideal two-band PR-QMFs are employed.

is repeated L times onto only the lower-spectral half component of the higher-level node in the tree. Figure 3.28 shows the composite analysis-synthesis structure for a three-level dyadic tree. It also shows the data rate at each point in the analysis tree. Note that the total data rate for the subband signals at the output of the analysis section equals f_s, the data rate of the source signal. Therefore this dyadic tree is also critically sampled.

This multiresolution (coarse-to-fine) signal decomposition idea was first proposed in 2D by Burt and Adelson (1983) for vision and image coding problems. This popular technique is called the *Laplacian pyramid*. In Chapter 5, we will show that the orthonormal wavelet transform also utilizes this dyadic subband tree. In that case, the coefficients in the multiresolution wavelet decomposition of a continuous-time signal, with proper initialization, are calculated using the discrete-time dyadic subband tree presented here. We will now briefly review the Laplacian pyramid signal decomposition technique and discuss its similarities with a dyadic tree-based PR filter bank structure.

3.4.5 Laplacian Pyramid for Signal Decomposition

A pyramid is a hierarchical data structure containing successively condensed information in a signal that is typically an image. Each layer of the pyramid represents a successively lower resolution (or blurred) representation of the image. The difference between the blurred representations at two adjacent levels is the detail at that level.

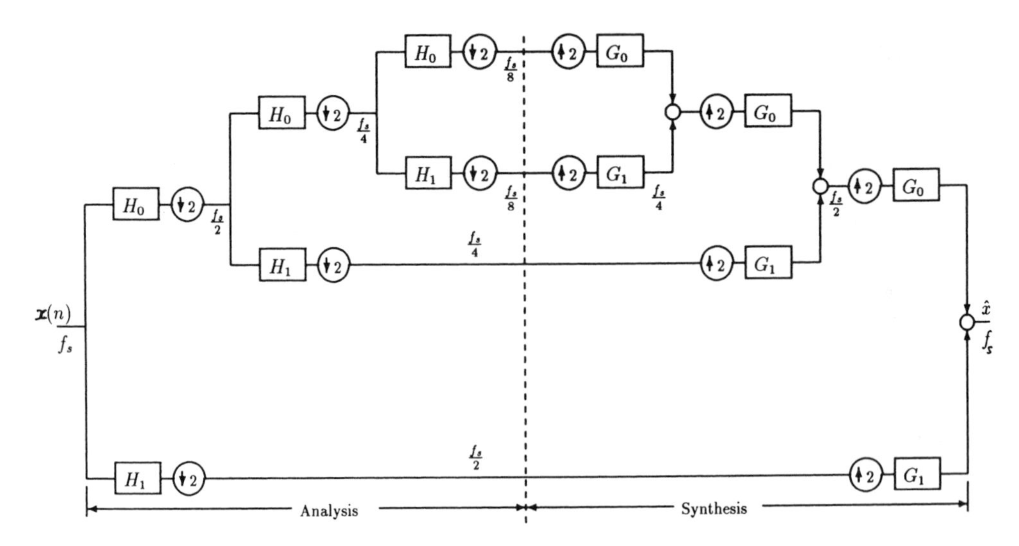

Figure 3.28: Dyadic analysis and synthesis tree.

Pyramid reconstitution of a signal may be done in a "progressive" manner. Starting with the *coarsest* approximation at the lowest level the signal is reassembled by adding on successively finer details at each resolution level until the original image is reconstituted. This reassembly can be performed either progressively in time or concurrently. This pyramid decomposition is of considerable interest in machine vision and image coding applications.

In its essence the Laplacian pyramid (Burt and Adelson, 1983) performs a dyadic tree-like spectral or subband analysis. The idea will be explained with a 1D example and the link to a dyadic PR subband tree established here.

In Fig. 3.29 the signal $x(n)$ is low-pass filtered and decimated by 2. Let us denote this signal as $x_1(n)$. Then $x_1(n)$ is up-sampled by 2 and interpolated to form interpolated signal $\hat{x}_0(n)$. The corresponding interpolation error, or high-resolution detail,

$$d_0(n) = x(n) - \hat{x}_0(n) \tag{3.52}$$

has a Laplacian-shaped pdf for most image sources. To obtain $x(n)$ perfectly one should sum the detail and the interpolated low-pass signal

$$x(n) = \hat{x}_0(n) + d_0(n). \tag{3.53}$$

Since $\hat{x}_0(n)$ is obtained from $x_1(n)$, then $d_0(n)$ and $x_1(n)$ are sufficient to represent $x(n)$ perfectly. The data rate of $x_1(n)$ is half of the data rate of $x(n)$. This provides

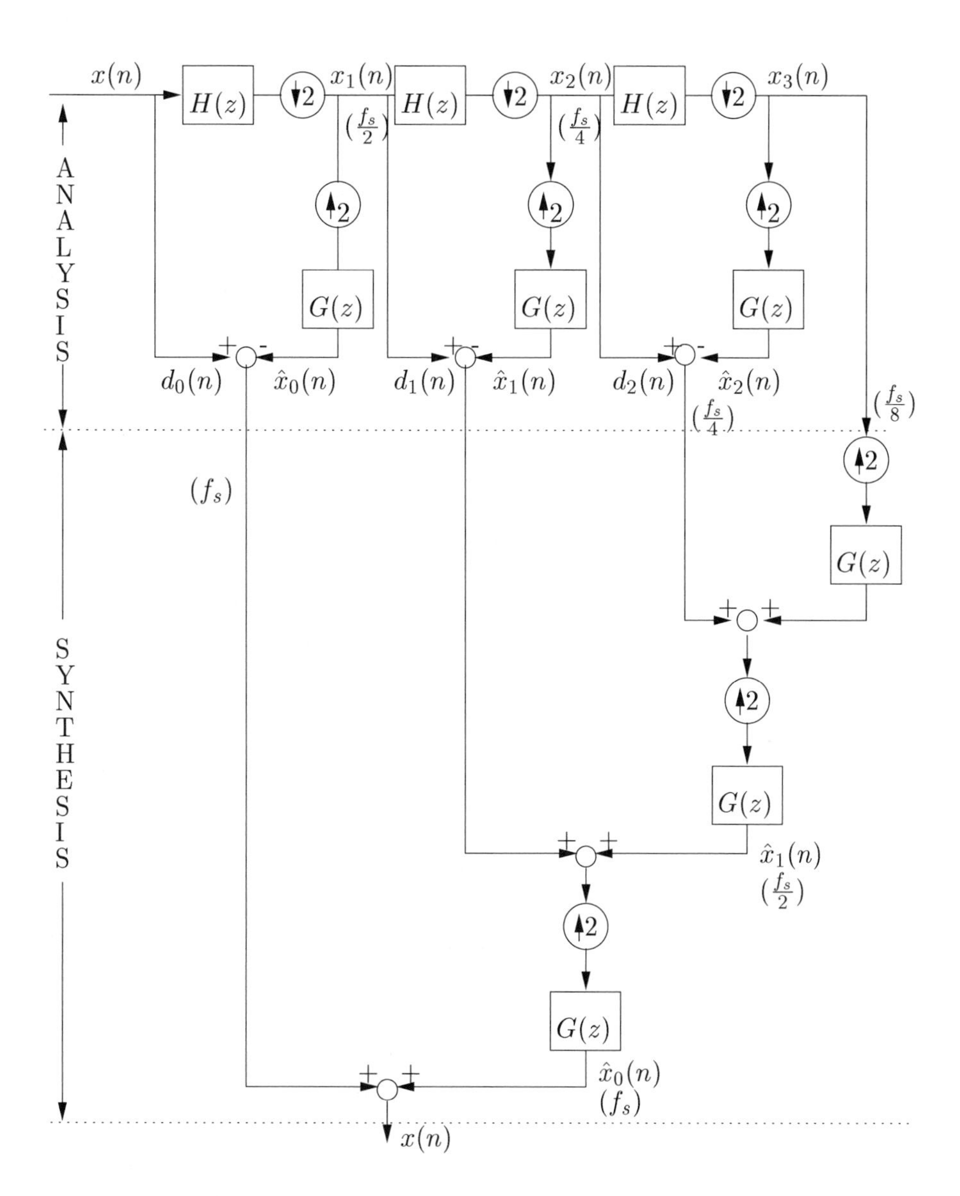

Figure 3.29: Analysis and synthesis structures for Laplacian pyramid.

a multiresolution or coarse-to-fine signal representation in time. The decimation and interpolation steps on the higher level low-pass signal are repeated until the desired level L of the dyadic-like tree structure is reached. Figure 3.29 displays the Laplacian pyramid and its frequency resolution for $L = 3$. It shows that $x(n)$ can be recovered perfectly from the coarsest low-pass signal $x_3(n)$ and the detail signals, $d_2(n)$, $d_1(n)$, and $d_0(n)$. The data rate corresponding to each of these signals is noted on this figure. The net rate is the sum of these or

$$f_s + \frac{f_s}{2} + \frac{f_s}{4} + \frac{f_s}{8} = \frac{15}{8} f_s,$$

which is almost double the data rate in a critically decimated PR dyadic tree.

This weakness of the Laplacian pyramid scheme can be fixed easily if the proper antialiasing and interpolation filters are employed. These filters, PR-QMFs, also provide the conditions for the decimation and interpolation of the high-frequency signal bands. This enhanced pyramid signal representation scheme is actually identical to the dyadic subband tree, resulting in critical sampling.

3.4.6 Modified Laplacian Pyramid for Critical Sampling

The oversampling nature of the Laplacian pyramid is clearly undesirable, particularly for signal coding applications. We should also note that the Laplacian pyramid does not put any constraints on the low-pass antialiasing and interpolation filters, although it decimates the signal by 2. This is also a questionable point in this approach.

In this section we modify the Laplacian pyramid structure to achieve critical sampling. In other words, we derive the filter conditions to decimate the Laplacian error signal by 2 and to reconstruct the input signal perfectly. Then we point out the similarities between the modified Laplacian pyramid and two-band PR-QMF banks.

Figure 3.30 shows one level of the modified Laplacian pyramid. It is seen from the figure that the error signal $D_0(z)$ is filtered by $H_1(z)$ and down- and up-sampled by 2 then interpolated by $G_1(z)$. The resulting branch output signal $\hat{X}_1(z)$ is added to the low-pass predicted version of the input signal, $\hat{X}_0(z)$, to obtain the reconstructed signal $\hat{X}(z)$.

We can write the low-pass predicted version of the input signal from Fig. 3.30 similar to the two-band PR-QMF case given earlier,

$$\hat{X}_0(z) = \frac{1}{2} G_0(z)[H_0(z)X(z) + H_0(-z)X(-z)], \tag{3.54}$$

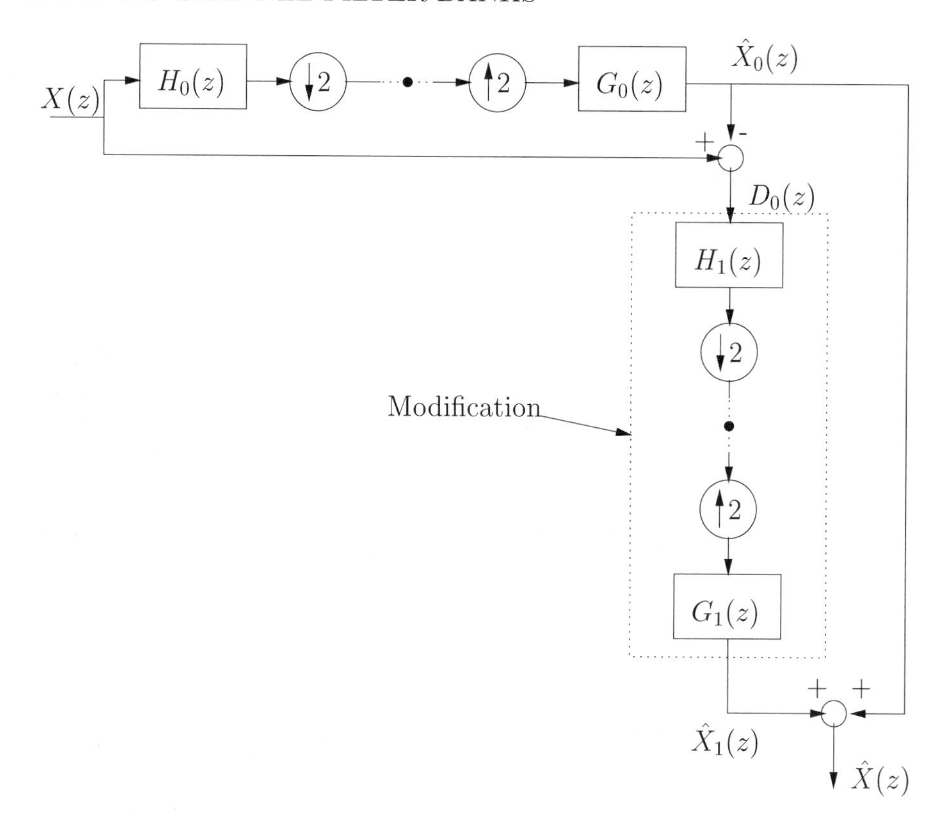

Figure 3.30: Modified Laplacian pyramid structure allowing perfect reconstruction with critical number of samples.

and the Laplacian or prediction error signal

$$D_0(z) = X(z) - \hat{X}_0(z) \tag{3.55}$$

is obtained. As stated earlier $D_0(z)$ has the full resolution of the input signal $X(z)$. Therefore this structure oversamples the input signal. Now, let us decimate and interpolate this error signal. From Fig. 3.30,

$$\hat{X}_1(z) = \frac{1}{2} G_1(z)[H_1(z)D_0(z) + H_1(-z)D_0(-z)]. \tag{3.56}$$

If we put Eqs. (3.54) and (3.55) in this equation, and then add $\hat{X}_0$ and $\hat{X}_1$, we get the reconstructed signal

$$\begin{aligned} \hat{X}(z) &= \hat{X}_0(z) + \hat{X}_1(z) \\ &= T(z)X(z) + S(z)X(-z), \end{aligned} \tag{3.57}$$

where

$$\begin{aligned} T(z) &= \frac{1}{2}[H_0(z)G_0(z) + H_1(z)G_1(z)] - \frac{1}{4}[H_0(z)G_0(z)H_1(z)G_1(z) \\ &\quad + H_0(z)G_0(-z)H_1(-z)G_1(z)] \end{aligned} \tag{3.58}$$

and

$$\begin{aligned} S(z) &= \frac{1}{2}[H_0(-z)G_0(z) + H_1(-z)G_1(z)] - \frac{1}{4}[H_0(-z)G_0(z)H_1(z)G_1(z) \\ &\quad + H_0(-z)G_0(-z)H_1(-z)G_1(z)]. \end{aligned} \tag{3.59}$$

If we choose the synthesis or interpolation filters as

$$\begin{aligned} G_0(z) &= -z^{-1}H_1(-z) \\ G_1(z) &= z^{-1}H_0(-z) \end{aligned} \tag{3.60}$$

the aliasing terms cancel and

$$\begin{aligned} S(z) &= 0 \\ T(z) &= \frac{1}{2}z^{-1}[-H_0(z)H_1(-z) + H_1(z)H_0(-z)] \end{aligned} \tag{3.61}$$

as in Eq. (3.37) except for the inconsequential z^{-1} factor. One way of achieving PR is to let $H_0(z)$, $H_1(z)$ be the paraunitary pair of Eq. (3.38), $H_1(z) = z^{-(N-1)}H_0(-z^{-1})$ and then solve the resulting Eq. (3.40), or Eq. (3.47) in the time-domain. This solution implies that all filters, analysis and synthesis, have the same length N. Furthermore, for $h(n)$ real, the magnitude responses are mirror images,

$$|H_1(e^{j\omega})| = |H_0(e^{j(\omega-\pi)})|, \tag{3.62}$$

implying equal bandwidth low-pass and high-pass filters. In the 2-band orthonormal PR-QMF case discussed in Section 3.5.4, we show that the paraunitary solution implies the time-domain orthonormality conditions

$$\begin{aligned} \sum_k h_i(k)h_i(k+2n) &= \delta(n), \quad i = 0, 1 \\ \sum_k h_1(k)h_0(k+2n) &= 0. \end{aligned} \tag{3.63}$$

These equations state that sequence $\{h_0(n)\}$ is orthogonal to its own even translates (except n=0), and orthogonal to $\{h_1(n)\}$ and its even translates.

Vetterli and Herley (1992), proposed the PR biorthogonal two-band filter bank as an alternative to the paraunitary solution. Their solution achieves zero aliasing by Eq. (3.60). The PR conditions for $T(z)$ is obtained by satisfying the following *biorthogonal* conditions (Prob. 3.28):

$$\begin{aligned} \sum_k h_0(k)\tilde{g}_0(k-2n) &= \delta_n \qquad \sum_k h_1(k)\tilde{g}_0(k-2n) = 0 \\ \sum_k h_1(k)\tilde{g}_1(k-2n) &= \delta_n \qquad \sum_k h_0(k)\tilde{g}_1(k-2n) = 0 \end{aligned} \tag{3.64}$$

where

$$\tilde{g}_i(n) \triangleq g_i(-n)$$

These biorthogonal filters also provide basis sequences in the design of biorthogonal wavelet transforms discussed in Section 6.4. The low- and high-pass filters of a two-band PR filter bank are not mirrors of each other in this approach. Biorthogonality provides the theoretical basis for the design of PR filter banks with linear-phase, unequal bandwidth low-high filter pairs.

The advantage of having linear-phase filters in the PR filter bank, however, may very well be illusory if we do not monitor their frequency behavior. As mentioned earlier, the filters in a multirate structure should try to realize the antialiasing requirements so as to minimize the spillover from one band to another. This suggests that the filters $H_0(z)$ and $H_1(z)$ should be equal bandwidth low-pass and high-pass respectively, as in the orthonormal solution.

This derivation shows that the modified Laplacian pyramid with critical sampling emerges as a biorthogonal two-band filter bank or, more desirably, as an orthonormal two-band PR-QMF bank based on the filters used. The concept of the modified Laplacian pyramid emphasizes the importance of the decimation and interpolation filters employed in a multirate signal processing structure.

3.4.7 Generalized Subband Tree Structure

The spectral analysis schemes considered in the previous sections assume a two-band frequency split as the main decomposition operation. If the signal energy is concentrated mostly around $\omega = \pi/2$, the binary spectral split becomes inefficient. As a practical solution for this scenario, the original spectrum should be split into three equal bands. Therefore a spectral division by 3 should be possible. The three-band PR filter bank is a special case of the M-band PR filter bank presented in Section 3.5. The general tree structure is a very practical and powerful spectral analysis technique. An arbitrary general tree structure and its frequency resolution are displayed in Fig. 3.26 for $L = 3$ with the assumption of ideal decimation and interpolation filters.

The irregular subband tree concept is very useful for time-frequency signal analysis-synthesis purposes. The irregular tree structure should be custom tailored for the given input source. This suggests that an adaptive tree structuring algorithm driven by the input signal can be employed. A simple tree structuring algorithm based on the energy compaction criterion for the given input is proposed in Akansu and Liu (1991).

We calculated the compaction gain of the Binomial QMF filter bank (Section 3.6.1) for both the regular and the dyadic tree configurations. The test results for a one-dimensional AR(1) source with $\rho = 0.95$ are displayed in Table 3.1 for four-, six-, and eight-tap filter structures. The term G_{TC}^{ub} is the upper bound for G_{TC} as defined in Eq. (2.97) using ideal filters. The table shows that the dyadic tree achieves a performance very close to that of the regular tree, but with fewer bands and hence reduced complexity.

Table 3.2 lists the energy compaction performance of several decomposition techniques for the standard test images: LENA, BUILDING, CAMERAMAN, and BRAIN. The images are of 256×256 pixels monochrome with 8 bits/pixel resolution. These test results are broadly consistent with the results obtained for AR(1) signal sources.

For example, the six-tap Binomial QMF outperformed the DCT in every case for both regular and dyadic tree configurations. Once again, the dyadic tree with fewer bands is comparable in performance to the regular or full tree. However, as we alluded to earlier, more levels in a tree tends to lead to poor band isolation. This aliasing could degrade performance perceptibly under low bit rate encoding.

3.5 M-Band Filter Banks

The results of the previous two-band filter bank are extended in two directions in this section. First, we pass from two-band to M-band, and second we obtain more general perfect reconstruction (PR) conditions than those obtained previously.

Our approach is to represent the filter bank by three equivalent structures, each of which is useful in characterizing particular features of the subband system. The conditions for alias cancellation and perfect reconstruction can then be described in both time and frequency domains using the polyphase decomposition and the alias component (AC) matrix formats. In this section, we draw heavily on the papers by Vaidyanathan (ASSP Mag., 1987), Vetterli and LeGall (1989), and Malvar (Elect. Letts., 1990) and attempt to establish the commonality of these

(a) 4-tap Binomial-QMF.

level	Regular Tree			Half Band Irre gular Tree		
	# of bands	G_{TC}	G_{TC}^{ub}	# of bands	G_{TC}	G_{TC}^{ub}
1	2	3.6389	3.9462	2	3.6389	3.9462
2	4	6.4321	7.2290	3	6.3681	7.1532
3	8	8.0147	9.1604	4	7.8216	8.9617
4	16	8.6503	9.9407	5	8.3419	9.6232

(b) 6-tap Binomial-QMF.

level	Regular Tree			Half Band Irre gular Tree		
	# of bands	G_{TC}	G_{TC}^{ub}	# of bands	G_{TC}	G_{TC}^{ub}
1	2	3.7608	3.9462	2	3.7608	3.9462
2	4	6.7664	7.2290	3	6.6956	7.1532
3	8	8.5291	9.1604	4	8.2841	8.9617
4	16	9.2505	9.9407	5	8.8592	9.6232

(c) 8-tap Binomial-QMF.

level	Regular Tree			Half Band Irre gular Tree		
	# of bands	G_{TC}	G_{TC}^{ub}	# of bands	G_{TC}	G_{TC}^{ub}
1	2	3.8132	3.9462	2	3.8132	3.9462
2	4	6.9075	7.2290	3	6.8355	7.1532
3	8	8.7431	9.1604	4	8.4828	8.9617
4	16	9.4979	9.9407	5	9.0826	9.6232

Table 3.1: Energy compaction performance of PR-QMF filter banks along with the full tree and upper performance bounds for AR(1) source of $\rho = 0.95$.

TEST IMAGE	LENA	BUILDING	CAMERAMAN	BRAIN
8×8 2D DCT	21.99	20.08	19.10	3.79
64 Band Regular 4-tap B-QMF	19.38	18.82	18.43	3.73
64 Band Regular 6-tap B-QMF	22.12	21.09	20.34	3.82
64 Band Regular 8-tap B-QMF	24.03	22.71	21.45	3.93
4×4 2D DCT	16.00	14.11	14.23	3.29
16 Band Regular 4-tap B-QMF	16.70	15.37	15.45	3.25
16 Band Regular 6-tap B-QMF	18.99	16.94	16.91	3.32
16 Band Regular 8-tap B-QMF	20.37	18.17	17.98	3.42
*10 Band Irregular 4-tap B-QMF	16.50	14.95	13.30	3 .34
*10 Band Irregular 6-tap B-QMF	18.65	16.55	14.88	3 .66
*10 Band Irregular 8-tap B-QMF	19.66	17.17	15.50	3 .75

* Bands used are $lllll - llllh - lllhl - lllhh - lllh - llhl - llhh - lh - hl - hh$.

Table 3.2: Compaction gain, G_{TC}, of several different regular and dyadic tree structures along with the DCT for the test images.

approaches, which in turn reveals the connection between block transforms, lapped transforms, and subbands.

3.5.1 The M-Band Filter Bank Structure

The M-band QMF structure is shown in Fig. 3.31. The bank of filters $\{H_k(z), k = 0, 1, ..., M-1\}$ constitute the analysis filters typically at the transmitter in a signal transmission system. Each filter output is subsampled, quantized (i.e., coded), and transmitted to the receiver, where the bank of up-samplers/synthesis filters reconstruct the signal.

In the most general case, the decimation factor L satisfies $L \leq M$ and the filters could be any mix of FIR and IIR varieties. For most practical cases, we would choose maximal decimation or "critical subsampling," $L = M$. This ensures that the total data rate in samples per second is unaltered from $x(n)$ to the set of subsampled signals, $\{v_k(n), \quad k = 0, 1, ..., M-1\}$. Furthermore, we will consider FIR filters of length N at the analysis side, and length N' for the synthesis filters. Also, for deriving PR requirements, we do not consider coding errors. Under these conditions, the maximally decimated M-band $FIR\,QMF$ filter bank structure has the form shown explicitly in Fig. 3.32. [The term QMF is a carryover from the two-band case and has been used, somewhat loosely, in the DSP community for the M-band case as well.]

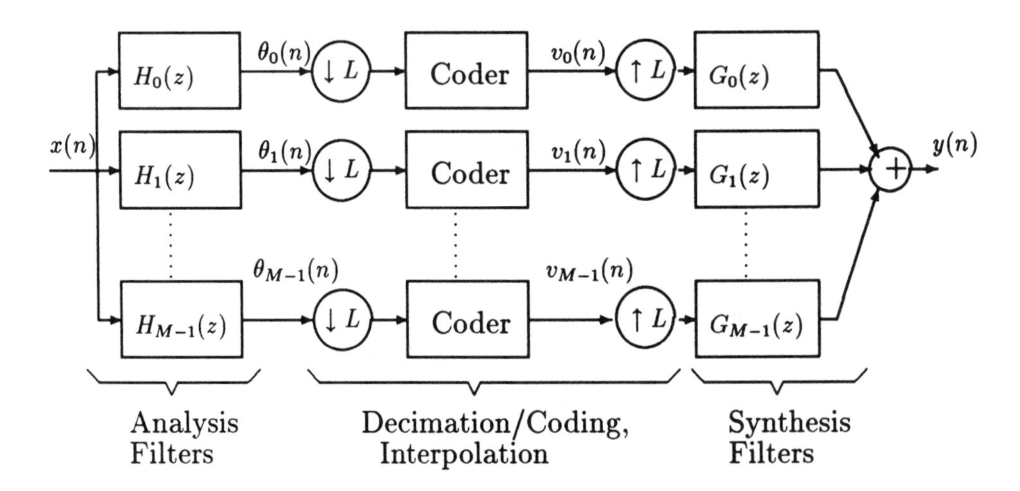

Figure 3.31: M-band filter bank.

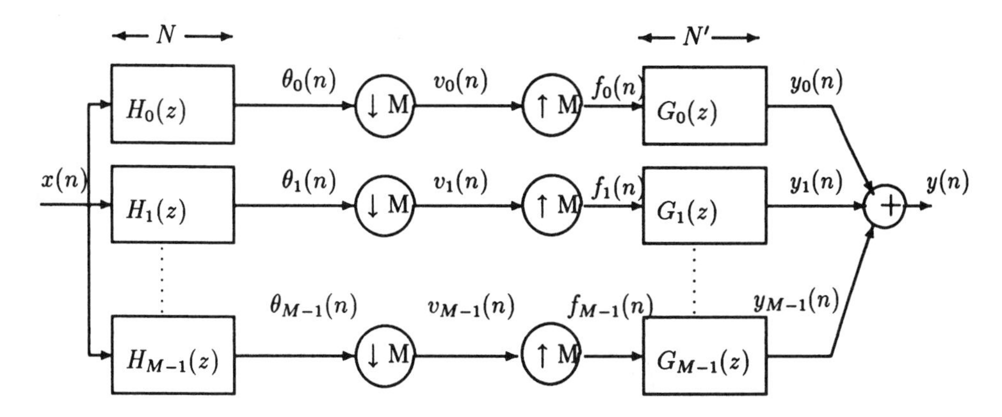

Figure 3.32: Maximally decimated M-band FIR QMF structures.

From this block diagram, we can derive the transmission features of this subband system. If we were to remove the up- and down-samplers from Fig. 3.32, we would have

$$Y(z) = \left(\sum_{k=0}^{M-1} G_k(z) H_k(z) \right) X(z) \tag{3.65}$$

and perfect reconstruction; i.e., $y(n) = x(n - n_0)$ can be realized with relative ease, but with an attendant M-fold increase in the data rate. The requirement is obviously

$$\sum_{k=0}^{M-1} G_k(z) H_k(z) = z^{-n_0}, \tag{3.66}$$

i.e., the composite transmission reduces to a simple delay.

Now with the samplers reintroduced, we have, at the analysis side,

$$\theta_k(z) = H_k(z) X(z), \qquad k = 0, 1, ..., M-1 \tag{3.67}$$

and

$$Y(z) = \sum_{k=0}^{M-1} G_k(z) F_k(z) \tag{3.68}$$

at the synthesis side.

The sampling bank is represented using Eqs. (3.12) and (3.9) in Section 3.1.1,

$$F_k(z) = V_k(z^M), \qquad \text{up-sampler} \tag{3.69}$$

$$V_k(z) = \frac{1}{M}\sum_{r=0}^{M-1} \theta_k(z^{1/M}W^r), \quad \text{down-sampler} \tag{3.70}$$

where $W = e^{-j2\pi/M}$. Combining these gives

$$\begin{aligned} Y(z) &= \sum_k G_k(z)V_k(z^M) \\ &= \sum_k G_k(z)\frac{1}{M}\sum_r \theta_k(zW^r) \\ &= \sum_k G_k(z)\frac{1}{M}\sum_r H_k(zW^r)X(zW^r) \\ &= \frac{1}{M}\sum_{r=0}^{M-1}\sum_{k=0}^{M-1} X(zW^r)H_k(zW^r)G_k(z). \end{aligned} \tag{3.71}$$

We can write this last equation more compactly as

$$\begin{aligned} Y(z) &= \frac{1}{M}[X(z), ..., X(zW^{M-1})] \times \\ &\begin{bmatrix} H_0(z) & ... & H_{M-1}(z) \\ H_0(zW) & ... & H_{M-1}(zW) \\ \vdots & \vdots & \vdots \\ H_0(zW^{M-1}) & ... & H_{M-1}(zW^{M-1}) \end{bmatrix} \begin{bmatrix} G_0(z) \\ G_1(z) \\ \vdots \\ G_{M-1}(z) \end{bmatrix} \\ &= \frac{1}{M}\underline{X}^T H_{AC}(z)\underline{g}(z), \end{aligned} \tag{3.72}$$

where $H_{AC}(z)$ is the *alias component*, or *AC* matrix.

The subband filter bank of Fig. 3.32 is linear, but time-varying, as can be inferred from the presence of the samplers. This last equation can be expanded as

$$Y(z) = \left(\frac{1}{M}\sum_{k=0}^{M-1} G_k(z)H_k(z)\right)X(z) + \underbrace{\frac{1}{M}\sum_{\substack{r \\ k\neq 0}}\sum_{\substack{k \\ r\neq 0}} H_k(zW^r)X(zW^r)G_k(z)}_{aliasing\ terms}. \tag{3.73}$$

Three kinds of errors or undesirable distortion terms can be deduced from this last equation.

(1) Aliasing error or distortion (ALD) terms. More properly, the subsampling is the cause of aliasing components while the up-samplers produce images.

The combination of these is still called *aliasing*. These aliasing terms in Eq. (3.73) can be eliminated if we impose

$$H_{AC}\underline{g}(z) = \begin{bmatrix} MT(z) \\ 0 \\ \vdots \\ 0 \end{bmatrix}. \qquad (3.74)$$

In this case, the input-output relation reduces to just the first term in Eq. (3.73), which represents the transfer function of a linear, time-invariant system:

$$\begin{aligned} Y(z) &= T(z)X(z) \\ T(z) &= \sum_k G_k(z)F_k(z). \end{aligned} \qquad (3.75)$$

(2) Amplitude and Phase Distortion. Having constrained $\{H_k, G_k\}$ to force the aliasing term to zero, we are left with classical magnitude (amplitude) and phase distortion, with

$$T(e^{j\omega}) = |T(e^{j\omega})|e^{j\phi(\omega)}. \qquad (3.76)$$

Perfect reconstruction requires $T(z) = z^{-n_0}$, a pure delay, or

$$T(e^{j\omega}) = e^{-jn_0\omega}. \qquad (3.77)$$

Deviation of $|T(e^{j\omega})|$ from unity constitutes amplitude distortion, and deviation of $\phi(\omega)$ from linearity is phase distortion. Classically, we could select an IIR all-pass filter to eliminate magnitude distortion, whereas a linear-phase FIR easily removes phase distortion.

When all three distortion terms are zero, we have perfect reconstruction:

$$T(z) = z^{-n_0}. \qquad (3.78)$$

The conditions for zero aliasing, and the more stringent PR, can be developed using the AC matrix formulation, and as we shall see, the polyphase decomposition that we consider next.

3.5.2 The Polyphase Decomposition

In this subsection, we formulate the PR conditions from a polyphase representation of the filter bank. Recall that from Eqs. (3.14) and (3.15), each analysis filter $H_r(z)$ can be represented by

$$H_r(z) = \sum_{k=0}^{M-1} z^{-k}H_{r,k}(z^M), \qquad r = 0, 1, ..., M-1 \qquad (3.79)$$

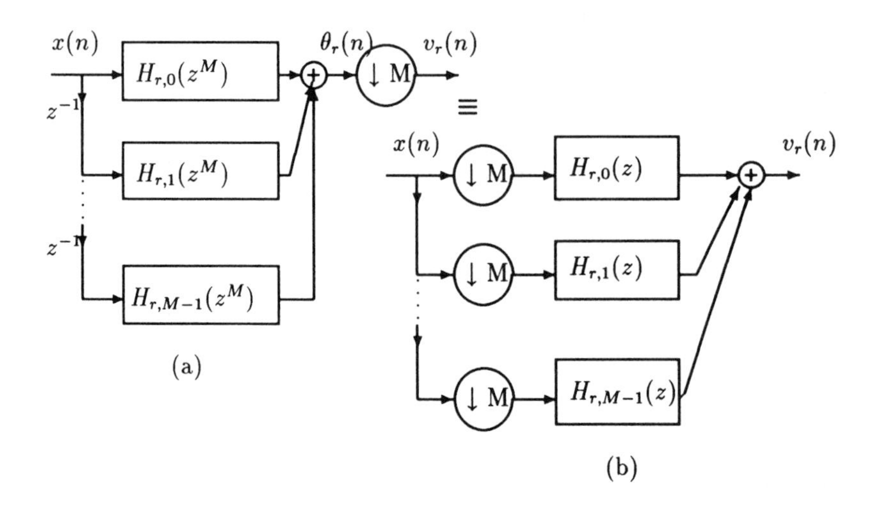

Figure 3.33: Polyphase decomposition of $H_r(z)$.

$$H_{r,k}(z) = \sum_{n=-\infty}^{\infty} h_r(k + Mn)z^{-n}, \qquad 0 \leq r, k \leq M - 1. \tag{3.80}$$

These are shown in Fig. 3.33.

When this is repeated for each analysis filter, we can stack the results to obtain

$$\begin{bmatrix} \theta_0(z) \\ \theta_1(z) \\ \vdots \\ \theta_{M-1}(z) \end{bmatrix} = \begin{bmatrix} H_{00}(z^M) & \ldots & H_{0,M-1}(z^M) \\ H_{10}(z^M) & \ldots & H_{1,M-1}(z^M) \\ \vdots & & \\ H_{M-1,0}(z^M) & \ldots & H_{M-1,M-1}(z^M) \end{bmatrix} \begin{bmatrix} 1 \\ z^{-1} \\ \vdots \\ z^{-(M-1)} \end{bmatrix} X(z)$$

or

$$\underline{\theta}(z) = [\mathcal{H}_p(z^M)\underline{Z}_M]X(z) = \underline{h}(z)X(z), \tag{3.81}$$

where $\mathcal{H}_p(z)$ is the polyphase matrix, and $\underline{Z}_M$ is a vector of delays

$$\underline{Z}_M^T = [1, z^{-1}, ..., z^{-(M-1)}] \tag{3.82}$$

and

$$\underline{h}^T(z) = [H_0(z), ..., H_{M-1}(z)].$$

Similarly, we can represent the synthesis filters by

$$G_s(z) \;\; = \;\; \sum_{k=0}^{M-1} z^{-k} G_{s,k}(z^M) = \sum_{k=0}^{M-1} z^{-(M-1-k)} G_{s,M-1-k}(z^M)$$

$$G_{s,k}(z) \;=\; \sum_{n=-\infty}^{\infty} g_s(k+Mn)z^{-n}. \tag{3.83}$$

This structure is shown in Fig. 3.34.

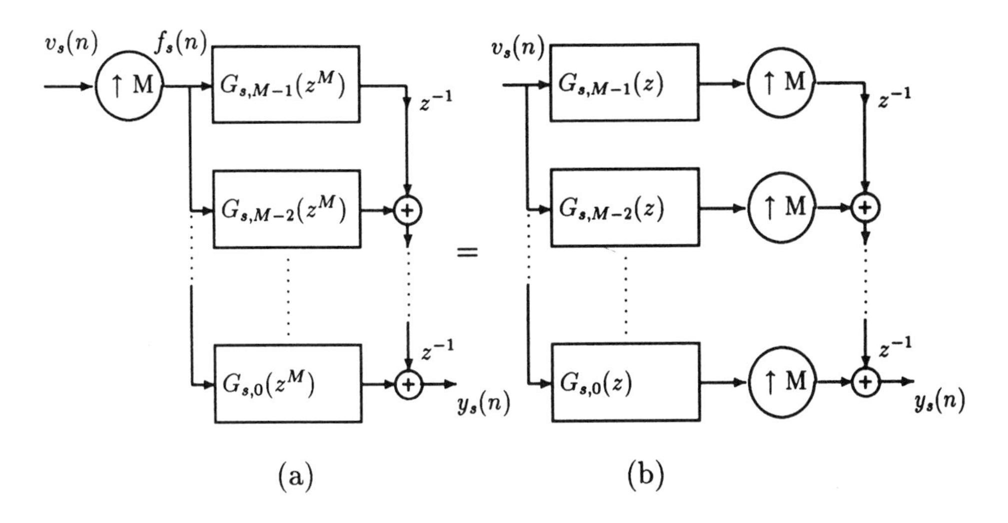

Figure 3.34: Synthesis filter decomposition.

In terms of the polyphase components, the output is

$$\begin{aligned}
Y(z) &= \sum_{s=0}^{M-1} Y_s(z) = \sum_{s=0}^{M-1} F_s(z)G_s(z) \\
&= \sum_{s=0}^{M-1} F_s(z) \sum_{k=0}^{M-1} G_{s,M-1-k}(z^M)z^{-(M-1-k)} \\
&= \sum_{k=0}^{M-1} z^{-(M-1-k)} \sum_{s=0}^{M-1} F_s(z)G_{s,M-1-k}(z^M) \\
&= \sum_{k=0}^{M-1} z^{-(M-1-k)} L_k(z).
\end{aligned} \tag{3.84}$$

The reason for rearranging the dummy indexing in these last two equations is to obtain a synthesis polyphase representation with delay arrows pointing down, as in Fig. 3.34(b). This last equation can now be written as

$$Y(z) = [L_0(z), L_1(z), ..., L_{M-1}(z)] \begin{bmatrix} z^{-(M-1)} \\ z^{-(M-2)} \\ \vdots \\ z^{-1} \\ 1 \end{bmatrix} = \underline{L}^T(z)\underline{Z}'_M, \tag{3.85}$$

where

$$\begin{bmatrix} L_0(z) \\ L_1(z) \\ \vdots \\ L_{M-1}(z) \end{bmatrix} = \begin{bmatrix} G_{0,M-1}(z^M) & .... & G_{M-1,M-1}(z^M) \\ G_{0,M-2}(z^M) & .... & G_{M-1,M-2}(z^M) \\ \vdots & & \\ G_{0,0}(z^M) & .... & G_{M-1,0}(z^M) \end{bmatrix} \begin{bmatrix} F_0(z) \\ F_1(z) \\ \vdots \\ F_{M-1}(z) \end{bmatrix}. \tag{3.86}$$

The synthesis polyphase matrix in this last equation has a row-column indexing different from $\mathcal{H}_p(z)$ in Eq. (3.81).

For consistency in notation, we introduce the *"counter-identity"* or interchange matrix J,

$$J = \begin{bmatrix} 0 & 0 & ... & 0 & 1 \\ 0 & 0 & ... & 1 & 0 \\ \vdots & & & & \\ 1 & 0 & ... & 0 & 0 \end{bmatrix}, \tag{3.87}$$

with the property that pre(post)multiplication of a matrix A by J interchanges the rows (columns) of A, i.e.,

$$\begin{aligned} [a_{ij}]J &= [a_{i,N-1-j}] \\ J[a_{ij}] &= [a_{N-1-i,j}]. \end{aligned} \tag{3.88}$$

Also note that

$$J[a_{ij}]J = [a_{N-1-i,N-1-j}] \tag{3.89}$$

and

$$J^2 = I. \tag{3.90}$$

We have already employed this notation, though somewhat implicitly, in the vector of delays:

$$\begin{bmatrix} z^{-(M-1)} \\ \vdots \\ z^{-1} \end{bmatrix} = \underline{Z}'_M = J\underline{Z}_M = J \begin{bmatrix} 1 \\ z^{-1} \\ \vdots \\ z^{-(M-1)} \end{bmatrix}. \tag{3.91}$$

With this convention, and with $\mathcal{G}_p(z)$ defined in the same way as $\mathcal{H}_p(z)$ of Eq. (3.81), i.e., by

$$\mathcal{G}_p(z) = \begin{bmatrix} G_{00}(z) & \dots & G_{0,M-1}(z) \\ \vdots & & \\ G_{M-1,0}(z) & \dots & G_{M-1,M-1}(z) \end{bmatrix}, \tag{3.92}$$

we recognize that the synthesis polyphase matrix in Eq. (3.86) is

$$J\mathcal{G}_p^T(z) \triangleq \mathcal{G}_p'(z). \tag{3.93}$$

This permits us to write the polyphase synthesis equation as

$$\begin{aligned} Y(z) &= (\underline{Z}_M')^T \mathcal{G}_p'(z^M)\underline{F}(z) \\ &= [\mathcal{G}_p(z^M)\underline{Z}_M]^T \underline{F}(z) \\ &= \underline{g}^T(z)\underline{F}(z). \end{aligned} \tag{3.94}$$

Note that we have defined the analysis and synthesis polyphase matrices in exactly the same way so as to result in

$$\begin{aligned} \underline{h}(z) &= \mathcal{H}_p(z^M)\underline{Z}_M \\ \underline{g}(z) &= \mathcal{G}_p(z^M)\underline{Z}_M. \end{aligned}$$

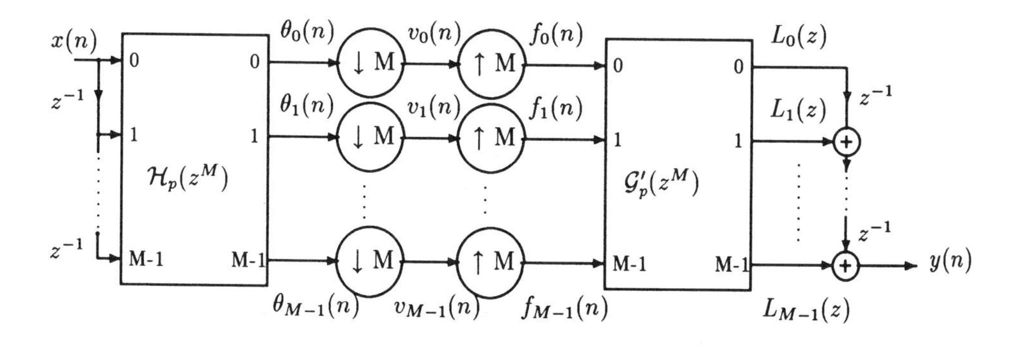

Figure 3.35: Polyphase representation of QMF filter bank.

Finally, we see that Eqs. (3.81) and (3.94) suggest the polyphase block diagram of Fig. 3.35. As explained in Section 3.1.2, we can shift the down-samplers to the left of the analysis polyphase matrix and replace z^M by z in the argument of

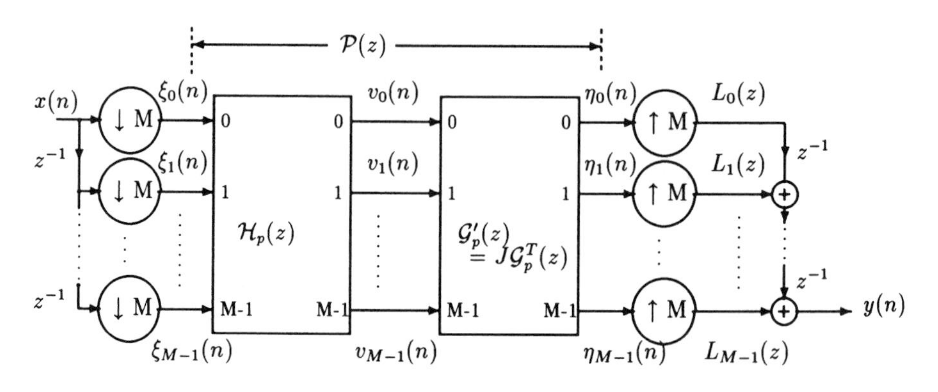

Figure 3.36: Equivalent polyphase QMF bank.

$\mathcal{H}_p(.)$. Similarly, we shift the up-samplers to the right of the synthesis polyphase matrix and obtain the structure of Fig. 3.36. These two polyphase structures are equivalent to the filter bank with which we started in Fig. 3.32.

We can obtain still another representation, this time with the delay arrows pointing up, by the following manipulations. From Eq. (3.81), noting that $J^2 = I$, we can write

$$\begin{aligned} \underline{\theta}(z) &= [\mathcal{H}_p(z^M)\underline{Z}_M]X(z) \\ &= [\mathcal{H}_p(z^M)J](J\underline{Z}_M)X(z) \\ &= [\mathcal{H}_p'(z^M)\underline{Z}_M']X(z). \end{aligned} \tag{3.95}$$

Similarly,

$$\begin{aligned} Y(z) &= (\underline{Z}_M')^T \mathcal{G}_p'(z^M)\underline{F}(z) \\ &= (\underline{Z}_M^T J)(J\mathcal{G}_p^T(z^M))\underline{F}(z) \\ &= \underline{Z}_M^T \mathcal{G}_p^T(z^M)\underline{F}(z). \end{aligned} \tag{3.96}$$

These last two equations define the alternate polyphase QMF representations of Figs. 3.37 and 3.38, where we are using

$$\underline{\eta}' = J\underline{\eta}, \quad \underline{\xi}' = J\underline{\xi}, \quad \text{and} \quad \underline{L}' = J\underline{L}. \tag{3.97}$$

It is now easy to show that (Prob. 3.14)

$$\mathcal{P}' = J\mathcal{P}J. \tag{3.98}$$

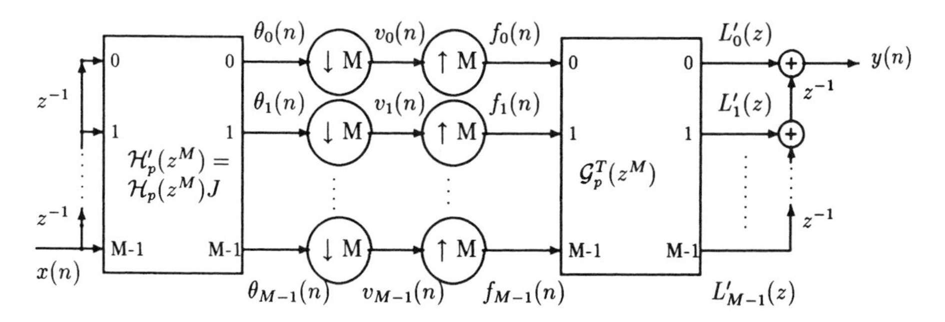

Figure 3.37: Alternative polyphase structure.

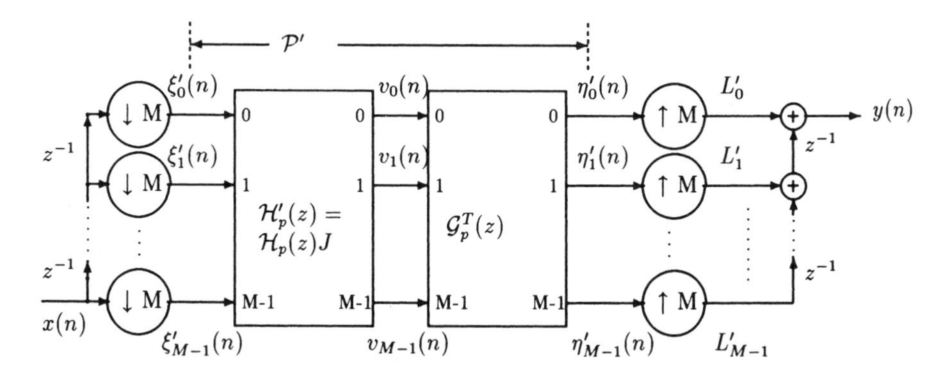

Figure 3.38: Alternative polyphase representation.

Either of the polyphase representations allow us to formulate the PR requirements in terms of the polyphase matrices. From Fig. 3.36, we have

$$\underline{\eta}(z) = \mathcal{G}'_p(z)\mathcal{H}_p(z)\underline{\xi}(z) = \mathcal{P}(z)\underline{\xi}(z), \tag{3.99}$$

which defines the composite structure of Fig. 3.39.

The condition for PR in Eq. (3.78) was $T(z) = z^{-n_0}$. It is shown by Vaidyanathan (April 1987) that PR is satisfied if

$$\mathcal{P}(z) = z^{k_1}\begin{bmatrix} 0 & I_{M-k_0} \\ z^{-1}I_{k_0} & 0 \end{bmatrix}$$

where I_m denotes the $m \times m$ identity matrix. This condition is very broadly stated. Detailed discussion of various special cases induced by imposing symmetries on

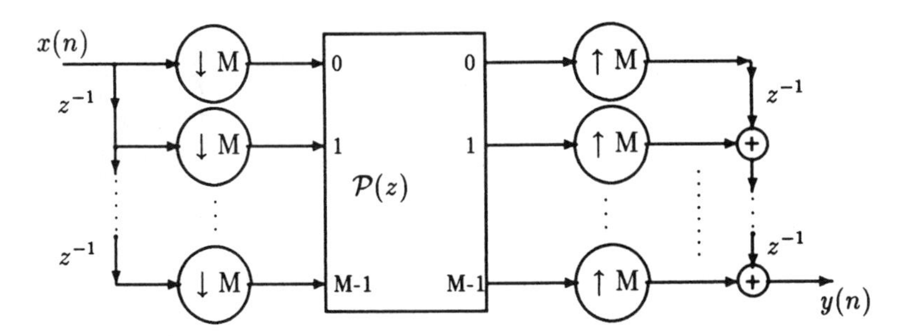

Figure 3.39: Composite M-band polyphase structure.

the analysis-synthesis filters can be found in Viscito and Allebach (1989). For our purposes we will only consider a *sufficient* condition for PR, namely,

$$\mathcal{P}(z) = \mathcal{G}'_p(z)\mathcal{H}_p(z) = z^{-\mu}I. \tag{3.100}$$

(This corresponds to the case where $k_0 = 0$.) For if this condition is satisfied, using the manipulations of Fig. 3.40, we can demonstrate that (Prob. 3.13)

$$y(n) = x(n - n_0), \qquad n_0 = \mu M + M - 1. \tag{3.101}$$

The bank of delays is moved to the right of the up-samplers, and then outside of the decimator-interpolator structure. It is easily verified that the signal transmission from point (1) to point (2) in Fig. 3.40(c) is just a delay of $M - 1$ units. Thus the total transmission from $x(n)$ to $y(n)$ is just $[(M-1)+M\mu]$ delays, resulting in $T(z) = z^{-n_0}$.

Thus we have two representations for the M-band filter bank, the AC matrix approach, and the polyphase decomposition. We next develop detailed PR filter bank requirements using each of these as starting points. The AC matrix provides a frequency-domain formulation, while the polyphase is useful for both frequency- and time-domain interpretations. We close this subsection by noting the relationship between the AC and polyphase matrices. From Eq. (3.72), we know that the AC matrix is

$$[H_{AC}(z)]_{r,k} = H_k(zW^r). \tag{3.102}$$

Substituting the polyphase expansion from Eq. (3.79) into this last equation gives

$$[H_{AC}(z)]_{r,k} \quad = \quad \sum_{l=0}^{M-1} (zW^r)^{-l} H_{k,l}((zW^r)^M)$$

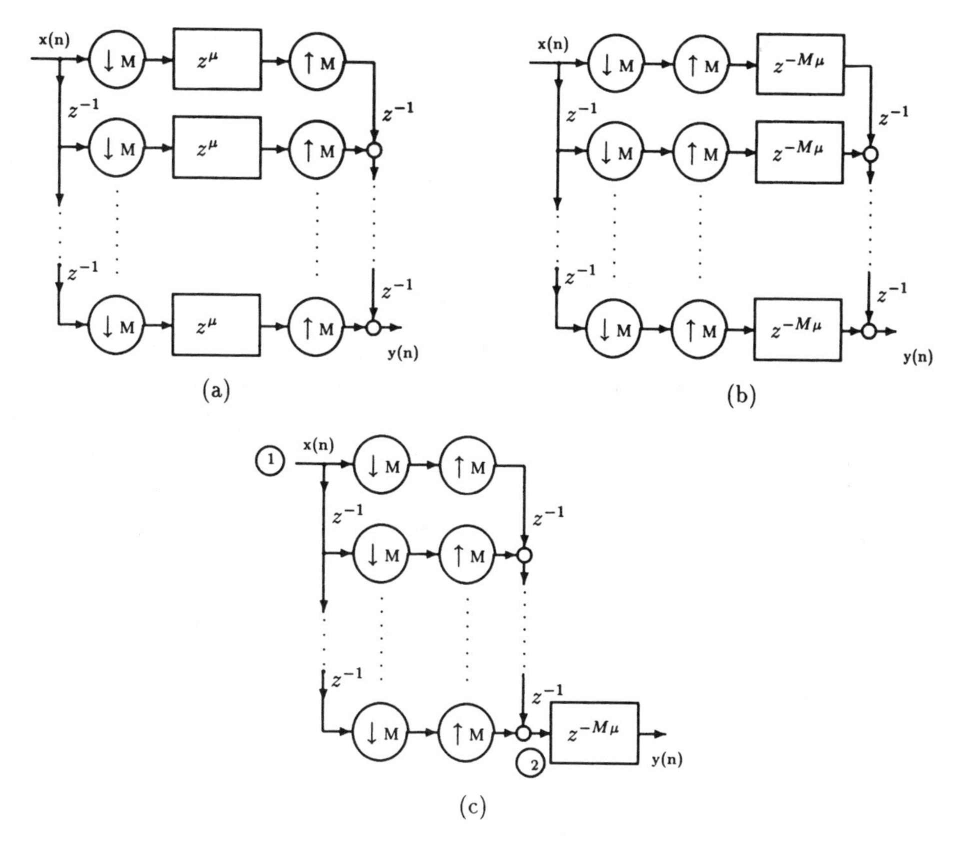

Figure 3.40: Polyphase implementation of PR condition of Eq. (3.100).

$$
\begin{aligned}
&= \sum z^{-l} W^{-rl} H_{k,l}(z^M W^{rM}) \\
&= \sum_{l=0}^{M-1} z^{-l} W^{-rl} H_{k,l}(z^M). \qquad (3.103)
\end{aligned}
$$

This last equation can be expressed as the product of three matrices,

$$
\begin{aligned}
H_{AC}(z) &= \left[W^{-rl}\right] \begin{bmatrix} 1 & 0 & \dots & 0 \\ 0 & z^{-1} & & \\ \vdots & & \ddots & \\ 0 & & \dots & z^{-(M-1)} \end{bmatrix} \left[H_{kl}(z^M)\right] \\
&= \mathcal{W} A(z) \mathcal{H}_p^T(z^M), \qquad (3.104)
\end{aligned}
$$

where $\mathcal{W}$ is the DFT matrix, and $A(z)$ is the diagonal matrix

$$A(z) = diag\{1, z^{-1}, ..., z^{-(M-1)}\}. \tag{3.105}$$

We can now develop filter bank properties in terms of either $H_{AC}(z)$ or $\mathcal{H}_p(z)$ or both.

3.5.3 PR Requirements for FIR Filter Banks

A simplistic approach to satisfying the PR condition in Eq. (3.100) is to choose $\mathcal{G}'_p(z) = z^{-\mu}\mathcal{H}_p^{-1}(z)$. Generally this implies that the synthesis filters would be IIR and possibly unstable, even when the analysis filters are FIR. Therefore, we want to impose conditions on the FIR $\mathcal{H}_p(z)$ that result in synthesis filters which are also FIR. Three conditions are considered (Vetterli and LeGall, 1989).

(1) Choose the FIR $\mathcal{H}_p(z)$ such that its determinant is a pure delay (i.e., $det\{\mathcal{H}_p(z)\}$ is a monomial),

$$det\{\mathcal{H}_p(z)\} = z^{-p}, \tag{3.106}$$

where p is an integer ≥ 0. Then we can satisfy Eq. (3.100) with an FIR synthesis bank. The sufficiency is established as follows. We want

$$\mathcal{P}(z) = \mathcal{G}'_p(z)\mathcal{H}_p(z) = z^{-\mu}I.$$

Multiply by $\mathcal{H}_p^{-1}(z)$ and obtain

$$\begin{aligned} \mathcal{G}'_p(z) &= z^{-\mu}\mathcal{H}_p^{-1}(z) \\ &= \frac{z^{-\mu}}{det\{\mathcal{H}_p(z)\}} adj\{\mathcal{H}_p(z)\} \\ &= z^{-(\mu-p)} adj\{\mathcal{H}_p(z)\}. \end{aligned} \tag{3.107}$$

The elements in the adjoint matrix are just cofactors of $\mathcal{H}_p(z)$, which are products and sums of FIR polynomials and thus FIR. Hence, each element of $\mathcal{G}'_p(z)$ is equal to the transposed FIR cofactor of $\mathcal{H}_p(z)$(within a delay). This approach generally leads to FIR synthesis filters that are considerably longer than the analysis filters.

(2) The second class consists of PR filters with equal length analysis and synthesis filters. Conditions for this using a time-domain formulation are developed in Section 3.5.5.

(3) Choose $\mathcal{H}_p(z)$ to be paraunitary or "lossless." This results in identical analysis and synthesis filters (within a time-reversal), which is the most commonly

stated condition. A lossless or paraunitary matrix is defined by the property

$$\begin{aligned} \tilde{\mathcal{H}}_p(z)\mathcal{H}_p(z) &= I \\ \tilde{\mathcal{H}}_p(z) &\triangleq \mathcal{H}_p^T(z^{-1}). \end{aligned} \tag{3.108}$$

And, if this condition is met, we can simply choose

$$\mathcal{G}_p'(z) = z^{-n_0}\mathcal{H}_p^T(z^{-1}) = z^{-n_0}\tilde{\mathcal{H}}_p(z). \tag{3.109}$$

This results in

$$\mathcal{G}_p'(z)\mathcal{H}_p(z) = z^{-n_0}\mathcal{H}_p^T(z^{-1})\mathcal{H}_p(z) = z^{-n_0}I.$$

The delay n_0 is selected to make $\mathcal{G}_p'(z)$ the polyphase matrix of a causal filter bank. The converse of this theorem is also valid.

We will return to review cases (1) and (2) from a time-domain standpoint. Much of the literature on PR structures deals with paraunitary solutions to which we now turn.

3.5.4 The Paraunitary FIR Filter Bank

We have shown that PR is assured if the analysis polyphase matrix is lossless (which also forces losslessness on the synthesis matrix). The main result is that the impulse responses of the paraunitary filter bank must satisfy a set of orthonormal constraints, which are generalizations of the $M = 2$ case dealt with in Section 3.4. (See also Prob. 3.17)

First, we note that the choice of $\mathcal{G}_p'(z)$ in Eq. (3.109) implies that each synthesis filter is just a time-reversed version of the analysis filter,

$$\begin{aligned} G_k(z) &= z^{-r}H_k(z^{-1}) = z^{-r}\tilde{H}_k(z) \\ g_k(n) &= h_k(r-n). \end{aligned} \tag{3.110}$$

To prove this, recall that the polyphase decomposition of the filter bank is

$$\underline{g}(z) = \mathcal{G}_p(z^M)\underline{Z}_M, \quad \text{and} \quad \underline{h}(z) = \mathcal{H}_p(z^M)\underline{Z}_M. \tag{3.111}$$

But, from Eq. (3.109), we had

$$\mathcal{G}_p'(z) = J\mathcal{G}_p^T(z) = z^{-n_0}\mathcal{H}_p^T(z^{-1})$$

or

$$\mathcal{G}_p(z)J = z^{-n_0}\mathcal{H}_p(z^{-1})$$

Now let's replace z by z^M, and multiply by $J\underline{Z}_M$ to obtain

$$\begin{aligned}
\underline{g}(z) &= \mathcal{G}_p(z^M)J^2\underline{Z}_M \\
&= z^{-Mn_0}\mathcal{H}_p(z^{-M})J\underline{Z}_M \\
&= z^{-[Mn_0+(M-1)]}\mathcal{H}_p(z^{-M})\begin{bmatrix} 1 \\ z \\ \vdots \\ z^{(M-1)} \end{bmatrix} \\
&= z^{-r}\underline{h}(z^{-1}),
\end{aligned} \tag{3.112}$$

where $r = [Mn_0+(M-1)]$. Thus $G_k(z) = z^{-r}\tilde{H}_k(z), k = 0, 1, ..., M-1$ as asserted in Eq. (3.110).

We can also write the paraunitary PR conditions in terms of elements of the AC matrix. In fact, we can show that lossless $\mathcal{H}_p(z)$ implies a lossless AC matrix and conversely, that is,

$$\tilde{\mathcal{H}}_p(z)\mathcal{H}_p(z) = I \longleftrightarrow \tilde{H}_{AC}(z)H_{AC}(z) = MI, \tag{3.113}$$

where

$$\tilde{H}_{AC}(z) \triangleq H_{AC*}^T(z^{-1}),$$

and the subscripted asterisk implies conjugation of coefficients in the matrix. The proof is straightforward. From Eq. (3.104)

$$H_{AC*}^T(z^{-1})H_{AC}(z) = \mathcal{H}_p(z^{-M})A(z^{-1})\mathcal{W}^*\mathcal{W}A(z)\mathcal{H}_p^T(z^M).$$

But

$$\mathcal{W}^*\mathcal{W} = MI$$

for a DFT matrix. Hence

$$H_{AC*}^T(z^{-1})H_{AC}(z) = M\mathcal{H}_p(z^{-M})\mathcal{H}_p^T(z^M) = MI.$$

The AC matrix approach will allow us to obtain the properties of filters in lossless structures. From Eq. (3.72), we had

$$Y(z) = \frac{1}{M}\underline{X}^T H_{AC}(z)\underline{g}(z),$$

where $H_{AC}(z)$ is the AC matrix. For zero aliasing, we had in Eq. (3.74)

$$H_{AC}(z)\underline{g}(z) = \begin{bmatrix} MT(z) \\ 0 \\ \vdots \\ 0 \end{bmatrix}.$$

Let us substitute successively $zW, zW^2, ..., zW^{M-1}$ for z in this last equation. Each substitution of zW in the previous equation induces a circular shift in the rows of H_{AC}. For example,

$$[H_{AC}(zW)\underline{g}(zW)]^T = [MT(zW), 0, ..., 0],$$

can be rearranged as

$$[H_{AC}(z)\underline{g}(zW)]^T = [0, MT(zW), 0, ..., 0].$$

This permits us to express the set of M equations as one matrix equation of the form

$$H_{AC}(z)G_{AC}^T(z) = M \quad diag\{T(z), T(zW), ..., T(zW^{M-1})\}, \tag{3.114}$$

where $G_{AC}^T(z)$ is the transpose of the AC matrix for the synthesis filters.

Equation (3.114) constitutes the requirements on the analysis and synthesis AC matrices for alias-free signal reconstructions in the broadest possible terms. If we impose the additional constraint of perfect reconstruction, the requirement becomes

$$H_{AC}(z)G_{AC}^T(z) = (Mz^{-n_0})diag\{1, W^{-n_0}, \cdots, W^{-(M-1)n_0}\}. \tag{3.115}$$

The PR requirements can be met by choosing the AC matrix to be lossless. The imposition of this requirement will allow us to derive time- and frequency-domain properties for the paraunitary filter bank. Thus, we want

$$\tilde{H}_{AC}(z)H_{AC}(z) = K(z) = MI$$

or

$$\frac{1}{M}\begin{bmatrix} H_0(z^{-1}) & \cdots & H_0(z^{-1}W^{-(M-1)}) \\ \vdots & \vdots & \vdots \\ H_{M-1}(z^{-1}) & \cdots & H_{M-1}(z^{-1}W^{-(M-1)}) \end{bmatrix}\begin{bmatrix} H_0(z) & \cdots & H_{M-1}(z) \\ \vdots & \vdots & \vdots \\ H_0(zW^{M-1}) & \cdots & H_{M-1}(zW^{M-1}) \end{bmatrix}$$

$$= [K_{rs}(z)] = I$$

or

$$K_{rs}(z) = \frac{1}{M}\sum_{j=0}^{M-1}\tilde{H}_r(zW^j)H_s(zW^j) = \delta_{rs}. \tag{3.116}$$

We will show that the necessary and sufficient conditions on filter banks satisfying the paraunitary condition are as follows. Let

$$\Phi_{rs}(z) \triangleq \tilde{H}_r(z)H_s(z) \leftrightarrow \rho_{rs}(n) = h_r(-n) * h_s(n). \tag{3.117}$$

Then

$$\rho_{rs}(Mn) = \begin{cases} 0, & r \neq s \\ \delta(n), & r = s. \end{cases} \tag{3.118}$$

We will first interpret these results, and then provide a derivation.

For $r = s$, we see that $\rho_{rr}(Mn) = \delta(n)$. Hence $\Phi_{rr}(z) = H_r(z^{-1})H_r(z)$ is the transfer function of an Mth *band* filter, Eq. (3.25), and $H_r(z)$ must be a spectral factor of $\Phi_{rr}(z)$. In the time-domain, the condition is

$$\rho_{rr}(Mn) = \sum_k h_r(k)h_r(k + Mn) = \delta(n), \tag{3.119}$$

which implies that the impulse response $h_r(n)$:

$$\begin{aligned} \sum_k |h_r(k)|^2 &= 1, \qquad \text{normalization} \\ \sum h_r(k)h_r(k + Mn) &= 0, \qquad n \neq 0. \end{aligned} \tag{3.120}$$

The latter asserts that $\{h_r(k)\}$ is orthogonal to its translates shifted by M. For $r \neq s$, we have $\rho_{rs}(Mn) = 0$, or

$$\sum_k h_r(k)h_s(k + Mn) = 0, \qquad \forall n. \tag{3.121}$$

This implies $\{h_r(k)\}$ is orthogonal to $\{h_s(k)\}$ and to all M translates of $\{h_s(k)\}$

$$\begin{aligned} \sum_k h_r(k)h_s(k) &= 0 \\ \sum_k h_r(k)h_s(k + Mn) &= 0. \end{aligned}$$

This condition corresponds to the off-diagonal terms in Eq. (3.116). It is a time-domain equivalent of aliasing cancellation.

The paraunitary requirement therefore imposes a set of orthonormality requirements on the impulse responses in the analysis filter bank and by Eq. (3.112) on the synthesis filters as well. Another version of this will be developed in Section 3.5.5 in conjunction with the polyphase matrix approach.

Another consequence of a paraunitary AC matrix is that the filter bank is *power complementary*, which means that

$$\begin{aligned} \frac{1}{M}\sum_{r=0}^{M-1} H_r(z)H_r(z^{-1}) &= 1, \quad \text{or} \\ \frac{1}{M}\sum |H_r(e^{j\omega})|^2 &= 1. \end{aligned} \tag{3.122}$$

To appreciate this, note that if $H_{AC}(z)$ is lossless, then $H_{AC}^T(z)$ is also lossless. Then $\tilde{H}_{AC}^T(z)H_{AC}(z) = MI$, and the first diagonal element is just

$$(\frac{1}{M})\sum_r H_r(z)H_r(z^{-1}).$$

Now for the proof of Eq. (3.118): First we define

$$\Phi_{rs}(zW^j) \triangleq \tilde{H}_r(zW^j)H_s(zW^j) = H_r(z^{-1}W^{-j})H_s(zW^j). \tag{3.123}$$

The following are Fourier transform pairs:

$$\begin{aligned} H_r(z) &\leftrightarrow h_r(n) \\ H_r(z^{-1}W^{-j}) &\leftrightarrow h_r(-n)W^{-jn} \\ H_s(zW^j) &\leftrightarrow h_s(n)W^{-jn}. \end{aligned}$$

The condition to be satisfied, Eq. (3.116), is

$$K_{rs}(z) = \frac{1}{M}\sum_{j=0}^{M-1}\Phi_{rs}(zW^j) = \delta_{rs}.$$

In the time-domain, this becomes

$$\frac{1}{M}\sum_{j=0}^{M-1} Z^{-1}\{\Phi_{rs}(zW^j)\} = \begin{cases} 0, & r \neq s \\ \delta(n), & r = s. \end{cases} \tag{3.124}$$

But

$$\begin{aligned} Z^{-1}\{\Phi_{rs}(zW^j)\} &= [h_r(-n)W^{-jn}] * [h_s(n)W^{-jn}] \\ &= \sum_k h_r(-k)W^{-jk}h_s(n-k)W^{-j(n-k)} \\ &= W^{-jn}\sum_k h_r(k)h_s(n+k) = W^{-jn}\rho_{rs}(n). \end{aligned} \tag{3.125}$$

Equation (3.124) becomes

$$\frac{1}{M}\sum_{j=0}^{M-1} W^{-jn}\rho_{rs}(n) = \frac{1}{M}\rho_{rs}(n)\left(\sum_{k=0}^{M-1} W^{kn}\right) = \begin{cases} 0, & r \neq s \\ \delta(n), & r = s \end{cases} \tag{3.126}$$

The sum in this last equation is recognized as the sampling function of Eq. (3.4)

$$\begin{aligned} \sum_{k=0}^{M-1} W^{-kn} &= \sum_{k=0}^{M-1}\left(e^{j\frac{2\pi}{M}n}\right)^k = \begin{cases} M, & n = 0, \pm M, \pm 2M, \cdots \\ 0, & \text{otherwise} \end{cases} \\ &= M\sum_{\nu=-\infty}^{\infty}\delta(n-\nu M). \end{aligned} \tag{3.127}$$

The product of this sampling function with $\rho_{rs}(n)$ leaves us with $\rho_{rs}(Mn)$ on the left-hand side of Eq. (3.126) which completes the proof.

On occasion, necessary conditions for a paraunitary filter bank are confused with sufficient conditions. Our solution, Eq. (3.118), implies a paraunitary filter bank. The Mth band filter requirement, Eq. (3.119), and the power complementary property of Eq. (3.122) are consequences of the paraunitary filter bank. Together they do *not* imply Eq. (3.116). The additional requirement of Eq. (3.121) must also be observed.

One can start with a prototype low-pass $H_0(z)$, satisfying the Mth band requirement $H_0(z)H_0(z^{-1}) = \Phi_{00}(z)$ and develop a bank of filters from

$$H_r(z) = z^{-n_0}H_0(z^{-1}W^{-r}).$$

This selection satisfies power complementarity and M^{th} band requirement, but is not necessarily paraunitary.

Another difficulty with this M^{th} band design is evident in this last equation. First, $H_r(z)$ can have complex coefficients resulting in complex subband signals. Secondly, as Vaidyanathan (April 1987) points out, the aliasing cancellation required by Eq. (3.116) for $r \neq s$ is difficult to realize when $H_0(z)$ is a sharp low-pass filter. It turns out that alias cancellation and sharp cutoff filters are largely incompatible in this design. For this reason we turn to alternate product-type realizations of lossless filter banks.

The Two-Band Case

To fix ideas, we particularize these results for the case $M = 2$ and demonstrate the consistency with the two-band paraunitary filter bank derived in Section 3.3. For alias cancellation from Eq. (3.114), we want (real coefficients are assumed throughout)

$$H_{AC}(z)G_{AC}^T(z) = \begin{bmatrix} H_0(z) & H_1(z) \\ H_0(-z) & H_1(-z) \end{bmatrix} \begin{bmatrix} G_0(z) & G_0(-z) \\ G_1(z) & G_1(-z) \end{bmatrix} = M \begin{bmatrix} T(z) & 0 \\ 0 & T(-z) \end{bmatrix}$$

or

$$\begin{aligned} H_0(z)G_0(z) + H_1(z)G_1(z) &= MT(z) \\ H_0(-z)G_0(z) + H_1(-z)G_1(z) &= 0. \end{aligned} \tag{3.128}$$

And, for perfect reconstruction, we set $T(z) = z^{-n_0}$.

The paraunitary analysis filters must obey

$$\begin{bmatrix} H_0(z^{-1}) & H_0(-z^{-1}) \\ H_1(z^{-1}) & H_1(-z^{-1}) \end{bmatrix} \begin{bmatrix} H_0(z) & H_1(z) \\ H_0(-z) & H_1(-z) \end{bmatrix} = 2 \begin{bmatrix} 1 & 0 \\ 0 & 1 \end{bmatrix}$$

$$H_\nu(z^{-1})H_\nu(z) + H_\nu(-z^{-1})H_\nu(-z) = 2, \qquad \nu = 0,1 \tag{3.129}$$

$$H_0(z^{-1})H_1(z) + H_0(-z^{-1})H_1(-z) = 0. \tag{3.130}$$

Let $\rho_\nu(n)$, $\Phi_\nu(z)$ be an autocorrelation function and spectral density function for $h_\nu(n)$

$$\Phi_\nu(z) \triangleq H_\nu(z)H_\nu(z^{-1}) \longleftrightarrow \rho_\nu(n) = h_\nu(n) * h_\nu(-n) \qquad \nu = 0,1. \tag{3.131}$$

Consequently $\rho_\nu(n)$ is an even function. The paraunitary condition becomes

$$\Phi_\nu(z) + \Phi_\nu(-z) = 2 \longleftrightarrow \rho_\nu(n) + (-1)^n \rho_\nu(n) = 2\delta(n) \tag{3.132}$$

or

$$[1 + (-1)^n] \rho_\nu(n) = 2\delta(n).$$

But for n odd, $[1 + (-1)^n] = 0$, which leaves us with

$$2\rho_\nu(n) = 2\delta(n) = \begin{cases} 2, & n = 0 \\ 0, & n = \pm 2, \pm 4, \cdots \end{cases} \tag{3.133}$$

Hence the first paraunitary requirement is stated succinctly as

$$\rho_\nu(2n) = \sum_k h_\nu(2n+k)h_\nu(k) = \delta(n) \tag{3.134}$$

or

$$\rho_\nu(0) = \sum_k |h_\nu(k)|^2 = 1$$

$$\sum_k h_\nu(2n+k)h_\nu(k) = 0, \qquad n \neq 0. \tag{3.135}$$

This last equation asserts that the impulse response of each filter $\{h_k(n)\}$ is orthogonal to its even translates and has unit norm—a general property for the two-band paraunitary filter bank. Also, we can see that the correlation function $\rho_\nu(n)$ with even samples (except $n = 0$) equal to zero is precisely a half-band filter defined in Eq. (3.25) with $M = 2$.

In a similar fashion, Eq. (3.130) can be expressed in the time-domain using the cross-correlation $\rho_{10}(n)$ and its transform $\Phi_{10}(z)$

$$\Phi_{10}(z) \triangleq H_0(z)H_1(z^{-1}) \longleftrightarrow \rho_{10}(n) = h_1(-n) * h_0(n) = \sum_k h_1(k)h_0(n+k). \tag{3.136}$$

The second requirement becomes

$$\Phi_{10}(z) + \Phi_{10}(-z) = 0 \longleftrightarrow \rho_{10}(n) + (-1)^n \rho_{10}(n) = 0. \tag{3.137}$$

Following a similar line of reasoning, we can conclude

$$\rho_{10}(2n) = 0, \qquad n = 0, \pm 1, \pm 2, \cdots \tag{3.138}$$

and in particular

$$\rho_{10}(0) = \sum_k h_0(k)h_1(k) = 0. \tag{3.139}$$

This demonstrates that the paraunitary impulse responses $\{h_0(k)\}$ and $\{h_1(k)\}$ are orthogonal to each other (and orthogonal to their even-indexed translates). Having selected $H_0(z)$ as an N-tap FIR filter (N even), we can then choose

$$H_1(z) = z^{-(N-1)}H_0(-z^{-1})$$

to satisfy the paraunitary requirement. Then the synthesis filters from Eq. (3.110) are

$$\begin{aligned} G_0(z) &= z^{-(N-1)}H_0(z^{-1}) \\ G_1(z) &= z^{-(N-1)}H_1(z^{-1}). \end{aligned} \tag{3.140}$$

These relationships are summarized in the block diagram of Fig. 3.41 and the time-domain sketches shown in Fig. 3.42. Note that $H_1(z)$ is quadrature to $G_0(z)$ and $H_0(z)$ quadrature to $G_1(z)$. In the time-domain, we have

$$\begin{aligned} h_1(n) &= (-1)^{n+1}h_0(N-1-n) \\ g_0(n) &= h_0(N-1-n) \\ g_1(n) &= (-1)^n h_0(n). \end{aligned} \tag{3.141}$$

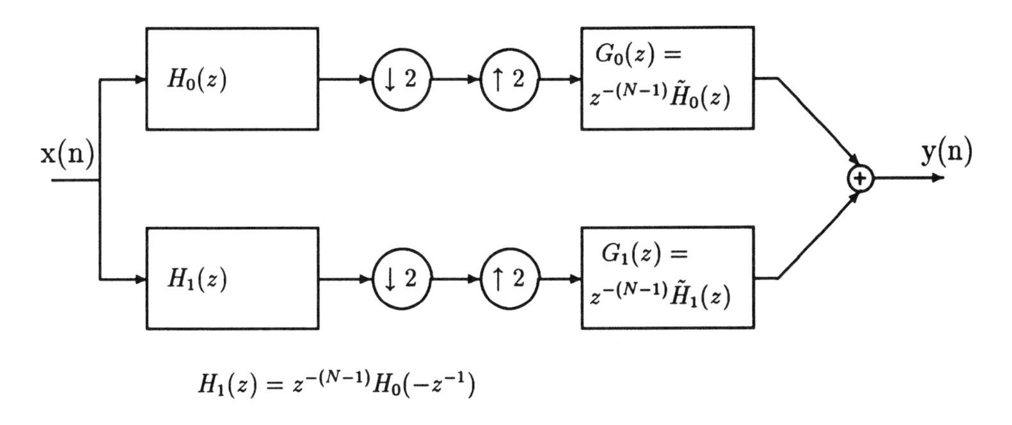

Figure 3.41: Two-band paraunitary filter bank.

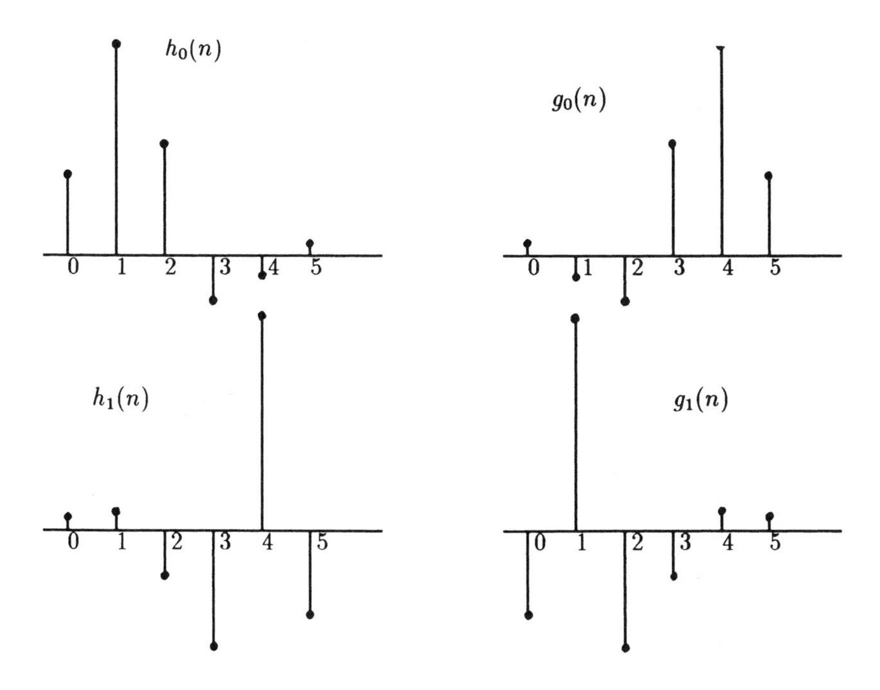

Figure 3.42: Filter responses for two-band, 6-tap Binomial PR-QMF.

The Binomial QMF bank presented in Section 4.1 is an example of a PR-QMF family. Impulse responses of this family for $N = 6$ are illustrated in Fig. 3.42.

Note that $H_0(z)$ and $G_0(z)$ are low-pass filters, whose magnitude responses are equal,

$$|H_0(e^{j\omega})| = |G_0(e^{j\omega})| = A_0(\omega).$$

Similarly $H_1(z)$ and $G_1(z)$ are high-pass filters with equal magnitude responses.

$$|H_1(e^{j\omega})| = |G_1(e^{j\omega})| = A_1(\omega).$$

But the low-pass and high-pass filters are also quadrature mirror filters, so that

$$A_1(\omega) = A_0(\omega - \pi).$$

But, of course, this is just the power complementary property of a paraunitary filter bank,

$$A_0^2(\omega) + A_1^2(\omega) = \text{constant}.$$

Finally we note that $\rho_\nu(2n) = 0$ for $n \neq 0$ is precisely the same as the half-band filter introduced in Eq. (3.23). Hence $H_0(z)$ is a spectral factor of $\Phi_0(z) = H_0(z)H_0(z^{-1})$, where $\Phi_0(z)$ is a half-band filter.

3.5.5 Time-Domain Representations

In this section we develop the properties of the PR filter bank in the time-domain and connect these with the extended lapped orthogonal transform. The works of Vetterli (1987), Malvar (Elec. Letts., 1990), and Nayebi, Barnwell, and Smith (1992) can then be viewed from a common, unified standpoint. In order to demonstrate the commonality of these seemingly disparate approaches, we need to introduce yet another equivalent structure for the maximally decimated QMF bank, as shown in Fig. 3.43.

The matrices P and Q are constant and of size $(NM \times M)$ and $(N'M \times M)$, respectively. The input data stream $\{x(n)\}$ is fed into a buffer. After $(NM - 1)$ clock pulses the input buffer is filled. It consists of N blocks of M data points each.

From the top to bottom at the input in Fig. 3.43, we define this input vector of NM points as $\underline{\psi}(n)$ and partition it into N vectors (or blocks) of M points

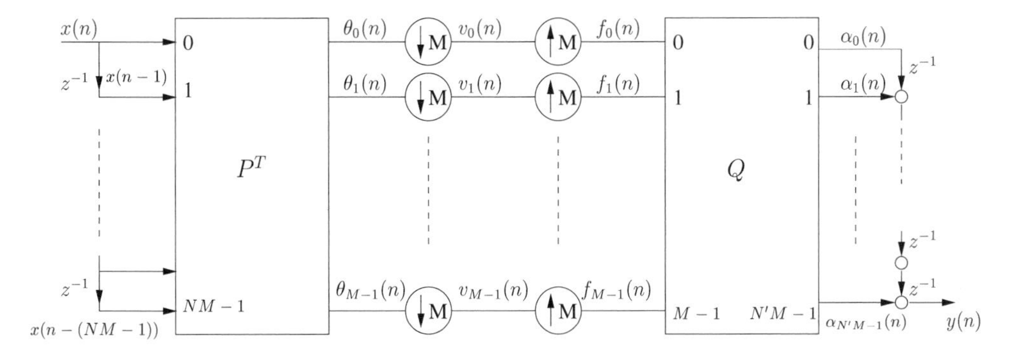

Figure 3.43: Multiblock transform representation.

each, as follows:

$$\underline{\psi}(n) = \begin{bmatrix} x(n) \\ x(n-1) \\ \vdots \\ x(n-(M-1)) \\ \cdots\cdots\cdots \\ x(n-M) \\ \vdots \\ x(n-(2M-1)) \\ \cdots\cdots\cdots \\ \vdots \\ \cdots\cdots\cdots \\ x(n-(N-1)M) \\ \vdots \\ x(n-(NM-1)) \end{bmatrix} = \begin{bmatrix} \underline{x}(n) \\ \underline{x}(n-M) \\ \vdots \\ \underline{x}(n-(N-1)M) \end{bmatrix} \tag{3.142}$$

where,

$$\underline{x}^T(n) \stackrel{\Delta}{=} [x(n), x(n-1), \cdots, x(n-(M-1))]. \tag{3.143}$$

This input vector is multiplied by P^T to give the coefficient vector $\underline{\theta}(n)$,

$$\underline{\theta}(n) = P^T \underline{\psi}(n) = \begin{bmatrix} P_0^T & P_1^T \cdots, P_{N-1}^T \end{bmatrix} \begin{bmatrix} \underline{x}(n) \\ \vdots \\ \underline{x}(n-(N-1)M) \end{bmatrix}, \tag{3.144}$$

where each P_k^T is an $(M \times M)$ block. This can be written as

$$\underline{\theta}(n) = \sum_{k=0}^{N-1} P_k^T \underline{x}(n - kM). \tag{3.145}$$

The subsampling by M transfers $\underline{\theta}$ to $\underline{v}$. During the interval until the next subsampling, M new input samples are entered into the stack buffer and the process repeated. We represent this by sampling $\underline{\theta}(n)$ at times Mn to get $\underline{v}(n)$

$$\underline{v}(n) = \underline{\theta}(Mn) = P^T \underline{\psi}(Mn). \tag{3.146}$$

From Fig. 3.36, we see that

$$\underline{x}(Mn) = \underline{\xi}(n). \tag{3.147}$$

Therefore, it follows that

$$\underline{v}(n) = P^T \underline{\psi}(Mn) = \sum_{k=0}^{N-1} P_k^T \underline{x}[M(n-k)] = \sum_{k=0}^{N-1} P_k^T \underline{\xi}(n-k). \tag{3.148}$$

This is like a block FIR filter, where the filter coefficients are the $(M \times M)$ matrices P_k^T and the signals are $(M \times 1)$ vectors $\underline{\xi}(k)$. After $(N-1)$ samples at the slow clock rate, there is an entirely new data vector in the input buffer. We can stack the successive outputs (starting at $n = 0$ for convenience):

$$\begin{bmatrix} \vdots \\ \underline{v}(0) \\ \underline{v}(1) \\ \vdots \\ \underline{v}(N-1) \\ \vdots \end{bmatrix} = \begin{bmatrix} \cdots & \cdot & \cdot & \cdots & \cdot & \cdot & \cdot & \cdots & \cdot & \cdots \\ \cdots & P_{N-1}^T & P_{N-2}^T & \cdots & P_0^T & 0 & 0 & \cdots & 0 & \cdots \\ \cdots & 0 & P_{N-1}^T & \cdots & P_1^T & P_0 & 0 & \cdots & 0 & \cdots \\ \vdots & \vdots & \vdots & & & & & & & \vdots \\ \cdots & 0 & 0 & \cdots & P_{N-1}^T P_{N-2}^T & \cdot & \cdot & \cdots & P_0^T & \cdots \end{bmatrix}$$

$$\times \begin{bmatrix} \vdots \\ \underline{\xi}(-(N-1)) \\ \vdots \\ \underline{\xi}(0) \\ \vdots \\ \underline{\xi}(N-1) \\ \vdots \end{bmatrix}. \tag{3.149}$$

The preceding $NM \times (2N-1)M$ transmission matrix is denoted as Π^T.

To relate this representation to the filter bank structure, we see from Fig. 3.43 and Eq. (3.143) that

$$\underline{\theta}(z) = P^T \begin{bmatrix} 1 \\ z^{-1} \\ \vdots \\ z^{-(NM-1)} \end{bmatrix} X(z) = (P^T \underline{Z}_{NM})X(z) \triangleq \underline{h}(z)X(z)$$

or

$$\underline{h}(z) = \begin{bmatrix} H_0(z) \\ \vdots \\ H_{M-1}(z) \end{bmatrix} = P^T \underline{Z}_{NM} \tag{3.150}$$

where the delay vector $\underline{Z}_{MN}$ as defined in Eq. (3.82) is

$$\underline{Z}_{MN}^T = \left[1, z^{-1}, \cdots, z^{-(NM-1)}\right]^T$$

and

$$H_r(z) = \sum_{k=0}^{NM-1} P_{rk}^T z^{-k}, \qquad r = 0, 1, ..., M-1. \tag{3.151}$$

This equivalence states that the block structure from $x(n)$ to $\underline{\theta}(n)$ in Fig. 3.43 can be replaced by the analysis filter bank $\underline{h}(z)$ as shown in Fig. 3.32, where now the length of each filter is NM. From the filter bank we then can obtain the polyphase representation of Fig. 3.35 or 3.37. From our previous derivation, we can now connect the extended block transform, the polyphase decomposition, and the filter bank by

$$\underline{h}(z) = P^T \underline{Z}_{NM} = \mathcal{H}_p(z^M)\underline{Z}_M. \tag{3.152}$$

Now using the polyphase representation of Fig. 3.36, we get

$$\underline{V}(z) = \mathcal{H}_p(z)\underline{\xi}(z). \tag{3.153}$$

Every element in the matrix $\mathcal{H}_p(z)$ is a polynomial in z^{-1} of degree $N-1$. Therefore we can expand $\mathcal{H}_p(z)$ as polynomials in z^{-1} with matrix coefficients. Substituting this expansion into Eq. (3.153) and converting to the time-domain,

$$\begin{aligned} \underline{V}(z) &= \mathcal{H}_p(z)\underline{\xi}(z) = \left(\sum_{k=0}^{N-1} \mathcal{H}_{p,k} z^{-k}\right)\underline{\xi}(z) \\ \underline{v}(n) &= \sum_{k=0}^{N-1} \mathcal{H}_{p,k}\underline{\xi}(n-k). \end{aligned} \tag{3.154}$$

Note that each $\mathcal{H}_{p,k}$ is a constant, $M \times M$ matrix.

Comparing this last expansion with that in Eq. (3.148) we see that the matrix coefficients in the polyphase expansion are simply related to the block transforms:

$$\mathcal{H}_{p,k} = P_k^T, \qquad k = 0, 1, \cdots, N-1. \tag{3.155}$$

In summary then, we have three equivalent representations for the analysis side of the subband coder: the filter bank, the polyphase decomposition, and the extended block transform, as related by Eqs. (3.152) and (3.155).

Next, we can develop a similar set of equivalences at the synthesis side. From Fig. 3.43 we partition $\underline{\alpha}(n)$, the $N'M$ sample output vector into N' blocks of M samples each, $\alpha_r(n) = [\alpha_{rM}(n), \alpha_{rM+1}(n), \ldots, \alpha_{rM+M-1}(n)]^T$, $r = 0, 1, \ldots, N' - 1$. This allows us to write

$$\underline{\alpha}(n) = \begin{bmatrix} \underline{\alpha}_0(n) \\ \underline{\alpha}_1(n) \\ \vdots \\ \underline{\alpha}_{N'-1}(n) \end{bmatrix} = \begin{bmatrix} Q_0 \\ Q_1 \\ \vdots \\ Q_{N-1} \end{bmatrix} \underline{f}(n), \tag{3.156}$$

where each Q_k is an $M \times M$ block. The up-sampler imposes

$$\underline{f}(n) = \begin{cases} \underline{v}(n/M), & n = 0, \pm M, \cdots \\ 0, & \text{otherwise} \end{cases} \tag{3.157}$$

Since $\underline{f}(n) = \underline{0}$ between up-sampled data points, we need only to evaluate $\underline{\alpha}(n)$ at times Mn to obtain

$$\underline{\alpha}(Mn) = \begin{bmatrix} \underline{\alpha}_0(Mn) \\ \underline{\alpha}_1(Mn) \\ \vdots \\ \underline{\alpha}_{N'-1}(Mn) \end{bmatrix} = \begin{bmatrix} Q_0 \\ Q_1 \\ \vdots \\ Q_{N'-1} \end{bmatrix} \underline{f}(Mn) = \begin{bmatrix} Q_0 \\ Q_1 \\ \vdots \\ Q_{N'-1} \end{bmatrix} \underline{v}(n). \tag{3.158}$$

Now let us examine the output sequence $\{y(n)\}$ in batches of M samples. Define

$$\underline{y}(n) = [y(n+M-1), \ldots, y(n+1), y(n)]^T. \tag{3.159}$$

Noting that $\underline{\alpha}(n) = 0$ except for integer multiples of M, we can express $\underline{y}(Mn)$ in terms of the $\underline{\alpha}$'s by

$$\begin{aligned} \underline{y}(Mn) &= \underline{\alpha}_{N'-1}(Mn) + \underline{\alpha}_{N'-2}[M(n-1)] + \ldots + \underline{\alpha}_0[M(n-(N'-1))]. \\ \textit{But} \ \ \underline{\alpha}_i(n) &= Q_i \underline{f}(n), \\ \textit{and} \ \ \underline{\alpha}_i(Mn) &= Q_i \underline{f}(Mn) = Q_i \underline{v}(n). \end{aligned}$$

Therefore,

$$\underline{y}(Mn) = \sum_{k=0}^{N'-1} Q_{N'-K-1}\underline{v}(n-k). \tag{3.160}$$

And stacking these for $n = 0, 1, \cdots, N'-1$, gives the synthesis block transmission form

$$\begin{bmatrix} \vdots \\ \underline{y}(0) \\ \underline{y}(M) \\ \vdots \\ \underline{y}(Mn) \\ \vdots \end{bmatrix} = \begin{bmatrix} \cdots & \cdot & \cdot & \cdots & \cdot & \cdot & & & & \\ & Q_0 & Q_1 & \cdots & Q_{N'-1} & 0 & 0 & \cdot & \cdots & 0 \\ & & Q_0 & \cdots & Q_{N'-2} & Q_{N'-1} & 0 & \cdot & \cdots & 0 \\ & & & & & & & & & \\ & & & & Q_0 & Q_1 & & Q_{N'-1} & \cdots & 0 \\ & & & & \cdot & \cdot & \cdot & \cdot & \cdots & \end{bmatrix} \times \begin{bmatrix} \vdots \\ \underline{v}(-(N'-1)) \\ \vdots \\ \underline{v}(0) \\ \vdots \\ \underline{v}(N'-1) \\ \vdots \end{bmatrix} \tag{3.161}$$

The preceding $N'M \times (2N'-1)M$ synthesis transmission matrix[1] is denoted by $\sum^T$.

To obtain the filter bank equivalent, we note that

$$Y(z) = \begin{bmatrix} z^{-(N'M-1)}, & \ldots, & z^{-1}, & 1 \end{bmatrix} \begin{bmatrix} \alpha_0(z) \\ \vdots \\ \alpha_{N'M-1}(z) \end{bmatrix} = \underline{\alpha}^T(z)\underline{Z}^T_{N'M}.$$

But

$$\underline{\alpha}(z) = Q\underline{F}(z)$$

Hence,

$$\begin{aligned} Y(z) &= \underline{\alpha}^T(z)\underline{Z}'_{N'M} = \underline{F}^T(z)[Q^T\underline{Z}'_{N'M}] = \underline{F}^T(z)\underline{g}(z) \\ \underline{g}(z) &= Q^T\underline{Z}'_{N'M}. \end{aligned} \tag{3.162}$$

[1]Some authors employ these transmission matrices Π^T and $\sum^T$ as fundamental analytic descriptors. We make no further use of them.

The connection with the polyphase representation is now evident. From Eq. (3.94),

$$\underline{g}(z) = \mathcal{G}_p(z^M)\underline{Z}_M = Q^T \underline{Z}'_{N'M} \tag{3.163}$$

$$G_r(z) = \sum_{k=0}^{N'M-1} Q^T_{rk} z^{-(N'M-1-k)}, \qquad r = 0, 1, \ldots, M-1.$$

Finally, from Fig. 3.36, we note that $\underline{\eta}(n) = \underline{y}(Mn)$, and also that

$$\underline{\eta}(z) = \mathcal{G}'_p(z)\underline{V}(z) = (\sum_{k=0}^{N'-1} \mathcal{G}'_{p,k} z^{-k})\underline{V}(z)$$

or

$$\underline{\eta}(n) = \sum_{k=0}^{N'-1} \mathcal{G}'_{p,k}\underline{v}(n-k) = \underline{y}(Mn). \tag{3.164}$$

Comparing this last equation with Eq. (3.160) shows that the synthesis blocks are the matrix coefficients in the $\mathcal{G}'_p(z)$ expansion

$$\mathcal{G}'_{p,k} = Q_{N'-1-k}. \tag{3.165}$$

This completes the equivalences at the synthesis side. We have a complete set of time-domain and transform domain representations. We can now turn to the prime objective, i.e., the PR conditions.

From Fig. 3.36 and Eqs. (3.154) and (3.164),

$$\begin{aligned}
\underline{\eta}(z) &= [\mathcal{G}'_p(z)\mathcal{H}_p(z)]\underline{\xi}(z) = \mathcal{P}(z)\underline{\xi}(z) \\
\mathcal{P}(z) &= \left(\sum_{k=0}^{N'-1} \mathcal{G}'_{p,k} z^{-k}\right)\left(\sum_{j=0}^{N-1} \mathcal{H}_{p,j} z^{-j}\right) = \sum_{k=0}^{N'-1}\sum_{j=0}^{N-1}\left(\mathcal{G}'_{p,k}\mathcal{H}_{p,j} z^{-(k+j)}\right) \\
&= \sum_{r=0}^{N+N'-2} \mathcal{P}_r z^{-r}.
\end{aligned} \tag{3.166}$$

Here, $\mathcal{P}_r$ is recognized as the convolution of the polyphase coefficient matrices (or the correlation of the block transforms P^T_k and Q_k)

$$\begin{aligned}
\mathcal{P}_r = \sum_{k=0}^{N'-1} \mathcal{G}'_{p,k}\mathcal{H}_{p,r-k} &= \sum_{k=0}^{N'-1} Q_{N'-1-k} P^T_{r-k} \\
&= \sum_{k=0}^{N'-1} Q_k P^T_{k+r-(N-1)}.
\end{aligned} \tag{3.167}$$

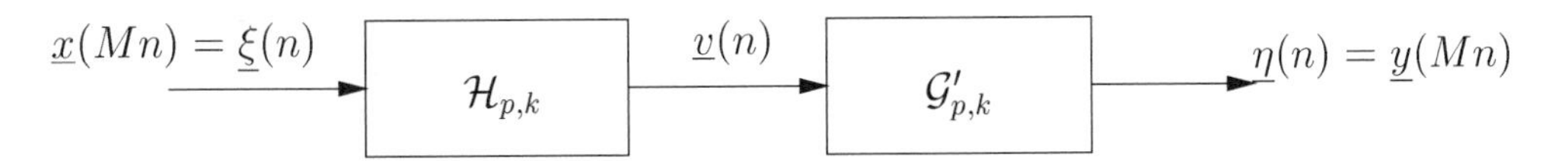

Figure 3.44: Time-domain equivalent representation.

Figure 3.44 shows the time-domain representation of the subband system. The terms $\{\mathcal{H}_{p,k}\}$ and $\{\mathcal{G}'_{p,k}\}$ are matrix weighting sequences for the analysis and synthesis banks, respectively. The convolution of these gives the overall matrix weighting sequence $\mathcal{P}_r$, which relates the block input and output sequences. Thus, from Eq. (3.166),

$$\underline{\eta}(n) = \sum_{r=0}^{N'+N-2} \mathcal{P}_r \underline{\xi}(n-r)$$

or

$$\underline{y}(Mn) = \sum_{r=0}^{N'+N-2} \mathcal{P}_r \underline{x}[(n-r)M]. \tag{3.168}$$

We saw in Fig. 3.40 and Eq. (3.101) that there is an inherent delay of $(M-1)$ fast-clock samples induced by the down-samplers and up-samplers, and additionally, a delay of $M\mu$ due to the filter bank. We have also accounted for the $(M-1)$ delay in the formulation of the input and output blocks, $\underline{y}(Mn)$ and $\underline{x}(Mn)$, as can be seen in Eqs. (3.143) and (3.159). Thus if $\mathcal{P}(z) = \mathcal{P}_0 = I$, we would have $\underline{y}(Mn) = \underline{x}(Mn)$, or

$$y(n) = x[n-(M-1)].$$

From Eq. (3.167) the sufficient condition for PR is

$$\mathcal{P}_r = I\delta(r-\mu) = \begin{cases} I & r = \mu \\ 0 & r \neq \mu \end{cases}. \tag{3.169}$$

There are several possibilities as to which $\mathcal{P}_r$ should be chosen as I. The condition developed here is for the case that $(N'-1)+(N-1) = (N'+N-2)$ is even. We can then choose the center term in the expansion to be the non-zero element; i.e.,

$$\mu = \frac{N'+N-2}{2}.$$

Then

$$\mathcal{P}(z) = z^{-\mu} I_{M\times M}$$

and

$$\underline{y}(Mn) = \underline{x}[Mn - M\mu] = \underline{x}[M(n-\mu)]$$

or

$$y(n) = y(n - n_0) \qquad n_0 = M\mu + (M-1).$$

The implementation of this selection is best illustrated by a simple example. Suppose $N = 3$, $N' = 5$, then

$$\mathcal{P}(z) = \mathcal{P}_0 + \mathcal{P}_1 z^{-1} + \ldots + \mathcal{P}_6 z^{-6}.$$

Setting the center term $\mathcal{P}_3(z) = I_{M\times M}$ and the others to zero in effect asserts that the correlation of Q_k and P_k^T is a delayed impulse $I\delta(n-3)$. This requirement is similar to the requirement for deconvolution of two weighting functions.

For the example chosen we can arrange the matrices as follows for $r = 0$.

$$\begin{array}{cccccccc} & & Q_4 & Q_3 & Q_2 & Q_1 & Q_0 \\ P_2^T & P_1^T & P_0^T & \rightarrow r & & & \end{array}$$

As r increases, the sequence $\{P_k^T\}$ advances to the right. Our requirement is that this correlation be zero for $r = 0, 1, 2$; equal to I for $r = 3$; and then equal to zero for $r = 4, 5, 6$. This leads to the set of equations

$$\begin{array}{lclclcl} \mathcal{P}_6 & = & Q_0 P_2^T = 0 & & \mathcal{P}_2 & = & Q_2 P_0^T + Q_3 P_1^T + Q_4 P_2^T = 0 \\ \mathcal{P}_5 & = & Q_1 P_2^T + Q_0 P_1^T = 0 & & \mathcal{P}_1 & = & Q_3 P_0^T + Q_4 P_1^T = 0 \\ \mathcal{P}_4 & = & Q_0 P_0^T + Q_1 P_1^T + Q_2 P_2^T = 0 & & \mathcal{P}_6 & = & Q_4 P_0^T = 0 \end{array}$$

$$\mathcal{P}_3 = Q_1 P_0^T + Q_2 P_1^T + Q_3 P_2^T = I_{M\times M}.$$

If $(N' + N - 2)$ is not even, we can choose either of the two center terms to be the identity matrix. For equal length analysis and synthesis filters, $N' = N$, we can satisfy Eq. (3.169) with $\mu = N - 1$, the middle term.

For the latter case, we can say the PR condition is realized if the synthesis matrix $\{Q_{N'-1-k}\}$ is orthogonal to block translates of the analysis matrix sequence $\{P_k^T\}$. Examples of a design based on this approach are given in Nayebi, Barnwell, and Smith (1990).

The foregoing conditions are very broad and, as such, are somewhat difficult to interpret or realize. We can, however, develop some special cases of the preceding and, in the process, obtain several results reported in the literature. The first simplification is one already mentioned, equal-length analysis and synthesis filters.

Next, we can impose orthogonality conditions on the analysis blocks themselves of the form

$$\sum_{k=0}^{N-1-r} P_{k+r}P_k^T = B\delta(r), \qquad r = 0, 1, \cdots, N-1, \tag{3.170}$$

where B is a symmetric positive definite matrix. This condition has been termed *orthogonality of the tails* by Vetterli and LeGall (1989). When this is imposed, the transmission matrix of Eq. (3.149) has the property that $\Pi\Pi^T$ is block diagonal

$$\Pi\Pi^T = diag\{B, \cdots, B\}.$$

Basically, this means that the impulse responses of the analysis filters are independent. Now we can select the synthesis blocks to be

$$Q_k = B^{-1}P_k. \tag{3.171}$$

This choice, along with Eq. (3.170), satisfies the PR requirement, since

$$\begin{aligned} \mathcal{P}_r &= \sum_{k=0}^{N-1} Q_{N-1-k}P_{r-k}^T = \sum_{k=0}^{N-1} B^{-1}P_{N-1-k}P_{r-k}^T \\ &= B^{-1}B\delta(N-1-r) = I\delta[r-(N-1)]. \end{aligned} \tag{3.172}$$

In the transform-domain, this choice implies that

$$\begin{aligned} [\mathcal{H}_p(z^{-1})]^T[\mathcal{H}_p(z)] &= \left(\sum_{j=0}^{N-1} P_j z^j\right)\left(\sum_{k=0}^{N-1} P_k^T z^{-k}\right) \\ &= \sum_{n=-(N-1)}^{N-1} z^n \Big(\sum_{k=0}^{N-1-n} P_{k+n}P_k^T\Big) = \sum_{n=-(N-1)}^{N-1} z^n B\delta(n) \\ &= B. \end{aligned} \tag{3.173}$$

Also, the synthesis polyphase matrix is

$$\begin{aligned} \mathcal{G}_p'(z) &= \sum_{k=0}^{N-1} z^{-k}Q_{N-1-k} = \sum_{k=0}^{N-1} z^{-k}B^{-1}P_{N-1-k} \\ &= z^{-(N-1)}B^{-1}\mathcal{H}_p^T(z^{-1}). \end{aligned} \tag{3.174}$$

Finally, let $B = I$. This makes the analysis bank paraunitary, since we now have

$$\mathcal{H}_p^T(z^{-1})\mathcal{H}_p(z) = I.$$

The lapped orthogonal transforms and extensions of it are a special case of the foregoing. For this case, Malvar (Elect. Letts., 1990) takes $N' = N$, chooses Q equal to P and imposes the orthogonality condition of Eq. (3.170). Hence, these lapped transforms are paraunitary filter banks. Finally, the LOT is itself a special case with $N = 2$. For this case, the paraunitary constraints are

$$P^T = \begin{bmatrix} P_0^T & P_1^T \end{bmatrix}, \qquad Q = P$$

$$P_0 P_0^T + P_1 P_1^T = I \quad \text{and} \quad P_0^T P_1 = 0. \tag{3.175}$$

In addition, other constraints were imposed (Malvar, ASSP 1990): M is even, and the filters have linear phase so there are $N/2$ symmetric and $N/2$ antisymmetric filters, as discussed in Chapter 2.

3.5.6 Modulated Filter Banks

These were originally proposed as a bank of frequency-translated filters fashioned from a low-pass prototype. Vetterli and LeGall (1989) and Malvar (ASSP 1990) point out that these can be made PR by the suitable choice of modulating function.

The bank of analysis filters is chosen to be

$$h_k(n) = h(n)\cos\{\omega_k(n - \frac{N-1}{2}) + \phi_k\} \qquad k = 0, 1, \cdots, M-1, \tag{3.176}$$

where ω_k, ϕ_k are respectively the modulating frequency and the phase shift in the kth band, and $h(n)$ is the prototype, low-pass filter. The frequency ω_k is chosen to evenly span the frequency band, and then ϕ_k is picked to eliminate aliasing:

$$\begin{aligned} \omega_k &= \frac{\pi}{M}(k + 1/2) \\ \phi_k &= \frac{\pi}{4} + k\frac{\pi}{2}. \end{aligned} \tag{3.177}$$

The synthesis filters $\{g_k(n)\}$ have the same form as $\{h_k(n)\}$, except that the phase is $-\phi_k$.

The Princen-Bradley filters (Section 4.7) can be obtained from this prototype for the special case that $N = 2M$ (in this respect similar to LOT in filter size). The requirements for PR are that the low-pass prototype satisfy

$$\begin{aligned} h(n) &= h(2M - n - 1), && \text{symmetry} \\ h^2(n) + h^2(n + M) &= 1, && \text{paraunitary.} \end{aligned} \tag{3.178}$$

Proof: For the unwindowed filter bank, $(h(n) = 1)$, the PR filter bank of length $2M$ must satisfy Eq. (3.175). In the present context, the P^T matrix is

$$P^T(n,k) = \cos\{\omega_k(n - \frac{N-1}{2}) + \phi_k\}$$

But the modulating matrix has the property

$$\begin{aligned} P_0 P_0^T &= \frac{1}{2}(I + J) \\ P_1 P_1^T &= \frac{1}{2}(I - J). \end{aligned} \tag{3.179}$$

Let us now insert the symmetric prototype $h(n)$ and define

$$D = diag\{h(0), h(1), \cdots, h(M-1)\}. \tag{3.180}$$

The modulated matrices become

$$(P_0')^T \triangleq P_0^T D \quad and \quad (P_1')^T \triangleq P_1^T(JDJ). \tag{3.181}$$

It is easy to show that

$$P_0'(P_1')^T = DP_0P_1^T(JDJ) = 0$$

and

$$\begin{aligned} P_0'(P_0')^T + P_1'(P_1')^T &= DP_0P_0^TD + (JDJ)P_1P_1^T(JDJ) \\ &= \frac{1}{2}D(I+J)D + \frac{1}{2}(JDJ)(I-J)(JDJ) \\ &= \frac{1}{2}[D^2 + JD^2J] = I, \end{aligned} \tag{3.182}$$

where D^2 is a diagonal matrix and JD^2J is the same as D^2 but with an interchange of elements [see Eq. (3.89)]. Hence

$$h^2(n) + h^2(M-1-n) = 2. \tag{3.183}$$

By symmetry, the latter can also be represented as

$$h^2(n) + h^2(n+M) = 2.$$

And we note that windowing has not destroyed the orthogonality of overlapping blocks. Equation (3.183) is a time-domain counterpart to the half-band filter property,

$$A_0^2(\omega) + A_0^2(\omega - \pi) = 2.$$

The modulated lapped transform (MLT)(Malvar, ASSP 1990) has essentially the same structure as the foregoing with the same requirements, Eq. (3.178). Additionally, a particular window function,

$$h(n) = \sin \frac{\pi}{2M}(n + 1/2)$$

is chosen for the prototype.

A key advantage to this modulated bank is that fast transform methods can be used to implement the cosine terms in a computationally efficient manner. More recently, Malvar has developed an extended version of the foregoing called the Extended MLT. This generalization—the filter lengths are $N = 2KM$—is essentially the same modulated filter bank structure as in Eq. (3.176).

For the extended modulated lapped transform, Malvar sets $N = 2KM$, and as before employs the paraunitary P, and sets $Q = P$. Thus instead of just one block overlap as in Eqs. (3.176)–(3.183), we now have a K block overlap.

In Eq. (3.181), each P_r block was represented as the product of a diagonal window matrix D and the modulation matrix. In the extended case, this is repeated for each of the $2K$ blocks P_r, such that

$$P_r = H_r \Phi_r, \tag{3.184}$$

where P_r, Φ_r are $M \times M$ matrices

$$\begin{aligned} \Phi_r(n,k) &= \sqrt{\frac{2}{M}} \cos\{\omega_k(n + rM + \frac{M-1}{2})\} \\ \omega_k &= \frac{\pi}{M}(k + \frac{1}{2}) \\ H_r &= diag\{h(rM), h(rM+1), ..., h(rM+M-1)\}. \end{aligned}$$

Applying the paraunitary constraint of Eq. (3.170)(with $B = I$), leads to

$$\sum_{i=0}^{2K-1-2m} H_i H_{i+2m} = \delta(m) I. \tag{3.185}$$

This can be viewed as a matrix generalization of the two-band paraunitary filter bank in Eq. (3.134). Here, the matrix impulse response $\{H_i\}$ is orthogonal to its even translates and has a unit norm. The scalar form of this orthogonality condition is the set of nonlinear equations, which can be solved for the window function $h(n)$.

$$\sum_{i=0}^{2K-1-2m} h(n + iM) h(n + iM + 2mM) = \delta(m). \tag{3.186}$$

It is reported that the *MLT* is superior to the *DCT* in several respects: the *MLT* is free of blocking effects, requires somewhat fewer memory locations, fewer coefficients, and is comparable in computational complexity, and independent of the sign of the correlation coefficient for an AR(1) source (Malvar, June 1990). Lattice realization of PR modulated filter bank is provided in the next section.

3.6 Cascaded Lattice Structures

In the previous section, we developed PR requirements based on a finite series expansion of the polyphase matrix as in Eq. (3.154). This led to constraints on the analysis and synthesis polyphase matrices, special cases of which resulted in paraunitary filter banks, modulated filter banks, and lapped orthogonal transforms. This may be regarded as a basically time- (or spatial) domain approach.

In this section we construct the polyphase matrix as a product of constituent modular sections, each of which has the desired properties—whatever they may be. For $M = 2$, this regular structure is recognized as a lattice structure, and for arbitrary M, as a generalized lattice.

We consider first the two-band paraunitary lattice, and then a two-band linear-phase lattice which is *PR* but *not* paraunitary. Then we extend these results to the case of arbitrary M and discover that we can obtain lattice structures that are *both* paraunitary and linear-phase.

Previously we expressed

$$\mathcal{H}_p(z) = \sum_{k=0}^{N-1} \mathcal{H}_{p,k} z^{-k}$$

to obtain *PR* requirements. In particular, we could obtain constraints to force $\mathcal{H}_p(z)$ to be *lossless*.

In the present instance, we can write

$$\mathcal{H}_p(z) = A_{N-1} D_{N-1}(z) A_{N-2} D_{N-2}(z) ... A_1 D_1(z) A_0, \tag{3.187}$$

where the $M \times M$ matrices A_k have full rank, and $D_k(z)$ are diagonal matrices of delays. According to Belevitch (1968), if $\mathcal{H}_p(z)$ is lossless, the product form always exists; conversely, if each A_k is unitary, then $\mathcal{H}_p(z)$ is paraunitary, which is to say that the product of lossless modules is itself lossless. This property allows us to synthesize individual paraunitary blocks, connect them in cascade, and have full confidence that the resulting structure is PR.

3.6.1 The Two-Band Lossless Lattice

To fix ideas, we consider the two-band case; let

$$\mathcal{H}_p(z) = A'_{N-1}D(z)A'_{N-2}...D(z)A'_0, \tag{3.188}$$

where A'_k is a constant 2×2 orthogonal matrix, and

$$D(z) = \begin{bmatrix} 1 & 0 \\ 0 & z^{-1} \end{bmatrix}. \tag{3.189}$$

This structure is evidently lossless. A particular orthogonal matrix is the rotation matrix

$$\begin{aligned} A'_k &= \begin{bmatrix} \cos\phi_k & \sin\phi_k \\ -\sin\phi_k & \cos\phi_k \end{bmatrix} = (\cos\phi_k)\begin{bmatrix} 1 & \alpha_k \\ -\alpha_k & 1 \end{bmatrix} \\ &= \frac{1}{\sqrt{1+\alpha_k^2}}A_k; \end{aligned} \tag{3.190}$$

since $\cos(\phi_k) = (1+\alpha_k^2)^{-1/2}$. Collecting the $(\cos\phi_k)$ product terms in a single normalization factor c gives

$$\begin{aligned} \mathcal{H}_p(z) &= cA_{N-1}DA_{N-2}...A_1DA_0 \\ &= c[\prod_{k=1}^{N-1} A_{N-k}D]A_0 \end{aligned} \tag{3.191}$$

and

$$c = \prod_{k=0}^{N-1}\left[\frac{1}{1+\alpha_k^2}\right]^{1/2}.$$

This structure is shown in Fig. 3.45, where each A_k is realized by the lattice structure. The output taps at the end of the lattice yield the analysis filters.

This realization is paraunitary for *any* α_k; the PR condition continues to hold even when $\{\alpha_k\}$ are quantized from their design values. In this sense, the lattice is PR robust. The paraunitary property is easily shown from

$$\mathcal{H}_p^{-1}(z) = [cA_{N-1}D...A_0]^{-1} = (\frac{1}{c})A_0^{-1}D^{-1}A_1^{-1}...A_{N-1}^{-1}.$$

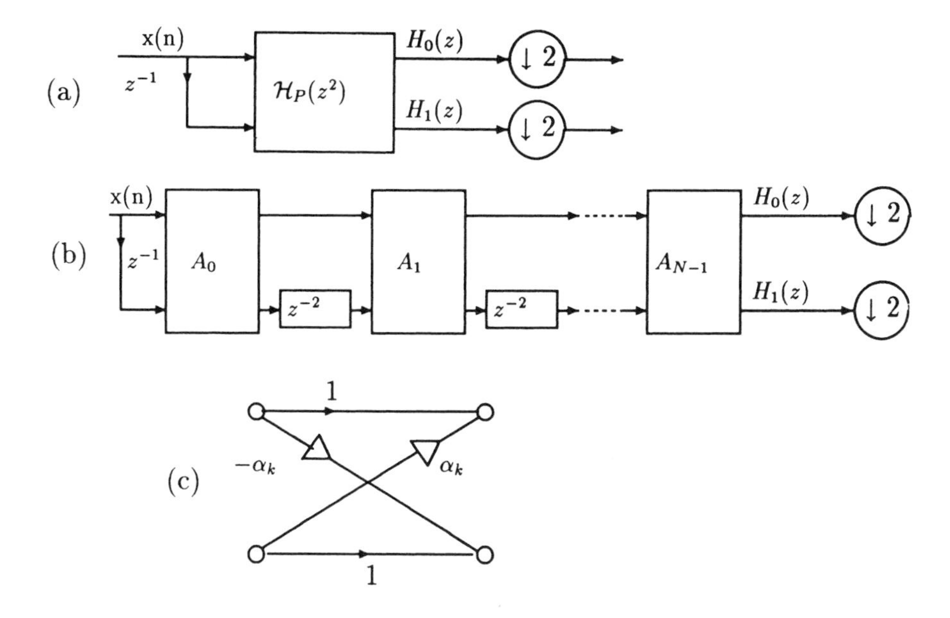

Figure 3.45: Two-band lattice structure (normalization suppressed): (a) polyphase representation of analysis stage; (b) realization as a cascaded structure; (c) individual lattice module.

But

$$A_k^{-1} = \frac{1}{1+\alpha_k^2} A_k^T, \quad and \quad D^{-1}(z) = D(z^{-1})$$

$$\begin{aligned} \mathcal{H}_p^{-1}(z) &= cA_0^T D(z^{-1}) A_1^T ... D(z^{-1}) A_{N-1}^T \\ &= [cA_{N-1} D(z^{-1}) A_{N-2} ... D(z^{-1}) A_0]^T \\ &= \mathcal{H}_p^T(z^{-1}). \end{aligned}$$

Of course, our two-band paraunitary lattice filters have the same properties as the two-band paraunitary filters discussed in Section 3.5.4. In particular, they are power complementary and share the filter relations of Eq. (3.141).

As in Eq. (3.109) we can choose the PR synthesis polyphase matrix

$$\mathcal{G}_p'(z) = z^{-(N-1)} \mathcal{H}_p^T(z^{-1}).$$

We can distribute the delay $z^{-(N-1)}$ among each of the $(N-1)$ blocks $D(z^{-1})$ to get the synthesis lattice

$$\mathcal{G}_p'(z) \;=\; cA_0^T D'(z) A_1^T ... D'(z) A_{N-1}^T$$

$$D'(z) = z^{-1}D(z^{-1}) = \begin{bmatrix} z^{-1} & 0 \\ 0 & 1 \end{bmatrix} = JD(z)J. \tag{3.192}$$

This structure is depicted in Fig. 3.46.

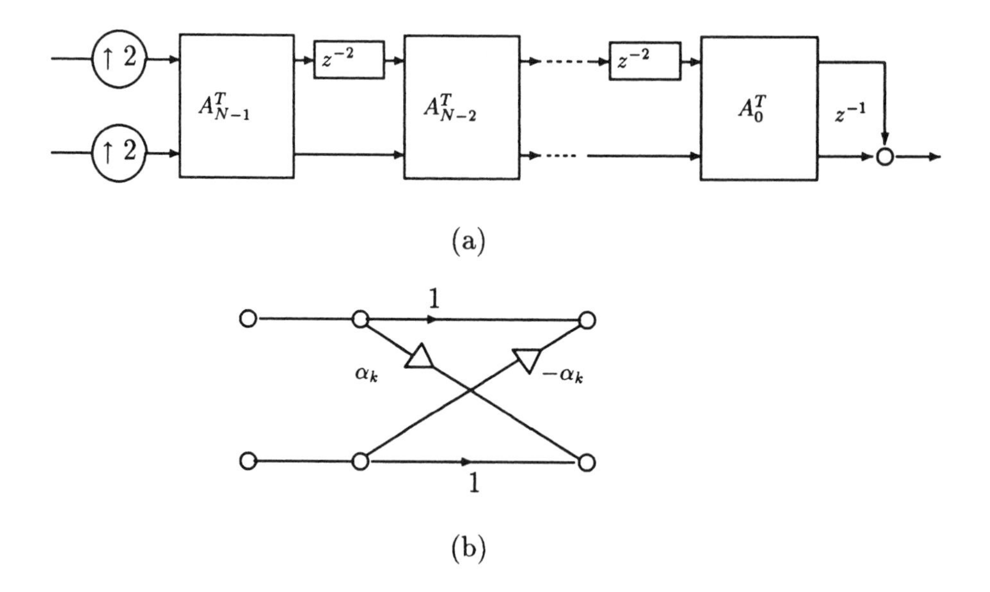

Figure 3.46: Paraunitary synthesis filter (unnormalized): (a) cascade structure; (b) individual structure.

So long as the same values of α_ks are used at the synthesis and analysis sides, this lattice realizes perfect reconstruction. Researchers (Vaidyanathan and Hoang, Jan. 1988; Delsarte, Macq, and Slock, 1992) have exploited this property by developing optimization procedures for selecting α_k, stage-by-stage, using the lattice as the basic element.

The desired behavior of the low-pass and high-pass filters is indicated in Fig. 3.47, where ω_p, and ω_s are, respectively, the passband and stopband cut-off frequencies.

The function to be minimized is the low-frequency filter energy that spills over into the stopband:

$$J = \int_{\omega_s}^{\pi} |H_0(e^{j\omega})|^2 d\omega$$

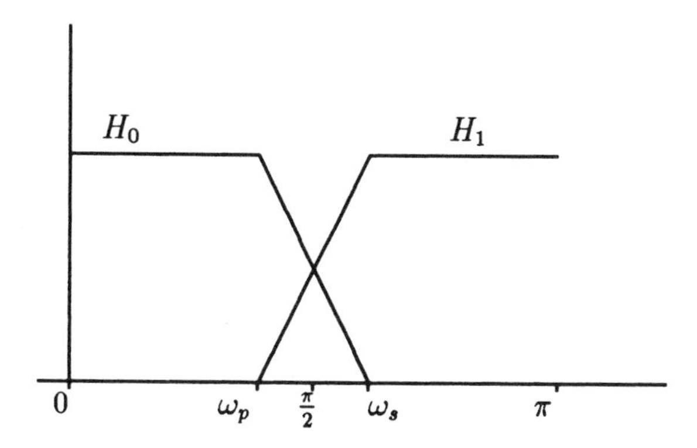

Figure 3.47: Frequency responses.

Because of the power complementarity of the paraunitary pair, that is,

$$|H_0(e^{j\omega})|^2 + |H_1(e^{j\omega})|^2 = \text{const.},$$

minimum J also ensures that $H_1(e^{j\omega})$ is a "good" high-pass filter with minimum energy in its stopband $[0, \omega_p]$.

The optimization algorithm and the lattice property have a *hierarchical* property, in the sense that a higher-order PR bank can be obtained from a lower-order one by adding more lattice sections (Vaidyanathan and Hoang, Jan. 1988). The design is based on iterations as we progress to the right in the analysis lattice. As k increases, α_k gets smaller, and the lattice frequency responses approach acceptable design specs. This $H_0(z)$ design based on lattice parameters results in filters with the maximum possible number of stopband zeros on the unit circle and a monotone decreasing peak error in the stopband.

3.6.2 The M-Band Paraunitary Lattice

A somewhat obvious M-band extension of Eq. (3.188) expresses $\mathcal{H}_p(z)$ as

$$\mathcal{H}_p(z) = R_{N-1}D_{N-1}(z)R_{N-2}.....R_1D_1(z)R_0, \tag{3.193}$$

where each R_k is an $M \times M$ orthogonal matrix, and $D_k(z)$ is a diagonal matrix whose elements are delays. This $\mathcal{H}_p(z)$ is lossless by construction, since $R_k^{-1} = R_k^T$, and $D_k^T(z^{-1}) = D_k^{-1}(z)$. Typically R_k is selected as a product of planar rotation matrices. For example, for $M = 3$,

$$R_k = \begin{bmatrix} \cos\alpha_k & \sin\alpha_k & 0 \\ -\sin\alpha_k & \cos\alpha_k & 0 \\ 0 & 0 & 1 \end{bmatrix} \begin{bmatrix} 1 & 0 & 0 \\ 0 & \cos\beta_k & \sin\beta_k \\ 0 & -\sin\beta_k & \cos\beta_k \end{bmatrix} \begin{bmatrix} \cos\gamma_k & 0 & \sin\gamma_k \\ 0 & 1 & 0 \\ -\sin\gamma_k & 0 & \cos\gamma_k \end{bmatrix}$$

and

$$D(z) = \begin{bmatrix} 1 & 0 & 0 \\ 0 & 1 & 0 \\ 0 & 0 & z^{-1} \end{bmatrix}.$$

Each modular section has three design parameters, the three rotation angles α_k, β_k, γ_k. Low-pass, band-pass, and high-pass filters $H_0(z)$, $H_1(z)$, and $H_2(z)$ can then be designed so as to minimize the sum of the *spillover* energies of each filter into its respective stopband (Vaidyanathan, April 1987),

$$J = \int_{\omega_0} |H_0(e^{j\omega})|^2 d\omega + \int_{\omega_1} |H_1(e^{j\omega})|^2 d\omega + \int_{\omega_2} |H_2(e^{j\omega})|^2 d\omega,$$

where ω_0, ω_1, ω_2 are the stopbands for H_0, H_1, H_2, respectively. This lattice formulation guarantees perfect reconstruction, for any filter order $(N-1)$.

Still another, less restrictive, product structure has been proposed. The polyphase matrix is lossless if and only if it can be written as the product

$$\begin{aligned} \mathcal{H}_p(z) &= V_{N-1}(z)V_{N-2}(z)...V_1(z)B \\ &= \left[\prod_{k=0}^{N-2} V_{N-k-1}\right] B \end{aligned} \tag{3.194}$$

where V_k is an $M \times M$ matrix

$$V_k = (I - \underline{v}_k\underline{v}_k^T + z^{-1}\underline{v}_k\underline{v}_k^T),$$

$\underline{v}_k$ is an $M \times 1$ column vector of unit norm, and B is unitary. The sufficiency is proved by demonstrating that each $V_k(z)$ is lossless, i.e., that

$$Q \triangleq V_k^T(z^{-1})V_k(z) = I.$$

By direct expansion, using $A = \underline{v}\underline{v}^T$ and noting that

$$A^2 = \underline{v}(\underline{v}^T\underline{v})\underline{v}^T = A,$$

we find that

$$Q = I - 2A + A(z + z^{-1}) + A^2(-2 - z^{-1} - z) = I.$$

Additionally, the unitary matrix B itself can be written as the product of $(M-1)$ orthogonal matrices of the form

$$(I - 2\underline{w}_n \underline{w}_n^T),$$

where $\underline{w}_n$ is a unit norm vector. The design approach—a direct extension of the 2 band case—involves optimization of the normalized vectors $\underline{v}_n$ and of the unitary matrix B so as to minimize the sum of the stop-band energies associated with each filter $H_k(e^{j\omega})$. The reader is referred to (Vaidyanathan) (July 1989) for a detailed description of this process, and sample filter designs. (See also Probs. 3.18, 3.19)

3.6.3 The Two-Band Linear-Phase Lattice

Here we demonstrate that linear-phase perfect reconstruction filters can be synthesized, but without the paraunitary and power complementary properties. Our approach is to build these filters using a cascade of modular linear-phase lattices.

We had previously noted that the PR requirement $\mathcal{G}_p'(z)\mathcal{H}_p(z) = z^{-n_0}I$ can be met by selecting $\mathcal{G}_p'(z) = z^{-n_0}\mathcal{H}_p^{-1}(z)$. For the two-band case,

$$\mathcal{H}_p^{-1}(z) = \begin{bmatrix} H_{00}(z) & H_{01}(z) \\ H_{10}(z) & H_{11}(z) \end{bmatrix}^{-1} = \frac{1}{\Delta(z)} \begin{bmatrix} H_{11}(z) & -H_{01}(z) \\ -H_{10}(z) & H_{00}(z) \end{bmatrix}$$

where

$$\Delta(z) \triangleq det\{\mathcal{H}_p(z)\}.$$

Therefore, from Eq. (3.93),

$$\mathcal{G}_p(z) = [\mathcal{G}_p'(z)]^T J = \frac{z^{-n_0}}{\Delta(z)} \begin{bmatrix} -H_{10}(z) & H_{11}(z) \\ H_{00}(z) & -H_{01}(z) \end{bmatrix}.$$

If we force $\Delta(z) = cz^{-n_0}$, the PR condition can be satisfied with FIR synthesis filters; from Eq. (3.94), these are

$$\begin{aligned} \begin{bmatrix} G_0(z) \\ G_1(z) \end{bmatrix} = \mathcal{G}_p(z^2) \begin{bmatrix} 1 \\ z^{-1} \end{bmatrix} &= \frac{1}{c} \begin{bmatrix} -H_{10}(z^2) & H_{11}(z^2) \\ H_{00}(z^2) & -H_{01}(z^2) \end{bmatrix} \begin{bmatrix} 1 \\ z^{-1} \end{bmatrix} \\ &= \frac{1}{c} \begin{bmatrix} -H_{10}(z^2) + z^{-1}H_{11}(z^2) \\ H_{00}(z^2) - z^{-1}H_{01}(z^2) \end{bmatrix} \\ &= \frac{1}{c} \begin{bmatrix} -H_1(-z) \\ H_0(-z) \end{bmatrix}. \end{aligned} \tag{3.195}$$

Thus Eq. (3.195) and $det\{\mathcal{H}_p(z)\} = cz^{-n_0}$ constitute the PR requirements in this case. Note that $G_0(z)$, $G_1(z)$ are mirror image filters of $H_1(z)$, $H_0(z)$, respectively, and that Eq. (3.195) satisfies the alias cancellation requirement of Eq. (3.36). If we make $H_0(z)$, and $H_1(z)$ linear-phase, then the synthesis filters are also linear-phase. But also observe that paraunitary and power complementary conditions are *not* met.

We now pause to review some properties of linear- phase filters. Let $H_0(z)$, $H_1(z)$ be symmetric and antisymmetric, respectively, and of length N,

$$\begin{aligned} h_0(n) &= h_0(N-1-n), \qquad n = 0, 1, \cdots, \frac{N}{2} - 1 \\ h_1(n) &= -h_1(N-1-n). \end{aligned} \tag{3.196}$$

Such filters can be written in the form

$$H_0(z) = A(z) + z^{-(N-1)}A(z^{-1}),$$
$$A(z) = \sum_{n=0}^{\frac{N}{2}-1} h_0(n) z^{-n}$$

$$H_1(z) = B(z) - z^{-(N-1)}B(z^{-1}),$$
$$B(z) = \sum_{n=0}^{\frac{N}{2}-1} h_1(n) z^{-n},$$

where $A(z)$, $B(z)$ are polynomials in z^{-1} of degree $(\frac{N}{2} - 1)$. More succinctly, we have

$$\begin{aligned} H_0(z) &= z^{-(N-1)}H_0(z^{-1}) \\ H_1(z) &= -z^{-(N-1)}H_1(z^{-1}). \end{aligned}$$

Next, we want to show that the polyphase matrix for such linear-phase filters has a particular structure. Since we wish to build our filters by an iteration process, we take N to be even. For this case,

$$\mathcal{H}_p(z) = \begin{bmatrix} H_{00}(z) & H_{01}(z) \\ H_{10}(z) & H_{11}(z) \end{bmatrix}$$

is constrained to satisfy (for N even)

$$\begin{aligned} H_{00}(z) &= z^{-(\frac{N}{2}-1)}H_{01}(z^{-1}) \\ H_{10}(z) &= -z^{-(\frac{N}{2}-1)}H_{11}(z^{-1}). \end{aligned} \tag{3.197}$$

To appreciate this, expand $H_0(z)$ into polyphase form:

$$H_{00}(z) = \sum_{n=0}^{\frac{N}{2}-1} h_0(2n) z^{-n} \tag{3.198}$$

$$H_{01}(z) = \sum_{n=0}^{\frac{N}{2}-1} h_0(2n+1) z^{-n}.$$

But with $h_0(n)$ even symmetric, Eq. (3.196), we have $h_0(2n) = h_0(N-1-2n)$. Substituting into $H_{00}(z)$ gives

$$\begin{aligned} H_{00}(z) &= \sum_{n=0}^{\frac{N}{2}-1} h_0(N-1-2n) z^{-n} \\ &= z^{-(\frac{N}{2}-1)} \sum_{n=0}^{\frac{N}{2}-1} h_0(2n+1) z^{n} \\ &= z^{-(\frac{N}{2}-1)} H_{01}(z^{-1}). \end{aligned}$$

Similarly, we can obtain the second constraint of Eq. (3.198) for the polyphase components of an antisymmetric filter.

The theorem to be proved is that, if $\mathcal{H}_p(z)$ is a polyphase matrix for a linear-phase filter pair, Eq. (3.196), then $\hat{\mathcal{H}}_p(z)$ given by

$$\hat{\mathcal{H}}_p(z) = \mathcal{H}_p \begin{bmatrix} 1 & 0 \\ 0 & z^{-1} \end{bmatrix} \begin{bmatrix} 1 & \alpha \\ \alpha & 1 \end{bmatrix} \tag{3.199}$$

also represents a linear-phase, symmetric and antisymmetric pair, $\hat{H}_0(z)$, $\hat{H}_1(z)$ of length $N+2$. (For an alternative version, see Prob. 3.20) Direct expansion yields

$$\hat{H}_{00}(z) = H_{00}(z) + \alpha z^{-1} H_{01}(z)$$

$$\hat{H}_{01}(z) = \alpha H_{00}(z) + z^{-1} H_{01}(z).$$

But $H_{00}(z)$,$H_{01}(z)$ must satisfy Eq. (3.198). Hence

$$\hat{H}_{00}(z) = H_{00}(z) + \alpha z^{-\frac{N}{2}} H_{00}(z^{-1})$$

$$\hat{H}_{01}(z) = \alpha H_{00}(z) + z^{-\frac{N}{2}} H_{00}(z^{-1}).$$

From the foregoing, we see that

$$\hat{H}_{00}(z) = z^{-\frac{N}{2}} \hat{H}_{01}(z^{-1})$$

Similarly, we can show

$$\hat{H}_{10}(z) = -z^{-\frac{N}{2}} \hat{H}_{11}(z^{-1})$$

and conclude that $\hat{H}_p(z)$ represents a polyphase matrix for linear phase filters of length $(N+2)$.

This theorem sets the stage for a cascaded formulation of the form

$$\mathcal{H}_p(z) = \begin{bmatrix} 1 & 1 \\ 1 & -1 \end{bmatrix} \prod_{k=1}^{N-1} \begin{bmatrix} 1 & 0 \\ 0 & z^{-1} \end{bmatrix} \begin{bmatrix} 1 & \alpha_{N-k-1} \\ \alpha_{N-k-1} & 1 \end{bmatrix}. \tag{3.200}$$

This polyphase cascade is initialized with the 2×2 Hadamard matrix

$$\mathcal{H}_p(z) = \begin{bmatrix} 1 & 1 \\ 1 & -1 \end{bmatrix};$$

and the lowest order filters are simply $H_0(z) = 1 + z^{-1}$ and $H_1(z) = 1 - z^{-1}$. The resulting lattice structure is the same as that for the paraunitary analysis lattice shown in Fig. 3.45 except that, in the present instance, because of Eq. (3.191) we use only positive α_k in the lattice and the Hadamard matrix at the right-hand end in place of A_{N-1}.

Now, it is a simple matter to obtain PR by choosing

$$\mathcal{G}_p'(z) = z^{-n_0} \mathcal{H}_p^{-1}(z), \quad n_0 = (N-1)$$

or

$$\begin{aligned} \mathcal{G}_p'(z) &= c' \begin{bmatrix} 1 & -\alpha_0 \\ -\alpha_0 & 1 \end{bmatrix} \begin{bmatrix} z^{-1} & 0 \\ 0 & 1 \end{bmatrix} \cdots \begin{bmatrix} z^{-1} & 0 \\ 0 & 1 \end{bmatrix} \begin{bmatrix} 1 & 1 \\ 1 & -1 \end{bmatrix} \\ &= c' \left\{ \prod_{k=0}^{N-2} \begin{bmatrix} 1 & -\alpha_k \\ -\alpha_k & 1 \end{bmatrix} \begin{bmatrix} z^{-1} & 0 \\ 0 & 1 \end{bmatrix} \right\} \begin{bmatrix} 1 & 1 \\ 1 & -1 \end{bmatrix} \\ c' &= \frac{1}{2} \prod_{k=0}^{N-2} (1 - \alpha_k^2)^{-1/2}. \end{aligned} \tag{3.201}$$

The resulting synthesis lattice is now fairly obvious. From $\mathcal{G}_p'(z) = z^{-(N-1)}\mathcal{H}_p^{-1}(z)$ and

$$\begin{aligned}\underline{h}(z) &= \mathcal{H}_p(z^M)\underline{Z}_M \\ \underline{g}(z) &= \mathcal{G}_p(z^M)\underline{Z}_M.\end{aligned}$$

We can show that the synthesis filters are also a linear-phase, modulated version of the analysis set:

$$\begin{aligned}G_0(z) &= c' H_1(-z) \\ G_1(z) &= -c' H_0(-z).\end{aligned}$$

The lattice structure permits a stage-by-stage design of the linear-phase lattice. We can use optimization procedures similar to those in the paraunitary lattice.

Linear-phase filters of odd length can also be realized. In this case $H_0(z)$ and $H_1(z)$ are *each symmetric* filters of lengths $(2N-1)$ and $(2N+1)$, respectively. The polyphase matrix is such that $det\{\mathcal{H}_p(z)\} = cz^{-N}$, and $H_{00}(z)$, $H_{11}(z)$ are of degree N, while $H_{01}(z)$, $H_{10}(z)$ are of degree $(N-1)$ and $(N+1)$, respectively. Additionally, each polyphase component is itself a symmetric polynomial of the form $z^{-k}p(z^{-1}) = p(z)$, as in Eq. (3.197). See Vetterli and LeGall (1989) for a detailed derivation.

We close this section by noting that all product forms of the type considered here yield linear-phase filters. But it is also noted that not all linear-phase filters can be factored into product form, except for $N \leq 8$.

3.6.4 M-Band PR Linear Phase Filter Bank

The linear-phase requirement for the case $M = 2$ was set down in Section 3.6.3. An alternate phrasing of these conditions is (Prob. 3.20)

$$z^{-K}\begin{bmatrix} 1 & 0 \\ 0 & -1 \end{bmatrix}\mathcal{H}_p(z^{-1})\begin{bmatrix} 0 & 1 \\ 1 & 0 \end{bmatrix} = \mathcal{H}_p(z),$$

where K is the highest degree in $\mathcal{H}_p(z)$. The extension to the M-band case has the form

$$z^{-K}\begin{bmatrix} I & 0 \\ 0 & -I \end{bmatrix}\mathcal{H}_p(z^{-1})J = \mathcal{H}_p(z). \tag{3.202}$$

Here M is even, the first $\frac{M}{2}$ filters are symmetric, and the bottom $\frac{M}{2}$ are antisymmetric. We will simply outline the M-band extensions.

The polyphase matrices satisfying the linear-phase conditions are generated by a generalization of Eq. (3.199). If $\mathcal{H}_p(z)$ is a polyphase matrix of a linear-phase filter bank, Eq. (3.202), then $\hat{\mathcal{H}}_p(z)$ generated by

$$\hat{\mathcal{H}}_p(z) = \mathcal{H}_p(z)D(z)R \tag{3.203}$$

is also a linear-phase polyphase matrix. In this case, $D(z)$ is a diagonal matrix of delay elements and R is unitary. The main theorem proved by Vetterli and LeGall (1989) is that $\hat{\mathcal{H}}_p(z)$ is the polyphase matrix of a linear-phase filter bank if and only if

$$\begin{aligned} z^{-l}JD(z^{-1})J &= D(z), \\ JRJ &= R \end{aligned} \tag{3.204}$$

where l is an integer such that $D(z)$ is causal. From Eq. (3.89), this implies that

$$R_{ij} = R_{M-1-i,M-1-j}. \tag{3.205}$$

A matrix R with this property is said to be *persymmetric.*

The point is, we may start with any unitary matrix R and $D(z)$ satisfying Eq. (3.205). A particular case is when R is symmetric Toeplitz, and $D(z)$ any one of the four forms (for $M = 4$)

$$\begin{array}{ll} \text{diag}\,(1, 1, z^{-1}, z^{-1}) & \text{diag}\,(1, z^{-1}, 1, z^{-1}) \\ \text{diag}\,(z^{-1}, z^{-1}, 1, 1) & \text{diag}\,(z^{-1}, 1, z^{-1}, 1). \end{array}$$

It can be shown that a symmetric Toeplitz matrix that satisfies the persymmetric form, Eq. (3.205), has the form

$$R = \begin{bmatrix} M_0 & M_1 \\ JM_1J & JM_0J \end{bmatrix}. \tag{3.206}$$

Each matrix is of size $\frac{M}{2} \times \frac{M}{2}$. Since R is unitary, $RR^T = I$ implies

$$\begin{aligned} M_0M_0^T + M_1M_1^T &= I \\ M_1JM_0^T + M_0JM_1^T &= 0. \end{aligned} \tag{3.207}$$

The first of these constitutes an orthonormality of the first $\frac{M}{2}$ rows of R, while the second asserts the orthogonality of M_1 and (M_0J). These bear a strong resemblance to the extended LOT orthonormalities of Section 3.5.5, Eqs. (3.175).

Another way of stating Eq. (3.207) is

$$\begin{aligned} \underline{r}_i^T \underline{r}_j &= \delta_{ij} \\ \underline{r}_i^T J \underline{r}_j &= 0 \qquad 0 \le i,j \le \frac{M}{2} - 1, \end{aligned}$$

where $\underline{r}_i^T$ is the ith row of R.

The choice of R determines the filter bank structure. An example is R as the product of two persymmetric matrices

$$R = \begin{bmatrix} c_1 & -s_1 & 0 & 0 \\ s_1 & c_1 & 0 & 0 \\ 0 & 0 & c_1 & s_1 \\ 0 & 0 & -s_1 & c_1 \end{bmatrix} \begin{bmatrix} c_2 & 0 & -s_2 & 0 \\ 0 & c_2 & 0 & s_2 \\ s_2 & 0 & c_2 & 0 \\ 0 & -s_2 & 0 & c_2 \end{bmatrix},$$

where

$$c_i = \cos\alpha_i, \qquad s_i = \sin\alpha_i.$$

Finally, we emphasize that for $M > 2$, we can have *both* a paraunitary and a linear-phase filter bank. That is, in addition to Eq. (3.202) we can simultaneously satisfy

$$\mathcal{H}_p^T(z^{-1})\mathcal{H}_p(z) = I.$$

As a consideration separate from linear-phase, it is sometimes desirable that a filter bank consist of mirror image filters, i.e., $H_{M-1-i}(z) = H_i(z)$, for $i = 0, 1, ..., \frac{M}{2} - 1$. This implies that the polyphase components are related by the relation $H_{M-1-i,k}(z) = (-1)^k H_{i,k}(z)$. That is, the even (odd) indexed polyphase components of $H_{M-1-i}(z)$, and $H_i(z)$ are equal (negative of each other). Polyphase matrices with this property can be generated recursively by product forms. In particular, postmultiplication by a delay matrix $D(z)$, and/or by a rotation matrix R, with $R_{ij} = 0$, i+j=even, results in a mirror-image filter bank.

As a case in point, suppose $M = 4$, and we start with $\mathcal{H}_p(z)$ as a row-shuffled 4×4 Hadamard matrix

$$\mathcal{H}_p(z) = R_0 = \frac{1}{2}\begin{bmatrix} 1 & 1 & 1 & 1 \\ 1 & -1 & -1 & 1 \\ 1 & 1 & -1 & -1 \\ 1 & -1 & 1 & -1 \end{bmatrix}.$$

The corresponding analysis filters are

$$H_0(z) = 1 + z^1 + z^{-2} + z^{-3} = (1 + z^{-2})(1 + z^{-1})$$

$$\begin{aligned} H_1(z) &= 1 - z^1 - z^{-2} + z^{-3} = (1 - z^{-1})^2(1 + z^{-1}) \\ H_2(z) &= 1 + z^1 - z^{-2} - z^{-3} = (1 + z^{-1})^2(1 - z^{-1}) \\ H_3(z) &= 1 - z^1 + z^{-2} - z^{-3} = (1 + z^{-2})(1 - z^{-1}). \end{aligned}$$

This filter bank is paraunitary, since $\mathcal{H}_p^T(z^{-1})\mathcal{H}_p(z) = R_0^T R_0 = I$. It is linear-phase, since $h_0(n)$ and $h_1(n)$ are symmetric, and $h_2(n)$, $h_3(n)$ are antisymmetric. And, the filter bank has mirror-image filters, $H_3(z) = H_0(-z)$ and $H_2(z) = H_1(-z)$.

(1) To build up a paraunitary structure, it is sufficient to write $\mathcal{H}_p$ as a product of paraunitary matrices as in Eq. (3.193), $\hat{\mathcal{H}}_p(z) = D_1(z)\mathcal{H}_p(z) = D_1(z)R_0$, or equally valid, $\hat{\mathcal{H}}_p(z) = R_0 D_1(z)$, since the product of paraunitary matrices is paraunitary.

(2) To construct a linear-phase structure, we want $\hat{\mathcal{H}}_p(z) = \mathcal{H}_p(z)D(z)R$, where $D(z)$, and R are constrained by Eq. (3.204).

(3) To continue the mirror-image property, we may postmultiply R_0 by a diagonal delay matrix $D(z)$, and/or the special rotation matrix with $R_{ij} = 0$ when i+j is odd. For example, we can satisfy all three properties with

$$\hat{\mathcal{H}}_p(z) = R_0 \begin{bmatrix} 1 & 0 & 0 & 0 \\ 0 & z^{-1} & 0 & 0 \\ 0 & 0 & z^{-1} & 0 \\ 0 & 0 & 0 & 1 \end{bmatrix} \begin{bmatrix} c & 0 & -s & 0 \\ 0 & c & 0 & s \\ s & 0 & c & 0 \\ 0 & -s & 0 & c \end{bmatrix},$$

where $c = \cos\alpha$, $s = \sin\alpha$.

This example demonstrates that one can recursively construct M-band paraunitary, linear-phase, mirror-image filter banks from elementary building blocks. The cascaded structure depends on initialization, choice of $D(z)$, and of rotation matrices, the angles of which can be recursively optimized.

3.6.5 Lattice Realizations of Modulated Filter Bank

The cosine modulated PR filter bank of Section 3.5.6 is revisited here, where in the present instance, each filter is of length $2lM$, l an integer ≥ 1. A polyphase expansion of each filter and imposition of PR constraints lead to a realization based on the two channel lossless lattice of Section 3.6.1. The latter structure leads to efficient design procedures (Koilpillai and Vaidyanathan, 1992).

The rth band analysis filter has the form

$$h_r(n) = h(n)c_{r,n} \leftrightarrow H_r(z) = \sum_{n=0}^{2lM-1} h(n)c_{r,n}z^{-n}, \quad 0 \leq r \leq M-1. \tag{3.208}$$

In Eq. (3.208) $h(n)$ is the linear-phase low-pass prototype of length $2lM$, and $c_{r,n}$ is the cosine modulation term,

$$c_{r,n} = cos\{\frac{(2r+1)\pi}{2M}(n - \frac{(N-1)}{2}) + \theta_r\} \tag{3.209}$$

$$where \quad N = 2lM, \quad and \quad \theta_r = (-1)^r \pi/4. \tag{3.210}$$

The modulation term has a half period of $2pM$ and a periodicity evidenced by

$$c_{r,n+pM} = (-1)^p c_{r,n}. \tag{3.211}$$

The associated synthesis filter $g_r(n) = h_r(N-1-n)$ has the same form as Eq. (3.208), except that the phase is $-\theta_r$.

The derivation of the lattice structure shown in Fig. 3.49 is based on evaluating the polyphase matrix $H_p(z)$ in terms of the polyphase components of $H_r(z)$, followed by imposition of PR constraints. The details are as follows:

(1) Expand the prototype $h(n)$, and band-pass $h_r(n)$, each of length $2lM$, in a polyphase expansion using base $2M$ (instead of the usual M), to obtain

$$H(z) = \sum_{n=0}^{2lM-1} h(n)z^{-n} = \sum_{k=0}^{2M-1} z^{-k}G_k(z^{2M}), \tag{3.212}$$

realizable by $2M$ parallel branches. From Equation (3.15),

$$\begin{aligned} G_k(z) &= h(k) + h(k+2M)z^{-1} + h(k+4M)z^{-2} + .. \\ &= \sum_{p=0}^{l-1} h(k+2Mp)z^{-p}. \end{aligned} \tag{3.213}$$

Equation (3.212) now becomes

$$H(z) = \sum_{k=0}^{2M-1} z^{-k} \sum_{p=0}^{l-1} h(k+2Mp)z^{-2pM}. \tag{3.214}$$

(2) Next, by comparison with Eqs. (3.212) and (3.214), the polyphase expansion of $H_r(z)$ is

$$H_r(z) = \sum_{n=0}^{2lM-1} [h(n)c_{r,n}]z^{-n} = \sum_{k=0}^{2M-1} z^{-k} \sum_{p=0}^{l-1} h(k+2Mp)c_{r,k+2Mp}z^{-2pM}. \tag{3.215}$$

Using the periodicity, Eq. (3.211), the last expression becomes

$$H_r(z) = \sum_{k=0}^{2M-1} z^{-k} c_{r,k} \underbrace{\sum_{p=0}^{l-1} h(k+2Mp)(-z^{2M})^p}_{G_k(-z^{2M})}.$$

Finally,

$$H_r(z) = \sum_{k=0}^{2M-1} [c_{r,k} G_k(-z^{2M}] z^{-k}. \tag{3.216}$$

(3) The vector of analysis filters $\underline{h}(z)$ is now of the form,

$$\underline{h}(z) = \begin{bmatrix} H_0(z) \\ \vdots \\ H_{M-1}(z) \end{bmatrix} = CG\underline{Z}_{2M} \tag{3.217}$$

$$\left.\begin{array}{c} \textit{where } C \textit{ is } M \times 2M \textit{ modulation matrix } = [c_{r,k}] \\ G \textit{ is a } 2M \times 2M \textit{ diagonal matrix} \\ G = diag\{G_0(-z^{2M}), \ldots, G_{2M-1}(-z^{2M})\} \\ \textit{and } \underline{Z}_{2M}^T = [1, z^{-1}, \ldots, z^{-(2M-1)}] \end{array}\right\}. \tag{3.218}$$

The next step is to impose PR conditions on the polyphase matrix defined by $\underline{h}(z) = H_p(z^M)\underline{z}_M$. To obtain this form, we partition C, G, and $\underline{Z}_{2M}$ into

$$C = [C_0 \vdots C_1], \ G = \begin{bmatrix} g_0(z^{2M}) & \phi \\ \phi & g_1(z^{2M}) \end{bmatrix}, \ \underline{Z}_{2M} = \begin{bmatrix} \underline{Z}_M \\ \cdots\cdots \\ z^{-M}\underline{Z}_M \end{bmatrix}, \tag{3.219}$$

where C_0, C_1, g_0, g_1 are each $M \times M$ matrices. Expanding Eq. (3.217) in terms of the partitional matrices leads to the desired form,

$$\underline{h}(z) = [C_0 g_0(z^{2M}) + z^{-M} C_1 g_1(z^{2M})]\underline{Z}_M = H_p(z^M)\underline{Z}_M. \tag{3.220}$$

In Eq. (3.220),

$$\begin{aligned} [C_0]_{k,l} = c_{k,l}, \quad [C_1]_{k,l} &= c_{k,l+M} \\ g_0(z) = diag\{G_0(-z), \ldots, G_{M-1}(-z)\} \\ g_1(z) = diag\{G_M(-z), \ldots, G_{2M-1}(-z)\}. \end{aligned} \tag{3.221}$$

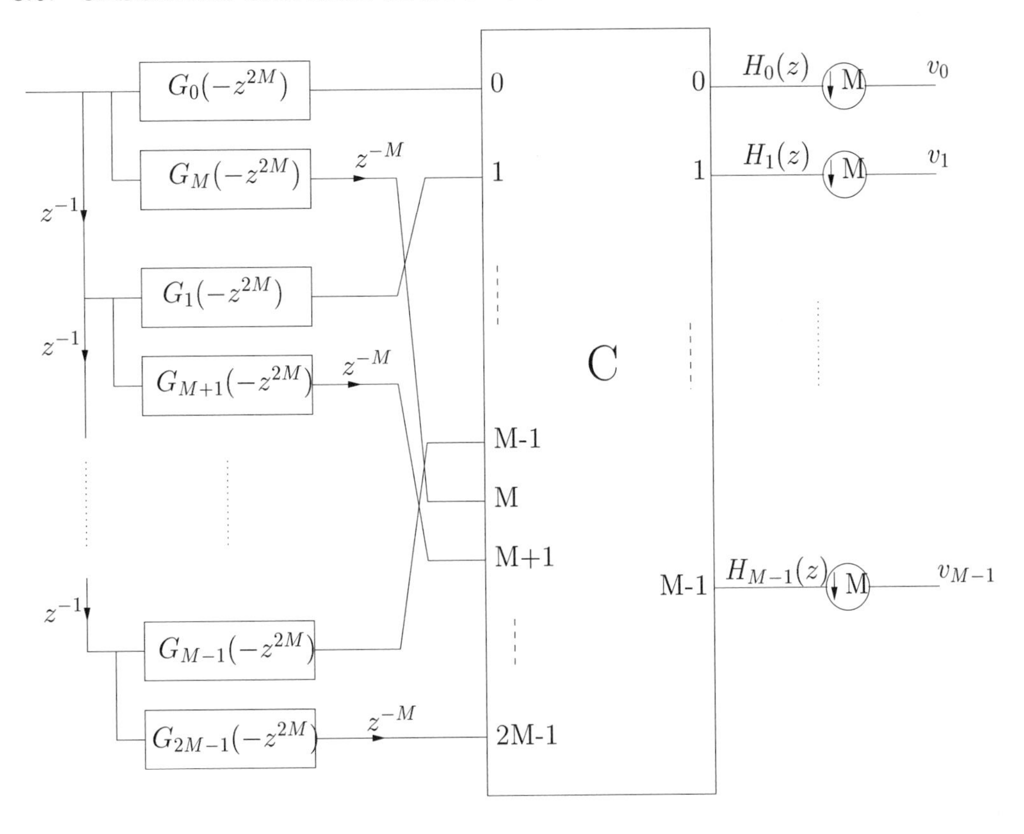

Figure 3.48: Structure of cosine-modulated filter bank. Each pair G_k and G_{k+M} is implemented as a two-channel lattice.

Equation (3.220) suggests that the polyphase components can be grouped into pairs, G_k and $z^{-M}G_{k+M}$, as shown in Fig. 3.48. Moving the down-samplers to the left in Fig. 3.48 then gives us the structure of Fig. 3.49.

Up to this point, the realization has been purely structural. By using the properties of C_0, C_1, g_0, g_1 in Eq. (3.220) and imposing $H_p^T(z^{-1})H_p(z) = I$, it is shown (Koilpillai and Vaidyanathan, 1992) (see also Prob. 3.31) that the necessary and sufficient condition for paraunitary perfect reconstruction is that the polyphase component filters $G_k(z)$ and $G_{k+M}(z)$ be pairwise power complementary, i.e.,

$$G_k(z^{-1})G_k(z) + G_{k+M}(z^{-1})G_{k+M}(z) = \frac{1}{2M}, \quad 0 \leq k \leq M-1,$$

$$or \quad |G_k(e^{j\omega}|^2 + |G_{k+M}(e^{j\omega})|^2 = \frac{1}{2M}. \qquad (3.222)$$

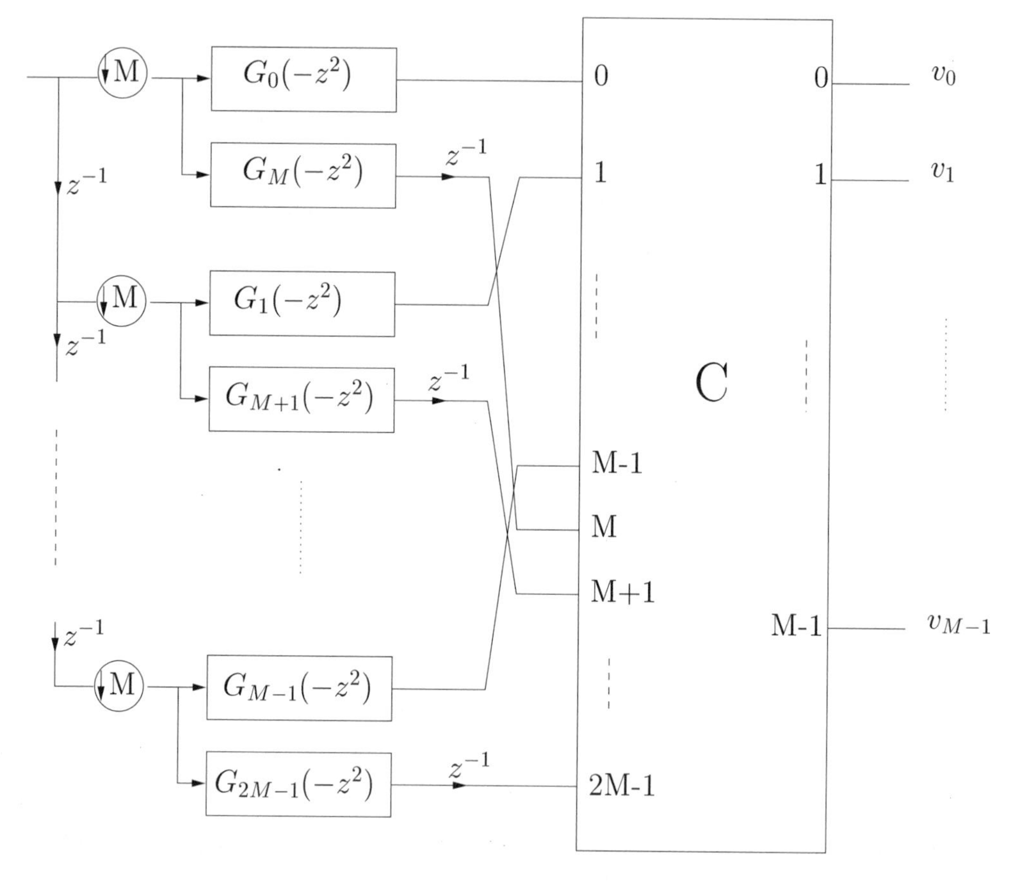

Figure 3.49: Alternate representation of cosine modulated filter bank.

For the case $l = 1$, all polyphase components are constant, $G_k(z) = h(k)$. This last equation then becomes $h^2(k)+h^2(k+M) = \frac{1}{2M}$, which corresponds to Eq. (3.178). Therefore the filters in Fig. 3.49 $\{G_k(-z^2), G_{k+M}(-z^2)\}$ can be realized by a two channel lossless lattice. We design $G_k(z)$ and $G_{k+M}(z)$ to be power complementary or lossless as in Section 3.6.1, Fig. 3.45 (with down-samplers shifted to the left), and then replace each delay z^{-1} by $-z^2$ in the realization. The actual design of each component lattice is described in Koilpillai and Vaidyanathan (1992). The process involves optimization of the lattice parameters.

In Nguyen and Koilpillai (1996), these results were extended to the case where the filter length is arbitrary. It was shown that Eq. (3.222) remains necessary and sufficient for paraunitary perfect reconstruction.

3.7 IIR Subband Filter Banks

Thus far, we have restricted our studies to FIR filter banks. The reason for this hesitancy is that it is extremely difficult to realize perfect reconstruction IIR analysis and synthesis banks. To appreciate the scope of this problem, consider the PR condition of Eq. (3.100):

$$\mathcal{G}_p'(z)\mathcal{H}_p(z) = z^{-\mu}I,$$

which requires

$$\begin{aligned} \mathcal{G}_p'(z) &= z^{-\mu}\mathcal{H}_p^{-1}(z) \\ &= z^{-\mu}\frac{1}{det\{\mathcal{H}_p(z)\}}\left(adj\mathcal{H}_p(z)\right). \end{aligned} \tag{3.223}$$

Stability requires the poles of $\mathcal{G}_p'(z)$ to lie within the unit circle of the Z-plane. From Eq. (3.223), we see that the poles of $\mathcal{G}_p'(z)$ are the uncancelled poles of the elements of the adjoint of $\mathcal{H}_p(z)$ and the *zeros* of $det(\mathcal{H}_p(z))$. Suppose $\mathcal{H}_p(z)$ consists of stable, rational IIR filters (i.e., poles within the unit circle). Then $adj\mathcal{H}_p(z)$ is also stable, since its common poles are poles of elements of $\mathcal{H}_p(z)$. Hence stability depends on the zeros of $det(\mathcal{H}_p(z))$, which must be minimum-phase—i.e., lie within the unit circle—a condition very difficult to ensure.

Next suppose $\mathcal{H}_p(z)$ is IIR lossless, so that $\mathcal{H}_p^T(z^{-1})\mathcal{H}_p(z) = I$. If $\mathcal{H}_p(z)$ is stable with poles inside the unit circle, then $\mathcal{H}_p(z^{-1})$ must have poles outside the unit circle, which cannot be stabilized by multiplication by z^{-n_0}. Therefore, we *cannot* choose $\mathcal{G}_p'(z) = \mathcal{H}_p^T(z^{-1})$ as we did in the FIR case. Thus, we cannot obtain a stable *causal* IIR lossless analysis-synthesis PR structure.

We will consider two alternatives to this impasse:

(1) It is possible, however, to obtain PR IIR structures if we operate the synthesis filters in a noncausal way. In this case, the poles of $\mathcal{G}_p'(z)$ outside the unit circle are the *stable* poles of an anticausal filter, and the filtering is performed in a noncausal fashion, which is quite acceptable for image processing. Two approaches for achieving this are described subsequently. In the first case, the signals are reversed in time and applied to causal IIR filters (Kronander, ASSP, Sept. 1988). In the second instance, the filters are run in both causal and anticausal modes (Smith, and Eddins, ASSP, Aug. 1990).

(2) We can still use the concept of *losslessness* if we back off from the PR requirement and settle for no aliasing and no amplitude distortion, but tolerate some phase distortion. This is achieved by *power complementary* filters synthesized from all-pass structures. To see this (Vaidyanathan, Jan. 1990), consider a lossless IIR polyphase analysis matrix expressed as

$$\mathcal{H}_p(z) = \frac{1}{d(z)}\mathcal{F}(z), \tag{3.224}$$

where $d(z)$ is the least common multiple of the denominators of the elements of $\mathcal{H}_p(z)$, and $\mathcal{F}(z)$ is a matrix of adjusted numerator terms; i.e., just polynomials in z^{-1}. We assume that $d(z)$ is stable. Now let

$$\mathcal{G}_p^{'}(z) = \frac{z^{-\mu}}{d(z)}\mathcal{F}^T(z^{-1}). \tag{3.225}$$

Therefore,

$$\begin{aligned} \mathcal{P}(z) &\triangleq \mathcal{G}_p^{'}(z)\mathcal{H}_p(z) = \frac{z^{-\mu}}{d(z)}\mathcal{F}^T(z^{-1})\frac{\mathcal{F}(z)}{d(z)} \\ &= z^{-\mu}\left[\frac{d(z^{-1})}{d(z)}\right]\mathcal{H}_p^T(z^{-1})\mathcal{H}_p(z) \\ &= z^{-\mu}\frac{d(z^{-1})}{d(z)}I. \end{aligned} \tag{3.226}$$

With this selection, $\mathcal{P}(z)$ is all-pass and diagonal, resulting in

$$T(z) = \frac{\hat{X}(z)}{X(z)} = z^{-\mu'}\frac{d(z^{-1})}{d(z)}. \tag{3.227}$$

Hence $|T(e^{j\omega})| = 1$, but the phase response is not linear. The phase distortion implicit in Eq. (3.227) can be reduced by all-pass phase correction networks.

A procedure for achieving this involves a modification of the product form of the M-band paraunitary lattice of Eq. (3.193). The substitution

$$z^{-1} \longrightarrow \frac{a_n^* + z^{-1}}{1 + a_n z^{-1}}, \qquad |a_n| < 1, \tag{3.228}$$

converts $\mathcal{H}_p(z)$ from a lossless FIR to a lossless IIR polyphase matrix. We can now select $\mathcal{G}_p^{'}(z)$ as in Eq. (3.225) to obtain the all-pass, stable $T(z)$.

To delve further into this subject, we pause to review the properties of all-pass filters.

3.7.1 All-Pass Filters and Mirror Image Polynomials

An all-pass filter is an IIR structure defined by

$$A(z) = \frac{z^{-p} + a_1 z^{-(p-1)} + \cdots + a_{p-1} z + a_p}{1 + a_1 z^{-1} + \cdots + a_p z^{-p}}. \tag{3.229}$$

This can also be expressed as

$$\begin{aligned} A(z) &= z^{-p} \frac{D(z^{-1})}{D(z)} \\ D(z) &= 1 + a_1 z^{-1} + \cdots + a_p z^{-p} \\ &= (1 - z_1 z^{-1}) \cdots (1 - z_p z^{-1}). \end{aligned} \tag{3.230}$$

From this last expression, we see that if poles of $A(z)$ are at $(z_1, z_2, \cdots, z_p)$, then the zeros are at reciprocal locations, $(z_1^{-1}, z_2^{-1}, \cdots, z_p^{-1})$, as depicted in Fig. 3.50. Hence $A(z)$ is a product of terms of the form $(1 - az)/(1 - az^{-1})$, each of which is all-pass. Therefore,

$$|A(e^{j\omega})| = 1$$

and note that the zeros of $A(z)$ are all non-minimum phase. Furthermore

$$A(z)A(z^{-1}) = 1 \tag{3.231}$$

These all-pass filters provide building blocks for lattice-type low-pass and high-pass power complementary filters. These are defined as the sum and difference of all-pass structures,

$$\begin{aligned} H_0(z) &= \frac{1}{2}[A_0(z) + A_1(z)] = \frac{N_0(z)}{D(z)} \\ H_1(z) &= \frac{1}{2}[A_0(z) - A_1(z)] = \frac{N_1(z)}{D(z)}, \end{aligned} \tag{3.232}$$

where $A_0(z)$, and $A_1(z)$ are all-pass networks with real coefficients.

Two properties can be established immediately:

(1) $N_0(z)$ is a mirror-image polynomial (even symmetric FIR), and $N_1(z)$ is an antimirror image polynomial (odd symmetric FIR).

(2) $H_0(z)$ and $H_1(z)$ are power complementary (Prob. 3.6):

$$\begin{aligned} H_0(z)H_0(z^{-1}) + H_1(z)H_1(z^{-1}) &= 1 \\ |H_0(e^{j\omega})|^2 + |H_1(e^{j\omega})|^2 &= 1. \end{aligned} \tag{3.233}$$

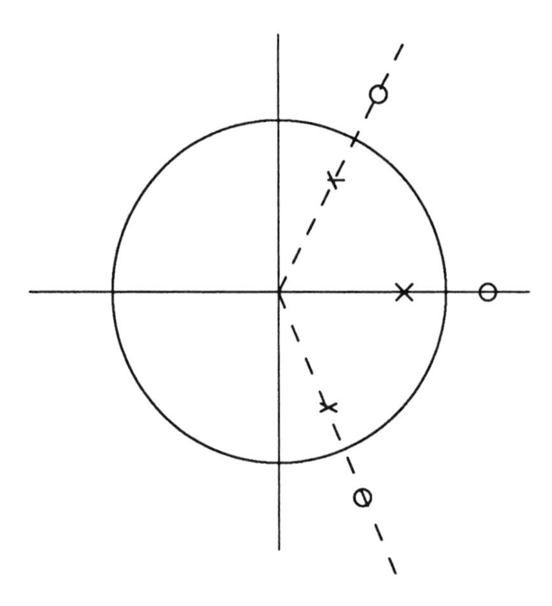

Figure 3.50: Pole-zero pattern of a typical all-pass filter.

A mirror image polynomial (or FIR impulse response with even symmetry) is characterized by Eq. (3.196) as

$$f_n = f_{N-1-n} \longleftrightarrow F(z) = z^{-(N-1)}F(z^{-1}). \tag{3.234}$$

The proof of this property is left as an exercise for the reader (Prob. 3.21). Thus if z_1 is a zero of $F(z)$, then z_1^{-1} is also a zero. Hence zeros occur in reciprocal pairs. Similarly, $F(z)$ is an antimirror image polynomial (with odd symmetric impulse response), then

$$f_k = -f_{N-1-k} \longleftrightarrow F(z) = -z^{-(N-1)}F(z^{-1}). \tag{3.235}$$

To prove property (1), let $A_0(z)$, $A_1(z)$ be all-pass of orders p_o and p_1, respectively. Then

$$A_0(z) = \frac{z^{-p_0}D_0(z^{-1})}{D_0(z)}, \qquad A_1(z) = \frac{z^{-p_1}D_1(z^{-1})}{D_1(z)}$$

and

$$H_0(z) = \frac{1}{2}[\frac{z^{-p_0}D_0(z^{-1})}{D_0(z)} + \frac{z^{-p_1}D_1(z^{-1})}{D_1(z)}] = \frac{N_0(z)}{D(z)},$$

where

$$D(z) = D_0(z)D_1(z)$$

and

$$N_0(z) = z^{-p_0} D_0(z^{-1}) D_1(z) + z^{-p_1} D_0(z) D_1(z^{-1}).$$

From this, it follows that

$$z^{-(p_0+p_1)} N_0(z^{-1}) = N_0(z).$$

Similarly, we can combine the terms in $H_1(z)$ to obtain the numerator

$$N_1(z) = z^{-p_0} D_0(z^{-1}) D_1(z) - z^{-p_1} D_0(z) D_1(z^{-1}),$$

which is clearly an antimirror image polynomial.

The power complementary property, Eq. (3.218), is established from the following steps: Let

$$Q(z) = D_0(z) D_1(z^{-1}).$$

Then

$$\begin{aligned} N_0(z) &= z^{-p_0} Q(z^{-1}) + z^{-p_1} Q(z) \\ N_1(z) &= z^{-p_0} Q(z^{-1}) - z^{-p_1} Q(z). \end{aligned}$$

But

$$\begin{aligned} W(z) &\triangleq H_0(z) H_0(z^{-1}) + H_1(z) H_1(z^{-1}) \\ &= \frac{N_0(z) N_0(z^{-1}) + N_1(z) N_1(z^{-1})}{D(z) D(z^{-1})} = \frac{N(z)}{D(z) D(z^{-1})}. \end{aligned}$$

By direct expansion and cancellation of terms, we find

$$\begin{aligned} N(z) = Q(z) Q(z^{-1}) &= D_0(z) D_1(z^{-1}) D_0(z^{-1}) D_1(z) \\ &= D(z) D(z^{-1}) \end{aligned}$$

and therefore, $W(z) = 1$, confirming the power complementary property.

These filters have additional properties:

(3)

$$|H_0(e^{j\omega}) \pm H_1(e^{j\omega})| = 1. \tag{3.236}$$

(4) There exists a simple lattice realization as shown in Fig. 3.51, and we can write

$$\begin{bmatrix} H_0(z) \\ H_1(z) \end{bmatrix} = \frac{1}{2} \begin{bmatrix} 1 & 1 \\ 1 & -1 \end{bmatrix} \begin{bmatrix} A_0(z) \\ A_1(z) \end{bmatrix}. \tag{3.237}$$

Observe that the lattice butterfly is simply a 2×2 Hadamard matrix.

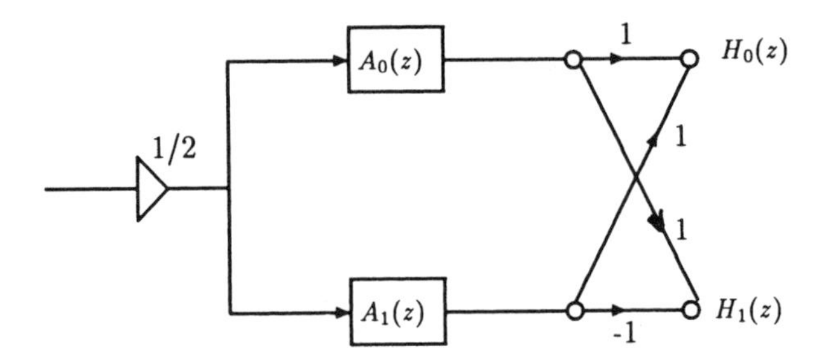

Figure 3.51: Lattice realization of power complementary filters; $A_0(z)$, $A_1(z)$ are all-pass networks.

3.7.2 The Two-Band IIR QMF Structure

Returning to the two-band filter structure of Fig. 3.20, we can eliminate aliasing from Eq. (3.36) by selecting $G_0(z) = H_1(-z)$ and $G_1(z) = -H_0(-z)$. This results in

$$T(z) = H_0(z)H_1(-z) - H_0(-z)H_1(z).$$

Now let $H_1(z) = H_0(-z)$, which ensures that $H_1(z)$ will be high-pass if $H_0(z)$ is low-pass. Thus

$$T(z) = H_0^2(z) - H_0^2(-z) = H_0^2(z) - H_1^2(z).$$

Finally, the selection of $H_0(z)$ and $H_1(z)$ by Eq. (3.232) results in

$$T(z) = \frac{1}{4}(A_0(z) + A_1(z))^2 - \frac{1}{4}(A_0(z) - A_1(z))^2 = A_0(z)A_1(z). \qquad (3.238)$$

Thus, $T(z)$ is the product of two all-pass transfer functions and, therefore, is itself all-pass. Some insight into the nature of the all-pass is achieved by the polyphase representations of the analysis filters,

$$\begin{aligned} H_0(z) = a_0(z^2) + z^{-1}a_1(z^2) &= A_0(z) + A_1(z) \\ H_1(z) = H_0(-z) = a_0(z^2) - z^{-1}a_1(z^2) &= A_0(z) - A_1(z). \end{aligned}$$

The all-pass networks are therefore

$$\begin{aligned} A_0(z) &= a_0(z^2) \\ A_1(z) &= z^{-1}a_1(z^2). \end{aligned} \qquad (3.239)$$

These results suggest the two-band lattice of Fig. 3.52, where $a_0(z)$ and $a_1(z)$ are both all-pass filters.

We can summarize these results with

$$\begin{cases} H_0(z) &= A_0(z) + A_1(z) = a_0(z^2) + z^{-1}a_1(z^2) = N_0(z)/D(z) \\ H_1(z) &= H_0(-z) = N_1(z)/D(z) \end{cases} \tag{3.240}$$

$$\begin{aligned} |H_0(e^{j\omega})|^2 + |H_1(e^{j\omega})|^2 &= 1 \\ T(z) &= A_0(z)A_1(z). \end{aligned}$$

In addition to the foregoing constraints, we also want the high-pass filter to have zero DC gain (and correspondingly, the low-pass filter gain to be zero at $\omega = \pi$). It can be shown that if the filter length N is even (i.e., filter order $N-1$ is odd), then $N_0(z)$ has a zero at $z = -1$ and $N_1(z)$ has a zero at $z = 1$. Pole-zero patterns for typical $H_0(z)$, $H_1(z)$ are shown in Fig. 3.53.

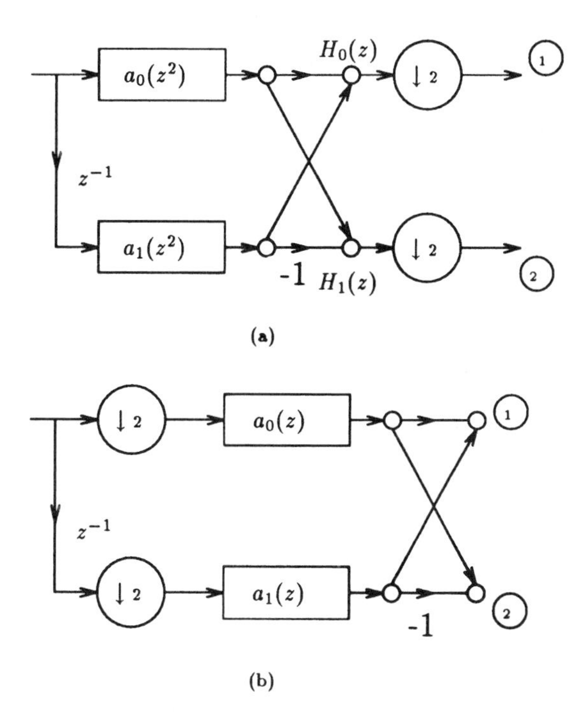

Figure 3.52: Two-band power complementary all-pass IIR structure.

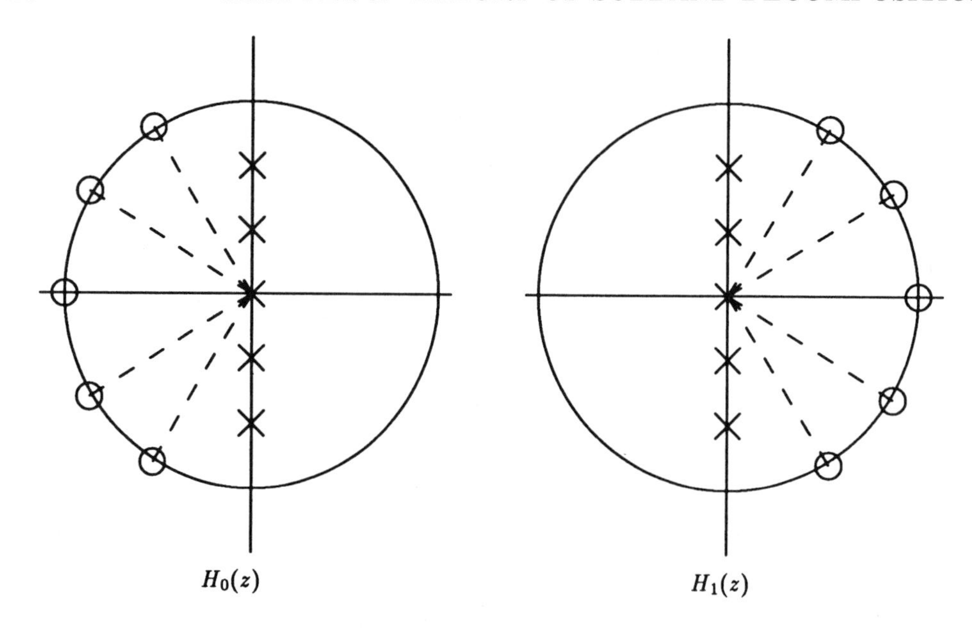

Figure 3.53: Typical IIR power complementary two-band filters.

A design procedure as described in Vaidyanathan (Jan. 1990) is as follows. Let the all-pass polyphase components $a_0(z)$, $a_1(z)$ have alternating real poles

$$a_0(z) = \prod_{k=\text{odd}} \frac{(p_k + z^{-1})}{(1 + p_k z^{-1})}, \quad a_1(z) = \prod_{k=\text{even}} \frac{(p_k + z^{-1})}{(1 + p_k z^{-1})}$$
$$0 < p_1 < p_2 < \cdots < p_n. \tag{3.241}$$

Then,

$$H_0(z) = a_0(z^2) + z^{-1}a_1(z^2) = \frac{N_0(z)}{\prod_k (1 + p_k z^{-2})}. \tag{3.242}$$

By construction, $N_0(z)$ is a mirror image polynomial of odd order $N - 1$, and the poles of $H_0(z)$ are all purely imaginary. The set $\{p_k\}$ can then be chosen to put the zeros of $N_0(z)$ on the unit circle as indicated in Fig. 3.53. Procedures for designing M-band power complementary filters are given in Vaidyanathan (Jan. 1990), and S. R. Pillai, Robertson, and Phillips (1991). (See also Prob. 3.31)

3.7.3 Perfect Reconstruction IIR Subband Systems

We know that physically realizable (i.e., causal) IIR filters cannot have a linear-phase. However, noncausal IIR filters can exhibit even symmetric impulse responses and thus have linear-phase, in this case, zerophase.

This suggests that noncausal IIR filters can be used to eliminate phase distortion as well as amplitude distortion in subbands. One procedure for achieving a linear-phase response uses the tandem connection of identical causal IIR filters separated by two time-reversal operators, as shown in Fig. 3.54.

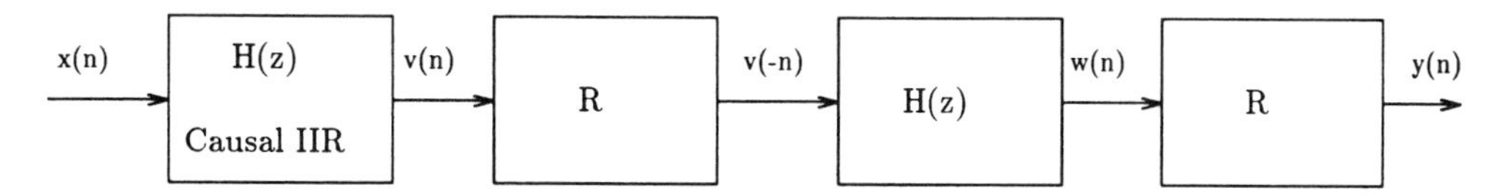

Figure 3.54: Linear-phase IIR filter configuration; R is a time-reversal.

The finite duration input signal $x(n)$ is applied to the causal IIR filter $H(z)$. The output $v(n)$ is lengthened by the impulse response of the filter and hence in principle is of infinite duration. In time, this output becomes sufficiently small and can be truncated with negligible error. This truncated signal is then reversed in time and applied to $H(z)$ to generate the signal $w(n)$; this output is again truncated after it has become very small, and then reversed in time to yield the final output $y(n)$.

Noting that the time-reversal operator induces

$$R\{v(n)\} = v(-n) \longleftrightarrow R\{V(z)\} \longleftrightarrow V(z^{-1}), \tag{3.243}$$

we can trace the signal transmission through Fig. 3.52 to obtain

$$V(z) = H(z)X(z), \qquad W(z) = H(z)V(z^{-1})$$

and

$$Y(z) = W(z^{-1}) = H(z^{-1})V(z) = [H(z^{-1})H(z)]X(z) = \Phi(z)X(z).$$

Hence,

$$Y(z) = \Phi(z)X(z) \longleftrightarrow y(n) = \rho(n) * x(n), \tag{3.244}$$

where

$$\Phi(z) = H(z)H(z^{-1}) \longleftrightarrow \rho(n) = h(n) * h(-n), \tag{3.245}$$
$$\Phi(e^{j\omega}) = |H(e^{j\omega})|^2.$$

The composite transfer function is $|H(e^{j\omega})|^2$ and has zero phase. The time reversals in effect cause the filters to behave like a cascade of stable causal and stable anticausal filters.

This analysis does not account for the inherent delays in recording and reversing the signals. We can account for these by multiplying $\Phi(z)$ by $z^{-(N_1+N_2)}$, where N_1 and N_2 represent the delays in the first and second time-reversal operators.

Kronander (ASSP, Sept. 88) employed this idea in his perfect reconstruction two-band structure shown in Fig. 3.55. Two time-reversals are used in each leg but these can be distributed as shown, and all analysis and synthesis filters are causal IIR.

Using the transformations induced by time-reversal and up- and down-sampling, we can calculate the output as

$$\begin{aligned}\hat{X}(z) &= T(z)X(z) + S(z)X(-z)\\ S(z) &= \frac{1}{2}[G_0(z^{-1})H_0(-z) - G_1(z)H_0(-z^{-1})]\\ T(z) &= \frac{1}{2}[G_0(z^{-1})H_0(z) + G_1(z)H_0(z^{-1})]. \end{aligned} \tag{3.246}$$

The aliasing term $S(z)$ can be eliminated, and a low-pass/high-pass split obtained by choosing

$$\begin{aligned}H_1(z) &= H_0(-z)\\ G_0(z) &= 2H_0(z)\\ G_1(z) &= 2H_1(z). \end{aligned} \tag{3.247}$$

This forces $S(z) = 0$, and $T(z)$ is simply

$$T(z) = [H_0(z^{-1})H_0(z) + H_0(-z)H_0(-z^{-1})]. \tag{3.248}$$

On the unit circle (for real $h_0(n)$), the PR condition reduces to

$$T(e^{j\omega}) = |H_0(e^{j\omega})|^2 + |H_0(e^{j(\omega-\pi)})|^2 = 1. \tag{3.249}$$

Hence, we need satisfy only the power complementarity requirement of causal IIR filters to obtain perfect reconstruction! We may regard this last equation as the culmination of the concept of combining causal IIR filters and time-reversal operators to obtain linear-phase filters, as suggested in Fig. 3.54.

The design of $\{H_0(z), H_1(z)\}$ IIR pair can follow standard procedures, as outlined in the previous section. We can implement $H_0(z)$ and $H_1(z)$ by the all-pass lattice structures as given by Eqs. (3.241) and (3.242) and design the constituent all-pass filters using standard tables (Gazsi, 1985).

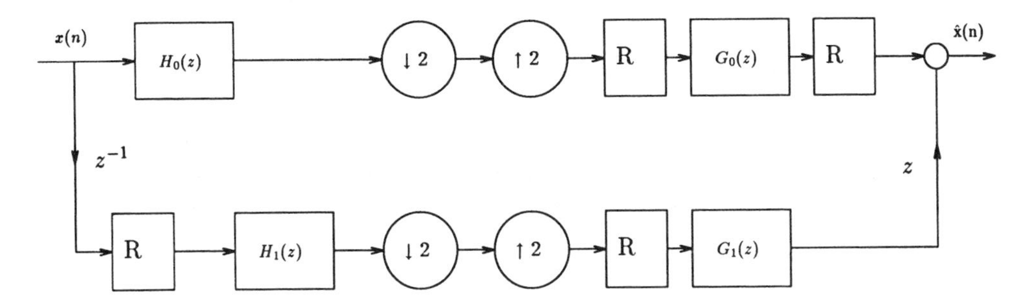

Figure 3.55: Two-band perfect reconstruction IIR configuration; R denotes a time-reversal operator.

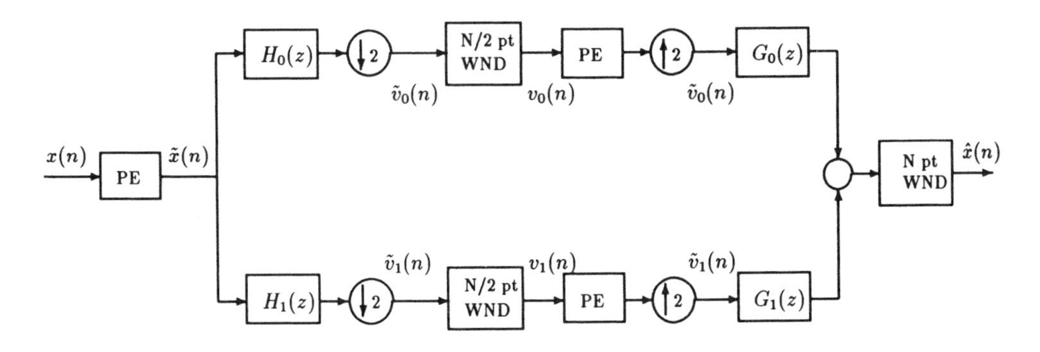

Figure 3.56: Two-channel IIR subband configuration. PE means periodic extension and WND is the symbol for window.

The second approach to PR IIR filter banks was advanced by Smith and Eddins (ASSP, Aug. 1990) for filtering finite duration signals such as sequences of pixels in an image. A continuing stream of sequences such as speech is, for practical purposes, infinite in extent. Hence each subband channel is maximally decimated at its respective Nyquist rate, and the total number of input samples equals the number of output samples of the analysis section. For images, however, the convolution of the spatially limited image with each subband analysis filter generates outputs whose lengths exceed the input extent. Hence the total of all the samples (i.e., pixels) in the subband output exceeds the total number of pixels in the image; the achievable compression is decreased accordingly, because of this overhead.

The requirements to be met by Smith and Eddins are twofold:

(1) The analysis section should not increase the number of pixels to be encoded.

(2) IIR filters with PR property are to be used.

The proposed configuration for achieving these objectives is shown in Fig. 3.56 as a two-band codec and in Fig. 3.57 in the equivalent polyphase lattice form.

The analysis section consists of low-pass and high-pass causal IIR filters, and the synthesis section of corresponding anticausal IIR filters. The key to the proposed solution is the conversion of the finite-duration input signal to a periodic one:

$$\tilde{x}(n) = \sum_{l=-\infty}^{\infty} x(n - lN), \tag{3.250}$$

and the use of circular convolution. In the analysis section the causal IIR filter is implemented by a difference equation running forward in time over the periodic signal; in the synthesis part, the anticausal IIR filter operates via a backward-running difference equation. Circular convolution is used to establish initial conditions for the respective difference equations.

These periodic repetitions are indicated by *tildes* on each signal. The length N input $x(n)$ is periodically extended to form $\tilde{x}(n)$ in accordance with Eq. (3.250). As indicated in Fig. 3.57 this signal is subsampled to give $\tilde{\xi}_0(n)$ and $\tilde{\xi}_1(n)$, each of period $N/2$ (N is assumed to be even). Each subsampled periodic sequence is then filtered by the causal IIR polyphase lattice to produce the $N/2$ point periodic sequences $\tilde{v}_0(n)$ and $\tilde{v}_1(n)$. These are then windowed by an $N/2$ point window prior to encoding. Thus, the output of the analysis section consists of two $N/2$ sample sequences while the input $x(n)$ had N samples. Maximal decimation is thereby preserved. Inverse operations are performed at the synthesis side using noncausal operators. Next, we show that this structure is indeed perfect reconstruction and describe the details of the operations.

For the two-band structure, the perfect reconstruction conditions were given by Eq. (3.74), which is recast here as

$$\begin{bmatrix} H_0(z) & H_1(z) \\ H_0(-z) & H_1(-z) \end{bmatrix} \begin{bmatrix} G_0(z) \\ G_1(z) \end{bmatrix} = \begin{bmatrix} 2 \\ 0 \end{bmatrix}. \tag{3.251}$$

The unconstrained solution is

$$\begin{bmatrix} G_0(z) \\ G_1(z) \end{bmatrix} = \frac{2}{\Delta} \begin{bmatrix} H_1(-z) \\ -H_0(-z) \end{bmatrix}, \tag{3.252}$$

where

$$\Delta \triangleq H_0(z)H_1(-z) - H_0(-z)H_1(z).$$

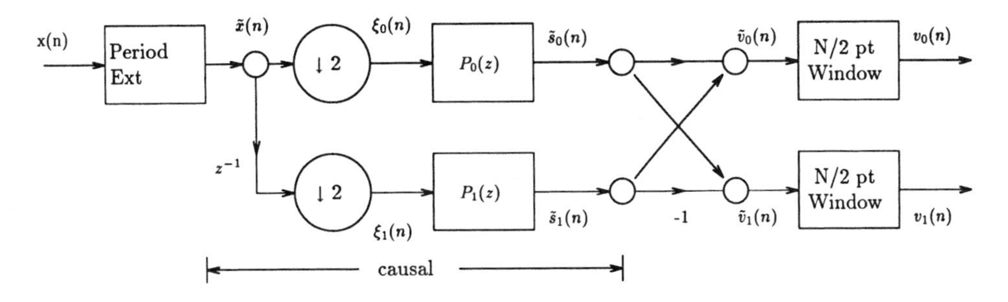

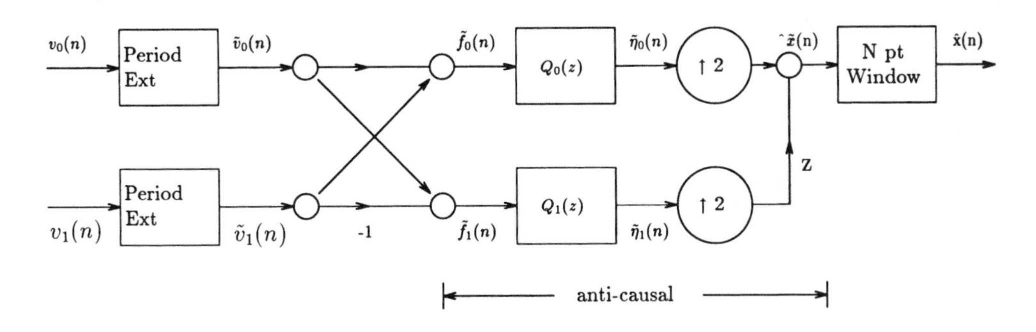

Figure 3.57: Two-channel polyphase lattice configuration with causal analysis and anticausal synthesis sections.

Let us construct the analysis filters from all-pass sections and constrain $H_1(z) = H_0(-z)$. Thus, we have the polyphase decomposition

$$\begin{aligned} H_0(z) &= P_0(z^2) + z^{-1}P_1(z^2) \\ H_1(z) &= P_0(z^2) - z^{-1}P_1(z^2); \end{aligned} \tag{3.253}$$

and $P_0(z)$, $P_1(z)$ are both all-pass. For this choice, Δ reduces to simply

$$\Delta = 4z^{-1}P_0(z^2)P_1(z^2). \tag{3.254}$$

The PR conditions are met by

$$G_0(z) = \frac{2H_1(-z)}{\Delta} = \frac{2[P_0(z^2) + z^{-1}P_1(z^2)]}{4z^{-1}P_0(z^2)P_1(z^2)} = \frac{1}{2}[\frac{1}{P_0(z^2)} + \frac{z}{P_1(z^2)}].$$

But, for an all-pass, $P_0(z)P_0(z^{-1}) = 1$. Hence

$$\begin{aligned} G_0(z) &= \frac{1}{2}[P_0(z^{-2}) + zP_1(z^{-2})] = \frac{1}{2}H_0(z^{-1}) \\ &\triangleq Q_0(z^2) + zQ_1(z^2). \end{aligned} \tag{3.255}$$

Similarly, we find

$$\begin{aligned} G_1(z) &= \frac{1}{2}[\frac{1}{P_0(z^2)} - \frac{z}{P_1(z^2)}] = -\frac{1}{2}H_1(z^{-1}) \\ &= Q_0(z^2) - zQ_1(z^2). \end{aligned} \tag{3.256}$$

Therefore, the PR conditions for the synthesis all-pass filters are simply

$$\begin{aligned} Q_0(z) &= \frac{1/2}{P_0(z)} = \frac{1}{2}P_0(z^{-1}) \\ Q_1(z) &= \frac{1/2}{P_1(z)} = \frac{1}{2}P_1(z^{-1}), \end{aligned} \tag{3.257}$$

which are recognized as anticausal, if the analysis filters are causal.

To illustrate the operation, suppose $P_0(z)$ is first-order:

$$\begin{aligned} P_0(z) = \left(\frac{a + z^{-1}}{1 + az^{-1}}\right) &\longleftrightarrow p_0(n) = \beta_0\delta(n) + \beta_1(-a)^n u(n) \\ &\beta_0 = 1/a, \quad \beta_1 = (a - 1/a). \end{aligned} \tag{3.258}$$

The difference equation is then (the subscript is omitted for simplicity)

$$\tilde{s}(n) = a[\tilde{\xi}(n) - \tilde{s}(n-1)] + \tilde{\xi}(n-1). \tag{3.259}$$

Since the $N/2$ point periodic sequence $\tilde{\xi}(n)$ is given, we can solve the difference equation recursively for $n = 0, 1, 2, ..., \frac{N}{2} - 1$. Use is made of the periodic nature of $\tilde{\xi}(n)$ so that terms like $\tilde{\xi}(-1)$ are replaced by $\tilde{\xi}(\frac{N}{2} - 1)$; *but* we need an initial condition $\tilde{s}(-1)$. This is obtained via the following steps. The impulse response $p_0(n)$ is circularly convolved with the periodic input.

$$\begin{aligned} \tilde{s}(n) &= \tilde{\xi}(n) * p_0(n) = \beta_0\tilde{s}(n) + \sum_{m=-\infty}^{\infty} \tilde{\xi}(n-m)\beta_1(-a)^m u(m) \\ &= \beta_0\tilde{\xi}(n) + \sum_{m=0}^{\infty} \tilde{\xi}(n-m)\beta_1(-a)^m. \end{aligned} \tag{3.260}$$

But $\sum_{m=0}^{\infty}\phi(m)$ can be written as $\sum_{k=0}^{\frac{N}{2}-1}\sum_{l=0}^{\infty}\phi(k+lN/2)$. The sum term becomes

$$\sum_{k=0}^{\frac{N}{2}-1}\sum_{l=0}^{\infty}\tilde{\xi}(n-(k+lN/2))\beta_1(-a)^{k+lN/2} = \sum_{k=0}^{\frac{N}{2}-1}\beta_1(-a)^k\tilde{\xi}(n-k)\sum_{l=0}^{\infty}[(-a)^{N/2}]^l$$

$$= \sum_{k=0}^{\frac{N}{2}-1}\tilde{\xi}(n-k)\left(\frac{\beta_1(-a)^k}{1-(-a)^{N/2}}\right) = \sum_{k=0}^{\frac{N}{2}-1}\tilde{\xi}(n-k)g(k).$$

Finally,

$$\tilde{s}(n) = \beta_0\tilde{\xi}(n) + \sum_{k=0}^{\frac{N}{2}-1}\tilde{\xi}(n-k)g(k) \tag{3.261}$$

This last equation is used to compute $\tilde{s}(-1)$, the initial condition needed for the difference equation, Eq. (3.259).

The synthesis side operates with the anticausal all-pass

$$Q_0(z) = \frac{1}{2}P_0(z^{-1}) = \frac{1}{2}\frac{(a+z)}{(1+az)}$$

or

$$2q_0(n) = p_0(-n) = \beta_0\delta(n) + \beta_1(-a)^{-n}u(-n).$$

The difference equation is

$$\tilde{\eta}(n) = a[\frac{1}{2}\tilde{f}(n) - \tilde{\eta}(n+1)] + \frac{1}{2}\tilde{f}(n+1)$$

which is iterated *backward* in time to obtain the sequence

$$\{\tilde{\eta}(0), \tilde{\eta}(-1), \cdots, \tilde{\eta}(2-\frac{N}{2})\}$$

with starting value $\tilde{\eta}(1)$ obtained from the circular convolution of $g_0(n)$ and $\tilde{f}(n)$. This can be shown to be

$$\tilde{\eta}(1) = \alpha_0\tilde{f}(1) + \sum_{k=0}^{\frac{N}{2}-1}g(k)\tilde{f}(1+k)$$

with

$$g(k) = \frac{\alpha_1(-\frac{1}{a})^{-k}}{1-(-\frac{1}{a})^{-N/2}}.$$

The classical advantages of IIR over FIR are again demonstrated in subband coding. Comparable magnitude performance is obtained for a first-order $P_0(z)$ (or fifth-order $H_0(z)$) and a 32-tap QMF structure The computational complexity is also favorable to the IIR structure, typically by factors of 7 to 14 (Smith and Eddins, 1990).

3.8 Transmultiplexers

The subband filter bank or codec of Fig. 3.32 is an analysis/synthesis structure. The front end or "analysis" side performs signal decomposition in such a way as to allow compression for efficient transmission or storage. The receiver or "synthesis" section reconstructs the signal from the decomposed components.

The transmultiplexer, depicted in Fig. 3.58, on the other hand, can be viewed as the dual of the subband codec. The front end constitutes the synthesis section wherein several component signals are combined to form a composite signal which can be transmitted over a common channel. This composite signal could be any one of the time-domain multiplexed (TDM), frequency-domain multiplexed (FDM), or code division multiplexed (CDM) varieties. At the receiver the analysis filter bank separates the composite signal into its individual components. The multiplexer can therefore be regarded as a synthesis/analysis filterbank structure that functions as the conceptual dual of the analysis/synthesis subband structure.

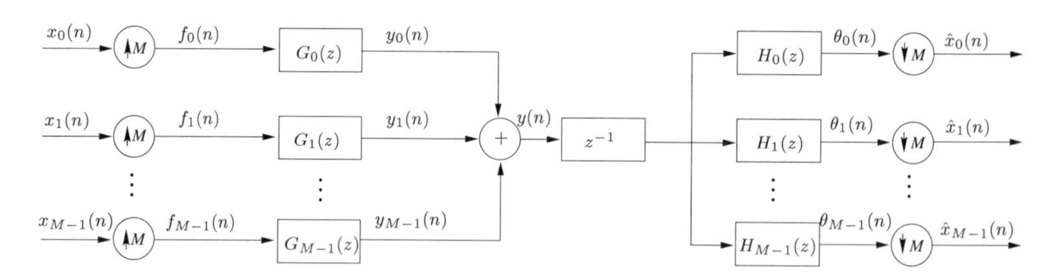

Figure 3.58: M-band multiplexer as a critically sampled synthesis/analysis multirate filterbank.

In this section we explore this duality between codec and transmux and show that perfect reconstruction and alias cancellation in the codec correspond to PR and cross-talk cancellation in the transmux.

3.8.1 TDMA, FDMA, and CDMA Forms of the Transmultiplexer

The block diagram of the M-band digital transmultiplexer is shown in Figure 3.58. Each signal $x_k(n)$ of the input set

$$\{x_0(n), x_1(n), ..., x_{M-1}(n)\}$$

is up-sampled by M, and then filtered by $G_k(z)$, operating at the fast clock rate. This signal $y_k(n)$ is then added to the other components to form the composite signal $y(n)$, which is transmitted over one common channel wherein a unit delay is introduced[2]. This is a multiuser scenario wherein the components of this composite signal could be TDM, FDM, or CDM depending on the filter used. The simplest case is that of the TDM system. Here each synthesis filter $(G_k(z) = z^{-k}, k = 0, 1, ..., M-1)$ is a simple delay so that the composite signal $y(n)$ is the interleaved signal

$$y(n) = \{..., x_0(0), x_1(0), ..., x_{M-1}(0), x_0(1), x_1(1), ...\}. \tag{3.262}$$

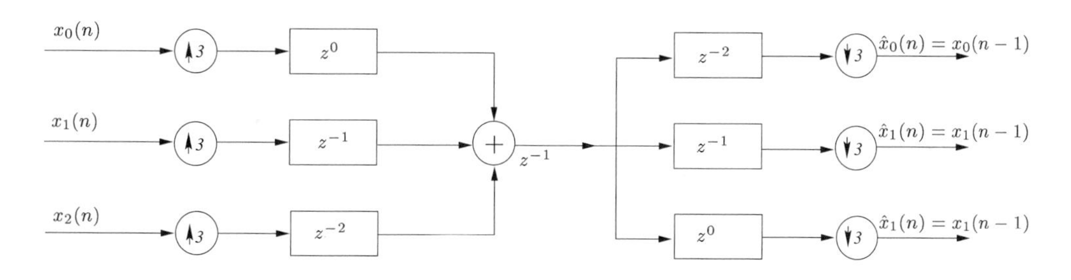

Figure 3.59: Three-band TDMA Transmultiplexer.

At the receiver (or "demux"), the composite TDM signal is separated into its constituent components. This is achieved by feeding the composite signal into a bank of appropriately chosen delays, and then down-sampling, as indicated in Fig. 3.59 for a three-band TDMA transmux. For the general case with $G_k(z) = z^{-k}, 0 \leq k \leq M-1$, the separation can be realized by choosing the corresponding analysis filter to be

$$z^{-1}H_k(z) = z^{-(rM-k)}, \quad 0 \leq k \leq M-1, \tag{3.263}$$

[2]Insertion of a delay z^{-1} (or more generally $z^{-(lM+1)}$ for l any integer) simplifies the analysis to follow and obviates the need for a shuffle matrix in the system transfer function matrix.

for r any integer. The simplest noncausal and causal cases correspond to $r = 0$ and $r = 1$, respectively. This reconstruction results in just a simple delay,

$$\hat{x}_k(n) = x_k(n - r), \quad 0 \leq k \leq M - 1, \tag{3.264}$$

as can be verified by a study of the selectivity provided by the upsampler-delay-down-sampler structure shown as Fig. 3.60. This is a linear time-invariant system whose transfer function is zero unless the delay r is a multiple of M, i.e.,

$$Y(z) = T(z)X(z), \quad T(z) = \begin{cases} z^{-l}, & r = lM; \\ 0 & \text{otherwise.} \end{cases} \tag{3.265}$$

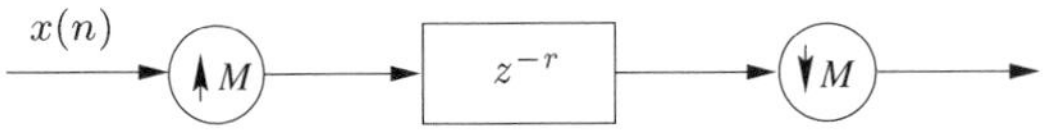

$x(n)$ → $\uparrow M$ → z^{-r} → $\downarrow M$ → $y(n) = \begin{cases} x(n-l), & if \quad r = lM, \\ 0 & \quad\;\; r \neq lM. \end{cases} \forall l$

Figure 3.60: Up-sampler-delay-down-sampler structure.

In essence, the TDMA transmux provides a kind of time-domain orthogonality across the channels. Note that the impulse responses of the synthesis filters

$$g_k(n) = \delta(n - k), \quad 0 \leq k \leq M - 1, \tag{3.266}$$

are orthonormal in time. Each input sample is provided with its own time slot, which does not overlap with the time slot allocated to any other signal. That is,

$$\sum_{n=0}^{M-1} g_k(n) g_l(n) = \delta_{k-l}. \tag{3.267}$$

This represents the rawest kind of orthogonality in time. From a time-frequency standpoint, the impulse response is the time-localized Kronecker delta sequence while the frequency response,

$$G_k(e^{j\omega}) = e^{-jk\omega}, \quad \text{or} \quad |G_k(e^{j\omega})| = 1, \tag{3.268}$$

has a flat, all-pass frequency characteristic with linear-phase. The filters all overlap in frequency but are absolutely non-overlapping in the time domain; this is a pure TDM⟶TDM system.

The second scenario is the TDM ⟶FDM system. In this case, the up-sampler compresses the frequency scale for each signal (see Fig. 3.61). This is followed by an ideal, "brick-wall," band-pass filter of width π/M, which eliminates the images and produces the FDM signal occupying a frequency band π/M. These FDM signals are then added in time or butted together in frequency with no overlap and transmitted over a common channel. The composite FDM signal is then separated into its component parts by band-pass, brick-wall filters in the analysis bank, and then down-sampled by M so as to occupy the full frequency band at the slow clock rate.

An example for an ideal 2-band FDM transmux is depicted in Figs. 3.61 and 3.62. The FDM transmux is the frequency-domain dual of the TDM transmux. In the FDM system, the band-pass synthesis filterbank allocates frequency bands or "slots" to the component signals. The FDM signals are distributed and overlap time, but occupy non-overlapping slots in frequency. On the interval $[0, \pi]$, the synthesis filters defined by

$$G_k(e^{j\omega}) = \begin{cases} 1, & \frac{k\pi}{M} \leq \omega \leq \frac{(k+1)\pi}{M}; \\ 0, & \text{else.} \end{cases} \quad ,k = 0, 1, ..., M-1, \tag{3.269}$$

are clearly orthogonal by virtue of non-overlap in frequency

$$\int_{-\pi}^{\pi} G_k(e^{j\omega}) G_l(e^{j\omega}) d\omega = \delta_{k-l}. \tag{3.270}$$

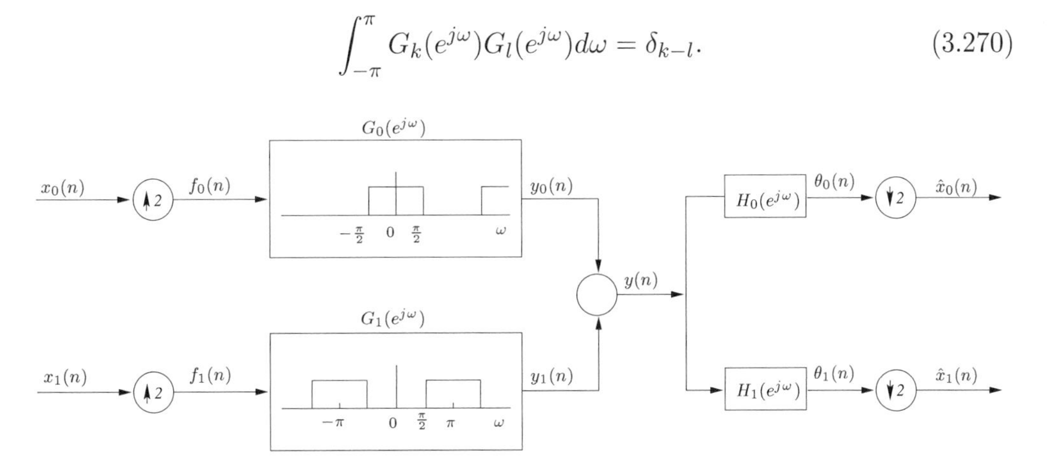

Figure 3.61: Ideal two-band TDM-FDM transmux. H_0 and H_1 are ideal low-pass and band-pass filters.

These filters are localized in frequency but distributed over time, The time-frequency duality between TDM and FDM transmultiplexers is summarized in Table 3.3. It should be evident at this point that the orthonormality of a transmultiplexer need not be confined to purely TDM or FDM varieties. The orthonormality and localization can be distributed over both time- and frequency-domains,

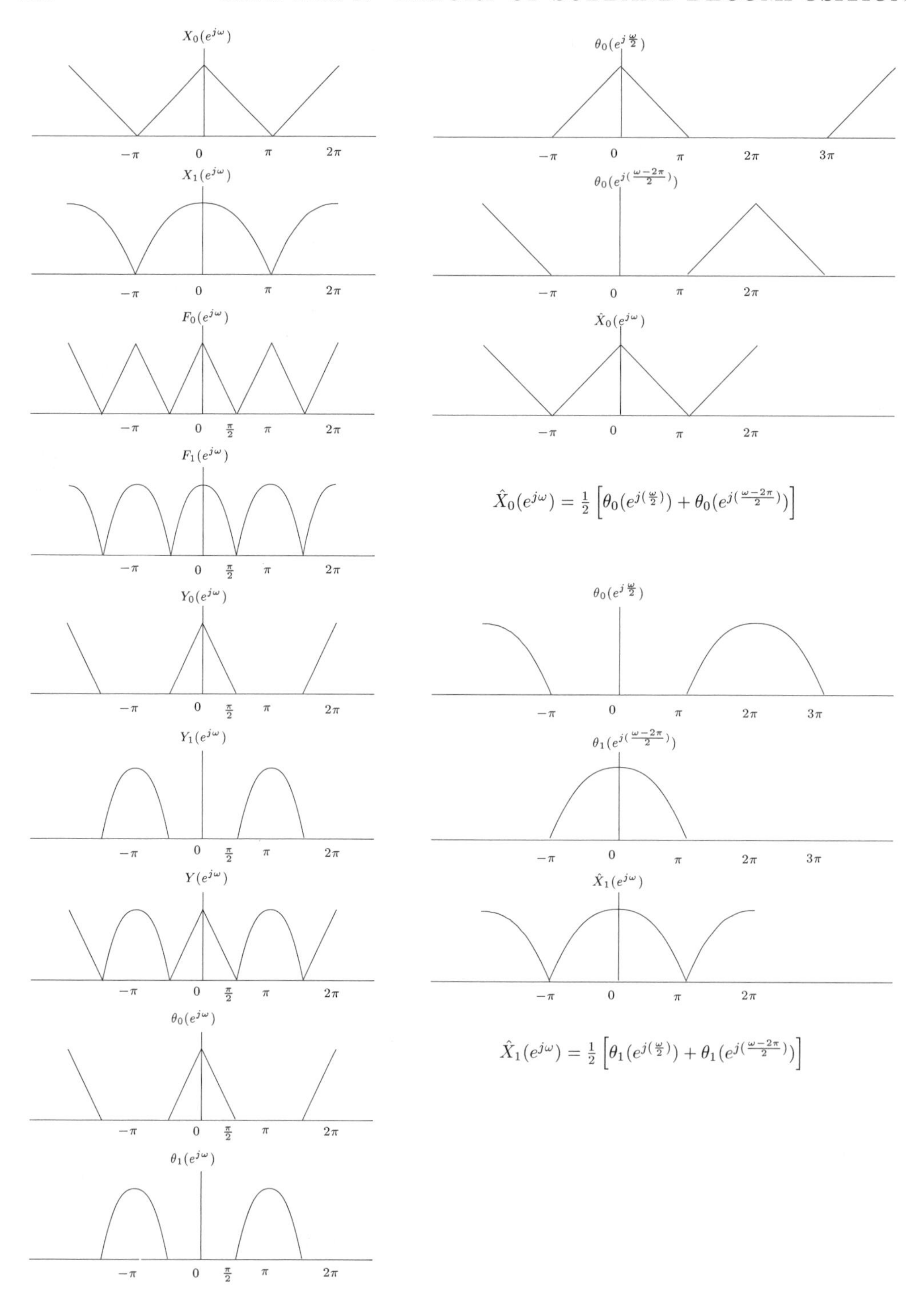

Figure 3.62: Signal transforms in ideal 2-band FDM system transmux of Fig. 3.61.

as in QMF filter banks. When this view is followed, we are led to a consideration of a broader set of orthonormality conditions which lead to perfect reconstruction. The filter impulse responses for this class are the code-division multiple access (CDMA) codes for a set of signals. These filter responses are also the same as what are known as orthogonal spread spectrum codes.

	Impulse response $g_k(n)$	Frequency response $G_k(e^{j\omega})$	Localization
TDM	$\delta(n-k)$	$e^{-jk\omega}$ all-pass	Time
FDM	$\frac{1}{M}\frac{\sin(n\pi/2M)}{n\pi/2M}\cos\{(k+\frac{1}{2})\frac{n\pi}{M}\}$	Eq. (3.269), band-pass	Frequency
CDM	Distributed over time and frequency		

Table 3.3: Time-frequency characteristics of TDM and FDM transmultiplexers.

3.8.2 Analysis of the Transmultiplexer

In this section we show that the conditions on the synthesis/analysis filters for perfect reconstruction and for cross-talk cancellation are identical to those for PR and alias cancellation in the QMF filterbank. Using the polyphase equivalences for the synthesis and analysis filter, we can convert the structure in Fig. 3.58 to the equivalent shown in Fig. 3.63 where the notation is consistent with that used in connection with Figs. 3.35 and 3.36. Examination of the network within the dotted lines shows that there is no cross-band transmission, and that within each band, the transmission is a unit delay, i.e.,

$$\xi_k(n) = \eta_k(n-1), \quad k = 0, 1, ..., M-1. \tag{3.271}$$

This is also evident from the theorem implicit in Fig. 3.60. Using vector notation and transforms, we have

$$\underline{\xi}(z) = z^{-1} I \underline{\eta}(z). \tag{3.272}$$

Therefore, at the slow clock rate, the transmission from $\underline{\eta}(z)$ to $\underline{\xi}(z)$ is just a diagonal delay matrix. The system within the dotted line in Fig. 3.63 can therefore be replaced by matrix $z^{-1}I$ as shown in Fig. 3.64. This diagram also demonstrates that the multiplexer from slow clock rate input $\underline{x}(n)$ to slow clock rate output $\hat{x}(n)$ is linear, time-invariant (LTI) for any polyphase matrices $\mathcal{G}'_p(z), \mathcal{H}_p(z)$, and hence is LTI for any synthesis/analysis filters. This should be compared with the analysis/synthesis codec which is LTI at the slow-clock rate (from $\underline{\xi}(n)$ to

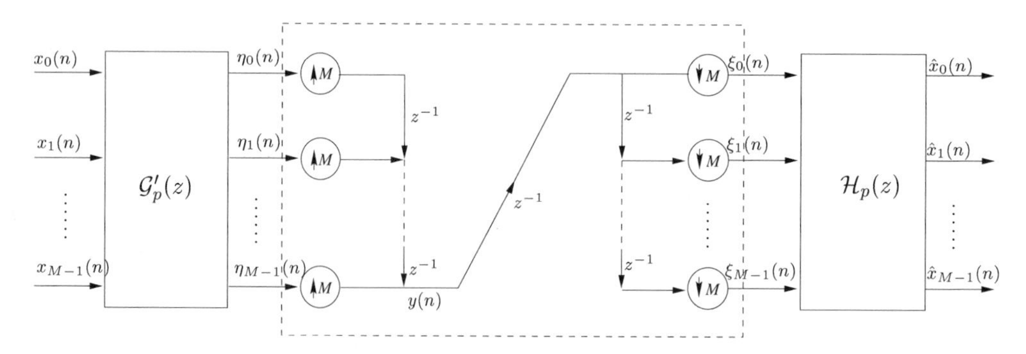

Figure 3.63: Equivalent polyphase transmultiplexer.

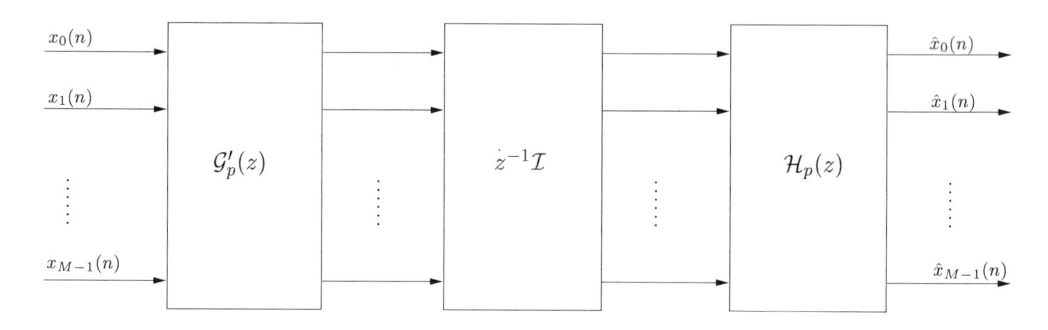

Figure 3.64: Reduced polyphase equivalent of transmultiplexer.

$\underline{\eta}(n)$ in Fig. 3.36), but is LTI at the fast clock rate (from $x(n)$ to $\hat{x}(n)$) *only if aliasing terms are cancelled.* The complete analysis of the transmultiplexer using polyphase matrices is quite straightforward. From Fig. 3.64, we see that

$$\underline{\hat{X}}(z) = \left[z^{-1}\mathcal{H}_p(z)\mathcal{G}'_p(z)\right] \underline{X}(z). \tag{3.273}$$

For PR with a unit (slow clock) delay, we want

$$\underline{\hat{x}}(n) = \underline{x}(n-1). \tag{3.274}$$

Hence, the necessary and sufficient condition for a PR transmultiplexer is simply

$$\mathcal{H}_p(z)\mathcal{G}'_p(z) = I. \tag{3.275}$$

From Eqs. (3.100) and (3.101), the corresponding condition for PR in the QMF filter bank is

$$\hat{x}(n) = x(n-(M-1)), \tag{3.276}$$

which is achieved iff

$$\mathcal{G}'_p(z)\mathcal{H}_p(z) = I. \tag{3.277}$$

Since $\mathcal{G}'_p(z)$ and $\mathcal{H}_p(z)$ are each square, then Eq. (3.275) implies Eq. (3.277), and conversely. An immediate consequence of this is that any procedure for designing PR codecs can be used to specify PR transmultiplexers. In particular, the orthonormal (or paraunitary) filter bank conditions in Section 3.5.4 carry over intact for the transmux.

The cross-talk cancellation condition obtains when there is no interference from one channel to another. This is secured iff $\mathcal{H}_p(z)\mathcal{G}'_p(z)$ is a diagonal matrix. This condition is satisfied if the progenitor codec is alias free. The argument in support of this contention is as follows:

In the codec of Fig. 3.36, let $\mathcal{P}(z) = \mathcal{G}'_p(z)\mathcal{H}_p(z)$ be diagonal

$$\mathcal{P}(z) = \mathcal{D}(z) = diag\{D_0(z), D_1(z), ..., D_{M-1}(z)\}.$$

Then Fig. 3.36 can be put into the form of Fig. 3.65(a), which in turn can be manipulated into Fig. 3.65(b) using the noble identities of Fig. 3.7. From Fig. 3.65(b) and (c), we can write

$$\hat{X}(z) = \sum_{r=0}^{M-1} X_r(z), \quad X_r(z) = D_r(z^M)V_r(z) \tag{3.278}$$

and

$$V_r(z) = \frac{z^{-(M-1)}}{M} \sum_{k=0}^{M-1} W^{-kr} X(zW^k). \tag{3.279}$$

Combining these, we obtain

$$\hat{X}(z) = \frac{z^{-(M-1)}}{M} \sum_{k=0}^{M-1} \left[\sum_{r=0}^{M-1} D_r(z^M) W^{-kr} \right] X(zW^k). \tag{3.280}$$

To eliminate aliasing, Eq. (3.280) must reduce to $\hat{X}(z) = T(z)X(z)$, the input-output relationship of a LTI system. This is achieved if

$$D_r(z) = D_s(z) \stackrel{\triangle}{=} D(z), \quad \forall\, r, s \tag{3.281}$$

for then the term in square brackets in Eq. (3.280) becomes

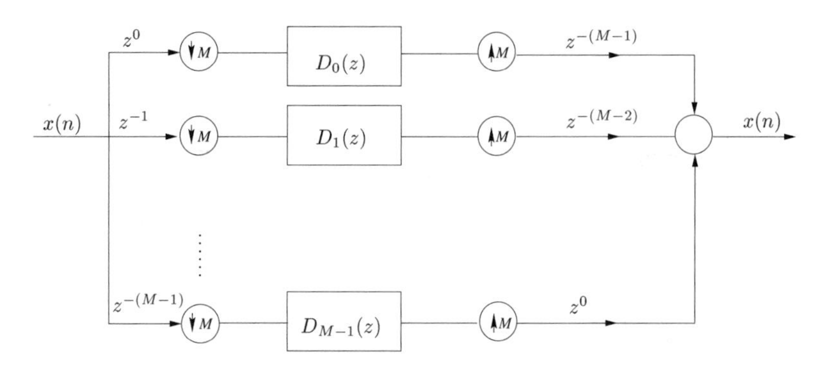

(a)

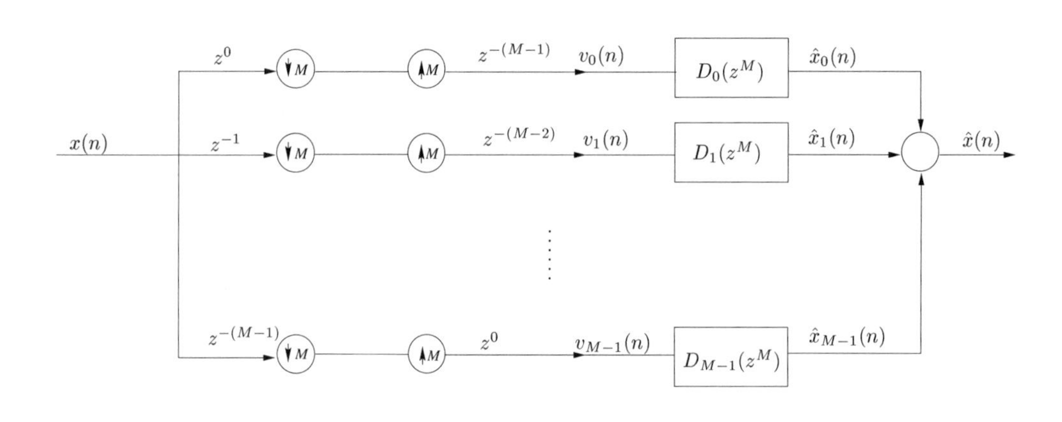

(b)

(c)

Figure 3.65: (a) Codec with diagonal $\mathcal{P}(z)$. (b) Equivalent representation. (c) rth channel of codec.

$$D(z)\sum_{r=0}^{M-1} W^{-kr} = D(z)M\delta_k = \begin{cases} MD(z), & k=0; \\ 0, & k=1,...,M-1. \end{cases} \tag{3.282}$$

Thus, Eq. (3.280) reduces to

$$\hat{X}(z) = z^{-(M-1)}D(z^M)X(z). \tag{3.283}$$

We conclude that the condition for alias cancellation in the codec is satisfied if $\mathcal{P}(z)$ is a diagonal matrix with equal elements, i.e.,

$$\mathcal{P}(z) = \mathcal{G}_p'(z)\mathcal{H}_p(z) = D(z)I. \tag{3.284}$$

Equation (3.284) in turn implies that $\mathcal{H}_p(z)\mathcal{G}_p'(z) = D(z)I$, which is a sufficient condition for cross-talk cancellation.

In summary, if the progenitor codec satisfies PR, then the transmux is also PR, and if the codec is designed to be alias free, then the paired transmux will enjoy cross-talk cancellation.

We can now apply the PR properties to the class of orthonormal transmultiplexers.

3.8.3 Orthogonal Transmultiplexer

The conventional orthogonal TMUX is an FDMA or a TDMA multiuser communication system wherein frequency or time slots are allocated to users. In a code-division multiple access (CDMA) system all users are equally entitled to all of the available time and frequency slots, with the ultimate goal of optimizing the overall throughput by maximizing the number of users in the same cell. The filter responses or user codes in a CDMA transmux are therefore spread both in time and frequency.

The equivalence between codecs and transmultiplexers developed in the preceding section enables us to design orthogonal CDMA codes using the design techniques developed for orthonormal, i.e., paraunitary, codecs.

The conditions for the paraunitary filter structure, defined by Eqs. (3.119) and (3.120), are succinctly restated here as

$$\sum_k h_r(k)h_s(k+Mn) = \delta_{r-s}\delta(n) = \begin{cases} 0, & s \neq r \\ \delta(n), & s=r \end{cases} \tag{3.285}$$

and

$$g_r(n) = h_r(n_0 - n) = h_r(M - 1 - n). \tag{3.286}$$

These same conditions must be satisfied for orthogonal CDMA codes. But as we've seen for the QMF filter banks, the filter impulse responses are not unique. There are many free parameters that can be used to optimize system performance. The CDMA user codes in an orthogonal transmultiplexer should therefore be spread out in time and frequency with minimum inter- and intracode correlations. Optimal design criteria incorporating these features were described in Akansu, Tazebay, and Haddad (1997). This feature is discussed in detail in Chapter 7.

3.9 Two-Dimensional Subband Decomposition

Except in the separable case the two-dimensional (2D) multirate filter bank is not a simple extension of the 1D case. The main complication arises from the subsampling lattice used in the decimator. In 1D, the decimator or down-sampler retains every M^{th} sample in the sequence, discards the rest, and then reindexes the time scale. In 2D, the down-sampler retains samples located on a *subsampling lattice*, which is represented by the subsampling matrix D, with integer elements. We will see that the decimation factor $M = |detD|$, so that one of M samples is retained. This implies an M-band filter bank for a maximally decimated system.

In this section, we develop 2D multirate filter bank theory as a generalization of the 1D theory presented earlier in this chapter. The prime reference for this section is Viscito and Allebach (1991) and supported also by Karlsson and Vetterli (1990).

3.9.1 2D Transforms and Notation

A 2D signal $x(n_1, n_2)$ is defined on a rectangular grid of points Λ, where $\{n_1, n_2\}$ is the set of all integers. Physically $\{n_1, n_2\}$ can refer to pixel locations in an image. The Z and Fourier transforms (Dudgeon and Mersereau, 1984) are

$$X(z_1, z_2) = \sum_{n_2=-\infty}^{\infty} \sum_{n_1=-\infty}^{\infty} x(n_1, n_2) z_1^{-n_1} z_2^{-n_2} \tag{3.287}$$

$$X(e^{j\omega_1}, e^{j\omega_2}) = \sum \sum x(n_1, n_2) e^{-j\omega_1 n_1} e^{-j\omega_2 n_2} \triangleq X'(\omega_1, \omega_2) \tag{3.288}$$

and inversely

$$x(n_1, n_2) = (\frac{1}{2\pi})^2 \int_{-\pi}^{\pi} \int_{-\pi}^{\pi} X'(\omega_1, \omega_2) e^{j(\omega_1 n_1 + \omega_2 n_2)} d\omega_1 d\omega_2. \tag{3.289}$$

The notation can be simplified and the relationship to the 1D counterpart made more evident by employing the following vector shorthand. The integer pair $\{n_1, n_2\}$ is represented as an *integer vector* $\underline{n}$,

$$\underline{n} \triangleq \begin{bmatrix} n_1 \\ n_2 \end{bmatrix}. \tag{3.290}$$

The transform variables are indicated by

$$\underline{\omega} \triangleq \begin{bmatrix} \omega_1 \\ \omega_2 \end{bmatrix}, \qquad \underline{z} \triangleq \begin{bmatrix} z_1 \\ z_2 \end{bmatrix}. \tag{3.291}$$

For any integer vector $\underline{n}$ and integer matrix

$$D = [\ \underline{d}_1 \quad \underline{d}_2\ ] = \begin{bmatrix} d_{11} & d_{12} \\ d_{21} & d_{22} \end{bmatrix}, \tag{3.292}$$

we define

$$\begin{aligned} \underline{z}^{\underline{n}} &\triangleq z_1^{n_1} z_2^{n_2} \\ \underline{z}^{D} &\triangleq \begin{bmatrix} \underline{z}^{\underline{d}_1} \\ \underline{z}^{\underline{d}_2} \end{bmatrix} = \begin{bmatrix} z_1^{d_{11}} z_2^{d_{21}} \\ z_1^{d_{12}} z_2^{d_{22}} \end{bmatrix} \\ (e^{j\underline{\omega}})^D &\triangleq e^{jD^T\underline{\omega}} = \begin{bmatrix} e^{j(\omega_1 d_{11} + \omega_2 d_{21})} \\ e^{j(\omega_1 d_{12} + \omega_2 d_{22})} \end{bmatrix}. \end{aligned} \tag{3.293}$$

Note that $\underline{z}^{D}$ is a 2×1 vector.

With this notation, the transforms can be written as

$$\begin{aligned} X(\underline{z}) &= \sum_{\underline{n} \in \Lambda} x(\underline{n}) \underline{z}^{-\underline{n}} \\ X(e^{j\underline{\omega}}) &= \sum_{\underline{n}} x(\underline{n}) e^{-j\underline{\omega}^T \underline{n}} \triangleq X'(\underline{\omega}) \\ x(\underline{n}) &= (\frac{1}{2\pi})^2 \int X(e^{j\underline{\omega}}) e^{j\underline{\omega}^T \underline{n}} d\underline{\omega}. \end{aligned} \tag{3.294}$$

Other notational definitions will be introduced as needed. (Prob. 3.22)

3.9.2 Periodic Sequences and the DFT

Let $\tilde{x}(\underline{n})$ be a 2D periodic sequence with periodicity matrix D such that

$$\tilde{x}(\underline{n}) = \tilde{x}(\underline{n} + D\underline{r}) \tag{3.295}$$

for any integer vector $\underline{r}$. Let I_M denote the region in the (n_1, n_2) plane containing exactly one period of this pattern. This *unit cell* contains $M = |detD|$ samples. Explicitly, Eq. (3.295) is

$$\tilde{x}(n_1, n_2) = \tilde{x}(n_1 + d_{11}r_1 + d_{12}r_2, n_2 + d_{21}r_1 + d_{22}r_2)$$

where $\{d_{ij}\}$ are integers. For the special case of rectangular periodicity, D is diagonal and $M = d_{11}d_{22}$. In this case, the 2D DFT is just

$$\tilde{X}(\underline{k}) \triangleq \tilde{X}(k_1, k_2) = \sum_{n_2=0}^{d_{22}-1} \sum_{n_1=0}^{d_{11}-1} \tilde{x}(n_1, n_2) e^{-j2\pi n_1 k_1/d_{11}} e^{-j2\pi n_2 k_2/d_{22}}$$

and inversely

$$\tilde{x}(\underline{n}) = \tilde{x}(n_1, n_2) = \frac{1}{d_{11}d_{22}} \sum_{k_2=0}^{d_{22}-1} \sum_{k_1=0}^{d_{11}-1} \tilde{X}(k_1, k_2) e^{j2\pi n_1 k_1/d_{11}} e^{j2\pi n_2 k_2/d_{22}}.$$

The unit cell I_M is the rectangular region $\{0 \leq n_1 \leq d_{11} - 1,\ 0 \leq n_2 \leq d_{22} - 1\}$ containing $M = d_{11}d_{22}$ points.

Now consider an arbitrary non-singular integer matrix D. This periodicity matrix defines a unit cell that is related to the subsampling lattice to be introduced shortly. Each point in this cell constitutes a vector. These vectors are called the *coset vectors* associated with D. There are exactly $M = |detD|$ of these, denoted by $\{\underline{k}_0, \underline{k}_1 \cdots \underline{k}_{M-1}\}$, with $\underline{k}_0 = \underline{0}$. For example,

$$D = \begin{bmatrix} 2 & 1 \\ 0 & 2 \end{bmatrix}$$

defines the periodic regions and the parallelogram-shaped unit cell $ABCD$ in Fig. 3.66. The coset vectors within this unit cell are

$$\underline{k}_0 = \begin{bmatrix} 0 \\ 0 \end{bmatrix}, \quad \underline{k}_1 = \begin{bmatrix} 1 \\ 0 \end{bmatrix}, \quad \underline{k}_2 = \begin{bmatrix} 1 \\ 1 \end{bmatrix}, \quad \underline{k}_3 = \begin{bmatrix} 2 \\ 1 \end{bmatrix}.$$

Note that other cells and coset vectors can be defined for the same D. For the diamond-shaped cell $ADEF$ in Fig. 3.66 the associated coset vectors are

$$\underline{k}_0 = \begin{bmatrix} 0 \\ 0 \end{bmatrix}, \quad \underline{k}_1 = \begin{bmatrix} 0 \\ 1 \end{bmatrix}, \quad \underline{k}_2 = \begin{bmatrix} 0 \\ 2 \end{bmatrix}, \quad \underline{k}_3 = \begin{bmatrix} 0 \\ 3 \end{bmatrix}.$$

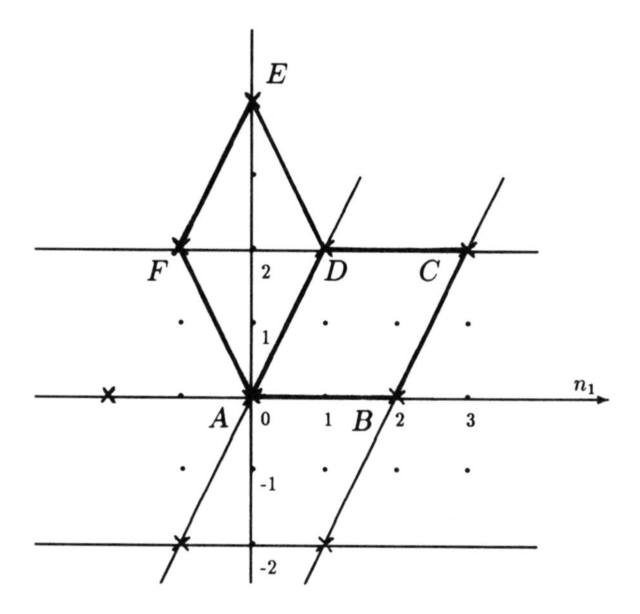

Figure 3.66: Periodic regions for , diamond and parallelogram-shaped unit cells.

Now for an arbitrary D, we can expand $\tilde{x}(\underline{n}) = \tilde{x}(\underline{n} + D\underline{r})$ in a discrete Fourier series,

$$\tilde{x}(\underline{n}) = \frac{1}{M} \sum_{l=0}^{M-1} \tilde{X}(\underline{k}) e^{j2\pi \underline{k}_l^T D^{-1} \underline{n}} = \tilde{x}(\underline{n} + D\underline{r}) \tag{3.296}$$

and inversely,

$$\tilde{X}(\underline{k}) = \sum_{l=0}^{M-1} \tilde{x}(\underline{n}) e^{-j2\pi \underline{k}^T D^{-1} \underline{n}_l} = \tilde{X}(\underline{k} + D^T \underline{r}) \tag{3.297}$$

where $M = |detD|$, and $\{\underline{k}_l\}$, $\{\underline{n}_l\}$ are coset vectors associated with D^T, and D, respectively, and the sum is taken over the respective unit cells in the spatial- and frequency-domains. These relations are valid since the complex exponentials $e^{j2\pi \underline{k}^T D^{-1} \underline{n}}$ are periodic in $\underline{k}$ and $\underline{n}$ with periodicity matrices D^T and D, respectively, and are orthogonal over the unit cell I_M specified by D. (Prob. 3.23)

As a special case of the foregoing, consider the periodic sampling function

$$i(\underline{n}) = \begin{cases} 1, & \text{if } \underline{n} \in \Lambda_D \\ 0, & \text{otherwise} \end{cases} \tag{3.298}$$

where Λ_D represents the sublattice generated by D. (In Fig. 3.66, Λ is the set of all grid points, Λ_D is the set of subsampled points indicated by the crosses.)

This sampling function can be expressed in Kronecker delta form,

$$i(\underline{n}) = \sum_{\underline{r}} \delta(\underline{n} - D\underline{r}) = \begin{cases} 1, & \underline{\mathrm{n}} = \mathrm{D}\underline{\mathrm{r}} \\ 0, & \text{otherwise}, \end{cases} \tag{3.299}$$

and $\{\underline{r}\}$ is the set of all integer vectors. The corresponding DFT is then

$$I(\underline{k}) = \sum_{l=0}^{M-1} i(\underline{n})e^{-j2\pi\underline{k}^T D^{-1}\underline{n}_l} = 1$$

so that

$$i(\underline{n}) = \frac{1}{M}\sum_{l=0}^{M-1} e^{j2\pi\underline{k}_l^T D^{-1}\underline{n}}. \tag{3.300}$$

Observe that Eqs. (3.299) and (3.300) are generalizations of the 1D version given in Eq. (3.4). This result will be used in deriving formulas for decimated and interpolated 2D signals.

3.9.3 Two-Dimensional Decimation and Interpolation

Let Λ be the set of integer vectors $\{\underline{n}\}$ and Λ_D the set of integer vectors $\{\underline{m}\}$ generated by $\underline{m} = D\underline{n}$. In Fig. 3.66, Λ is the set of grid points at all integer values, and Λ_D the lattice subset indicated by crosses. The coset is the set of points within a unit cell indicated previously. Note that a given sublattice can be described by more than one D matrix. For example,

$$D_1 = \begin{bmatrix} 1 & 1 \\ -2 & 2 \end{bmatrix}, \quad \text{and} \quad D_2 = \begin{bmatrix} 2 & 1 \\ 0 & 2 \end{bmatrix}$$

define the same sublattice. The matrices are related by $D_2 = D_1 \begin{bmatrix} 1 & 0 \\ 1 & 1 \end{bmatrix}$, i.e., postmultiplication by an integer matrix with determinant equal to ± 1. (Prob. 3.24)

The down-sampler and up-sampler shown schematically in Fig. 3.67 are defined by

$$v(\underline{n}) = x(D\underline{n}) \tag{3.301}$$

$$y(\underline{n}) = \begin{cases} v(D^{-1}\underline{n}), & D^{-1}\underline{n} \text{ an integer vector} \\ 0, & \text{otherwise} \end{cases} \tag{3.302}$$

$$y(\underline{n}) = \begin{cases} x(\underline{n}), & \underline{n} \in \Lambda_D \\ 0, & \text{otherwise.} \end{cases} \tag{3.303}$$

The down-sampler accepts samples lying on the sublattice Λ_D, discards the others, and reindexes the spatial axes.

The up-sampler takes points $v(\underline{n})$ on a rectangular lattice and maps them onto the sublattice Λ_D. Equation (3.303) combines the operations of down-sampling followed by up-sampling. This operation is equivalent to modulating the input $x(\underline{n})$ by the periodic sampling function $i(\underline{n})$ of Eq. (3.299), and it will prove to be the key to unveiling the connection among the transforms of these three signals.

For the up-sampler of Eq. (3.302), the output transform is

$$\begin{aligned} Y(\underline{z}) &= \sum_{\underline{n}\in\Lambda_D} y(\underline{n})\underline{z}^{-\underline{n}} = \sum_{\underline{n}\in\Lambda_D} v(D^{-1}\underline{n})\underline{z}^{-\underline{n}} \\ &= \sum v(\underline{n})\underline{z}^{-D\underline{n}} = \sum v(\underline{n})(\underline{z}^D)^{-\underline{n}} \\ &= V(\underline{z}^D). \end{aligned} \tag{3.304}$$

In the Fourier domain, this becomes

$$Y(e^{j\underline{\omega}}) = V(e^{jD^T\underline{\omega}}). \tag{3.305}$$

Using the simpler notation of Eq. (3.288) the exponent is suppressed, and the last equation is written as

$$Y^{'}(\underline{\omega}) = V^{'}(D^T\underline{\omega}), \tag{3.306}$$

which is a generalization of the 1D version of Eq. (3.13). In the 1D case, the baseband region for $X(e^{j\omega})$, $(-\pi,\pi)$, is mapped into $(\frac{-\pi}{M},\frac{\pi}{M})$ for $Y(e^{j\omega})$. In addition there are $(M-1)$ contiguous images on $[-\pi,\pi]$. See, for example, Fig. 3.4, which is redrawn here as Fig. 3.68, for $M=4$. In 2D, Eq. (3.305) implies that the rectangular frequency region $\{-\pi \le \omega_1 \le \pi, -\pi \le \omega_2 \le \pi\}$ is mapped into the baseband parallelogramshaped region

$$\{-\pi \le d_{11}\omega_1 + d_{21}\omega_2 \le \pi, -\pi \le d_{12}\omega_1 + d_{22}\omega_2 \le \pi\}. \tag{3.307}$$

The $(M-1)$ images of this cell are mapped into regions surrounding this baseband by shifting the baseband by

$$2\pi D^{-T}\underline{k}_l,$$

$x(\underline{n})$ → ↓D → $v(\underline{n})$ → ↑D → $y(\underline{n})$

Figure 3.67: Representation of two-dimensional down-sampling and up-sampling.

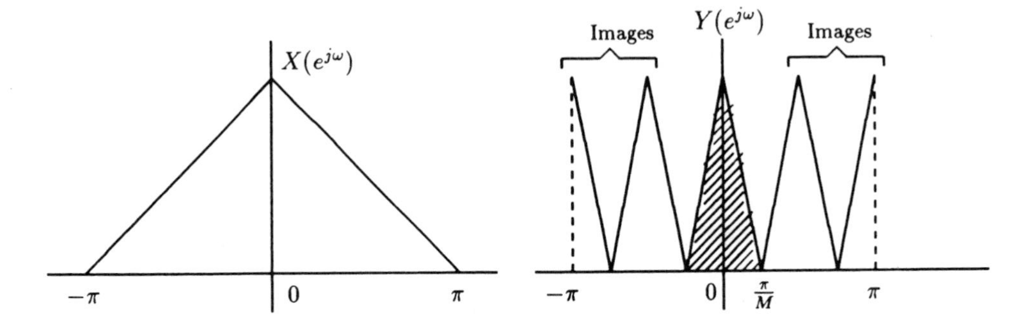

Figure 3.68: One-dimensional frequency axis compression due to interpolation shown for $M = 4$.

where $\underline{k}_l$ is the coset vector, and D^{-T} is an abbreviation for $(D^{-1})^T$. An example of the sublattice structure, the coset vectors, and the resulting frequency bands are illustrated in Fig. 3.69 for

$$D = \begin{bmatrix} 2 & 2 \\ 1 & -1 \end{bmatrix} = [\underline{d}_1 \quad \underline{d}_2]. \tag{3.308}$$

The subsampling lattice is constructed by first drawing vectors $\underline{d}_1, \underline{d}_2$ and lines parallel to these at the spacing indicated by $\underline{m} = D\underline{n}$. The sublattice points lie at the intersection of these lines. In Fig. 3.69, the unit cell is the parallelogram formed by $\underline{d}_1$ and $\underline{d}_2$. The coset vectors are the four points contained within this cell, excluding two of the boundaries. These are

$$\underline{k}_0 = \begin{bmatrix} 0 \\ 0 \end{bmatrix}, \quad \underline{k}_1 = \begin{bmatrix} 1 \\ 0 \end{bmatrix}, \quad \underline{k}_2 = \begin{bmatrix} 2 \\ 0 \end{bmatrix}, \quad \underline{k}_3 = \begin{bmatrix} 3 \\ 0 \end{bmatrix}. \tag{3.309}$$

The shaded diamond-shaped baseband in Fig. 3.69(b) is obtained from Eq. (3.309)

$$\{-\pi \le 2\omega_1 + \omega_2 \le \pi\} \bigcap \{-\pi \le 2\omega_1 - \omega_2 \le \pi\}.$$

The three other image bands are obtained by translating the baseband by $2\pi D^{-T}\underline{k}_l$. Figure 3.69(b) is therefore a 2D generalization of the 1D case. The bands are compressed in extent and skewed in orientation. This capability of forming the shape and location of these subbands is the foundation for the *fan filters* discussed subsequently. (Prob. 3.32)

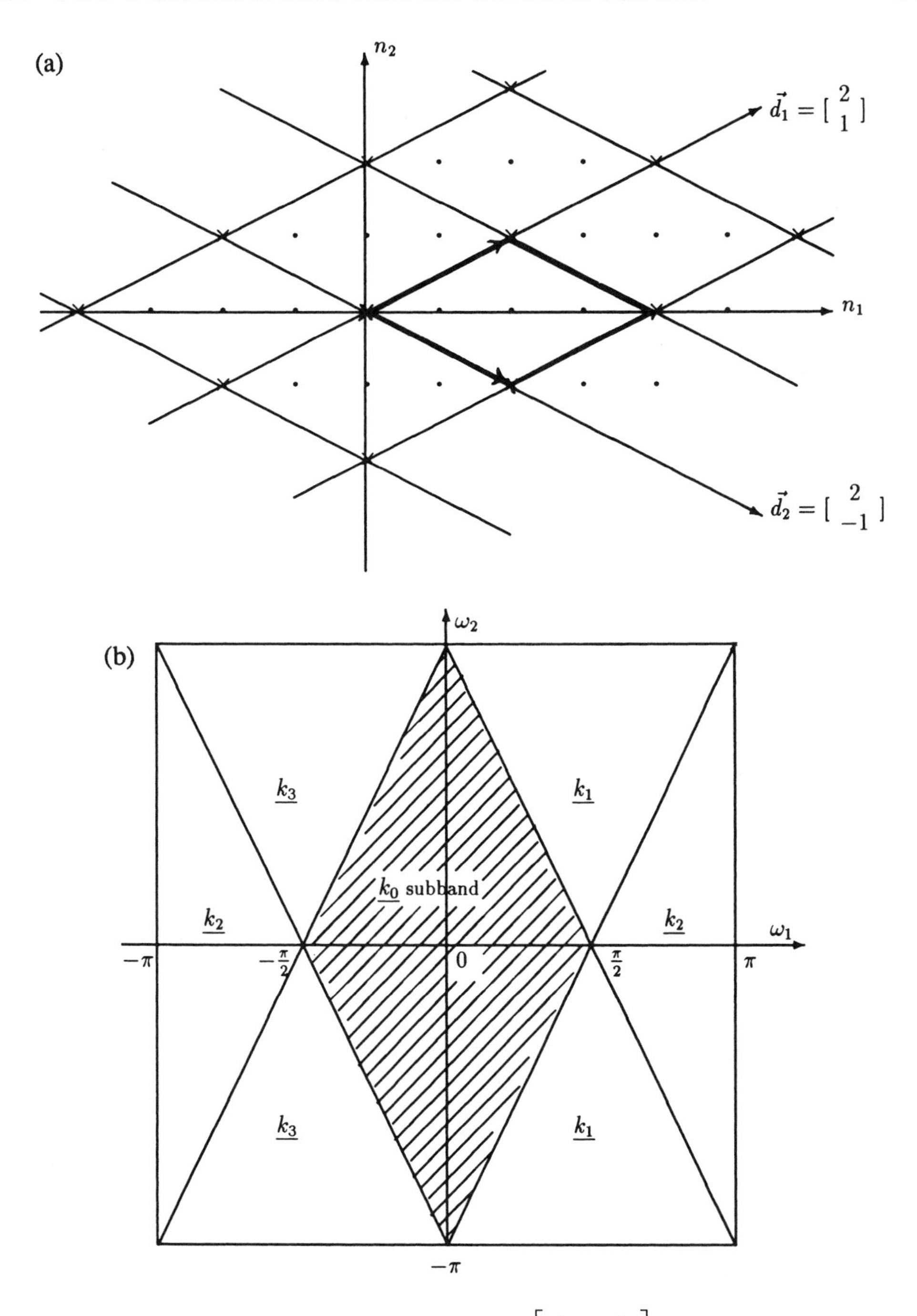

Figure 3.69: (a) Subsampling lattice for $D = \begin{bmatrix} 2 & 2 \\ 1 & -1 \end{bmatrix}$; (b) corresponding subbands.

The combined operations of down-sampling and up-sampling can be represented as a modulation of $x(\underline{n})$ by $i(\underline{n})$. Hence

$$y(\underline{n}) = x(\underline{n})i(\underline{n}) = \frac{1}{M}\sum_{l=0}^{M-1} x(\underline{n})e^{j2\pi \underline{k}_l^T D^{-1}\underline{n}}. \tag{3.310}$$

Consequently,

$$\begin{aligned} Y(\underline{z}) &= \sum_{\underline{n}\in\Lambda} y(\underline{n})\underline{z}^{-\underline{n}} = \frac{1}{M}\sum_{l=0}^{M-1}\sum_{\underline{n}} x(\underline{n})e^{j2\pi \underline{k}_l^T D^{-1}\underline{n}}\underline{z}^{-\underline{n}} \\ &= \frac{1}{M}\sum_{l=0}^{M-1}\sum_{\underline{n}} x(\underline{n})(\underline{z}e^{-j2\pi D^{-T}\underline{k}_l})^{-\underline{n}} \\ &= \frac{1}{M}\sum_{l=0}^{M-1} X(\underline{z}e^{-j2\pi D^{-T}\underline{k}_l}). \end{aligned} \tag{3.311}$$

The argument in this last equation involves the product of two complex vectors of the form $\underline{zw}$, which is defined as follows

$$\underline{zw} = \begin{bmatrix} z_1 \\ z_2 \end{bmatrix}\begin{bmatrix} w_1 \\ w_2 \end{bmatrix} = \begin{bmatrix} z_1 w_1 \\ z_2 w_2 \end{bmatrix}.$$

And the frequency characterization for the down- and up-sampling combination is

$$Y(e^{j\underline{\omega}}) = \frac{1}{M}\sum_{l=0}^{M-1} X(e^{j(\underline{\omega}-2\pi D^{-T}\underline{k}_l)})$$

or

$$Y'(\underline{\omega}) = \frac{1}{M}\sum_{l=0}^{M-1} X'(\underline{\omega} - 2\pi D^{-T}\underline{k}_l). \tag{3.312}$$

Next, since $Y(\underline{z}) = V(\underline{z}^D)$, then

$$V(\underline{z}) = Y(\underline{z}^{D^{-1}}) = \frac{1}{M}\sum_{l=0}^{M-1} X(\underline{z}^{D^{-1}}e^{-j2\pi D^{-T}\underline{k}_l}) \tag{3.313}$$

But

$$\underline{z}^{D^{-1}}\big|_{\underline{z}=e^{j\underline{\omega}}} = e^{jD^{-T}\underline{\omega}}.$$

This leaves us with the down-sampler characterization

$$V'(\underline{\omega}) = V(e^{j\underline{\omega}}) = \frac{1}{M}\sum X'[D^{-T}(\underline{\omega} - 2\pi\underline{k}_l)]. \tag{3.314}$$

The up-sampling, down-sampling, and composite operation are therefore given by Eqs. (3.306), (3.314), and (3.312), respectively.

3.9.4 The 2D Filter Bank

The 2D M-band filter bank is shown in Fig. 3.70. The structure is maximally decimated if the number of channels equals $|detD|$, which is the case at hand. In this subsection, we will develop the AC matrix, the polyphase representation, and the conditions for perfect reconstruction as generalizations of the 1D version of Section 3.5.

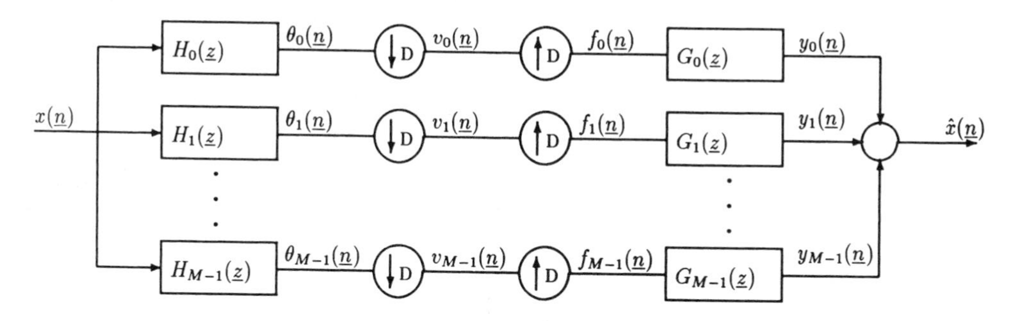

Figure 3.70: M-band maximally decimated 2D filter bank.

From Fig. 3.70 and Eq. (3.311), we can obtain

$$\hat{X}(\underline{z}) = \sum_{r=0}^{M-1} G_r(\underline{z})F_r(\underline{z}), \qquad \theta_r(\underline{z}) = H_r(\underline{z})X(\underline{z})$$

$$F_r(\underline{z}) = \frac{1}{M}\sum_{l=0}^{M-1} \theta_r(\underline{z}e^{-j2\pi D^{-T}\underline{k}_l}).$$

Combining these last equations gives

$$\hat{X}(\underline{z}) = \frac{1}{M}\underline{X}^T(\underline{z})H_{AC}(\underline{z})\underline{g}(\underline{z}). \tag{3.315}$$

The separate terms in Eq. (3.315) are extensions of those in Eq. (3.72). For notational simplicity, we will use $e^{-j2\pi D^{-T}\underline{k}_l} \triangleq W^{\underline{k}_l}$. Then

$$\begin{aligned} \underline{X}^T(\underline{z}) &\triangleq [X(\underline{z})X(\underline{z}W^{\underline{k}_1})...X(\underline{z}W^{\underline{k}_{M-1}})] \\ \underline{g}^T(\underline{z}) &\triangleq [G_0(\underline{z})......G_{M-1}(\underline{z})] \end{aligned} \tag{3.316}$$

$$H_{AC}(\underline{z}) \triangleq \begin{bmatrix} H_0(\underline{z}) & \cdots & \cdots & H_{M-1}(\underline{z}) \\ H_0(\underline{z}W^{\underline{k}_1}) & \cdots & \cdots & H_{M-1}(\underline{z}W^{\underline{k}_1}) \\ \vdots & \vdots & \vdots & \vdots \\ H_0(\underline{z}W^{\underline{k}_{M-1}}) & \cdots & \cdots & H_{M-1}(\underline{z}W^{\underline{k}_{M-1}}) \end{bmatrix}. \tag{3.317}$$

The conditions for zero-aliasing and perfect reconstruction are extensions of the 1D case:

$$\begin{aligned} \frac{1}{M} H_{AC}(\underline{z}) \underline{g}(\underline{z}) &= \begin{bmatrix} T(\underline{z}) \\ 0 \\ \vdots \\ 0 \end{bmatrix}, \quad \text{no aliasing} \\ T(\underline{z}) &= \underline{z}^{-\underline{m}}, \quad \text{PR.} \end{aligned} \tag{3.318}$$

The necessary and sufficient condition for PR is given formally by

$$\begin{aligned} \frac{1}{M} \underline{g}(\underline{z}) &= H_{AC}^{-1}(\underline{z}) \begin{bmatrix} 1 \\ 0 \\ \vdots \\ 0 \end{bmatrix} \underline{z}^{-\underline{m}} \\ &= [1^{st} \text{ column of } H_{AC}^{-1}(\underline{z})] \underline{z}^{-\underline{m}}. \end{aligned} \tag{3.319}$$

This condition, as it stands, is of little practical use since it involves inversion of the AC matrix. The resulting synthesis filters will likely be a high-order IIR and possibly unstable. The polyphase approach is more amenable to design.

In 1D, the coset vectors are the points $\{k\}$ on the interval $[0, M-1]$ defined by the decimation factor M. To get the polyphase component we shifted $x(n)$ by r and subsampled the translated $x(n+r)$ by M to get $x_r(n) = x(r+Mn) \leftrightarrow X_r(z)$. Repeating this for each $r \in [0, M-1]$ gave us

$$X(z) = \sum_{r=0}^{M-1} z^{-r} X_r(z^M).$$

To obtain the 2D polyphase expansion, we perform the following steps:
(1) Select $\{\underline{k}_0, \underline{k}_1, ..., \underline{k}_{M-1}\}$, the coset vectors associated with D
(2) Shift $x(\underline{n})$ by $\underline{k}_l$, and down-sample by D to get

$$x_l(\underline{n}) = x(\underline{k}_l + D\underline{n}) \leftrightarrow X_l(\underline{z}) = \sum_{\underline{n} \in \Lambda} x(\underline{k}_l + D\underline{n}) \underline{z}^{-\underline{n}} \tag{3.320}$$

(3) Combine the polyphase components to obtain (Prob. 3.16)

$$X(\underline{z}) = \sum_{\underline{n}} x(\underline{n}) \underline{z}^{-\underline{n}} = \sum_{l=0}^{M-1} \sum_{\underline{n}} x(\underline{k}_l + D\underline{n}) \underline{z}^{-(\underline{k}_l + D\underline{n})}$$

$$= \sum_{l=0}^{M-1} \underline{z}^{-\underline{k}_l} \sum_{\underline{n}} x(\underline{k}_l + D\underline{n}) \underline{z}^{-D\underline{n}}$$

$$= \sum_{l=0}^{M-1} \underline{z}^{-\underline{k}_l} X_l(\underline{z}^D) \tag{3.321}$$

This polyphase decomposition can now be applied to each analysis filter:

$$H_i(\underline{z}) = \sum_{l=0}^{M-1} \underline{z}^{-\underline{k}_l} H_{i,l}(\underline{z}^D), \qquad i = 0, 1, ...M-1. \tag{3.322}$$

Following the steps in the 1D case, we can obtain

$$\underline{\theta}(\underline{z}) = [\mathcal{H}_p(\underline{z}^D)\underline{Z}_M]X(\underline{z}) = \underline{h}(\underline{z})X(\underline{z}), \tag{3.323}$$

where

$$\begin{aligned} \underline{h}^T(\underline{z}) &\triangleq [H_0(\underline{z}) \cdots H_{M-1}(\underline{z})] \\ \underline{Z}_M^T &\triangleq [\underline{z}^{-\underline{k}_0}, \underline{z}^{-\underline{k}_1}, \cdots, \underline{z}^{-\underline{k}_{M-1}}] \\ \mathcal{H}_p(\underline{z}) &= \begin{bmatrix} H_{0,0}(\underline{z}) & \cdots & \cdots & H_{0,M-1}(\underline{z}) \\ H_{1,0}(\underline{z}) & \cdots & \cdots & H_{1,M-1}(\underline{z}) \\ \vdots & \vdots & \vdots & \vdots \\ H_{M-1,0}(\underline{z}) & \cdots & \cdots & H_{M-1,M-1}(\underline{z}) \end{bmatrix}. \end{aligned} \tag{3.324}$$

The bank of analysis filters $\underline{h}(\underline{z})$ in Fig. 3.70 can be replaced by $\mathcal{H}_p(\underline{z}^D)\underline{Z}_M$. Following the 1D argument, the decimators can be moved to the left of the polyphase matrix and the argument in $\mathcal{H}_p(.)$ is changed from $\underline{z}^D$ to $\underline{z}$. This gives the front end of the equivalent polyphase structure shown in Fig. 3.71(a).

For the 2D case we need to define the synthesis polyphase expansion using *positive* exponents. The reason for this departure from the 1D case will soon become evident. Using

$$\sum_{\underline{n}\in\Lambda} \alpha(\underline{n}) = \sum_{l=0}^{M-1} \sum_{\underline{n}\in\Lambda} \alpha(D\underline{n} - \underline{k}_l)$$

we can show (see Prob. 3.16) that

$$G_s(\underline{z}) = \sum_{l=0}^{M-1} \underline{z}^{\underline{k}_l} G_{s,l}(\underline{z}^D), \tag{3.325}$$

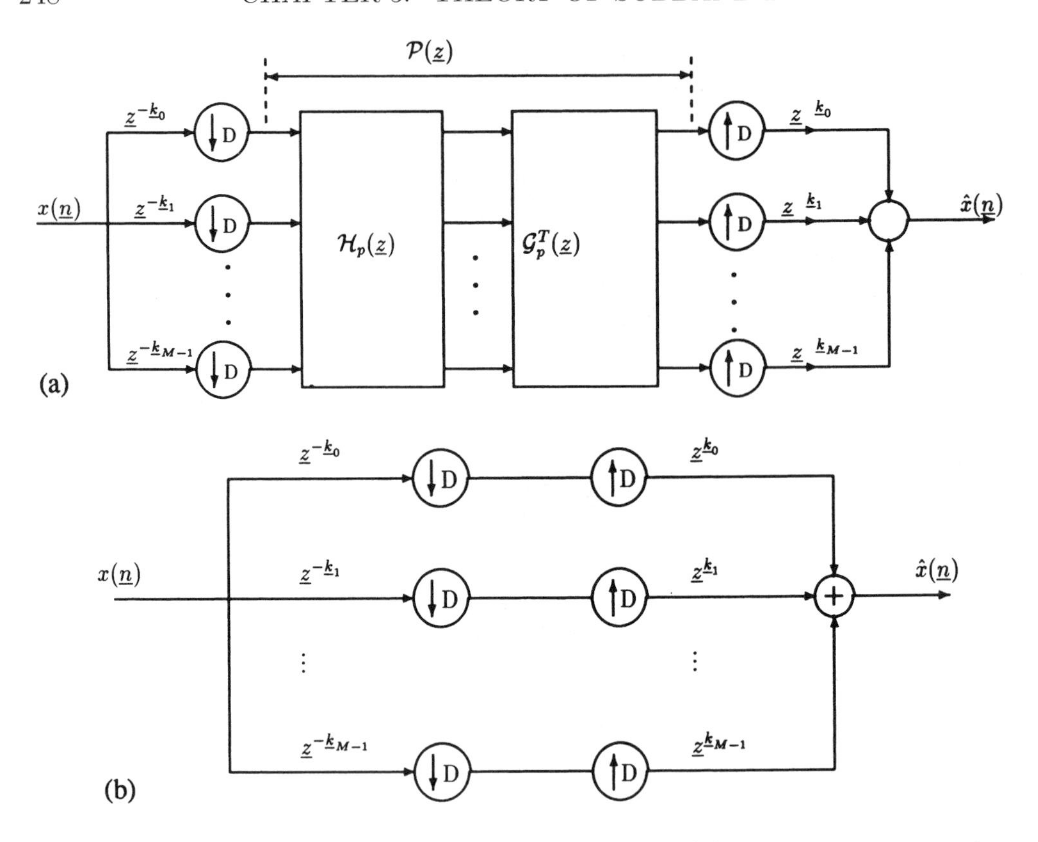

Figure 3.71: (a) The polyphase form of the filter bank; (b) reduced structure when $\mathcal{P}(z) = I$.

where

$$G_{s,l}(\underline{z}) = \sum_{\underline{n}\in\Lambda} g(D\underline{n} - \underline{k}_l)\underline{z}^{-\underline{n}}.$$

Next we define the synthesis polyphase matrix by

$$\mathcal{G}_p(\underline{z}) = \begin{bmatrix} G_{0,0}(\underline{z}) & G_{0,1}(\underline{z}) & \cdots & G_{0,M-1}(\underline{z}) \\ G_{1,0}(\underline{z}) & G_{1,1}(\underline{z}) & \cdots & G_{1,M-1}(\underline{z}) \\ \vdots & \vdots & \vdots & \vdots \\ G_{M-1,0}(\underline{z}) & G_{M-1,1}(\underline{z}) & \cdots & G_{M-1,M-1}(\underline{z}) \end{bmatrix} \tag{3.326}$$

such that

$$\underline{g}(\underline{z}) = \mathcal{G}_p(\underline{z}^D)\hat{\underline{Z}}_M \tag{3.327}$$

where

$$\hat{\underline{Z}}_M^T = [\underline{z}^{\underline{k}_0}, \underline{z}^{\underline{k}_1}, \cdots, \underline{z}^{\underline{k}_{M-1}}].$$

The bank of synthesis filters $\underline{g}(\underline{z})$ in Fig. 3.70 is replaced by $\mathcal{G}_p(\underline{z}^D)\hat{\underline{Z}}_M$. Moving the up-sampler to the right changes the argument from $\underline{z}^D$ to $\underline{z}$. The final equivalent synthesis polyphase structure is shown in Fig. 3.71(a). Note that in this case the transpose of $\mathcal{G}_p(z)$ emerges as the synthesis polyphase matrix followed by positive exponent shift vectors. Now we have

$$\mathcal{P}(\underline{z}) = \mathcal{G}_p^T(\underline{z})\mathcal{H}_p(\underline{z}). \tag{3.328}$$

In this representation, the polyphase components of $H_k(z)$ and $G_k(z)$ are on the kth row of $\mathcal{H}_p(\underline{z})$, and the kth column of $\mathcal{G}_p^T(\underline{z})$.

Continuing in this fashion, we can show that the relationship between H_{AC} and $\mathcal{H}_p$, as might be expected from Eq. (3.104), is (see Prob. 3.25)

$$H_{AC}(\underline{z}) = \mathcal{W}A(\underline{z})\mathcal{H}_p^T(\underline{z}^D), \tag{3.329}$$

where

$$\begin{aligned} A(\underline{z}) &= diag\{\underline{z}^{-\underline{k}_0}, \underline{z}^{-\underline{k}_1}, \cdots, \underline{z}^{-\underline{k}_{M-1}}\} \\ \mathcal{W} &= [\mathcal{W}_{rs}] = [e^{j2\pi \underline{k}_r^T D^{-1} \underline{n}_s}]. \end{aligned} \tag{3.330}$$

The $\mathcal{W}$ matrix is a DFT-type matrix associated with the lattice. The element $\mathcal{W}_{rs}$ corresponds to an evaluation of $e^{j\underline{\omega}^T \underline{n}}$ at $\underline{\omega} = 2\pi D^{-T}\underline{k}$. Such a matrix has the property

$$\mathcal{W}^{-1} = \frac{1}{M}\mathcal{W}^H \triangleq \frac{1}{M}(\mathcal{W}^*)^T. \tag{3.331}$$

We can now determine the conditions for perfect reconstruction in terms of the polyphase matrices. From Fig. 3.71, we see that a *sufficient* condition for PR is simply

$$\mathcal{P}(\underline{z}) \triangleq \mathcal{G}_p^T(\underline{z})\mathcal{H}_p(\underline{z}) = I. \tag{3.332}$$

Figure 3.71(b) shows this condition. Observe that this has been structured such that the product of the delays along any path is the same and is equal to $\underline{z}^{\underline{0}} = 1$. Also, for 2D image processing causality is not a constraint so that positive indexed shifts are acceptable. However, we can obtain *necessary* and *sufficient* conditions as follows. From Eqs. (3.315), (3.327), and (3.329),

$$\begin{aligned} \hat{X}(\underline{z}) &= [\frac{1}{M}\underline{X}^T\mathcal{W}A(\underline{z})\mathcal{H}_p^T(\underline{z}^D)\mathcal{G}_p(\underline{z}^D)\hat{\underline{Z}}_M]^T \\ &= [\frac{1}{M}\hat{\underline{Z}}_M^T\mathcal{G}_p^T(\underline{z}^D)\mathcal{H}_p(\underline{z}^D)A(\underline{z})\mathcal{W}^T]\underline{X} \\ &\triangleq \underline{B}^T(\underline{z})\underline{X}. \end{aligned} \tag{3.333}$$

Now if we can make

$$\underline{B}^T(\underline{z}) = T(\underline{z})[1, 0, ..., 0] \tag{3.334}$$

then Eq. (3.333) reduces to $\hat{X}(\underline{z}) = T(\underline{z})X(\underline{z})$ and aliasing is eliminated. Now post multiply both sides of Eq. (3.334) by $\mathcal{W}^*$ and note that $\mathcal{W}^T\mathcal{W}^* = MI$. Also, since the first row and column of $\mathcal{W}^*$ have unity entries, then

$$[\ 1\ \ 0\ \ \ldots\ \ 0\ ]\mathcal{W}^* = [\ 1\ \ 1\ \ \ldots\ \ 1\ ].$$

We have the intermediate result

$$\hat{\underline{Z}}_M^T[\mathcal{G}_p^T(\underline{z}^D)\mathcal{H}_p(\underline{z}^D)]A(\underline{z}) = T(\underline{z})[\ 1\ \ 1\ \ \ldots\ \ 1\ ].$$

Multiplying by $A^{-1}(\underline{z})$, noting that $[11\cdots 1]A^{-1}(\underline{z}) = \hat{\underline{Z}}_M^T$, and transposing leads to

$$[\mathcal{H}_p^T(\underline{z}^D)\mathcal{G}_p(\underline{z}^D)]\hat{\underline{Z}}_M = T(\underline{z})\hat{\underline{Z}}_M$$

or

$$\mathcal{P}^T(\underline{z}^D)\hat{\underline{Z}}_M = T(\underline{z})\hat{\underline{Z}}_M. \tag{3.335}$$

This last equation is of the form $A\underline{x} = \lambda\underline{x}$. Hence, aliasing is eliminated if and only if $T(\underline{z})$ is an eigenvalue of $\mathcal{P}^T(\underline{z}^D)$ and $\hat{\underline{Z}}_M$ is the associated eigenvector. For PR, $T(\underline{z}) = \underline{z}^{-\underline{r}}$, for some integer vector $\underline{r}$. For convenience, we can choose $\underline{r} = \underline{0}$, leaving us with

$$\mathcal{P}^T(\underline{z}^D)\hat{\underline{Z}}_M = \hat{\underline{Z}}_M. \tag{3.336}$$

Clearly, a sufficient condition is $\mathcal{P}(\underline{z}) = I$.

We can, of course, continue to exploit the resemblance to the 1D case. For a paraunitary solution, we can impose (see also Prob. 3.26)

$$\mathcal{H}_p^T(\underline{z}^{-1})\mathcal{H}_p(\underline{z}) = I \tag{3.337}$$

and obtain $\mathcal{P}(\underline{z}) = \underline{z}^{-\underline{n}_0}I$ by choosing

$$\mathcal{G}_p(\underline{z}) = \underline{z}^{-\underline{n}_0}\mathcal{H}_p(\underline{z}^{-1}). \tag{3.338}$$

Several other avenues can be explored. Karlsson and Vetterli (1990) construct a paraunitary cascaded structure as a 2D version of the M-band paraunitary lattice mentioned in Section 3.5.9 (Prob. 3.27). They also go on to describe PR designs based on a state-space description for both paraunitary and nonparaunitary constraints and conditions for PR linear-phase structures.

This completes our generalization of the 1D PR filter bank. The reader can consult the references cited for a detailed treatment of this subject. Application of these concepts will be demonstrated in the next subsection dealing with fan filters.

3.9.5 Two-Band Filter Bank with Hexagonal or Quincunx Sampling

Figure 3.72 shows a two-band filter bank, which serves as an example for the application of the theory just developed. This example, patterned after the papers by Ansari and Lau (1987) and Ansari (1987), is chosen to demonstrate the connection between the shape of the desired band split and the decimation lattice and to illustrate how the subbands propagate through the configuration.

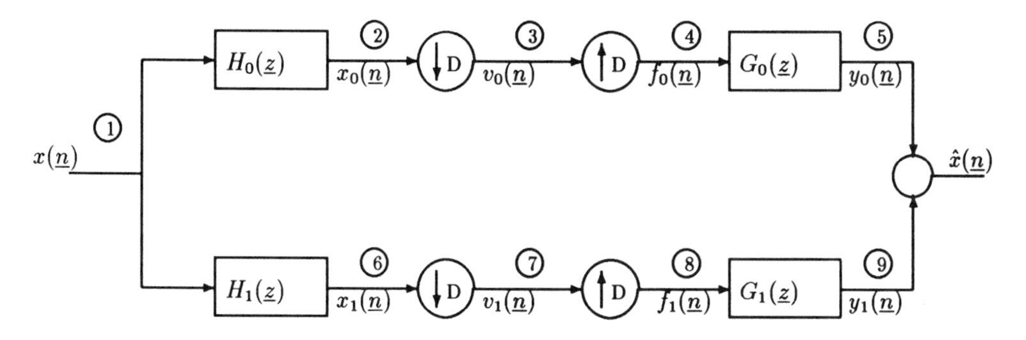

Figure 3.72: Two-band filter bank.

Suppose that the desired band split is that shown in Fig. 3.73 where the low-frequency region B_0 is the shaded interior of the diamond-shaped region, and the high-frequency subband B_1 is the complement of B_0 as indicated, such that

$$B_0 \cap B_1 = 0, \qquad B_0 \cup B_1 = B = \{|\omega_1| < \pi, |\omega_2| < \pi\}.$$

To obtain this split, let $H_0(\underline{z})$, and $H_1(\underline{z})$ be ideal low-pass and high-pass filters such that

$$H_0(e^{j\underline{\omega}}) \triangleq H_0'(\underline{\omega}) = \begin{cases} 1, & \underline{\omega} \in B_0 \\ 0, & \underline{\omega} \in B_1 \end{cases}$$

$$H_1(e^{j\underline{\omega}}) \triangleq H_1'(\underline{\omega}) = \begin{cases} 1, & \underline{\omega} \in B_1 \\ 0, & \underline{\omega} \in B_0. \end{cases} \tag{3.339}$$

Figure 3.74 shows the spectral bands at the various nodes in the subband structure. For illustrative purposes, the input signal spectrum is represented by eight bands, four belonging to B_0, and four to B_1. The ideal filter $H_0'(\underline{\omega})$ passes bands 1, 2, 3, and 4 to give the spectrum at node (2). Similarly, $H_1'(\underline{\omega})$ passes bands 5 through 8, yielding the high-pass spectrum at point (6). Next, we select a subsampling lattice

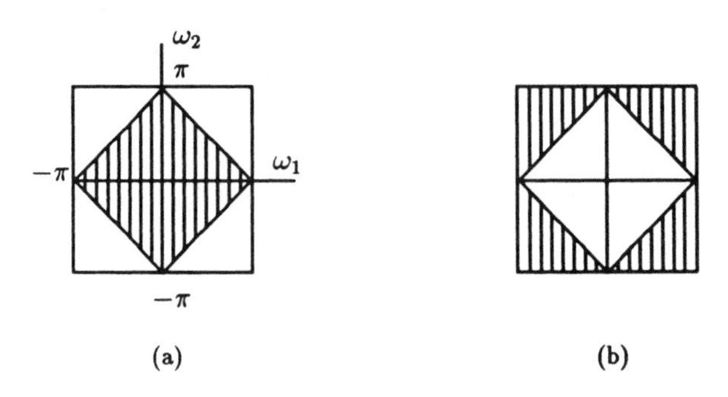

Figure 3.73: Ideal two-band split: (a) low frequency; (b) high frequency.

compatible with the subband split. In this respect, we want $D^T\underline{\omega}$ to partition B into the diamond-shaped region. The decimation matrix and coset vectors that achieve this are

$$D = \begin{bmatrix} 1 & -1 \\ 1 & 1 \end{bmatrix}, \quad \underline{k}_0 = \begin{bmatrix} 0 \\ 0 \end{bmatrix}, \quad \underline{k}_1 = \begin{bmatrix} 0 \\ 1 \end{bmatrix}. \tag{3.340}$$

Then

$$D^T\underline{\omega} = \begin{bmatrix} (\omega_1 + \omega_2) \\ (-\omega_1 + \omega_2) \end{bmatrix}$$

resulting in the partition

$$-\pi \leq \omega_1 + \omega_2 \leq \pi \qquad -\pi \leq -\omega_1 + \omega_2 \leq \pi.$$

The down-sampled and up-sampled spectra at nodes (3) and (4) are obtained by particularizing Eqs. (3.313) and (3.311). For the case at hand,

$$D^{-1} = \frac{1}{2}\begin{bmatrix} 1 & 1 \\ -1 & 1 \end{bmatrix}, \qquad D^{-T} = \frac{1}{2}\begin{bmatrix} 1 & -1 \\ 1 & 1 \end{bmatrix}$$

$$\underline{z}^{D^{-1}} = \begin{bmatrix} z_1^{1/2} & z_2^{-1/2} \\ z_1^{1/2} & z_2^{1/2} \end{bmatrix}, \quad 2\pi D^{-T}\underline{k}_0 = \begin{bmatrix} 0 \\ 0 \end{bmatrix}, \quad 2\pi D^{-T}\underline{k}_1 = \begin{bmatrix} -\pi \\ \pi \end{bmatrix}$$

$$\begin{aligned} \underline{z}^{D^{-1}} e^{-j2\pi D^{-T}\underline{k}_0} &= \underline{z}^{D^{-1}} \\ \underline{z}^{D^{-1}} e^{-j2\pi D^{-T}\underline{k}_1} &= \begin{bmatrix} z_1^{1/2} & z_2^{-1/2} \\ z_1^{1/2} & z_2^{1/2} \end{bmatrix} \begin{bmatrix} e^{-j\pi} \\ e^{j\pi} \end{bmatrix} \\ &= \begin{bmatrix} -z_1^{1/2} & z_2^{-1/2} \\ -z_1^{1/2} & z_2^{1/2} \end{bmatrix}. \end{aligned}$$

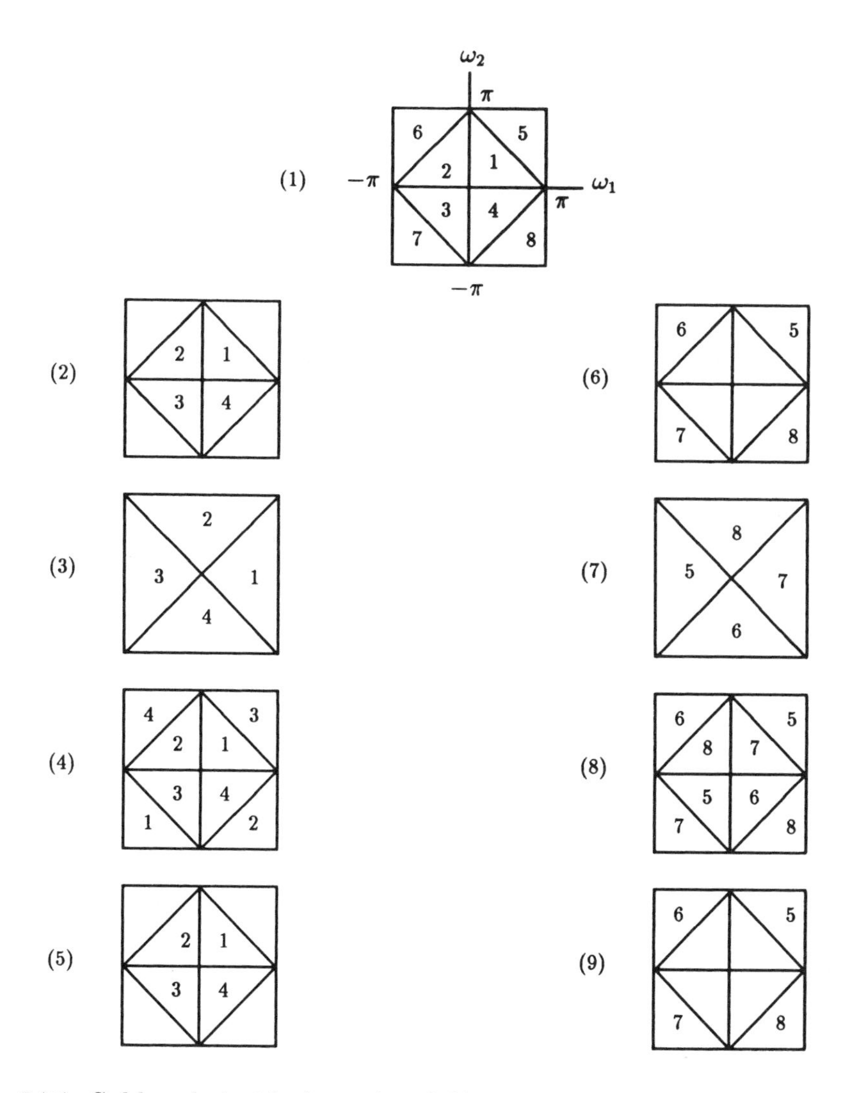

Figure 3.74: Subbands in ideal two-band filter bank, corresponding to signals in Fig. 3.72.

The down-sampled signal at node (3) is

$$\begin{aligned} V_0(\underline{z}) &= \frac{1}{2}\sum_{l=0}^{1} X_0(\underline{z}^{D^{-1}} e^{-j2\pi D^{-T}\underline{k}_l}) \\ &= \frac{1}{2}[X_0(z_1^{1/2}z_2^{-1/2}, z_1^{1/2}z_2^{1/2}) + X_0(-z_1^{1/2}z_2^{-1/2}, -z_1^{1/2}z_2^{1/2})] \\ V_0'(\underline{\omega}) &= \frac{1}{2}[X_0'(\frac{\omega_1-\omega_2}{2}, \frac{\omega_1+\omega_2}{2}) + X_0'(\frac{\omega_1-\omega_2}{2}+\pi, \frac{\omega_1+\omega_2}{2}+\pi)] \end{aligned} \quad (3.341)$$

This spectrum is shown in Fig. 3.74(3). Note that the subbands 1 through 4 occupy the full band, and because of the ideal filters, there is no aliasing at this point. Also note the rotation and stretching induced by D. The up-sampling compresses the spectrum $V_0'(\underline{\omega})$ and creates the images outside the diamond as shown in Fig. 3.74(4). This is also evident from

$$\begin{aligned} F_0(\underline{z}) &= \frac{1}{2}\sum_{l=0}^{1} X_0(\underline{z}e^{-j2\pi D^{-T}\underline{k}_l}) \\ &= \frac{1}{2}[X_0(z_1, z_2) + X_0(-z_1, -z_2)] \\ &= V_0(\underline{z}^D) \end{aligned}$$

and

$$F_0'(\underline{\omega}) = \frac{1}{2}[X_0'(\omega_1, \omega_2) + X_0'(\omega_1 - \pi, \omega_2 - \pi)] \quad (3.342)$$

The images are due to the term $X_0'(\omega_1 - \pi, \omega_2 - \pi)$. The ideal synthesis filter $G_0(\underline{z}) = H_0(\underline{z})$ removes these images, leaving us with the subbands shown at Fig. 3.74(5). In a similar way, we can trace the signals through the lower branch of the two-band structure. These spectra are also illustrated in Fig. 3.74, (6) through (9). Finally adding the signals at points (5) and (9) gives $\hat{x}(\underline{n}) = x(\underline{n})$, or perfect reconstruction. A detailed discussion of admissible passbands and their relationship to the subsampling lattice is provided by Viscito and Allebach (1991).

In this example the aliasing did not have to be cancelled out in the synthesis section. The ideal filters eliminated aliasing at inception. The design of realizable filters for perfect reconstruction is based on the theory previously developed—the polyphase approach of Viscito and Allebach, and Vetterli, among others. Various other techniques are possible. We will next describe a variation of an approach suggested by Ansari (1987) in the design of a diamondshaped subband filter.

(1) Start with a 1D filter $F(z)$ that approximates an ideal low-pass filter on $|\omega| < \pi/2$, as in Fig. 3.75(a).

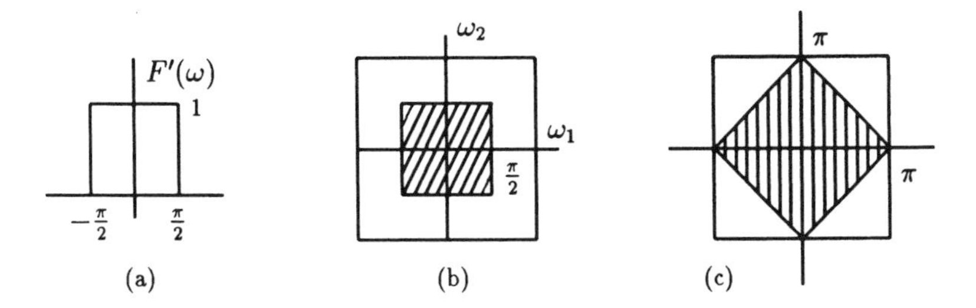

Figure 3.75: (a) 1D low-pass filter; (b) 2D rectangular low-pass filter; (c) diamond-shaped subband filter obtained by transformation of (b).

(2) Let $D(z_1, z_2) = F(z_1)F(z_2)$. This symmetric, *separable* filter approximates a subband filter with the pass band shown in Fig. 3.75(b).

(3) We can obtain a diamond-shaped subband filter by rotating and expanding the rectangle in Fig. 3.75(b). This is achieved by independent variable transformation, using the substitutions

$$z_1 \rightarrow z_1^{1/2} z_2^{-1/2} \qquad z_2 \rightarrow z_1^{1/2} z_2^{1/2}$$

$$H(z_1, z_2) = D(z_1^{1/2} z_2^{-1/2}, z_1^{1/2} z_2^{1/2}) = F(z_1^{1/2} z_2^{-1/2})F(z_1^{1/2} z_2^{1/2})$$

or

$$H'(\omega_1, \omega_2) = F'(\frac{\omega_1 - \omega_2}{2})F'(\frac{\omega_1 + \omega_2}{2}).$$

Specific design procedures for perfect reconstruction can be found in the references cited.

Before closing this subsection, we want to describe a polyphase design given in Viscito and Allebach (1991) for a two-band decomposition, Fig. 3.76, using the subsampling lattice D of Eq. (3.340) in the preceding example. In this instance,

$$\begin{aligned} \underline{Z}_M^T &= [\underline{z}^{-\underline{k}_0}, \underline{z}^{-\underline{k}_1}] = [1, z_2^{-1}] \\ \underline{z}^D &= \begin{bmatrix} z_1 z_2 \\ z_1^{-1} z_2 \end{bmatrix}. \end{aligned} \tag{3.343}$$

The analysis filters are given by Eq. (3.323)

$$\underline{h}(\underline{z}) = \begin{bmatrix} H_0(\underline{z}) \\ H_1(\underline{z}) \end{bmatrix} = \begin{bmatrix} H_{00}(\underline{z}^D) & H_{01}(\underline{z}^D) \\ H_{10}(\underline{z}^D) & H_{11}(\underline{z}^D) \end{bmatrix} \begin{bmatrix} 1 \\ \underline{z}^{-\underline{k}_1} \end{bmatrix} \tag{3.344}$$

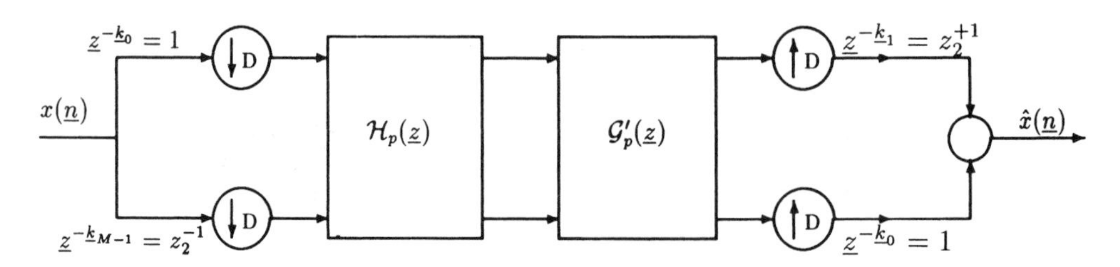

Figure 3.76: Two-band polyphase structure.

It is now convenient to impose symmetry constraints on the analysis filters. First, we can choose $H_0(\underline{z})$, and $H_1(\underline{z})$ to have quadrature mirror symmetry, which implies

$$h_1(\underline{n}) = (-1)^{n_1+n_2} h_0(\underline{n}) \leftrightarrow H_1(\underline{z}) = H_0(-\underline{z})$$

or

$$H_1^{'}(\omega_1, \omega_2) = H_0^{'}(\omega_1 - \pi, \omega_2 - \pi).$$

This implies that the H_0 frequency components inside the diamond-shaped passband B_0 are mapped into B_1, the other subband. Next, we can require $h_1(\underline{n})$ to be the spatial mirror of $h_0(\underline{n})$; i.e.,

$$h_1(\underline{n}) = h_0(-\underline{n}) \leftrightarrow H_1(\underline{z}) = H_0(\underline{z}^{-I}).$$

Combining this frequency and spatial symmetry with a shift $\underline{z}^{-\underline{p}}$ leads to the constraint

$$H_1(\underline{z}) = \underline{z}^{\underline{p}} H_0(-\underline{z}^{-I}). \tag{3.345}$$

Expanding $H_0(\underline{z})$, noting that $(-\underline{z}^{-I})^D = \underline{z}^{-D}$, $(-\underline{z}^{-I})^{-\underline{k}_1} = -\underline{z}^{\underline{k}_1}$, and choosing $\underline{p} = -\underline{k}_1$ leads to

$$\begin{aligned} H_1(\underline{z}) &= -H_{01}(\underline{z}^{-D}) + \underline{z}^{-\underline{k}_1} H_{00}(\underline{z}^{-D}) \\ &= H_{10}(\underline{z}^D) + \underline{z}^{-\underline{k}_1} H_{11}(\underline{z}^D). \end{aligned} \tag{3.346}$$

Thus the symmetry constraints on $H_1(\underline{z})$ imply that the polyphase components are related by

$$\begin{aligned} H_{10}(\underline{z}^D) &= -H_{01}(\underline{z}^{-D}) \\ H_{11}(\underline{z}^D) &= H_{10}(\underline{z}^{-D}). \end{aligned}$$

The analysis polyphase matrix becomes

$$\mathcal{H}_p(\underline{z}) = \begin{bmatrix} H_{00}(\underline{z}) & H_{01}(\underline{z}) \\ -H_{01}(\underline{z}^{-I}) & H_{00}(\underline{z}^{-I}) \end{bmatrix}. \tag{3.347}$$

Perfect reconstruction is assured by choosing $\mathcal{G}_p'(\underline{z})\mathcal{H}_p(\underline{z}) = I$:

$$\mathcal{G}_p'(\underline{z}) = \frac{1}{\Delta}\begin{bmatrix} H_{00}(\underline{z}^{-I}) & -H_{01}(\underline{z}) \\ H_{01}(\underline{z}^{-I}) & H_{00}(\underline{z}) \end{bmatrix}, \tag{3.348}$$

where

$$\Delta \triangleq det\mathcal{H}_p(\underline{z}).$$

Next, we set $\Delta = 1$, or

$$H_{00}(\underline{z})H_{00}(\underline{z}^{-I}) + H_{01}(\underline{z})H_{01}(\underline{z}^{-I}) = 1$$

or explicitly

$$H_{00}(z_1, z_2)H_{00}(z_1^{-1}, z_2^{-1}) + H_{01}(z_1, z_2)H_{01}(z_1^{-1}, z_2^{-1}) = 1. \tag{3.349}$$

With this choice, the synthesis filters will be FIR whenever the analysis filters are FIR. The final form of the synthesis filters can now be determined. Noting that

$$\begin{bmatrix} G_0(\underline{z}) \\ G_1(\underline{z}) \end{bmatrix} = \underline{g}(\underline{z}) = \mathcal{G}_p(\underline{z}^D)\underline{Z}_M$$

and

$$\mathcal{G}_p(\underline{z}) = (\mathcal{G}_p')^T J = [\mathcal{H}_p^{-1}(\underline{z})]^T J,$$

we can combine these last two equations with Eq. (3.308) to eventually arrive at

$$\begin{aligned} G_0(\underline{z}) &= \underline{z}^{-\underline{k}_1} H_0(\underline{z}^{-I}) \\ G_1(\underline{z}) &= \underline{z}^{-\underline{k}_1} H_1(\underline{z}^{-I}). \end{aligned} \tag{3.350}$$

In summary, the analysis and synthesis filters are expressed in terms of $H_0(\underline{z})$ by Eqs. (3.345) and (3.350). The perfect reconstruction requirement is then specified in terms of the polyphase components of $H_0(\underline{z})$ by Eq. (3.349), the satisfaction of which constitutes the two-band filter design problem.

3.9.6 Fan Filter Banks

Filter banks with wedge-shaped subbands have potential applications in several signal processing areas (Bamberger and Smith, 1992). The structure of a two-band, tree-structured configuration is examined here. Our focus is on the generation of the subbands and the transmission of these subbands through the filter bank. Therefore, we will assume ideal filters throughout.

The spectrum is to be partitioned into the subband wedges as shown in Fig. 3.77(d). We will consider two cases: the two-band, analysis-synthesis structure shown in Fig. 3.77(a), and the four-band analysis subband tree of Fig. 3.79.

In the first instance, our objective is to isolate the two bands consisting of wedges (1, 4, 5, 8), and (2, 3, 6, 7). We need to select antialiasing analysis filters that are compatible with the subsampling lattice chosen. Let $H_0(\underline{z})$ and $H_1(\underline{z})$ be the ideal hourglass-shaped filters indicated in Fig. 3.77, and let the subsampling

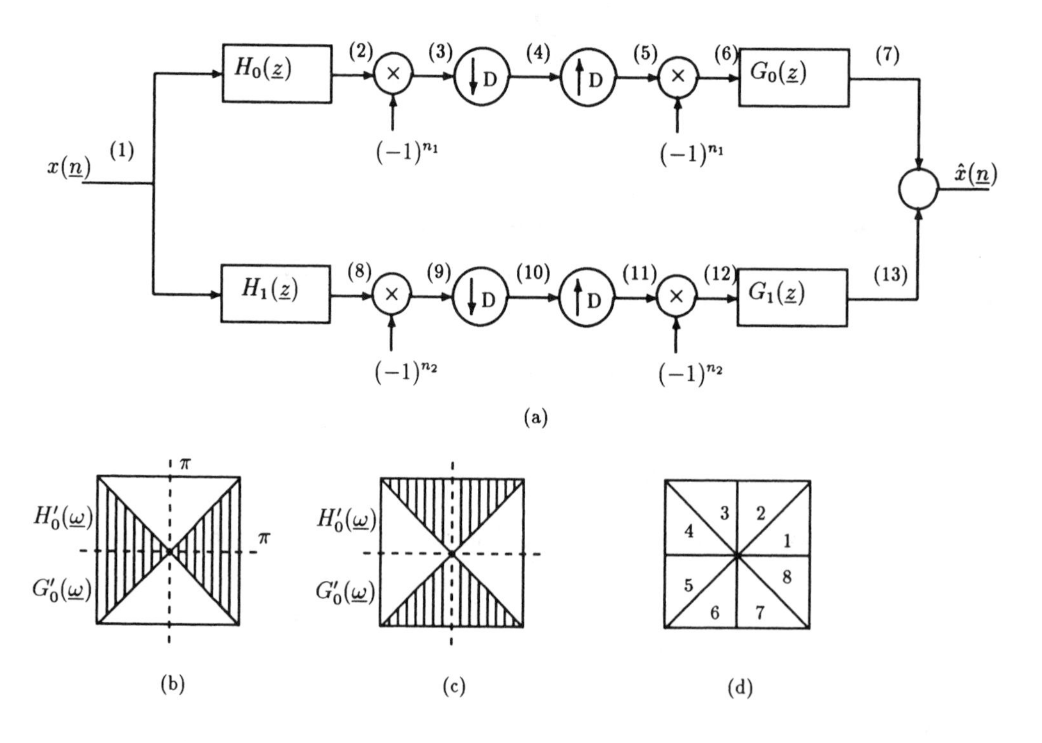

Figure 3.77: (a) Two-band directional subband structure; (b) and (c) hourglass-shaped filters; (d) fan-shaped subbands.

lattice be the familiar

$$D = \begin{bmatrix} 1 & -1 \\ 1 & 1 \end{bmatrix}.$$

The spectrum at point (2) is shown in Fig. 3.78(2). We can make this spectrum compatible with D if we shift it by π in the ω_1 direction. Thus modulation by $(-1)^{n_1}$ provides the shifted spectrum at point (3), now located within the diamond region as indicated in Fig. 3.78(3). This is the key trick; the rest is mere commentary! Down-sampling and up-sampling then produce the subbands in Fig. 3.78(4) and (5). We then modulate again by $(-1)^{n_1}$ to shift the spectrum to the wedge-shaped subbands (with images) in Fig. 3.78(6). The ideal hourglass filter $G_0(\underline{z})$ removes the images and reconstructs the subbands (1, 4, 5, 8) in their original positions. The signals in the lower branch can be similarly traced out. In this case, the shift is in the ω_2 direction requiring modulation by $(-1)^{n_2}$. The spectra at various nodes in the lower branch are shown in Figs. 3.78(8)–(13). For the ideal case considered, the reconstituted signal is exactly $x(\underline{n})$.

Extension of this idea to the four-band tree structure is quite straightforward and shown in Fig. 3.79. The signals in the first level are the same as those in Fig. 3.77. The spectra at points (5)–(10) in the two top branches are displayed explicitly in Fig. 3.80. The spectra at points (14)–(19) in the lower two branches can be worked out in a similar way. The result is the four-band split with no aliasing shown in Fig. 3.79.

Our description of fan or directional filter banks relied on ideal hourglass filters to eliminate all aliasing. This is satisfactory since our intent is to demonstrate how a filter structure with desired subband partitions can be configured with compatible filters and sampling lattices. We can, of course, back off from ideal and develop FIR and IIR filters for perfect reconstruction. In this regard, the papers by Bamberger and Smith (1992) and Ansari (1987) are noteworthy.

3.10 Summary

In this chapter, we developed the theory of subband filter banks from first principles. Starting with the fundamental operations of decimation and interpolation, we analyzed the two-band filter bank and derived and interpreted the unitary PR conditions, which is called the PR-QMF.

Using the tree structure expansion of the two-band PR-QMF, we were able to define a hierarchy of M-band filter banks with a variety of subband splits. The connection between the classical oversampled Laplacian pyramid decomposition and the critically sampled dyadic subband tree was explored and studied. In

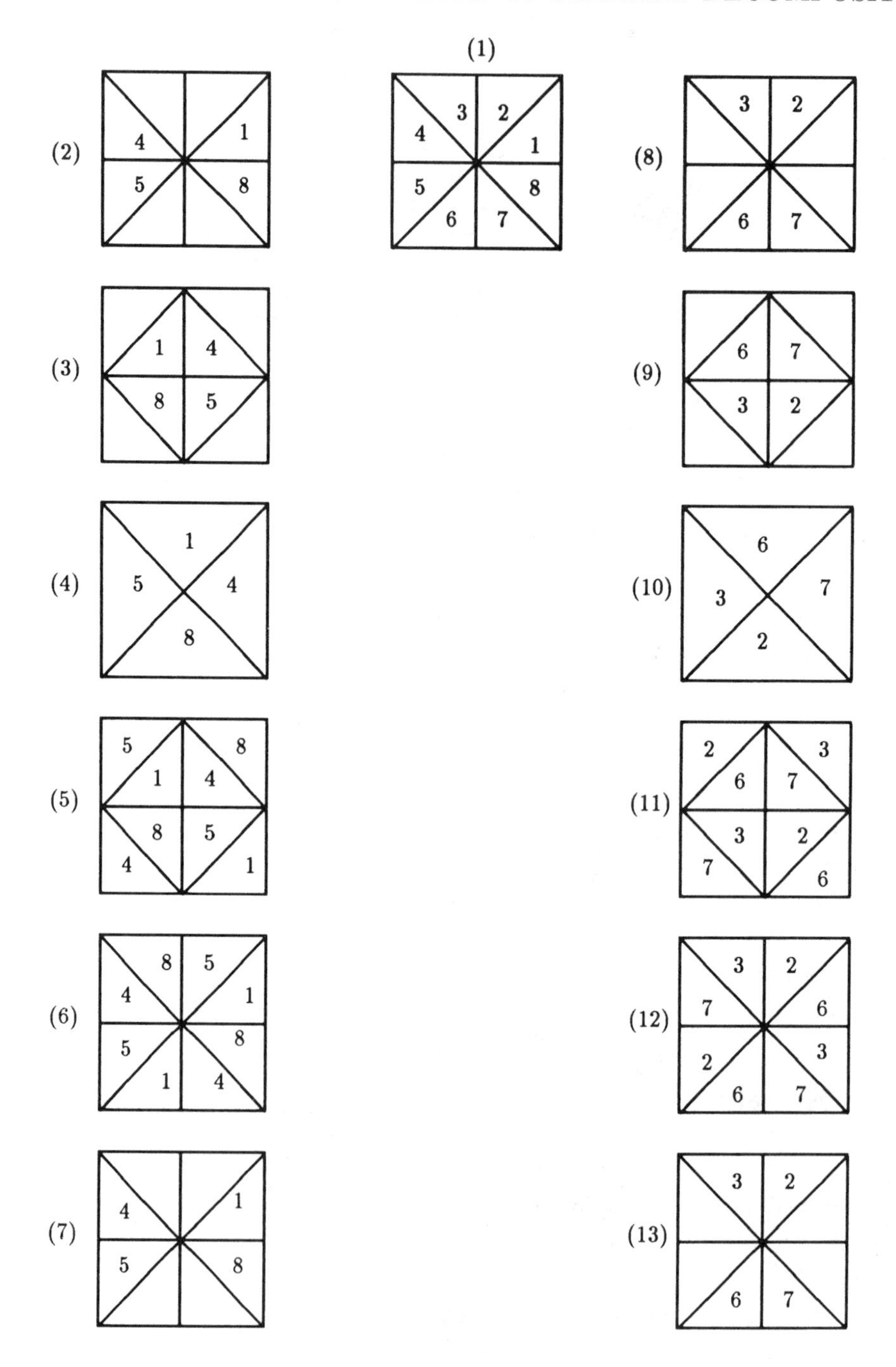

Figure 3.78: Subbands corresponding to nodes in Fig. 3.77.

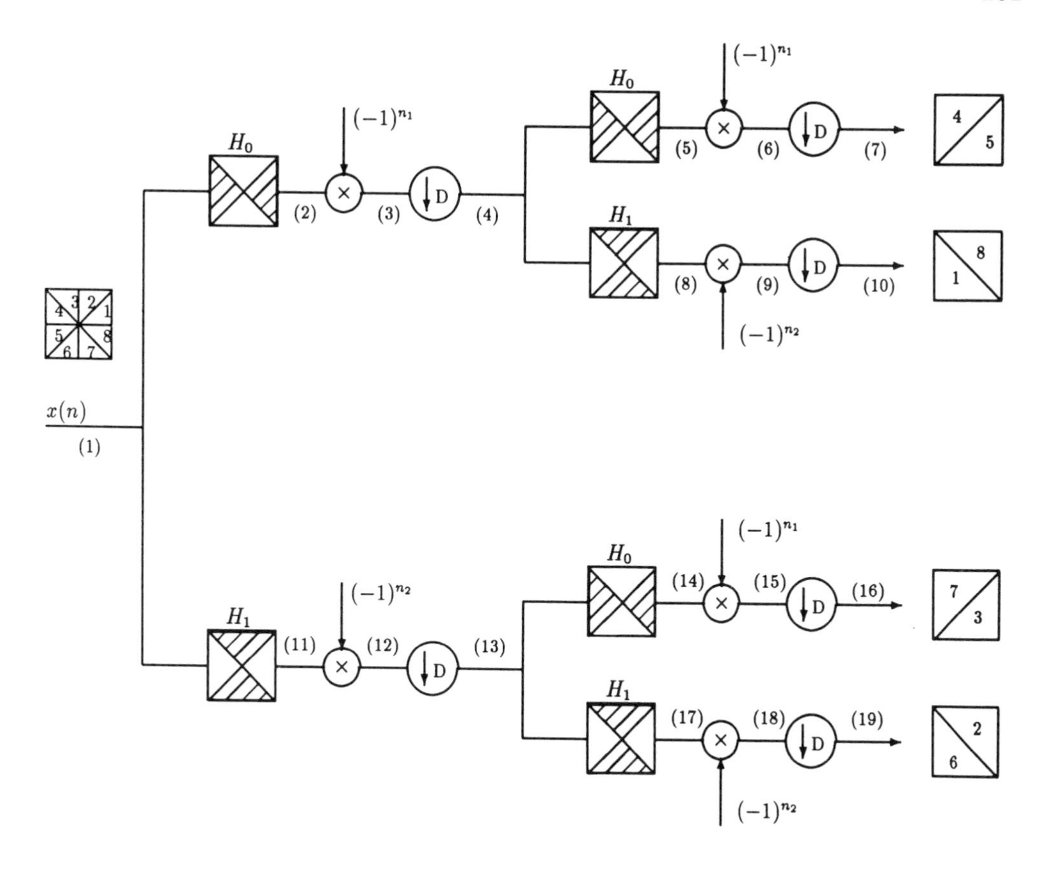

Figure 3.79: Four-subband directional filter bank.

Chapter 6, we will show this dyadic tree to be a precursor of the orthonormal wavelet transform.

Using the AC matrix and the polyphase decomposition, we were able to formulate general conditions for PR in the M-band filter structure. This led to a general time-domain formulation of the analysis-synthesis subband system that unifies critically sampled block transforms, LOTs, and critically sampled subband filter banks. The paraunitary filter bank provided an elegant solution in terms of the polyphase matrix.

The focus on the two-dimensional subband filter bank was the subsampling or decimation lattice. We showed how the 1D results could be generalized to 2D, but in a nontrivial way.

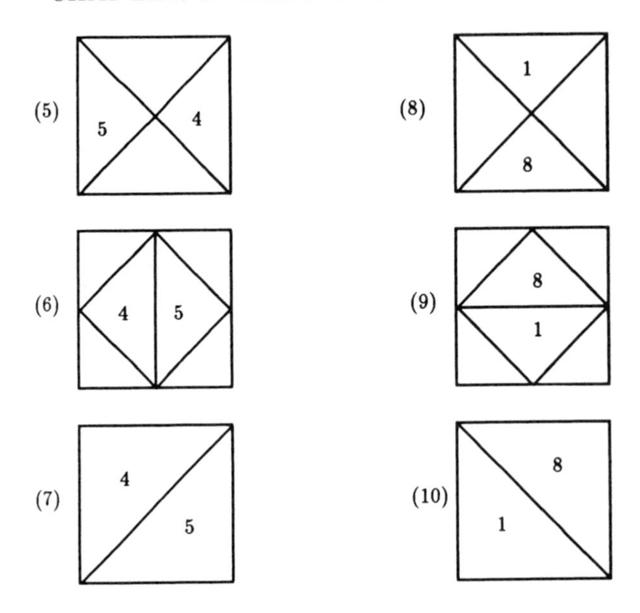

Figure 3.80: Subbands corresponding to nodes (5)–(10) in Fig. 3.79.

References

E. H. Adelson, E. Simoncelli, and R. Hingorani, "Orthogonal Pyramid Transforms for Image Coding," Proc. SPIE Visual Communication and Image Processing," pp. 50–58, 1987.

A. N. Akansu, and Y. Liu, "On Signal Decomposition Techniques," Optical Engineering, pp. 912–920, July 1991.

A. N. Akansu, R. A. Haddad, and H. Caglar, "Perfect Reconstruction Binomial QMF-Wavelet Transform," Proc. SPIE Visual Communication and Image Processing, Vol. 1360, pp. 609–618, Oct. 1990.

A. N. Akansu, R. A. Haddad, and H. Caglar, "The Binomial QMF-Wavelet Transform for Multiresolution Signal Decomposition," IEEE Trans. on Signal Processing, Vol. 41, No. 1, pp. 13–20, Jan. 1993.

R. Ansari, "Efficient IIR and FIR Fan Filters," IEEE Trans. Circuits and Systems, Vol. CAS-34, pp. 941–945, Aug. 1987.

R. Ansari and C. Guillemot, "Exact Reconstruction Filter Banks Using Diamond FIR Filters," Proc. BILCON, pp. 1412–1424, July 1990.

R. Ansari and C. L. Lau, "Two-Dimensional IIR Filters for Exact Reconstruction in Tree-Structured Subband Decomposition," Electronics Letters, Vol. 23, pp.

633–634, June, 1987.

R. Ansari and S. H. Lee, "Two-Dimensional Non-rectangular Interpolation, Decimation, and Filter Banks," presented at ICASSP, 1988.

R. Ansari and B. Liu, "Efficient Sample Rate Alteration Using Recursive(IIR) Digital Filters," IEEE Trans. ASSP, Vol. ASSP-32, pp. 1366–1373, Dec. 1983.

R. Ansari and B. Liu, "A Class of Low-noise Computationally Efficient Recursive Digital Filters with Applications to Sampling Rate Alterations," IEEE Trans. ASSP, Vol. ASSP-33, pp. 90–97, Feb. 1985.

R. Ansari, C. Guillemot, and J. F. Kaiser, "Wavelet Construction Using Lagrange Halfband Filters," IEEE Trans. Circuits and Systems, Vol. CAS-38, pp. 1116–1118, Sept. 1991.

M. Antonini, M. Barlaud, P. Mathieu, I. Daubechies, "Image Coding Using Vector Quantization in the Wavelet Transform Domain," Proc. ICASSP, pp. 2297–2300, 1990.

R. H. Bamberger and M. J. T. Smith, "A Filter Bank for the Directional Decomposition of Images: Theory and Design," IEEE Trans. on Signal Processing, Vol. 40, pp. 882–893, April 1992.

T. P. Barnwell III, "Subband Coder Design Incorporating Recursive Quadrature Filters and Optimum ADPCM Coders," IEEE Trans. ASSP, Vol. ASSP-30, pp. 751–765, Oct. 1982.

V. Belevitch, *Classical Network Theory.* Holden-Day, 1968.

M. G. Bellanger, "Computation Rate and Storage Estimation in Multirate Digital Filtering with Half-band Filters," IEEE Trans. ASSP, Vol. ASSP-25, pp. 344–346, Aug. 1977.

M. G. Bellanger and J. L. Daguet, "TDM-FDM Transmultiplexer: Digital Polyphase and FFT," IEEE Trans. Communications, Vol. COM-22, pp. 1199–1204, Sept. 1974.

M. G. Bellanger, J. L. Daguet, and G. P. Lepagnol, "Interpolation, Extrapolation, and Reduction of Computation Speed in Digital Filters," IEEE Trans. ASSP, Vol. ASSP-22,, pp. 231–235, Aug. 1974.

M. G. Bellanger, G. Bonnerot, and M. Coudreuse, "Digital Filtering by Polyphase Network: Application to Sample-rate Alteration and Filter Banks," IEEE Trans. ASSP, Vol. ASSP-24, pp. 109–114, April 1976.

P. J. Burt and E. H. Adelson, "The Laplacian Pyramid as a Compact Image Code," IEEE Trans. Comm., pp. 532–540, April 1983.

T. Caelli and M. Hubner, "Coding Images in the Frequency Domain: Filter Design and Energy Processing Characteristics of the Human Visual System," IEEE Trans. on Systems, Man and Cybernetics, pp. 1018–1021, May 1980.

B. Chitprasert and K. R. Rao, "Discrete Cosine Transform Filtering," Proc. IEEE ICASSP, pp. 1281–1284, 1990.

L. Chen, T. Q. Nguyen, and K-P Chan, "Symmetric extension methods for M-channel linear-phase perfect reconstruction filter banks", IEEE Trans. on Signal Processing, Vol. 13, No. 12, pp. 2505–2511, Dec. 1995.

P. L. Chu, "Quadrature Mirror Filter Design for an Arbitrary Number of Equal Bandwidth Channels," IEEE Trans. ASSP, Vol. ASSP-33, pp. 203–218, Feb. 1985.

A. Cohen, I. Daubechies, and J. C. Feauveau, "Biorthogonal Bases of Compactly Supported Wavelets," Technical Memo., #11217–900529–07, AT&T Bell Labs., Murray Hill.

R. E. Crochiere and L. R. Rabiner, "Optimum FIR Digital Filter Implementations for Decimation, Interpolation, and Narrow-band Filtering," IEEE Trans. ASSP, Vol. ASSP-23, pp. 444–456, Oct. 1975.

R. E. Crochiere and L. R. Rabiner, "Further Considerations in the Design of Decimators and Interpolators," IEEE Trans. ASSP, Vol. ASSP-24, pp. 296–311, Aug. 1976.

R. E. Crochiere and L. R. Rabiner, "Interpolation and Decimation of Digital Signals-A Tutorial Review," Proc. IEEE, Vol. 69, pp. 300–331, March 1981.

R. E. Crochiere and L. R. Rabiner, *Multirate Digital Signal Processing.* Prentice-Hall, 1983.

R. E. Crochiere, S. A.Weber, and J. L. Flanagan, "Digital Coding of Speech Subbands," Bell Syst. Tech. J., Vol. 55, pp. 1069–1085, 1976.

A. Croisier, D. Esteban, and C. Galand, "Perfect Channel Splitting by use of Interpolation/Decimation/Tree Decomposition Techniques," Int'l Conf. on Information Sciences and Systems, Patras, 1976.

A. W. Crooke and J. W. Craig, "Digital Filters for Sample-Rate Reduction," IEEE Trans. Audio and Electroacous., Vol. AU-20, pp. 308–315, Oct. 1972.

P. Delsarte, B. Macq, and D. T.M. Slock, "Signal-Adapted Multiresolution Transform for Image Coding," IEEE Trans. Information Theory, Vol. 38, pp. 897–904, March 1992.

R. L. de Queiroz, T. Q. Nguyen, and K. R. Rao, "The GenLOT: generalized linear-phase lapped orthogonal transform", IEEE Trans. on Signal Processing, Vol. 44, No. 3, pp. 497–507, March 1996.

D. E. Dudgeon and R. M. Mersereau, *Multidimensional Digital Signal Processing.* Prentice-Hall, 1984.

D. Esteban and C. Galand, "Application of Quadrature Mirror Filters to Split-band Voice Coding Schemes," Proc. ICASSP, pp. 191–195, 1977.

A. Fettweis, J. A. Nossek, and K. Meerkotter, "Reconstruction of Signals after Filtering and Sampling Rate Reduction," IEEE Trans. ASSP, Vol. ASSP-33, pp. 893–902, Aug. 1985.

S. W. Foo and L. F. Turner, "Design of Nonrecursive Quadrature Mirror Filters," IEE Proc., Vol. 129, part G, pp. 61–67, June 1982.

R. Forchheimer and T. Kronander, "Image Coding—From Waveforms to Animation," IEEE Trans. ASSP, Vol. ASSP-37, pp. 2008–2023, Dec. 1989.

D. Gabor, "Theory of Communications," Proc. IEE, pp. 429–461, 1946.

C. R. Galand and H. J. Nussbaumer, "New Quadrature Mirror Filter Structures," IEEE Trans. ASSP, Vol. 32, pp. 522–531, June 1984.

L. Gazsi, "Explicit Formulas for Lattice Wave Digital Filters," IEEE Trans. Circuits and Systems, Vol. CAS-32, pp. 68–88, Jan. 1985.

H. Gharavi and A. Tabatabai, "Subband Coding of Digital Image Using Two-Dimensional Quadrature Mirror Filtering," Proc. SPIE Visual Communication and Image Processing, pp. 51–61, 1986.

H. Gharavi and A. Tabatabai, "Sub-band Coding of Monochrome and Color Images," IEEE Trans. on Circuits and Systems, Vol. CAS-35, pp. 207–214, Feb. 1988.

R. C. Gonzales and P. Wintz, *Digital Image Processing.* 2nd ed. Addison-Wesley, 1987.

D. J. Goodman and M. J. Carey, "Nine Digital Filters for Decimation and Interpolation," IEEE Trans. ASSP, Vol. ASSP-25, pp. 121–126, Apr. 1977.

R. A. Gopinath and C. S. Burrus, "On Upsampling, Downsampling and Rational Sampling Rate Filter Banks," Tech. Rep., CML TR-91-25, Rice Univ., Nov. 1991.

R. A. Haddad and T. W. Parsons, Digital Signal Processing: Theory, Applications and Hardware. Computer Science Press, 1991.

R. A. Haddad and A. N. Akansu, "A Class of Fast Gaussian Binomial Filters for Speech and Image Processing," IEEE Trans. on Signal Processing, Vol. 39, pp. 723–727, March 1991.

J. H. Husoy, *Subband Coding of Still Images and Video.* Ph.D. Thesis, Norwegian Institute of Technology, 1991.

A. Ikonomopoulos and M. Kunt, "High Compression Image Coding via Directional Filtering," Signal Processing, Vol. 8, pp. 179–203, 1985.

V. K. Jain and R. E. Crochiere, "Quadrature Mirror Filter Design in the Time Domain," IEEE Trans. ASSP, Vol. ASSP-32, pp. 353–361, April 1984.

J. D. Johnston, "A Filter Family Designed for Use in Quadrature Mirror Filter Banks," Proc. ICASSP, pp. 291–294, 1980.

G. Karlsson and M. Vetterli, "Theory of Two-Dimensional Multirate Filter Banks," IEEE Trans. ASSP, Vol. 38, pp. 925–937, June 1990.

C. W. Kim and R. Ansari, "FIR/IIR Exact Reconstruction Filter Banks with Applications to Subband Coding of Images," Proc. Midwest CAS Symp., 1991.

R. D. Koilpillai and P. P. Vaidyanathan, "Cosine modulated FIR filter banks satisfying perfect reconstruction", IEEE Trans. on Signal Processing, Vol. 40, No. 4, pp. 770–783, Apr., 1992.

T. Kronander, *Some Aspects of Perception Based Image Coding.* Ph.D. Thesis, Linkoping University, 1989.

T. Kronander, "A New Approach to Recursive Mirror Filters with a Special Application in Subband Coding of Images," IEEE Trans. ASSP, Vol. 36, pp. 1496–1500, Sept. 1988.

M. Kunt, A. Ikonomopoulos, and M. Kocher, "Second Generation Image Coding Techniques," Proc. IEEE, Vol. 73, pp. 549–574, April 1985.

H. S. Malvar, "The LOT: A Link Between Block Transform Coding and Multirate Filter Banks," Proc. IEEE ISCAS, pp. 781–784, 1988.

H. S. Malvar, "Modulated QMF Filter Banks with Perfect Reconstruction," Electronics Letters, Vol. 26, pp. 906–907, June 1990.

H. S. Malvar, "Lapped Transforms for Efficient Transform/Subband Coding," IEEE Trans. ASSP, Vol. 38, pp. 969–978, June 1990.

H. S. Malvar, "Efficient Signal Coding with Hierarchical Lapped Transforms," Proc. ICASSP, pp. 1519–1522, 1990.

H. S. Malvar, *Signal Processing with Lapped Transforms.* Artech House, 1992.

H. S. Malvar, "Extended lapped transforms: properties, applications and fast algorithms", IEEE Trans. on Signal Processing, Vol. 40, No. 11, pp. 2703–2714, Nov. 1992.

H. S. Malvar and D. H. Staelin, "Reduction of Blocking Effects in Image Coding with a Lapped Orthogonal Transform," Proc. IEEE ICASSP, pp. 781–784, April 1988.

H. S. Malvar and D. H. Staelin, "The LOT: Transform Coding without Blocking Effects," IEEE Trans. ASSP, Vol.37, pp. 553–559, Apr. 1989.

D. F. Marshall, W. K. Jenkins, and J. J. Murphy, "The Use of Orthogonal Transforms for Improving Performance of Adaptive Filters," IEEE Trans. Circuits and Systems, Vol. 36, pp. 474–484, April 1989.

J. Masson and Z. Picel, "Flexible Design of Computationally Efficient Nearly Perfect QMF Filter Banks," Proc. IEEE ICASSP, pp. 541–544, 1985.

P. C. Millar, "Recursive Quadrature Mirror Filters-Criteria Specification and Design Method," IEEE Trans. ASSP, Vol. ASSP-33, pp. 413–420, 1985.

F. Mintzer, "On Half-band, Third-band and Nth-band FIR Filters and Their Design," IEEE Trans. on ASSP, Vol. ASSP-30, pp. 734–738, Oct. 1982.

F. Mintzer, "Filters for Distortion-free Two-band Multirate Filter Banks," IEEE Trans. ASSP, Vol.33, pp. 626–630, June, 1985.

F. Mintzer and B. Liu, "The Design of Optimal Multirate Bandpass and Bandstop Filters," IEEE Trans. ASSP, Vol. ASSP-26, pp. 534–543, Dec. 1978.

F. Mintzer and B. Liu, "Aliasing Error in the Design of Multirate Filters," IEEE Trans. IEEE, Vol. ASSP-26, pp. 76–88, Feb. 1978.

T. Miyawaki and C. W. Barnes, "Multirate Recursive Digital Filters: A General Approach and Block Structures," IEEE Trans. ASSP, Vol. ASSP-31, pp. 1148–1154, Oct. 1983.

K. Nayebi, T. P. Barnwell III, and M. J. T. Smith, "The Time Domain Analysis and Design of Exactly Reconstructing FIR Analysis/Synthesis Filter Banks," Proc. IEEE ICASSP, pp. 1735–1738, April 1990.

K. Nayebi, T. P. Barnwell III, and M. J. T. Smith, "Time Domain Filter Bank Analysis: A New Design Theory," IEEE Trans. Signal Processing, Vol. 40, No. 6, pp. 1412–1429, June 1992.

K. Nayebi, T. P. Barnwell III, and M. J. T. Smith, "Nonuniform Filter Banks: A Reconstruction and Design Theory," IEEE Trans. Signal Processing, Vol. 41, No. 3, pp. 1114–1127, March 1993.

K. Nayebi, T. P. Barnwell III, and M. J. T. Smith, "The Design of Nonuniform Band Filter Banks," Proc. IEEE ICASSP, pp. 1781–1784, 1991.

T. Q. Nguyen and R. D. Koilpillai, "Theory and Design of Arbitrary Length Cosine Modulated Filter Banks and Wavelets Satisfying Perfect Reconstruction", IEEE Trans. on Signal Processing, Vol. 44, No.3, pp. 473–483, March 1996.

T. Q. Nguyen and P. P. Vaidyanathan, "Two-channel Perfect-reconstruction FIR QMF Structures Which Yield Linear-phase Analysis and Synthesis Filters," IEEE Trans. ASSP, Vol. ASSP-37, No. 5, pp. 676–690, May, 1989.

T. Q. Nguyen and P. P. Vaidyanathan, "Structure for M-channel Perfect Reconstruction FIR QMF Banks which Yield Linear-Phase Analysis and Synthesis Filters," IEEE Trans. ASSP, Vol. 38, pp. 433–446, March 1990.

H. J. Nussbaumer, "Pseudo-QMF Filter Bank," IBM Tech. Disclosure Bull., Vol. 24, pp. 3081–3087, Nov. 1981.

H. J. Nussbaumer and M. Vetterli, "Computationally Efficient QMF Filter Banks," Proc. IEEE ICASSP, pp. 11.3.1–11.3.4, 1984.

A. Papoulis, *Probability, Random Variables, and Stochastic Processes*, 3rd Edition, pp. 119–123. McGraw Hill, 1991.

S. R. Pillai, W. Robertson, W. Phillips, "Subband Filters Using All-pass Structures," Proc. IEEE ICASSP, pp. 1641–1644, 1991.

J. P. Princen and A. B. Bradley, "Analysis/Synthesis Filter Bank Design Based on Time Domain Aliasing Cancellation," IEEE Trans. ASSP, Vol. ASSP-34, pp. 1153–1161, Oct. 1986.

J. P. Princen, A. W. Johnson, and A. B. Bradley, "Sub-band/Transform Coding Using Filter Bank Designs Based on Time Domain Aliasing Cancellation," Proc. IEEE ICASSP, pp. 2161–2164, April 1987.

T. A. Ramstad, "Digital Methods for Conversion between Arbitrary Sampling Frequencies," IEEE Trans. ASSP, Vol. ASSP-32, pp. 577–591, June 1984.

T. A. Ramstad, "Analysis-Synthesis Filter Banks with Critical Sampling," Proc. Int'l Conf. Digital Signal Processing, pp. 130–134, Sept. 1984.

T. A. Ramstad, "IIR Filter Bank for Subband Coding of Images," Proc. ISCAS, pp. 827–830, 1988.

T. A. Ramstad and J. P. Tanem, "Cosine Modulated Analysis-Synthesis Filter Bank with Critical Sampling and Perfect Reconstruction, Proc. IEEE Int. Conf. Acoust., Speech, Signal Processing, pp. 1789–1792, May 1991.

T. A. Ramstad, S. O. Aase and J. H. Husoy, *Subband Compression of Images: Principles and Examples.* Elsevier, 1995.

A. Rosenfeld, Ed., *Multiresolution Techniques in Computer Vision.* Springer-Verlag, New York, 1984.

J. H. Rothweiler, "Polyphase Quadrature Filters—A New Subband Coding Technique," Proc. IEEE ICASSP, pp. 1280–1283, 1983.

R. W. Schafer and L. R. Rabiner, "A Digital Signal Processing Approach to Interpolation," Proc. IEEE, Vol. 61, pp. 692–702, June 1973.

R. R. Shively, "On Multistage FIR Filters with Decimation," IEEE Trans. ASSP, Vol. ASSP-23, pp. 353–357, Aug. 1975.

E. Simoncelli, *Orthogonal Sub-band Image Transforms.* M. S. Thesis, Massachusetts Institute of Technology, May 1988.

M. J. T. Smith and T. P. Barnwell, "A Procedure for Designing Exact Reconstruction Filter Banks for Tree-Structured Sub-band Coders," Proc. IEEE ICASSP, pp. 27.1.1–27.1.4, 1984.

M. J. T. Smith and T. P. Barnwell, "Exact Reconstruction Techniques for Tree-Structured Subband Coders," IEEE Trans. ASSP, pp. 434–441, 1986.

M. J. T. Smith and T. P. Barnwell, "A New Filter Bank Theory for Time-Frequency Representation," IEEE Trans. ASSP, Vol. ASSP-35, No.3, pp. 314–327, March, 1987.

M. J. T. Smith and S. L. Eddins, "Analysis/Synthesis Techniques for Subband Image Coding," IEEE Trans. ASSP, Vol. ASSP-38, pp. 1446–1456, Aug. 1990.

A. K. Soman, P. P. Vaidyanathan, and T. Q. Nguyen, "Linear Phase Paraunitary Filter Banks: Theory, Factorization, and Design", IEEE Trans. on Signal Processing, Vol. 41, No. 12, pp. 3480–3496, Dec. 1993.

N. Uzun and R. A. Haddad, "Modeling and Analysis of Quantization Errors in Two Channel Subband Filter Structures," Proc. SPIE Visual Comm. and Image Processing, Nov. 1992.

P. P. Vaidyanathan, "Quadrature Mirror Filter Banks, M-Band Extensions and Perfect Reconstruction Techniques," IEEE ASSP Magazine, pp. 4–20, July 1987.

P. P. Vaidyanathan, "Theory and Design of M-channel Maximally Decimated Quadrature Mirror Filters with Arbitrary M, Having the Perfect Reconstruction Property," IEEE Trans. ASSP, pp. 476–492, April 1987.

P. P. Vaidyanathan, *Multirate Systems and Filterbanks.* Prentice-Hall, 1993.

P. P. Vaidyanathan, "Multirate Digital Filters, Filter Banks, Polyphase Networks, and Applications: A Tutorial," Proc. IEEE, Vol. 78, pp. 56–93, Jan. 1990.

P. P. Vaidyanathan, and Z. Doganata, "The Role of Lossless Systems in Modern Digital Signal Processing," IEEE Trans. Education, Special Issue on Circuits and Systems, Vol. 32, No. 3, pp. 181–197, Aug. 1989.

P. P. Vaidyanathan, and P. Q. Hoang, "Lattice Structures for Optimal Design and Robust Implementation of Two-band Perfect Reconstruction QMF Banks," IEEE Trans. ASSP, Vol. ASSP-36, No.1, pp. 81–94, Jan. 1988.

P. P. Vaidyanathan, T. Q. Nguyen, Z. Doganata, and T. Saramaki, "Improved Technique for Design of Perfect Reconstruction FIR QMF Banks with Lossless Polyhase Matrices," IEEE Trans. ASSP, pp. 1042–1056, July 1989.

L. Vandendorpe, "Optimized Quantization for Image Subband Coding," Signal Processing, Image Communication, Vol. 4, No. 1, pp. 65–80, Nov. 1991.

M. Vetterli and C. Herley, "Wavelets and Filter Banks: Theory and Design," IEEE Trans. Signal Processing, Vol. 40, No. 9, pp. 2207–2232, Sept. 1992.

M. Vetterli, "Multi-dimensional Sub-band Coding: Some Theory and Algorithms," Signal Processing, pp. 97–112, 1984.

M. Vetterli, "A Theory of Multirate Filter Banks," IEEE Trans. ASSP, Vol. ASSP-35, pp. 356–372, March 1987.

M. Vetterli, and D. LeGall, "Perfect Reconstruction FIR Filter Banks: Some Properties and Factorizations," IEEE Trans. ASSP, pp. 1057–1071, July 1989.

E. Viscito and J. P. Allebach, "The Analysis and Design of Multidimensional FIR Perfect Reconstruction Filter Banks for Arbitrary Sampling Lattices," IEEE Trans. Circuits and Systems, Vol. CAS-38, pp. 29–41, Jan. 1991.

E. Viscito and J. Allebach, "The Design of Tree-structured M-channel Filter Banks Using Perfect Reconstruction Filter Blocks," Proc. IEEE ICASSP, pp. 1475–1478, 1988.

E. Viscito and J. Allebach, "The Design of Equal Complexity FIR Perfect Reconstruction Filter Banks Incorporating Symmetries," Tech. Rep., TR-EE-89–27, Purdue University, May 1989.

G. Wackersreuther, "On Two-Dimensional Polyphase Filter Banks," IEEE Trans. ASSP, Vol. ASSP-34, pp. 192–199, Feb. 1986.

P. H. Westerink, J. Biemond, and D. E. Boekee, "Scalar Quantization Error Analysis for Image Subband Coding Using QMF's," IEEE Trans. Signal Processing, pp. 421–428, Feb. 1992.

P. H. Westerink, *Subband Coding of Images.* Ph.D. Thesis, Delft University, 1989.

P. H. Westerink, J. Biemond, D. E. Boekee, and J. W. Woods, "Sub-band Coding of Images Using Vector Quantization," IEEE Trans. Communications, Vol. COM-36, pp. 713–719, June 1988.

J. W. Woods and T. Naveen, "A Filter Based Bit Allocation Scheme for Subband Compression of HDTV," IEEE Trans. Image Processing, Vol. 1, No. 3, pp. 436–440, July 1992.

J. W. Woods, Ed., *Subband Image Coding.* Kluwer, 1991.

J. W. Woods, S. D. O'Neil, "Subband Coding of Images," IEEE Trans. ASSP, Vol. ASSP-34, No. 5, Oct. 1986.

Chapter 4

Filter Bank Families: Design and Performance

This chapter deals with the description, listing of coefficients, and comparative performance evaluation of various filter families. We then connect and compare these evaluations with those of the block transforms and LOTs of Chapter 2.

Next, we describe a method for the optimal design of filters using extended performance measures that include not only the standard unitary PR, but also constraints embodying a mix of criteria such as linear-phase, compaction and source signal statistics, aliasing energy and cross-correlation of subband signals, and certain time and frequency constraints. Tables of optimized filter coefficients are provided along with performance comparisons with filters designed conventionally.

The chapter includes an analysis of the distribution of aliasing energy among the subbands, in terms of an energy matrix. Then we define a single parameter that measures this distribution. Finally, tables are provided that compare this matrix and parameter for a sample block transform, PR-QMF, and the most regular wavelet filter.

The chapter concludes with a section dealing with rigorous modeling of quantization effects and optimum design of quantized M-band filter banks.

4.1 Binomial QMF-Wavelet Filters

The Binomial sequences were introduced in Section 2.3.2 as a family of orthogonal sequences that can be generated with remarkable simplicity—no multiplications are necessary. We saw that the modified Hermite transform is a computationally efficient unitary transform based on the Binomial-Hermite sequences. While this transform is inferior to the DCT for most coding applications, such is decidedly *not* the case for subband coders. We will show that the Binomial QMFs are the maximally flat magnitude square PR paraunitary filters with good compression capability. In Chapter 5, these are shown to be wavelet filters as well.

The PR conditions for the two-band PR paraunitary filter bank are given by Eq. (3.47). It is now a straightforward matter to impose these conditions on the Binomial family. The 8-tap Binomial frequency responses[1] are shown in Fig. 2.8(b). The first four frequency responses have energies distributed primarily over $[0, \pi/2]$, and the lower four over $[\pi/2, \pi]$ for $N+1=8$. This suggests that we take as the low-pass half bandwidth filter a superposition of the lower half Binomial sequences. Therefore, we let

$$h_0(n) \triangleq h(n) = \sum_{r=0}^{\frac{N-1}{2}} \theta_r x_r(n) \tag{4.1}$$

where $x_r(n) \leftrightarrow X_r(z)$ is defined by Eqs. (2.134) and (2.139). Then

$$H(z) = \sum_{r=0}^{\frac{N-1}{2}} \theta_r (1+z^{-1})^{N-r}(1-z^{-1})^r = (1+z^{-1})^{(N+1)/2} F(z) \tag{4.2}$$

where $F(z)$ is a polynomial in z^{-1} of order $(N-1)/2$. For convenience, we take $\theta_0 = 1$, and later impose the required normalization. The correlation sequence, Eq. (3.43), becomes

$$\begin{aligned}
\rho(n) &= \left(\sum_{r=0}^{\frac{N-1}{2}} \theta_r x_r(n) \right) \odot \sum_{s=0}^{\frac{N-1}{2}} \theta_s x_s(n) \\
&= \sum_{r=0}^{\frac{N-1}{2}} \sum_{s=0}^{\frac{N-1}{2}} \theta_r \theta_s [x_r(n) \odot x_s(n)] \\
&= \sum_{r=0}^{\frac{N-1}{2}} \sum_{s=0}^{\frac{N-1}{2}} \theta_r \theta_s \rho_{rs}(n) \\
&= \sum_{r=0}^{\frac{N-1}{2}} \theta_r^2 \rho_{rr}(n) + \sum_{\substack{r=0 \\ r \neq s}}^{\frac{N-1}{2}} \sum_{s=0}^{\frac{N-1}{2}} \theta_r \theta_s \rho_{rs}(n)
\end{aligned} \tag{4.3}$$

[1]For the Binomial filters, the length is designated as $N+1$, where N is the order of the filter.

where

$$\rho_{rs}(n) = x_r(n) * x_s(-n) = \sum_k x_r(k)x_s(n+k) = \rho_{sr}(-n). \tag{4.4}$$

But from the properties of the Binomial sequences, Eqs. (2.143)–(2.146), we can show that

$$\rho_{rs}(n) = (-1)^{s-r}\rho_{sr}(n) \tag{4.5}$$

and

$$\rho_{N-r,N-r}(n) = (-1)^n \rho_{rr}(n). \tag{4.6}$$

Equation (4.5) implies that the second summation in Eq. (4.3) has terms only where the indices differ by an even integer. Therefore, the autocorrelation for the Binomial half-bandwidth low-pass filter is

$$\rho(n) = \sum_{n=0}^{\frac{N-1}{2}} \theta_r^2 \rho_{rr}(n) + 2\sum_{l=1}^{\frac{N-3}{2}} \sum_{\nu=0}^{\frac{N-1}{2}-2l} \theta_\nu \theta_{\nu+2l} \rho_{\nu,\nu+2l}(n) \tag{4.7}$$

Finally, the PR requirement is $\rho(2n) = \delta(n)$, or

$$\rho(0) = 1 \qquad \rho(n) = 0 \quad n = 2, 4, \ldots, N-1 \tag{4.8}$$

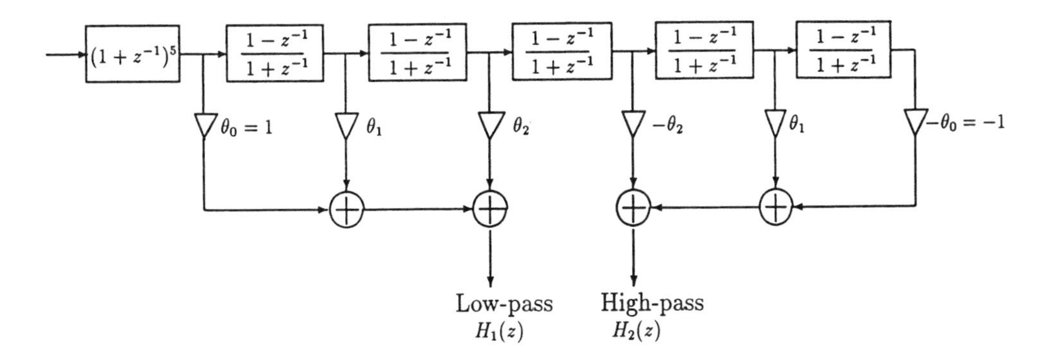

Figure 4.1: Low-pass and high-pass QMF filters from Binomial network.

This condition gives a set of $\frac{N-1}{2}$ nonlinear algebraic equations, in the $\frac{N-1}{2}$ unknowns $\theta_1, \theta_2, \ldots, \theta_{\frac{N-1}{2}}$ (Akansu, Haddad, and Caglar, 1990). The implementation of these half-bandwidth filters is trivially simple and efficient using either the purely FIR structure or the pole-zero cancellation configuration. The latter is shown in Fig. 4.1 for $N = 5$, wherein both low-pass and high-pass filters are

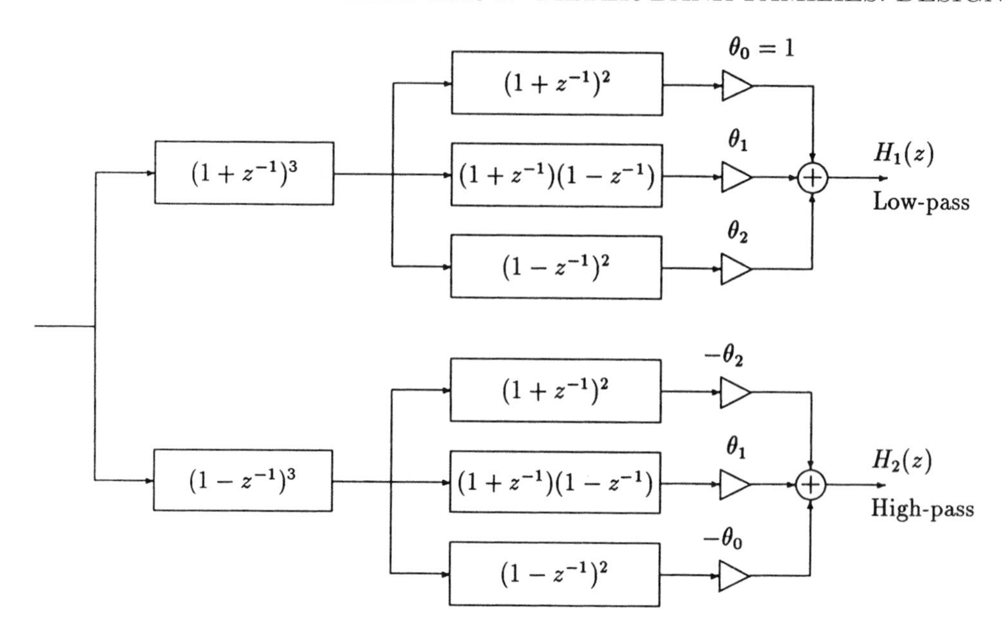

Figure 4.2: Low-pass and high-pass QMFs using direct form Binomial structure.

simultaneously realized. Figure 4.2 shows the QMF bank using the direct form. Coefficient θ_0 can be taken equal to unity, leaving only θ_1 and θ_2 as tap weights. These are the only multiplications needed when using the Binomial network as the PR-QMF rather than the six $h(n)$ weights in a transversal structure. The values of θ_r, for $N = 3, 5, 7$ (corresponding to 4-, 6-, 8-tap filters respectively) are given in Table 4.1 (where $\theta_0 = 1$).

As seen, there is more than one filter solution for a given N. For example, with $N = 3$, one obtains $\theta_1 = \sqrt{3}$ and also $\theta_1 = -\sqrt{3}$. The positive θ_1 corresponds to a minimum-phase solution, while the negative θ_1 provides a nonminimum-phase filter. The magnitude responses of both filters are identical. Although in our derivation, no linear-phase constraint on $h(n)$ was imposed, it is noteworthy that the phase responses are almost linear, the nonminimum-phase filters even more so. The magnitude and phase responses of these minimum-phase Binomial QMFs are given in Fig. 4.3 for the cases $N = 3, 5, 7$. Table 4.2 provides the normalized 4-,6-,8-tap Binomial QMF filter coefficients for a transversal realization for both minimum- and nonminimum-phase cases. We may recognize that these filters are the unique, maximally flat magnitude square PR-QMF solutions. In fact, it will be shown in Section 4.2 that the PR requirements are satisfied if we choose the θ_r

N=3		
θ_r	set 1	set 2
θ_0	1	1
θ_1	$\sqrt{3}$	$-\sqrt{3}$

N=5		
θ_r	set 1	set 2
θ_0	1	1
θ_1	$\sqrt{2\sqrt{10}+5}$	$-\sqrt{2\sqrt{10}+5}$
θ_2	$\sqrt{10}$	$\sqrt{10}$

N=7				
θ_r	set 1	set 2	set 3	set 4
θ_0	1	1	1	1
θ_1	4.9892	-4.9892	1.0290	-1.0290
θ_2	8.9461	8.9461	-2.9705	-2.9705
θ_3	5.9160	-5.9160	-5.9160	5.9160

Table 4.1: θ_r coefficients of Binomial QMF.

coefficients to satisfy maximally flat requirements at $\omega = 0$ and $\omega = \pi$. Explicitly, with $R(\omega) = |H(e^{j\omega})|^2$, we can set θ_r to satisfy

$$R(0) = 1 \qquad R(\pi) = 0$$

$$\left.\frac{d^k R(\omega)}{d\omega^k}\right|_{\substack{\omega = 0 \\ \omega = \pi}} = 0, \qquad k = 1, 2, ..., N. \tag{4.9}$$

Herrmann (1971) provides the unique maximally flat function on the interval [0,1]. This function can be easily mapped onto Z-plane to obtain the maximally flat magnitude square function $R(z)$. Now, one can obtain the corresponding $H(z)$ from $R(z)$ via factorization. This approach extends Herrmann's solution to the PR-QMF case.

n	h(n)			
	Mini Phase	Non-Minimum Phase		
	4 tap	4 tap		
0	0.48296291314453	-0.1294095225512		
1	0.83651630373780	0.2241438680420		
2	0.22414386804201	0.8365163037378		
3	-0.12940952255126	0.4829629131445		
	6 tap	6 tap		
0	0.33267055439701	0.0352262935542		
1	0.80689151040469	-0.0854412721235		
2	0.45987749838630	-0.1350110232992		
3	-0.13501102329922	0.4598774983863		
4	-0.08544127212359	0.8068915104046		
5	0.03522629355424	0.3326705543970		
	8 tap	8 tap	8 tap	8 tap
0	0.23037781098452	-0.0105973984294	-0.0757657137833	0.0322230981272
1	0.71484656725691	0.0328830189591	-0.0296355292117	-0.0126039690937
2	0.63088077185926	0.0308413834495	0.4976186593836	-0.0992195317257
3	-0.02798376387108	-0.1870348133969	0.8037387521124	0.2978578127957
4	-0.18703481339693	-0.0279837638710	0.2978578127957	0.8037387521124
5	0.03084138344957	0.6308807718592	-0.0992195317257	0.4976186593836
6	0.03288301895913	0.7148465672569	-0.0126039690937	-0.0296355292117
7	-0.01059739842942	0.2303778109845	0.0322230981272	-0.0757657137833

Table 4.2: Binomial QMF coefficients.

4.1.1 Binomial QMF and Orthonormal Wavelets

As shown in Chapter 5, the theory of orthonormal wavelet transforms is strongly linked with orthonormal PR-QMF filter banks. It develops that the convergence and differentiability of the continuous wavelet function, a property known as *regularity*, is related implicitly to the number of zeros of the discrete wavelet filter at $\omega = \pi$. From Eq. (4.2), this feature is seen to be inherent in the Binomial QMF. In fact, the Binomial QMFs developed here are *identical* to the wavelet filters proposed by Daubechies (1988).

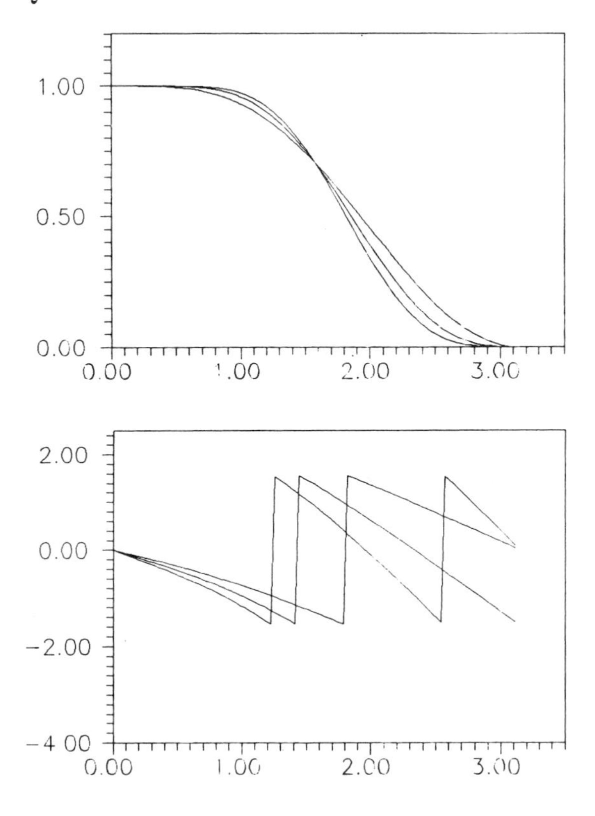

Figure 4.3: Amplitude and phase responses of minimum-phase Binomial QMFs for $N = 3, 5, 7$.

As shown in Chapter 5, wavelet regularity is related to the number of zeros at $\omega = \pi$ of a low-pass "inter-scaling" sequence of length $N + 1$, which is

$$H'(z) = (1 + z^{-1})^k F'(z) \quad 1 \leq k \leq \frac{N+1}{2}. \tag{4.10}$$

If $k = \frac{N+1}{2}$; the maximum number of zeros of $H'(z)$ is located at $\omega = \pi$; in this case $F'(z)$ is of degree $\frac{N-1}{2}$ and

$$H'(z) = (1 + z^{-1})^{\frac{N+1}{2}} F'(z).$$

Imposing PR requirement on $H'(z)$ forces the function $F'(z)$ to be equal to the Binomial $F(z)$, or

$$F'(z) = F(z) = \sum_{r=0}^{\frac{N-1}{2}} \theta_r (1+z^{-1})^{\frac{N-1}{2}-r}(1-z^{-1})^r \tag{4.11}$$

where $\{\theta_r\}$ are the binomial weights that satisfy Eq. (4.8). But it can be shown that the corresponding "spectral density" function is given *explicitly* by

$$V(z) = F(z)F(z^{-1}) = z^{\frac{N-1}{2}} \sum_{l=0}^{\frac{N-1}{2}} (-1)^l \binom{N}{l} (1+z^{-1})^{N-1-2l}(1-z^{-1})^{2l}. \tag{4.12}$$

This in fact is the polynomial used by Daubechies (1988), whose spectral factorization yields the orthonormal wavelet filters.

Furthermore, the magnitude square function (the transform of $\rho(n)$) is

$$\begin{aligned} R(z) &= H(z)H(z^{-1}) \\ &= \frac{z^N(1+z^{-1})^{N+1}}{4^{N+1}} \sum_{l=0}^{\frac{N-1}{2}} (-1)^l \binom{N}{l} (1+z^{-1})^{N-1-2l}(1-z^{-1})^{2l} \end{aligned} \tag{4.13}$$

This magnitude square function of the Binomial QMF solution is the *unique* maximally flat function obtained by Herrmann (1971).

4.2 Maximally Flat Filters

The class of maximally flat low-pass filters in a PR filter bank is defined by the number of vanishing derivatives of the magnitude square function at $\omega = 0$ and $\omega = \pi$. In this section, we develop explicit formulas for the design of such filters and relate these to Binomial QMFs.

Let $h(n)$ be a length $2N$ low-pass filter with the system function

$$H(z) = \sum_{n=0}^{2N-1} h(n)z^{-n}. \tag{4.14}$$

Its magnitude square function is

$$H(z)H(z^{-1}) \leftrightarrow \rho(n) = h(n) * h(-n)$$

or

$$\begin{aligned} |H(e^{j\omega})|^2 &= H(z)H(z^{-1}) \\ &= \rho(0) + 2\sum_{n=1}^{2N-1} \rho(n)cos(n\omega). \end{aligned} \tag{4.15}$$

We want to choose sequence $\rho(n)$ to satisfy the conditions

$$\left|H(e^{j\omega})\right|^2_{\omega=0} = 1 \tag{4.16}$$

$$\frac{d^\nu}{d\omega^\nu}\left|H(e^{j\omega})\right|^2_{\omega=0} = 0, \qquad \nu = 1, 2, ..., 2(2N-k)-1 \tag{4.17}$$

$$\frac{d^\mu}{d\omega^\mu}\left|H(e^{j\omega})\right|^2_{\omega=\pi} = 0, \qquad \mu = 0, 1, ..., 2k-1, \tag{4.18}$$

where k is an integer to be chosen arbitrarily within the limits $1 \leq k \leq 2N-1$. The parameter k defines the degree, ν, of flatness of the magnitude square function at $\omega = 0$ and μ, at $\omega = \pm\pi$. Note that all odd-ordered derivatives are zero at $\omega = 0$, and at $\omega = \pi$ for any $\{\rho(n)\}$. This effectively reduces the number of boundary conditions to $2N$, which just matches the number of samples $\{\rho(n)\}$.

If one defines the transform

$$cos\omega = 1 - 2x$$

where

$$x = \frac{1}{2}(1 - cos\omega)$$

$|H(e^{j\omega})|^2$ can be transformed into a simple polynomial of degree $2N-1$ as

$$P_{2N-1,k}(x) = \sum_{\nu=0}^{2N-1} a_\nu x^\nu \tag{4.19}$$

with an approximation interval $0 \leq x \leq 1$ and the following boundary conditions:
(1) $P_{2N-1,k}(x)$ has zeros of order k at $x = 1$,
(2) $P_{2N-1,k}(x)$ has zeros of order $2N-k$ at $x = 0$.

This is a special case of Hermite interpolation problem and can be solved by using the Newton interpolation formula (Miller, 1972). The set of $\{a_\nu\}$ that satisfy these constraints is given explicitly by (Herrmann, 1971)

$$\begin{aligned} P_{2N-1,k}(x) &= (1-x)^k \frac{1}{(1-k)!}\frac{d^{k-1}}{dx^{k-1}} \sum_{\nu=0}^{2N-2} x^\nu \\ &= (1-x)^k \sum_{\nu=0}^{2N-k-1} \binom{k+\nu-1}{\nu} x^\nu. \end{aligned} \tag{4.20}$$

Now we can inversely map $\{x, a_\nu\}$ into $\{\omega, \rho(n)\}$ and obtain (Rajagopaland and Dutta Roy, 1987)

$$\rho(0) = \frac{1}{2} \sum_{k=0}^{\lceil \frac{2N-1}{2} \rceil} \left[2^{-2k} \binom{2k}{k} \sum_{i=2k}^{2N-1} 2^{-i} \binom{i}{2k} a_i \right] \tag{4.21}$$

and

$$\rho(l) = \sum_{k=0}^{\lceil \frac{2N-1-l}{2} \rceil} \left[2^{-(2k+l)} \binom{2k+l}{k} \sum_{i=2k+l}^{2N-1} 2^{-i} \binom{i}{2k+l} a_i \right] \quad l = 1, 2, ..., 2N-1 \tag{4.22}$$

where $\lceil x \rceil$ means the integer part of x.

Equations (4.20)–(4.22) constitute the formulas for designing magnitude square function with prescribed flatness at $\omega = 0$ and $\omega = \pi$. On the other hand, the PR-QMF must satisfy

$$H(z)H(z^{-1}) + H(-z)H(-z^{-1}) = 1.$$

By inspection we see that a maximally flat PR-QMF requires its magnitude square function have maximum number of zeros at $\omega = 0$ and $\omega = \pm\pi$ *equally*, implying symmetry around $\omega = \pi/2$. This is expressed as

$$\left. \frac{d^\mu |H(e^{j\omega})|^2}{d\omega^\mu} \right|_{\omega=0} = 0 \tag{4.23}$$

$$\left. \frac{d^\mu |H(e^{j\omega})|^2}{d\omega^\mu} \right|_{\omega=\pm\pi} = 0 \quad \mu = 1, 2, ..., 2N-1.$$

Therefore, for this case, $P_{2N-1,k}(x)$ becomes

$$P_{2N-1,k}(x) = (1-x)^N \sum_{\nu=0}^{N-1} \binom{N+\nu-1}{\nu} x^\nu, \tag{4.24}$$

which maps into the magnitude square function of the Binomial QMF of Eq. (4.13) (except that N in Eq. (4.13) is replaced by $2N$ in the present context.)

Maximally flat filters will also provide the transition to the PR-QMF filters based on Bernstein polynomials, as described in the next section.

4.3 Bernstein QMF-Wavelet Filters

The Bernstein polynomials (Lorentz, 1953; Davis, 1963; Cheney, 1981) provide a ripple-free approximation to a set of points on the interval $[0,1]$. These parameterized polynomials and the mapping induced by them generate a magnitude square function that satisfies the PR-QMF conditions. We will show that several well-known orthonormal wavelet filters including the Binomial or Daubechies and Coiflet families emerge as special cases of this technique (Caglar and Akansu, 1992).

Let $\{f(\frac{k}{N})\}$ be $(N+1)$ uniformly spaced samples of a function $f(x)$ defined on the interval $[0,1]$. The Nth order Bernstein polynomial approximation to $f(x)$ is

$$B_N(f;x) = \sum_{k=0}^{N} f(\frac{k}{N}) \left(\begin{array}{c} N \\ k \end{array} \right) x^k (1-x)^{N-k} \tag{4.25}$$

Some features of this interpolation are (Davis, 1963):
(1) If $f(x)$ is differentiable, the approximation is also valid for its differentials. That implies

$$B_N(f;x) \to f(x) \quad \text{and} \quad B'_N(f;x) \to f'(x)$$

where the prime means the derivative. This feature also holds for higher derivatives. Therefore the Bernstein polynomials provide simultaneous approximations of a function and its derivatives.
(2) A monotonic and convex function is approximated by a monotonic and convex approximant. Hence it is ripple-free.

Consider now a low-pass function $f(x)$, $0 \leq x \leq 1$, which satisfies the PR-QMF conditions on $[0,1]$

$$f(x) + f(1-x) = 1, \qquad f(x) \geq 0. \tag{4.26}$$

Suppose $f(x)$ has sample values

$$f(\frac{i}{2N-1}) = \begin{cases} 1 & 0 \leq i \leq N-1 \\ 0 & N \leq i \leq 2N-1. \end{cases} \tag{4.27}$$

The Bernstein approximation is then

$$B_{2N-1}(f;x) = \sum_{i=0}^{N-1} \left(\begin{array}{c} 2N-1 \\ i \end{array} \right) x^i (1-x)^{2N-1-i} = (1-x)^N \sum_{i=0}^{N-1} \left(\begin{array}{c} N+i-1 \\ i \end{array} \right) x^i. \tag{4.28}$$

This last equation corresponds to a maximally flat symmetrical function around $1/2$ within $0 \leq x \leq 1$. It is precisely the magnitude square function of the Binomial QMF wavelet transform filter in x, Eq. (4.24). Using the inverse mappings,

$$x = \frac{1}{2}(1 - cos\omega), \qquad cos\omega = \frac{1}{2}(z + z^{-1}),$$

we obtain the Binomial QMF magnitude square function in the z domain of Eq. (4.13) (with N replaced by $2N$) whose solution was described in Section 4.1.

We can extend this technique to obtain a broad family of smooth PR-QMFs defined by a set of approximation parameters. Again assume Eq. (4.26), and let the set of nonincreasing samples be

$$f\left(\frac{i}{2N-1}\right) = \begin{cases} 1 & i = 0 \\ 1 - \alpha_i & 1 \leq i \leq N-1 \\ \alpha_i & N \leq i \leq 2(N-1) \\ 0 & i = 2N-1 \end{cases} \tag{4.29}$$

where $\alpha_i = \alpha_{2N-1-i}$ and $0 \leq \alpha_i < 0.5$ with $1 \leq i \leq N-1$. Then the Bernstein polynomial approximation is expressed as

$$\begin{aligned} B(f;x) &= \sum_{i=0}^{N-1} \binom{2N-1}{i} x^i (1-x)^{2N-1-i} \\ &- \sum_{i=1}^{N-1} \alpha_i \left(2N - 1i \right) x^i (1-x)^{2N-1-i} \\ &+ \sum_{i=N}^{2(N-1)} \alpha_i \binom{2N-1}{i} x^i (1-x)^{2N-1-i}. \end{aligned} \tag{4.30}$$

After applying the inverse mappings, we obtain $R(z)$ the corresponding magnitude square function in z domain,

$$\begin{aligned} R(z) &= z^{2N-1}\frac{(1+z^{-1})^{2N}}{4^{2N-1}} \left\{ \sum_{i=0}^{N-1} (-1)^i \binom{2N-1}{i} (1+z^{-1})^{2(N-1-i)}(1-z^{-1})^{2i} \right. \\ &- \sum_{i=1}^{N-1} (-1)^i \alpha_i \binom{2N-1}{i} (1+z^{-1})^{2(N-1-i)}(1-z^{-1})^{2i} \\ &+ \left. \sum_{i=N}^{2(N-1)} (-1)^i \alpha_i \binom{2N-1}{i} (1+z^{-1})^{2(N-1-i)}(1-z^{-1})^{2i} \right\}. \end{aligned} \tag{4.31}$$

Example: Consider the design of a 6-tap smooth PR-QMF with the constraints defined as

$$f\left(\frac{i}{2N-1}\right)=\begin{cases}1 & 0\leq i\leq 1\\ 1-\alpha & i=2\\ \alpha & i=3\\ 0 & 4\leq i\leq 5\end{cases} \tag{4.32}$$

where $0 \leq \alpha < 0.5$. This set of constraints actually corresponds to a filter function $h(n)$, with two vanishing moments for $\alpha > 0$ and three vanishing moments for $\alpha = 0$. The corresponding magnitude square function is

$$\begin{aligned} R(z) &= z^5\frac{(1+z^{-1})^6}{4^5}\left\{\sum_{i=0}^{2}(-1)^i\binom{5}{i}(1+z^{-1})^{2(2-i)}(1-z^{-1})^{2i}\right.\\ &\left. -\alpha\binom{5}{2}(1-z^{-1})^4-\alpha\binom{5}{3}(1+z^{-1})^{-2}(1-z^{-1})^6\right\}. \end{aligned} \tag{4.33}$$

At this point, any factorization technique can be used to obtain the corresponding PR-QMF $H(z)$. Figure 4.4 displays $f(i)$, $B(f;x)$ and $R(z)$ functions of Bernstein polynomial approximation for the 6-tap case with $\alpha = 0.25$.

We can relate the moments of a filter impulse response $h(n)$ to the derivatives of $H(e^{j\omega})$. With $H(e^{j\omega})$ and $H(e^{j(\omega+\pi)})$ the low-pass and high-pass filter pairs, respectively, $H(e^{j\omega}) = \sum h(n)e^{-jn\omega}$, and $H(e^{j(\omega+\pi)}) = \sum(-1)^n h(n)e^{-jn\omega}$, we obtain

$$\frac{d^\nu H(e^{j\omega})}{d\omega^\nu}=(-j)^\nu\sum n^\nu h(n)e^{-jn\omega}, \qquad \nu=0,1,....,N.$$

At $\omega = 0$, and $\omega = \pi$, these become

$$\frac{d^\nu H(e^{j\omega})}{d\omega^\nu}|_{\omega=0}=(-j)^\nu\sum n^\nu h(n)=\frac{d^\nu H(e^{j(\omega+\pi)})}{d\omega^\nu}|_{\omega=\pi}$$

$$\frac{d^\nu H(e^{j\omega})}{d\omega^\nu}|_{\omega=\pi}=(-j)^\nu\sum(-1)^n n^\nu h(n)=\frac{d^\nu H(e^{j(\omega+\pi)})}{d\omega^\nu}|_{\omega=0}.$$

For a given ν, we can choose to have the derivatives of the low-pass frequency function vanish either at $\omega = 0$, or at $\omega = \pi$, but not both. The maximally flat magnitude square (low-pass) Binomial-QMF has all derivatives vanish at $\omega = \pi$. Other PR-QMFs distribute these vanishing derivatives differently.

Any $\alpha_i \neq 0$ of the proposed approach decreases the number of vanishing moments of the high-pass filter by 1. The magnitude functions of several known smooth or regular 6-tap QMFs and their α values are given in Fig. 4.5.

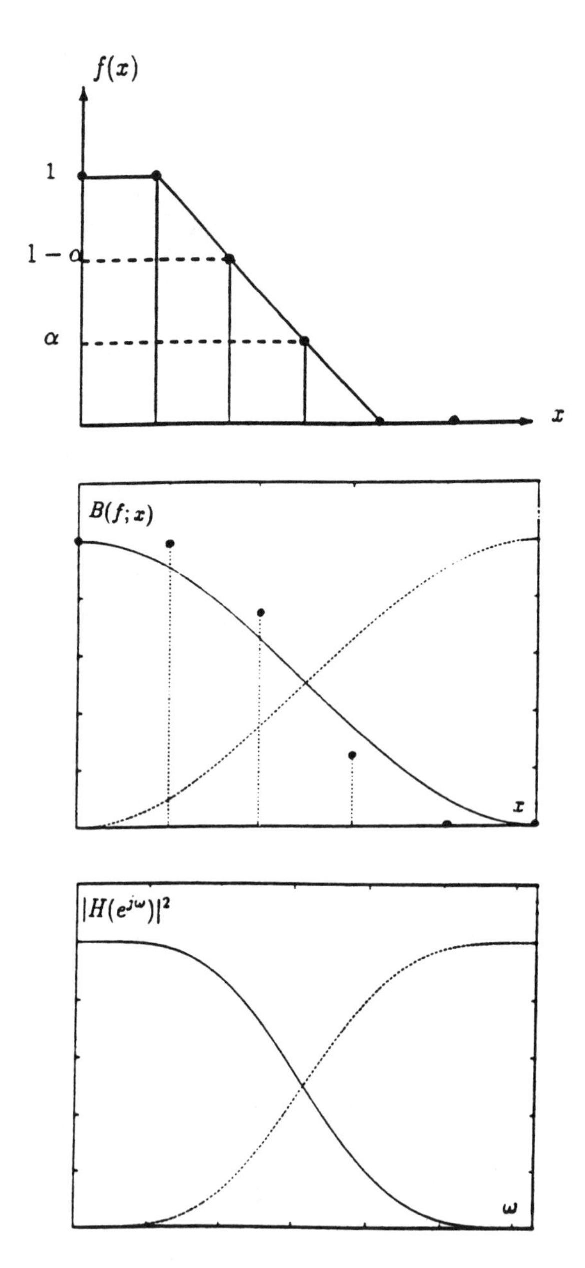

Figure 4.4: The functions $f(i)$, $B(f;x)$, and $R(z)$ of Bernstein polynomial approximation for $\alpha = 0.25$.

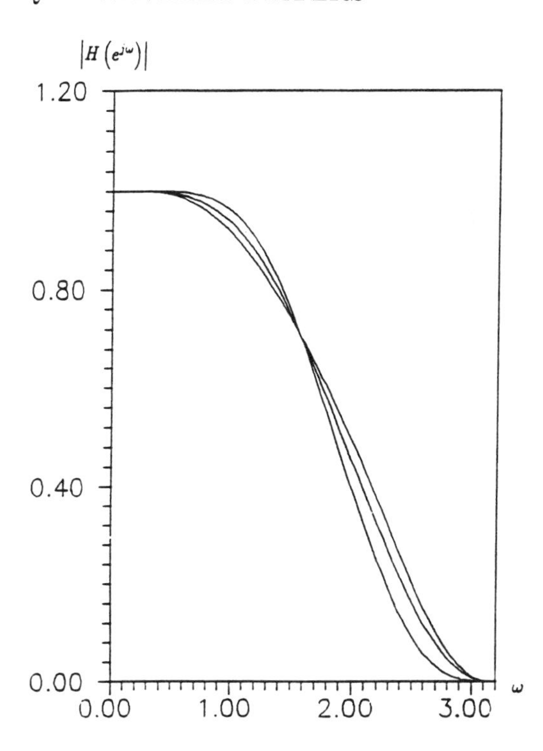

Figure 4.5: Magnitude functions of three different 6-tap PR-QMFs: maxflat ($\alpha = 0$). Coiflet (Daubechies) ($\alpha = 0.2708672$), and $\alpha = 0.480$.

The following special cases are worthy of note:

(1) $\alpha = 0$ gives the Binomial QMF-wavelet filter with three vanishing high-pass moments.

(2) $\alpha = 0.2708672$ corresponds to the 6-tap Coiflet filter, which is presented in Section 5.5.3.

(3) $\alpha = 0.0348642$ yields the 6-tap PR-QMF of the most regular wavelet solution (Daubechies, private communication). The coefficients of 6-tap most regular orthonormal wavelet filters follow:

n	0	1	2
$h(n)$	0.374328416	1.090933960	0.786941229
n	3	4	5
$h(n)$	-0.146269859	-0.161269645	0.055335898

Coefficients of most regular orthonormal wavelet filter.

The Bernstein polynomial approach provides a flexible parameterized method for designing FIR PR-QMF filter banks and can be easily extended to design orthonormal wavelet bases with compact support.

4.4 Johnston QMF Family

Johnston QMFs were the earliest popular filters used in the literature. These symmetrical filters constitute a non-PR-QMF bank. Their non-PR characteristics, particularly for longer duration filters, do not present practical significance for subband image coding. Design of these filters is based on two criteria:

- Ripple in the system or ripple energy

$$E_r = 2\sum_{\omega=0}^{\pi/2}[|H(\omega)|^2 + |H(\omega-\pi)|^2 - 1]^2. \tag{4.34}$$

- Out of band or stopband energy

$$E_s = \sum_{\omega=f_{SB}}^{\pi} |H(\omega)|^2 \tag{4.35}$$

where f_{SB} is the stopband edge frequency. The optimization procedure for this QMF family tries to minimize the objective function

$$E = E_r + \alpha E_s(f_{SB}) \tag{4.36}$$

where α is the weight of the stopband in the objective function. It is seen that this design approach tries to approximate the PR conditions with the first variable, E_r, of the objective function while minimizing the aliasing or stopband energy in E_s. As stated earlier, linear-phase two-band PR-QMF solution is not possible.

Johnston designed several sets of QMFs based on filter lengths, transition bands, and stopband weighting parameters. Johnston QMF coefficients, 8-, 12-, 16-, 24-, 32-tap, with their design parameters, are given in Tables 4.3 and 4.4.

4.5 Smith-Barnwell PR-CQF Family

Smith and Barnwell (1984) and Mintzer (1985) were the first to show that perfect reconstruction in a two-band filter bank is possible if the linear-phase requirement is relaxed. The Smith-Barnwell filters were called conjugate quadrature filters

8 TAP		
0.48998080		
0.06942827		
-0.07065183		
0.00938715		
12 TAP(A)	**12 TAP(B)**	
0.48438940	0.4807962	
0.08846992	0.09808522	
-0.08469594	-0.0913825	
-0.002710326	-0.00758164	
0.01885659	0.02745539	
-0.003809699	-0.006443977	
16 TAP(A)	**16 TAP(B)**	**16 TAP(C)**
0.4810284	0.4773469	0.4721122
0.09779817	0.1067987	0.1178666
-0.09039223	-0.09530234	-0.0992955
-0.009666376	-0.01611869	-0.0262756
0.0276414	0.03596853	0.04647684
-0.002589756	-0.001920936	0.00199115
-0.005054526	-0.009972252	-0.02048751
0.001050167	0.002898163	0.006525666
24 TAP(B)	**24 TAP(C)**	**24 TAP(D)**
0.4731289	0.4686479	0.4654288
0.1160355	0.1246452	0.1301121
-0.09829783	-0.09987885	-0.09984422
-0.02561533	-0.03464143	-0.04089222
0.04423976	0.05088162	0.05402985
0.003891522	0.01004621	0.01547393
-0.01901993	-0.02755195	-0.03295839
0.001446461	-0.0006504669	-0.004013781
0.006485879	0.01354012	0.0197638
-0.001373861	-0.002273145	-0.001571418
-0.001392911	-0.005182978	-0.010614
0.0003833096	0.002329266	0.004698426

Table 4.4: Johnston QMF coefficients (coefficients are listed from center to end) [J.D. Johnston, ©1980, IEEE].

Transition Code Letter	Normalized Transition Band
A	0.14
B	0.1
C	0.0625
D	0.043
E	0.023

Table 4.3: Normalized transition bands and their code letters for Johnston QMFs [J.D. Johnston, ©1980, IEEE].

32 TAP(C)	32 TAP(D)	32 TAP(E)
0.46640530	0.46367410	0.45964550
0.12855790	0.13297250	0.13876420
-0.099802430	-0.099338590	-0.097683790
-0.039348780	-0.044524230	-0.051382570
0.052947450	0.054812130	0.055707210
0.014568440	0.019472180	0.026624310
-0.031238620	-0.034964400	-0.038306130
-0.0041874830	-0.0079617310	-0.014569000
0.017981450	0.022704150	0.028122590
-0.0001303859	0.0020694700	0.0073798860
-0.0094583180	-0.014228990	-0.021038230
0.0014142460	0.00084268330	-0.0026120410
0.0042341950	0.0081819410	0.015680820
-0.0012683030	-0.0019696720	-0.00096245920
-0.0014037930	-0.0039715520	-0.011275650
0.00069105790	0.0022551390	0.0051232280

Table 4.4 (continued): Johnston QMF coefficients (coefficients are listed from center to end) [J.D. Johnston, ©1980, IEEE].

(CQFs). In this section we briefly discuss the highlights of their design procedure and provide 8-, 16-, and 32-tap PR-CQF coefficients in Table 4.5.

The CQF solution is essentially the same as the two-band paraunitary solution of Eqs. (3.140)–(3.141). The design was reduced to finding $H_0(z)$ such that

$$H_0(z)H_0(z^{-1}) + H_0(-z)H_0(-z^{-1}) = 2. \tag{4.37}$$

Let

$$R_0(z) = H_0(z)H_0(z^{-1}) \leftrightarrow \rho(n). \tag{4.38}$$

We have seen that $R_0(z) + R_0(-z) = 2$ implies that $R_0(z)$ is a half-band filter satisfying

$$\rho_0(2n) = \delta(n). \tag{4.39}$$

But the half-band filter needed in Eq. (4.37) must also satisfy

$$R_0(e^{j\omega}) = |H(e^{j\omega})|^2 \geq 0. \tag{4.40}$$

The steps in the filter design are as follows:
(1) Start with the design of a zero-phase, half-band filter $R_0(z)$ that necessarily satisfies Eq. (4.39), but $R_0(e^{j\omega})$ can go negative. Let $R_0(e^{j\omega}) = -\epsilon$. An equiripple half-band filter is shown in Fig. 4.6.
(2) We can make $R_0^{'}(e^{j\omega})$ positive semidefinite by

$$R_0^{'}(z) = aR_0(z) + b$$

or

$$\rho_0^{'}(n) = a\rho_0(n) + b\delta(n). \tag{4.41}$$

We can now choose a and b to make R_0 look like $R_0^{'}$ in Fig. 4.6. The parameter b raises the level in the frequency response and a renormalizes to make the pass-band gain equal to unity, or $\rho^{'}(0) = 1$. It is easily verified that

$$a = \frac{1}{(1+\epsilon)}, \qquad b = \frac{1}{(1+\epsilon)}$$

will do the trick. Note that $\rho_0^{'}(2n) = \delta(n)$.
(3) Evaluate the spectral factors $H_0(z)$ in

$$H_0(z)H_0(z^{-1}) = R_0^{'}(z).$$

The PR-QMF and the PR-CQF are the same except for a possible difference at the phase responses. Both satisfy the magnitude square condition as expected. Table 4.5 displays 8-, 16-, and 32-tap PR-CQF coefficients with 40 dB stopband attenuation.

4.6 LeGall-Tabatabai PR Filter Bank

These filters are the typical examples of the PR, unequal bandwidth and length, two-band filter banks. These filters have linear-phase responses and consequently are not paraunitary. They are computationally very efficient and can be implemented without multipliers. However, the frequency behavior is poor for decimation by 2. The low-pass filter length is 5 while the high-pass filter has length 3.

8 TAP	16 TAP	32 TAP
0.0348975582178515	0.02193598203004352	0.00849437247823317
-0.01098301946252854	0.001578616497663704	-0.0000961781687347404
-0.06286453934951963	-0.06025449102875281	-0.008795047132402801
0.223907720892568	-0.0118906596205391	0.000708779549084502
0.556856993531445	0.137537915636625	0.01220420156035413
0.357976304997285	0.05745450056390939	-0.001762639314795336
-0.02390027056113145	-0.321670296165893	-0.01558455903573820
-0.07594096379188282	-0.528720271545339	0.004082855675060479
	-0.295779674500919	0.01765222024089335
	0.0002043110845170894	-0.003835219782884901
	0.02906699789446796	-0.01674761388473688
	-0.03533486088708146	0.01823906210869841
	-0.006821045322743358	0.005781735813341397
	0.02606678468264118	-0.04692674090907675
	0.001033363491944126	0.05725005445073179
	-0.01435930957477529	0.354522945953839
		0.504811839124518
		0.264955363281817
		-0.08329095161140063
		-0.139108747584926
		0.03314036080659188
		0.09035938422033127
		-0.01468791729134721
		-0.06103335886707139
		0.006606122638753900
		0.04051555088035685
		-0.002631418173168537
		-0.02592580476149722
		0.0009319532350192227
		0.01535638959916169
		-0.0001196832693326184
		-0.01057032258472372

Table 4.5: The 8-, 16-, and 32-tap PR-CQF coefficients with 40 dB stopband attenuation [M.J.T. Smith and T.P. Barnwell, ©1986, IEEE].

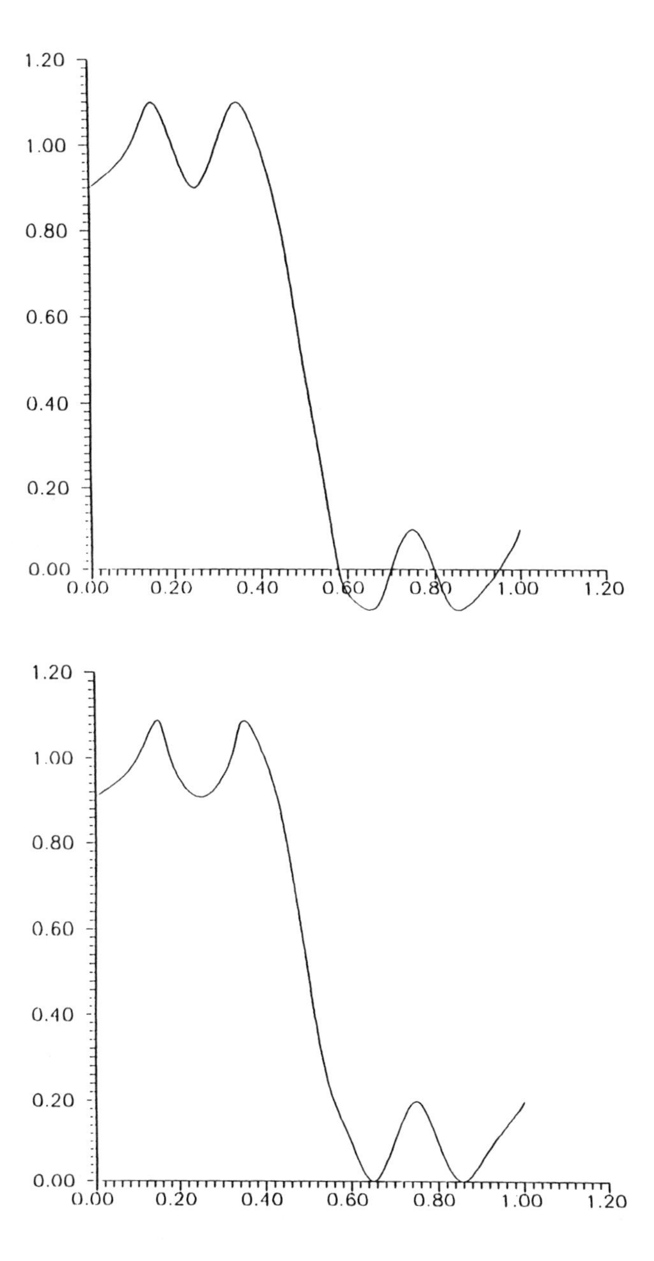

Figure 4.6: (a) $R_0(e^{j\omega})$ equiripple half-bandwidth filter; (b) $R_0'(e^{j\omega})$ product filter with double zeros on unit circle.

The coefficients of these filters are given in Table 4.6.

n	$h_L(n)$	$h_H(n)$
0	-1	1
1	2	-2
2	6	1
3	2	
4	-1	

Table 4.6: Low- and high-pass LeGall-Tabatabai filter coefficients (LeGall and Tabatabai, 1988).

4.7 Princen-Bradley QMF

The Princen-Bradley QMF can be viewed as an M-band PR-modulated filter bank of the type described in Section 3.5.6. It is equivalent to a bank of M filters

$$h_k(n) = h(n)cos\{\omega_k(n - \frac{2M-1}{2}) + \phi_k\}$$

$$\omega_k = \frac{\pi}{M}(k + \frac{1}{2}), \qquad \phi_k = \frac{\pi}{4} + k\frac{\pi}{2}$$

where $h(n)$ is a low-pass window function of length $2M$. A key advantage of such modulated systems is the implementation using fast DCT algorithms in the analysis and synthesis sections. A typical window design for 16- and 32-band filter banks is given in Princen and Bradley (1986). It is reported that this design has properties falling somewhere between subband coding and transform coding.

4.8 Optimal PR-QMF Design for Subband Image Coding

Subband filter banks with impulse responses of arbitrary length are significantly more flexible than block transforms with their fixed length basis sequences. Among the latter, the signal dependent KLT is the optimal block transform, providing complete inter-coefficient decorrelation and the best energy compaction. This feature is shown by the relation

$$E\{\theta_i(k)\theta_j(k)\} = \sigma_i^2\delta_{i-j}$$

where $\theta_i(k)$ and $\theta_j(k)$ are the ith and jth coefficients of the kth transform block for the given source.

The KLT (and all block transforms for that matter), however, does not address the issue of interband energy leakage or aliasing in the frequency domain. Furthermore, there is no freedom to adjust the joint time-frequency localization in block transforms.

Filter bank theory provides the means to assess and improve these features of block transforms. The ideal filter banks with infinite duration basis functions provide the best frequency and decorrelation properties. The interband correlation in the ideal filter bank for zero-mean WSS inputs is

$$E\{\theta_i(k)\theta_j(l)\} = \sigma_i^2 \delta_{i-j}\delta_{k-l} \tag{4.42}$$

where k and l are the sample locations in subbands i and j, respectively. It is seen from this relation that all coefficients or samples of different subbands except the ones coinciding in location are uncorrelated. This ideal solution provides an alias-free frequency split in a filter bank structure with perfect interband decorrelation, uncorrelated random processes rather than uncorrelated random variables. But the time functions are of infinite duration.

The common performance measures for a filter bank — compaction and perfect reconstruction — are, in fact, only partial descriptors. The reconstruction is "perfect" only in the absence of encoding quantization, transmission errors, and infinitely precise coefficient values and arithmetic. It is clear, then, that we need to expand and reformulate appropriate performance measures that can help to account for non-ideal behavior in "perfect" reconstruction filter banks. In this section we define an objective function that weights several performance measures, implying the time- and frequency-domain behaviors, and then we compare filter banks designed to optimize this criterion. Viscito and Allebach (1989) also proposed a statistical filter design approach. They treated the filters in an M-channel filter bank as the linear minimum mean-square error (MMSE) estimator of a hypothetical input signal in the presence of noise.

4.8.1 Parameters of Optimization

The design of optimal PR-QMFs should consider several parameters of practical significance. These parameters — namely the energy compaction, aliasing energy, unit step response, zero-mean high-pass filter, uncorrelated subband signals, constrained nonlinear-phase response, and input source statistics — are combined to define the objective function of the optimization problem. The optimal PR-QMF design approach presented in this section is a continuation and enhancement of earlier work in the field, particularly for image coding applications. The following performance measures are included in the design of optimal two-band PR-QMFs (Caglar, Liu, and Akansu, 1991).

(1) Orthonormal PR Requirement This set of requirements is included in the design to satisfy the unitary perfect reconstruction condition. The high-pass filter is assumed to be the mirror of the low-pass filter $\{h(n)\}$ of length $2N$ which is expressed in the vector form $\underline{h}$. The orthonormality condition can be written in vector product form as

$$\underline{h}^T \underline{h} = 1. \tag{4.43}$$

From Section 3.5.4 the perfect reconstruction condition of the orthonormal two-band PR-QMF is

$$\sum_n h(n)h(n+2k) = \delta(k). \tag{4.44}$$

Equations (4.43) and (4.44) can now be jointly expressed in the matrix form

$$\underline{h}^T C_i \underline{h} = 0 \qquad i = 1, 2, ..., N-1 \tag{4.45}$$

where C_i are the proper *filter coefficient shuffling matrices* as

$$C_1 = \begin{bmatrix} 0 & 0 & 1 & 0 & ... & 0 \\ 0 & 0 & 0 & 1 & ... & 0 \\ 1 & 0 & 0 & 0 & ... & 0 \\ . & . & & & & . \\ . & . & & & & . \\ 0 & 0 & 1 & 0 & ... & 1 \\ 0 & 0 & 0 & 1 & ... & 0 \\ 0 & 0 & 0 & 0 & ... & 0 \end{bmatrix}, ..., \quad C_{N-1} = \begin{bmatrix} 0 & 0 & 0 & ... & 1 & 0 \\ 0 & 0 & 0 & ... & 0 & 1 \\ 0 & 0 & 0 & ... & 0 & 0 \\ 0 & 0 & 0 & ... & 0 & 0 \\ . & . & & & . & . \\ . & . & & & . & . \\ 1 & 0 & 0 & ... & 0 & 0 \\ 0 & 1 & 0 & ... & 0 & 0 \end{bmatrix}. \tag{4.46}$$

(2) Energy Compaction Let σ_x^2 and R_{xx} be the variance and covariance matrix of the zero-mean input. For the two-band case, let σ_L^2, σ_H^2 be variances of the low-pass and high-pass outputs, respectively. For a paraunitary transformation, the Parseval theorem states the energy constraint

$$\sigma_x^2 = \frac{1}{2}(\sigma_L^2 + \sigma_H^2) \tag{4.47}$$

where

$$\sigma_L^2 = \underline{h}^T R_{xx} \underline{h} \tag{4.48}$$

and the compaction measure derived in Section 2.2 is

$$G_{TC} = \frac{\sigma_x^2}{\left[\sigma_L^2 \sigma_H^2\right]^{1/2}}. \tag{4.49}$$

It is clear that the maximization of σ_L^2 in Eq. (4.48) is sufficient for the constrained maximization of G_{TC}.

(3) Aliasing Energy All orthonormal signal decomposition techniques satisfy the conditions for alias cancellation. In practice, since all the decomposition bands or coefficients are not used in the synthesis, or because of different levels of quantization noise in the subbands, noncancelled aliasing energy components exist in the reconstructed signal. It is known that the aliasing energy causes annoying patterns in encoded images at low bit rates.

The aliasing energy component at the low-pass filter output in the two-band PR-QMF bank for the given input spectral density function $S_{xx}(e^{j\omega})$ is

$$\sigma_A^2 = \frac{1}{2\pi}\int_{-\pi}^{\pi} |H(e^{j\omega})|^2 S_{xx}(e^{j(\omega+\pi)})|H(e^{j(\omega+\pi)})|^2 d\omega. \tag{4.50}$$

The time-domain counterpart of this relation is expressed as

$$\sigma_A^2 = \sum_k [\rho(n) * (-1)^n \rho(n)] R_{xx}(k) \tag{4.51}$$

where $\rho(n)$ is the autocorrelation sequence of the filter coefficients $h(n)$ and defined as

$$\rho(n) = h(n) * h(-n)$$

and $R_{xx}(k)$ is the autocorrelation function of the input. The optimal solution should minimize this aliasing energy component.

(4) Step Response The representation of edges in images is a crucial problem. The edge structures are localized in time; therefore they should be represented by time-localized basis functions. Otherwise, the ringing artifacts occur in encoded images. An edge can be crudely considered as a step. Therefore, the step responses of the low-pass filter in the filter bank should be considered during the design procedure.

The *uncertainty principle* states that a signal cannot be localized perfectly in both time and frequency. The human visual system is able to resolve the time-frequency plane. Therefore, a joint time-frequency localization or behavior should be considered in a filter bank design. The trade-off between the time and frequency resolutions is reflected basically in the aliasing and step response performance of the designed filter.

The unit step response of the filter $h(n)$ can be written as

$$a(n) = h(n) * u(n) = \sum_{i=0}^{n} h(i)$$

where $u(n)$ is the unit step sequence. The difference energy between the unit step response $a(n)$ of the filter and the unit step sequence $u(n)$ is expressed as

$$E_s = \sum_{k=0}^{2N-1} [\sum_{n=0}^{k} h(n) - 1]^2. \tag{4.52}$$

The value of E_s should be minimized for the optimal filter solution. The optimization variable E_s here does not consider the symmetry of the unit step response around the step point. The ringing problem in image coding may be caused by an overshoot or an undershoot. This point is addressed later in the constrained nonlinear -phase condition of the desired filter.

(5) Zero Mean High-Pass Filter Most of the energy of practical signal sources is concentrated around the DC frequency. Therefore, practical signal decomposition techniques should be able to represent the DC component within only one basis function. Following this argument, we should constrain the high-pass QMF impulse response $h_1(n) = (-1)^n h(n)$ to have zero mean ,

$$\sum_n (-1)^n h(n) = 0. \tag{4.53}$$

This requirement implies that there should be at least one zero of the low-pass filter $H(e^{j\omega})$ at $\omega = \pi$. As we will see in Chapter 5, this condition is necessary to satisfy the regularity requirement in the design of wavelets.

(6) Uncorrelated Subband Signals Any good signal decomposition technique of coding applications should provide uncorrelated transform coefficients or subband signals. A performance demerit is the cross-correlation of the two subband signals for the given input

$$E\{y_L(m)y_H(m)\} = R_{LH}(0) = \sum_n [\sum_l h(l)(-1)^l h(n-l)] R_{xx}(n), \qquad \text{for all } m. \tag{4.54}$$

(7) Constrained Nonlinearity in Phase Response Linear-phase and PR are mutually exclusive in the orthonormal two-band QMF design. But severe phase nonlinearities are known to create undesired degradations in image and video applications. Therefore, a measure that indicates the level of nonlinearity in the filter-phase response is included as a parameter in the optimal filter design. Nonlinear-phase is related to the asymmetry of the impulse response. A measure is

$$E_p = \sum_n [h(n) - h(2N-1-n)]^2. \tag{4.55}$$

(8) Given Input Statistics The input spectral density function is needed for the optimal filter design variables discussed earlier. We assume an autoregressive,

AR(1) source model with correlation coefficient $\rho = 0.95$, which is a crude approximation to the real-world still frame images. The correlation function of this source is

$$R_{xx}(m) = \rho^{|m|} \qquad m = 0, \pm 1, \pm 2, \ldots \quad . \tag{4.56}$$

4.8.2 Optimal PR-QMF Design: Energy Compaction

This section deals with the optimization problem which consists of the PR and energy compaction for an AR(1) source.

The objective function J to be maximized is

$$\max\{J\} = \underline{h}^T R_{xx} \underline{h} + \lambda_0[1 - \underline{h}^T \underline{h}] + \lambda_1[\underline{h}^T C_1 \underline{h}] + \ldots + \lambda_i[\underline{h}^T C_i \underline{h}]. \tag{4.57}$$

Hence,

$$\frac{\partial J}{\partial \underline{h}} = 0.$$

Therefore,

$$R_{xx}\underline{h} + \lambda_1 C_1 \underline{h} + \ldots + \lambda_i C_i \underline{h} = \lambda_0 \underline{h}. \tag{4.58}$$

If the terms in the left side of the equation are combined as

$$R\underline{h} = \lambda_0 \underline{h} \tag{4.59}$$

where

$$R = R_{xx} + \lambda_1 C_1 + \ldots + \lambda_i C_i;$$

eq. (4.59) looks like a classical eigenvalue problem, but here the matrix R has unknown parameters $\{\lambda_i\}$ in it. The vector $\underline{h}$ that satisfies Eq. (4.59) is the optimal low-pass PR-QMF.

4.8.3 Optimal PR-QMF Design: Extended Set of Variables

The objective function of Eq. (4.57), which implies only the frequency-domain behavior of the filter, can be augmented to include the other performance measures described. The optimization problem is now set as

$$\begin{aligned} \max\{J\} &= \underline{h}^T R_{xx} \underline{h} - \alpha \sum_k [\rho(n) * (-1)^n \rho(n)] R_{xx}(k) \\ &\quad - \beta \sum_{k=0}^{2N-1} [\sum_{n=0}^{k} h(n) - 1]^2 - \gamma \sum_n [h(n) - h(2N-1-n)]^2 \end{aligned} \tag{4.60}$$

with the unitary PR, and zero-mean high-pass filter constraints

$$\begin{aligned}\sum_n h(n)h(n+2k) &= \delta(k) \\ \sum_n (-1)^n h(n) &= 0.\end{aligned} \tag{4.61}$$

This is a very general optimization problem. It simultaneously considers the time and frequency features of the filter. There are a set of parameters in the objective function that should be fine tuned for the application at hand. Therefore, this optimal filter design technique should be supported with experimental studies. The significance of the optimization variables in the objective function should be quantified for the human visual system. The following section presents examples of problem definition and performance of optimal filters.

4.8.4 Samples of Optimal PR-QMFs and Performance

The set of parameters in the optimization problem defined earlier admit many possible filter solutions. Therefore this section presents the interrelations among the performance parameters. Figure 4.7(a) shows the relationship between G_{TC} and aliasing energy σ_A^2 for an 8-tap two-band PR-QMF with AR(1) source, $\rho = 0.95$. As seen from the figure, this relation is linear-like and the energy compaction increases as the aliasing energy decreases. This trend is easily justified. The optimal PR-QMF solutions obtained are also consistent with this figure.

Figure 4.7(b) displays energy compaction versus interband correlations, $R_{LH}(0)$, again for the same source model. Although in block transforms these two variables merge in the unique optimal solution, KLT, this is not true for the filter banks. In other words, there is more than one possible solution. One should pick the solution that maximizes the objective function.

Figure 4.7(c) shows the relationship between energy compaction and phase nonlinearities.

Figure 4.7(d) plots energy compaction versus unit-step response error measure. This plot indicates that whenever the step response approaches the unit step the energy compaction decreases. This relation, time-domain vs frequency-domain, calls into question the practical merit of the energy compaction measure. Although the energy compaction may be optimal, the subjective coding performance of the corresponding filter may not be necessarily optimal.

Table 4.7 provides the coefficients of 4-, 6-, 8-, 12-, and 16-tap optimal PR-QMFs based on energy compaction with a zero-mean high-pass constraint. Similarly, Table 4.8. gives the optimal PR-QMF coefficients based on minimized aliasing energy with zero-mean high-pass. Table 4.9 has the optimal PR-QMFs similar

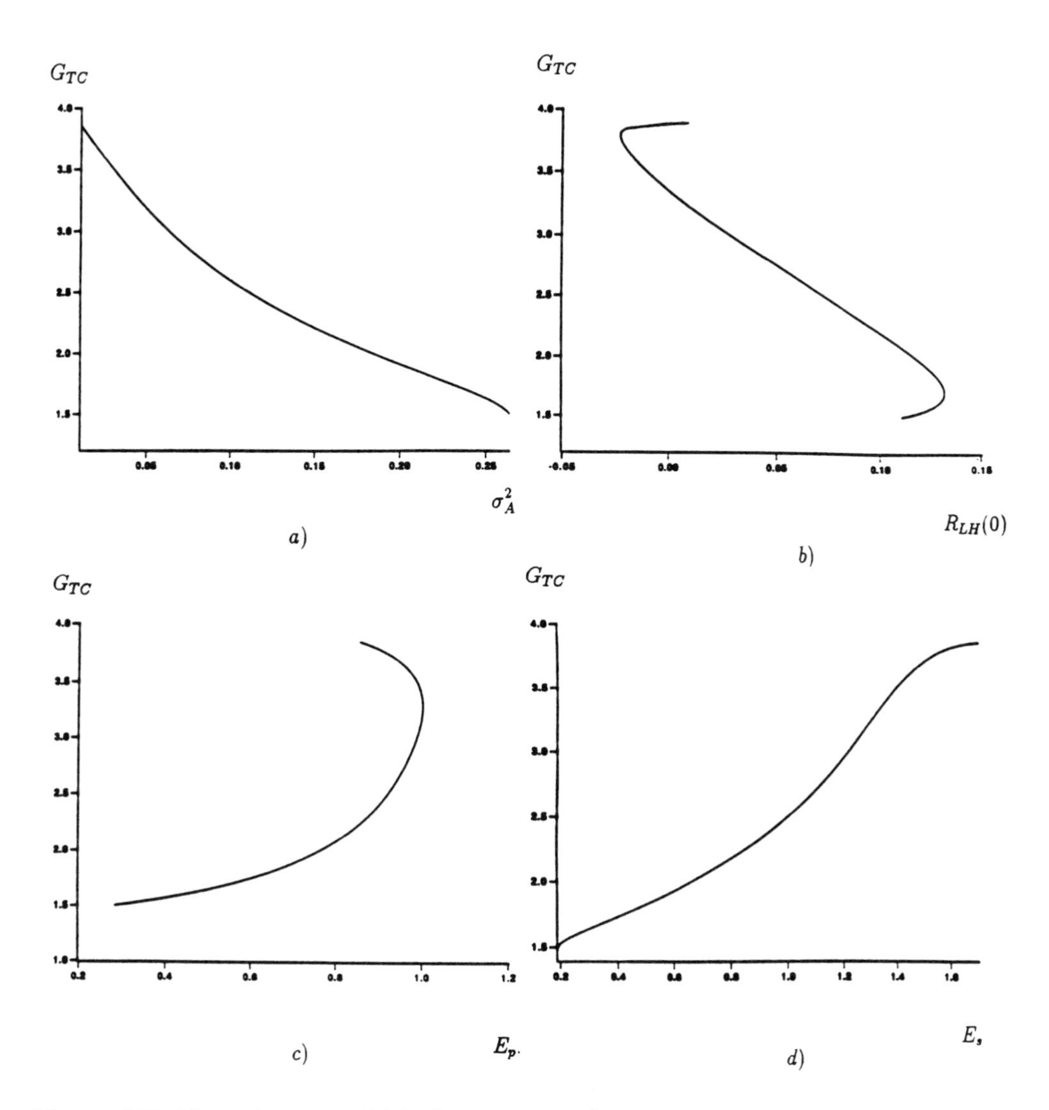

Figure 4.7: The relations of (a) G_{TC} versus σ_A^2, (b) G_{TC} versus $R_{LH}(0)$, (c) G_{TC} versus E_p, (d) G_{TC} versus E_s of 8-tap two-band PR-QMFs for AR(1), $\rho = 0.95$ source.

to Table 4.7 but additionally providing uncorrelated subbands, or $R_{LH}(0) = 0$. Table 4.10 also adds this constraint to the conditions of Table 4.8. Table 4.11 gives the optimal filters based on augmented objective function of Eq. (4.60). The filter solutions displayed in this section were obtained by using the IMSL FORTRAN Library(NCONF). The package solves a general nonlinear constrained minimization problem using the successive quadratic programming algorithm and a finite difference gradient.

n	$h(n)$	$h(n)$	$h(n)$	$h(n)$	$h(n)$
0	0.201087342	0.244206457	0.317976535	0.385659639	0.482962940
1	0.600007520	0.664513457	0.748898833	0.796281177	0.836516297
2	0.665259025	0.629717438	0.534939876	0.428145720	0.224143841
3	0.198773686	0.089423027	-0.058836349	-0.140851286	-0.129409515
4	-0.233790239	-0.251577216	-0.205817322	-0.106698578	
5	-0.153612998	-0.072467574	0.042523091	0.051676890	
6	0.118834741	0.134086583	0.060007692		
7	0.101350938	0.031916868	-0.025478793		
8	-0.074934374	-0.076499461			
9	-0.061434875	0.003706982			
10	0.053218300	0.027172980			
11	0.029837627	-0.009985979			
12	-0.037981695				
13	-0.002649357				
14	0.015413680				
15	-0.005165762				
G_{TC}	3.9220	3.9038	3.8548	3.7961	3.6426
σ_A^2	0.0056	0.0075	0.0115	0.0153	0.0239
$R_{LH}(0)$	0.0040	-0.1601	-0.0140	-0.0160	-0.0422
mean	0.0000	0.0000	0.0000	0.0000	0.0000
E_p	1.0622	1.0320	0.8566	1.2506	0.7500
E_s	3.3613	2.5730	1.7493	1.3059	0.8365

Table 4.7: A set of optimal PR-QMF filter coefficients and their performance. The optimality is based on energy compaction with zero mean high-pass filter.

n	$h(n)$	$h(n)$	$h(n)$	$h(n)$	$h(n)$
0	0.239674169	0.276769143	0.339291195	0.398655794	0.482962940
1	0.641863878	0.689345705	0.753812779	0.792728512	0.836516297
2	0.628941341	0.592445147	0.510688095	0.420459801	0.224143841
3	0.136154317	0.054082233	-0.062731472	-0.141949922	-0.129409515
4	-0.241530316	-0.247471430	-0.210405609	-0.112008814	
5	-0.123175317	-0.059746881	0.046422128	0.056328191	
6	0.128959373	0.138373438	0.067533100		
7	0.088433853	0.031525301	-0.030396654		
8	-0.083586814	-0.088498729			
9	-0.058991180	0.006149179			
10	0.061697343	0.035489212			
11	0.033431236	-0.014248756			
12	-0.050042508				
13	-0.002023897				
14	0.022994193				
15	-0.008586110				
G_{TC}	3.9189	3.9002	3.8513	3.7935	3.6426
σ_A^2	0.0054	0.0073	0.0113	0.0152	0.02397
$R_{LH}(0)$	0.0001	-0.1849	-0.0170	-0.0138	-0.0422
mean	0.0000	0.0000	0.0000	0.0000	0.0000
E_p	1.0521	0.9933	0.8450	1.2520	0.7500
E_s	3.2836	2.5202	1.7232	1.2930	0.8365

Table 4.8: A set of optimal PR-QMF filter coefficients and their performance. The optimality is based on minimized aliasing energy with a zero-mean high-pass filter.

n	$h(n)$	$h(n)$	$h(n)$	$h(n)$	$h(n)$
0	0.224159871	-0.106117265	0.240118698	0.312656005	0.00000000 0
1	0.629151335	-0.041624773	0.688564034	0.754045521	0.70710678 1
2	0.642510825	0.444275957	0.638286732	0.543768338	0.70710678 1
3	0.158071546	0.761031030	0.017567002	-0.108851490	0.00000000 0
4	-0.240893371	0.427762258	-0.235301591	-0.149317562	
5	-0.133127916	-0.066013158	0.023295098	0.061912751	
6	0.128098122	-0.107784207	0.064002943		
7	0.090074845	0.085537312	- 0.022319352		
8	-0.081998711	0.051558425			
9	-0.055306473	-0.038422405			
10	0.058081519	-0.002588387			
11	0.026452620	0.006598776			
12	-0.040400680				
13	-0.001956582				
14	0.017549205				
15	-0.006252594				
G_{TC}	3.9207	3.8935	3.8408	3.7661	3.2025
σ_A^2	0.0055	0.0083	0.0126	0.0167	0.0487
$R_{LH}(0)$	0.0000	0.0000	0.0000	0.0000	0.0000
mean	0.0000	0.0000	0.0000	0.0000	0.0000
E_p	1.0531	0.8694	0.9011	1.3048	0.0000
E_s	3.3117	4.4123	1.8685	1.3968	1.4289

Table 4.9: A set of optimal PR-QMF filter coefficients and their performance based on energy compaction with zero mean high-pass and uncorrelated subband signals.

n	$h(n)$	$h(n)$	$h(n)$	$h(n)$	$h(n)$
0	0.240173769	-0.121396419	0.249509936	0.348319026	0.00000000 0
1	0.642454295	-0.035246082	0.688584306	0.758774508	0.70710678 1
2	0.628348271	0.467924401	0.632097530	0.510327483	0.7071067 81
3	0.135389521	0.751312762	0.015778256	-0.121232755	0.00000000 0
4	-0.241606760	0.412397276	-0.240993887	-0.151539728	
5	-0.122763195	-0.062892458	0.026838168	0.069565029	
6	0.129125126	-0.109012591	0.066493202		
7	0.088184458	0.093200632	- 0.024093948		
8	-0.083719165	0.059816603			
9	-0.058849491	-0.048300585			
10	0.061801498	-0.002622488			
11	0.033339516	0.009032511			
12	-0.050088120				
13	-0.002023074				
14	0.023072163				
15	-0.008625249				
G_{TC}	3.9188	3.8897	3.8399	3.7611	3.2025
σ_A^2	0.0054	0.0083	0.0126	0.0165	0.0487
$R_{LH}(0)$	0.0000	0.0000	0.0000	0.0000	0.0000
mean	0.0000	0.0000	0.0000	0.0000	0.0000
E_p	1.0518	0.8667	0.8941	1.3052	0.0000
E_s	3.2826	4.4280	1.8564	1.3539	1.4289

Table 4.10: A set of optimal PR-QMF filter coefficients and their performance based on minimized aliasing energy with zero mean high-pass and uncorrelated subband signals.

Multiplier-free PR-QMFs

Multiplier-free filter algorithms are of great practical interest because of their computational efficiency. The optimal PR-QMF design introduced in this section can be modified for suboptimal multiplier free filters that have only the allowed coefficient values

$$h(n) = \pm 2^{k_n} \pm 1 \qquad\qquad n = 0, 1, 2, \cdots, 2N-1,$$

where k_n is an integer. Therefore, any filter coefficient $h(n)$ can be expressed as a binary shift and/or an addition. The 4-, 6-, 8-, and 10-tap examples of multiplier-free suboptimal paraunitary low-pass PR-QMF are found as in Table A (Akansu, 1992).

These multiplier-free suboptimal solutions are based on the criteria of orthonormality and energy compaction for an AR(0.95) source. The frequency behaviors of these filters are comparable with those of the Binomial QMF-wavelet filters of the same duration. The extensions of optimal PR-QMFs and the multiplier-free suboptimal PR-QMFs are the topics of current research.

	$\alpha = 0.5,\ \beta = 0.01,\ \gamma = 0.01$				
n	$h(n)$	$h(n)$	$h(n)$	$h(n)$	$h(n)$
0	0.349996497	0.360838504	0.377995233	0.442766931	0.466675669
1	0.731063819	0.744306049	0.768367237	0.805049213	0.840588657
2	0.505852096	0.490757098	0.462086554	0.352529377	0.240431112
3	-0.010803415	-0.036047928	-0.86013220	-0.146445561	-0.133481 875
4	-0.229358399	-0.222383198	-0.194919256	-0.088189527	
5	-0.029975411	-0.005408341	0.055225994	0.048503129	
6	0.134362313	0.128127832	0.061944250		
7	0.026991307	0.000007678	- 0.030473229		
8	-0.089102151	-0.079675397			
9	-0.017502278	0.018522733			
10	0.062860841	0.029441941			
11	0.006564367	-0.014273411			
12	-0.045242724				
13	0.009260600				
14	0.017738308				
15	-0.008492207				
G_{TC}	3.8950	3.8809	3.8432	3.7829	3.6407
σ_A^2	0.0065	0.0080	0.0115	0.0158	0.0240
$R_{LH(0)}$	-0.0084	-0.2052	-0.0196	-0.01970	-0.0437
mean	0.0000	0.0000	0.0000	0.0000	0.0000
E_p	0.9859	0.9439	0.8432	1.2022	0.7303
E_s	3.1015	2.395	1.6745	1.2436	0.8503

Table 4.11: Optimal PR-QMF filter solutions and their performance based on Eq. (4.60) and only the weight of the phase response variable is changed.

	h(n)			
n	10 tap	8 tap	6 tap	4 tap
0	-1	-8	4	2
1	-3	8	16	6
2	9	64	16	3
3	33	64	0	-1
4	32	8	-4	
5	4	-8	1	
6	-9	1		
7	1	1		
8	3			
9	-1			

Table A: 4-, 6-, 8-, and 10-tap examples of multiplier-free suboptimal paraunitary low-pass PR-QMF.

4.9 Performance of PR-QMF Families

In this section we compare the objective performance of several well-known PR-QMF families. Our broader aim is to lay the foundation for comparison of any orthonormal signal decomposition technique, block transforms, and filter banks.

Additionally, we extend the energy compaction measure in this section to different subband tree structures and quantify the objective performance of irregular tree structures, which are simpler to implement than regular subband trees.

(1) Compaction and Bit Allocation In Section 2.2.2 we derived formulas for compaction gain and bit allocation for an N-band orthonormal transform coder. Under the same assumptions as were used there (the same pdf at all points in the coder, and pdf-optimized quantizers), we can extend those formulas to an orthonormal subband tree with N_1 bands at the first level, each of which feeds N_2 bands at the second level.

Orthonormality ensures that the sum of the variances at the $N_1 \times N_2$ band outputs equals the input variance

$$\sigma_x^2 = \frac{1}{N_1 N_2} \sum_{k_1=0}^{N_1-1} \sum_{k_2=0}^{N_2-1} \sigma_{k_1 k_2}^2. \tag{4.62}$$

An orthonormal transform ensures that the average of the quantization errors in subbands is equal to the reconstruction error

$$\sigma_r^2 = \sigma_q^2 = \frac{1}{N_1 N_2} \sum_{k_1=0}^{N_1-1} \sum_{k_2=0}^{N_2-1} \sigma_{q k_1 k_2}^2. \tag{4.63}$$

From Section 2.2.2 the band distortions can be expressed as

$$\sigma_{q k_1 k_2}^2 = \epsilon_{k_1 k_2}^2 2^{-B_{k_1 k_2}} \sigma_{k_1 k_2}^2 \tag{4.64}$$

where $B_{k_1 k_2}$is the average bit rate for band $k_1 k_2$, and $\epsilon_{k_1 k_2}^2$ is the quantizer correction factor for that band. The same pdf type for all the bands implies

$$\epsilon^2 = \epsilon_{k_1 k_2}^2 \qquad \begin{array}{l} k_1 = 0, 1, ..., (N_1 - 1) \\ k_2 = 0, 1, ..., (N_2 - 1). \end{array}$$

Hence, the average distortion is

$$\sigma_q^2 = \frac{1}{N_1 N_2} \sum_{k_1=0}^{N_1-1} \sum_{k_2=0}^{N_2-1} \epsilon^2 2^{-B_{k_1 k_2}} \sigma_{k_1 k_2}^2. \tag{4.65}$$

The optimization problem is now to find the bit allocations of $(N_1 \times N_2)$ bands such that the average distortion σ_q^2 is minimized, subject to the constraint

$$B = \frac{1}{N_1 N_2} \sum_{k_1=0}^{N_1-1} \sum_{k_2=0}^{N_2-1} B_{k_1 k_2} = \text{constant} \tag{4.66}$$

Using the Lagrange multiplier method the optimum bit allocation is easily shown as

$$B_{k_1 k_2} = B + \frac{1}{2} log_2 \frac{\sigma_{k_1 k_2}^2}{\left[\prod_{k_1=0}^{N_1-1} \prod_{k_2=0}^{N_2-1} \sigma_{k_1 k_2}^2 \right]^{\frac{1}{N_1 N_2}}}. \tag{4.67}$$

Here, $B_{k_1 k_2}$ are not restricted to be nonnegative. In practice, they are truncated to zero if they become negative. A negative bit allocation result implies that if that band were completely discarded, its reconstruction error contribution would still be less than the corresponding distortion for the given rate. The resulting quantization error variance using this optimum bit allocation is

$$min\{\sigma_q^2\} = \epsilon^2 2^{-2B} \left[\prod_{k_1=0}^{N_1-1} \prod_{k_2=0}^{N_2-1} \sigma_{k_1 k_2}^2 \right]^{\frac{1}{N_1 N_2}}. \tag{4.68}$$

Assuming the same pdf type also for the input signal, the distortion for PCM at the same rate is

$$\sigma_{PCM}^2 = \epsilon^2 2^{-2B} \sigma_x^2$$

and the optimized compaction gain is therefore

$$max\{G_{TC}\} = \frac{\sigma_{qPCM}^2}{min\{\sigma_q^2\}} = \frac{\frac{1}{N_1 N_2} \sum_{k_1=0}^{N_1-1} \sum_{k_2=0}^{N_2-1} \sigma_{k_1 k_2}^2}{\left[\prod_{k_1=0}^{N_1-1} \prod_{k_2=0}^{N_2-1} \sigma_{k_1 k_2}^2 \right]^{\frac{1}{N_1 N_2}}}. \tag{4.69}$$

Similar expressions can be derived for regular trees with L levels and N_i bands at each level and for irregular subband tree structures.

(2) Energy Compaction of Ideal Filter Banks The upper bounds of G_{TC} for orthonormal block filter banks or transforms are set by the performance of KLT for the given N-band decomposition. On the other hand, the upper bounds of G_{TC} with zero aliasing are defined by the performance of the ideal filter banks. The ideal filter banks are optimal since they provide perfect interband decorrelation for any signal source as well as alias-free frequency characteristics for multirate signal processing. But, this perfect frequency localization implies infinite duration time functions. The poor time localization is not desired in some applications such as image coding. For known input power spectral density function $S_{xx}(e^{j\omega})$, the band variances of the N-band ideal filter bank are simply

$$\sigma_i^2 = \frac{1}{\pi} \int_{\frac{\pi}{N}i}^{\frac{\pi}{N}(i+1)} S_{xx}(e^{j\omega}) d\omega, \qquad i = 0, 1, ..., N-1. \tag{4.70}$$

The performance upper bound G_{TC}^{ub} is now calculated using these variances in the G_{TC} formula. A similar approach provides the performance upper bounds for irregular, unequal-bandwidth tree structures with the assumption of ideal filters.
(3) Performance Results G_{TC} results for several different cases are presented in this section. First, the decomposition schemes assume an AR(1) input signal, $\rho = 0.95$, with power spectral density function

$$S_{xx}(e^{j\omega}) = \frac{1-\rho^2}{1+\rho^2 - 2\rho cos\omega}. \tag{4.71}$$

Table 3.1 displays the compaction results of Binomial-QMF banks, which are identical to the orthonormal wavelet filters studied in Chapter 5, for 4-tap, 6-tap, and 8-tap cases. These results are for octave band or dyadic tree structures as well as for corresponding regular trees, along with the ideal filter bank cases. The levels of trees are limited to $L = 4$ here. It is observed from these tables that even the 5-octave band irregular tree with 4-tap filter has a better performance than the 16-band block filter bank. It is clear that the irregular tree structures reduce the computational burden of the subband filter banks and make them practical competitors to block filter banks or transforms.

These results suggest that an efficient algorithm to define an irregular subband tree structure, based on the input spectrum, is of practical importance. A simple algorithm based on the input statistics and energy compaction criterion is examined in Akansu and Liu (1991).

Table 4.12 displays the compaction performance of several different 6-tap orthonormal wavelet filters, namely Binomial-QMF, most regular wavelet filter, and Coiflet filters (Daubechies, Tech. Memo), which will be introduced in Chapter 5, for 2-, 4-, and 8-band signal decompositions along with the KLT and an ideal filter bank. These results indicate that the most regular filter does not perform the best even for highly correlated signal source. Although the mathematical interpretation of regularity in wavelets is meaningful, as we will see in Chapter 5,

	2-Bands	4-Bands	8-Bands
6-Tap Maxregular Filter	3.745	6.725	8.464
6-Tap Coiflet	3.653	6.462	8.061
6-Tap Binomial QMF(Maxflat)	3.760	6.766	8.529
KLT	3.202	5.730	7.660
Ideal Filter Bank	3.946	7.230	9.160

Table 4.12: Energy compaction performance of several 6-tap wavelet filters along with the KLT and ideal filter bank for an AR(1) source, $\rho = 0.95$.

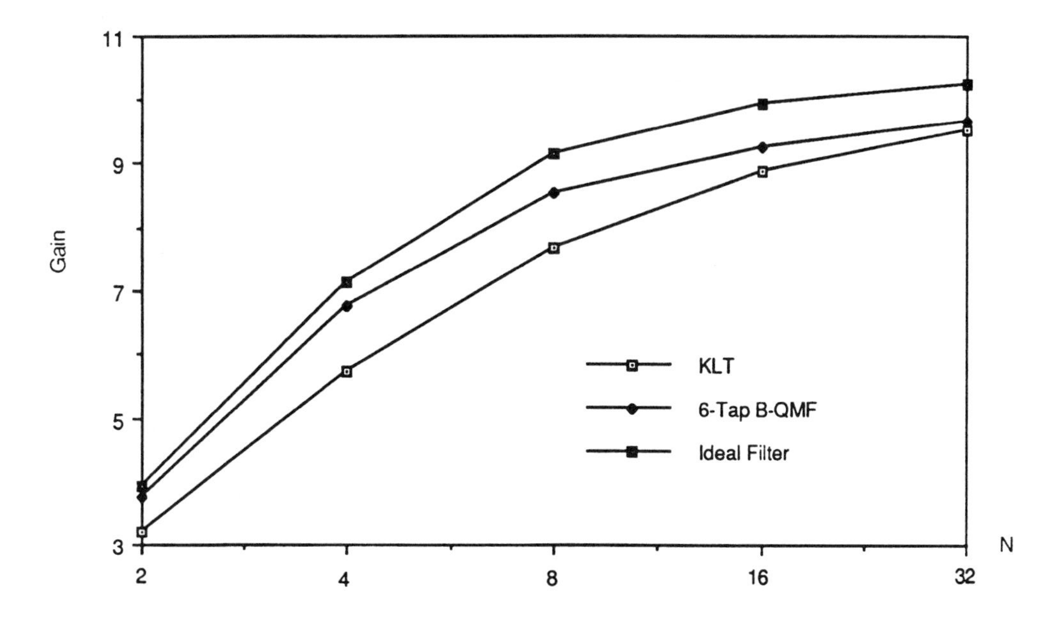

Figure 4.8: Graph of G_{TC} versus N for a six-tap Binomial-QMF, KLT, and ideal filter bank assuming an AR(1) source with $\rho = 0.95$.

its practical significance in signal processing is limited to imposing the obvious zero-mean high-pass filter condition.

Figure 4.8 compares the G_{TC} results of the KLT, ideal filter bank, and 6-tap Binomial-QMF for different resolution regular subband trees or block sizes. It is seen from this figure that, when the number of bands or transform size increases, the slope of the KLT and Binomial-QMF compaction curves get closer, since the aliasing energy or interband leakage becomes very significant. This phenomenon will be examined in Section 4.10.

Table 3.2 displays energy compaction performance of several decomposition tools and subband tree structures for the standard test images: LENA, BUILDING, CAMERAMAN, and BRAIN. These are monochrome, 256×256 size, 8 bits/pixel images. The test results displayed in Table 3.2 are broadly consistent with the results obtained for AR(1) sources. Again these results show that the irregular subband tree achieves a compaction performance very close to that of the regular tree, but with fewer bands and reduced computational burden. It must be remembered, however, that the data rate is the same for all tree structures in critically sampled systems.

4.10 Aliasing Energy in Multiresolution Decomposition

In this section, we present an analysis of signal energy distribution in PR multirate systems and evaluate the effects of aliasing. We also define a performance measure called the *nonaliasing energy ratio* (NER) for evaluation of decomposition techniques. The merit of the new measure is examined with respect to the block transforms and two-band PR-QMF based filter banks. We show that there is inverse relationship between G_{TC} and the new measure NER with respect to the number of bands or transform size.

4.10.1 Aliasing Effects of Decimation/Interpolation

Here we are evaluating the aliasing and nonaliasing energy components at the output of the ith branch of an M-band filter structure as shown in Fig. 4.9. From the decimator input to the output of the upsampler, we have

$$S_{u_i}(e^{j\omega}) = \frac{1}{M^2} \sum_{n=0}^{M-1} S_{y_i}(e^{j(\omega+\frac{2\pi}{M}n)}). \tag{4.72}$$

The filters impose

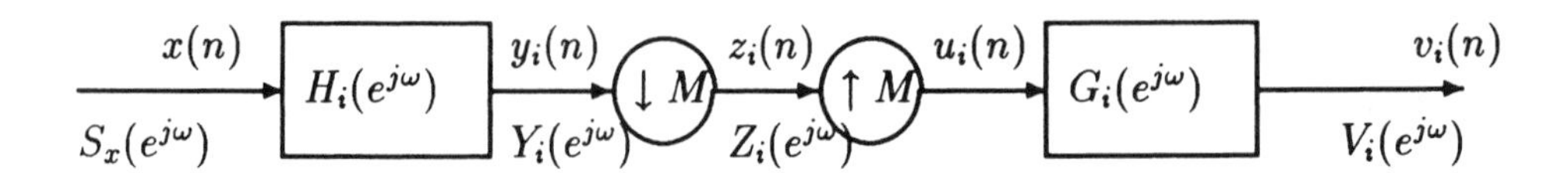

Figure 4.9: A decimation and interpolation branch.

$$S_{y_i}(e^{j\omega}) = |H_i(e^{j\omega})|^2 S_x(e^{j\omega})$$
$$S_{v_i}(e^{j\omega}) = \frac{1}{M}|G_i(e^{j\omega})|^2 S_{z_i}(e^{j\omega M}). \tag{4.73}$$

Combining these gives

$$S_{v_i}(e^{j\omega}) = \frac{1}{M^2}|G_i(e^{j\omega})|^2 \sum_{n=0}^{M-1} |H_i(e^{j(\omega+\frac{2\pi}{M}n)})|^2 S_x(e^{j(\omega+\frac{2\pi}{M}n)})) \tag{4.74}$$

which can be rewritten as

$$S_{v_i}(e^{j\omega}) = S_i^N(e^{j\omega}) + S_i^A(e^{j\omega}), \tag{4.75}$$

where

$$S_i^N(e^{j\omega}) = \frac{1}{M^2}|G_i(e^{j\omega})|^2 |H_i(e^{j\omega})|^2 S_x(e^{j\omega}) \tag{4.76}$$

and

$$S_i^A(e^{j\omega}) = \frac{1}{M^2}|G_i(e^{j\omega})|^2 \sum_{n=1}^{M-1} |H_i(e^{j(\omega+\frac{2\pi}{M}n)})|^2 S_x(e^{j(\omega+\frac{2\pi}{M}n)}). \tag{4.77}$$

It is seen that $S_i^N(e^{j\omega})$ consists of the nonaliasing component of the branch output spectral density while $S_i^A(e^{j\omega})$ consists of $(M-1)$ aliasing energy density terms caused by down- and up-sampling. We view these terms as somewhat misplaced energy components in frequency.

Finally, the branch output energy or variance for a zero mean input is

$$\begin{aligned} \sigma_i^2 &= \frac{1}{2\pi}\int_{-\pi}^{\pi} S_{V_i}(e^{j\omega})d\omega \\ &= \sigma_{i_N}^2 + \sigma_{i_A}^2 \end{aligned} \tag{4.78}$$

where

$$\begin{aligned} \sigma_{i_N}^2 &= \frac{1}{2\pi}\int_{-\pi}^{\pi} S_i^N(e^{j\omega})d\omega \\ \sigma_{i_A}^2 &= \frac{1}{2\pi}\int_{-\pi}^{\pi} S_i^A(e^{j\omega})d\omega. \end{aligned} \tag{4.79}$$

Hence, we can separate the branch output energy into its nonaliasing and aliasing components.

Figure 4.10 displays the spectra at different points in the decimation/interpolation branch for a two-band, 4-tap Binomial QMF with AR(1) source, $\rho = 0.5$.

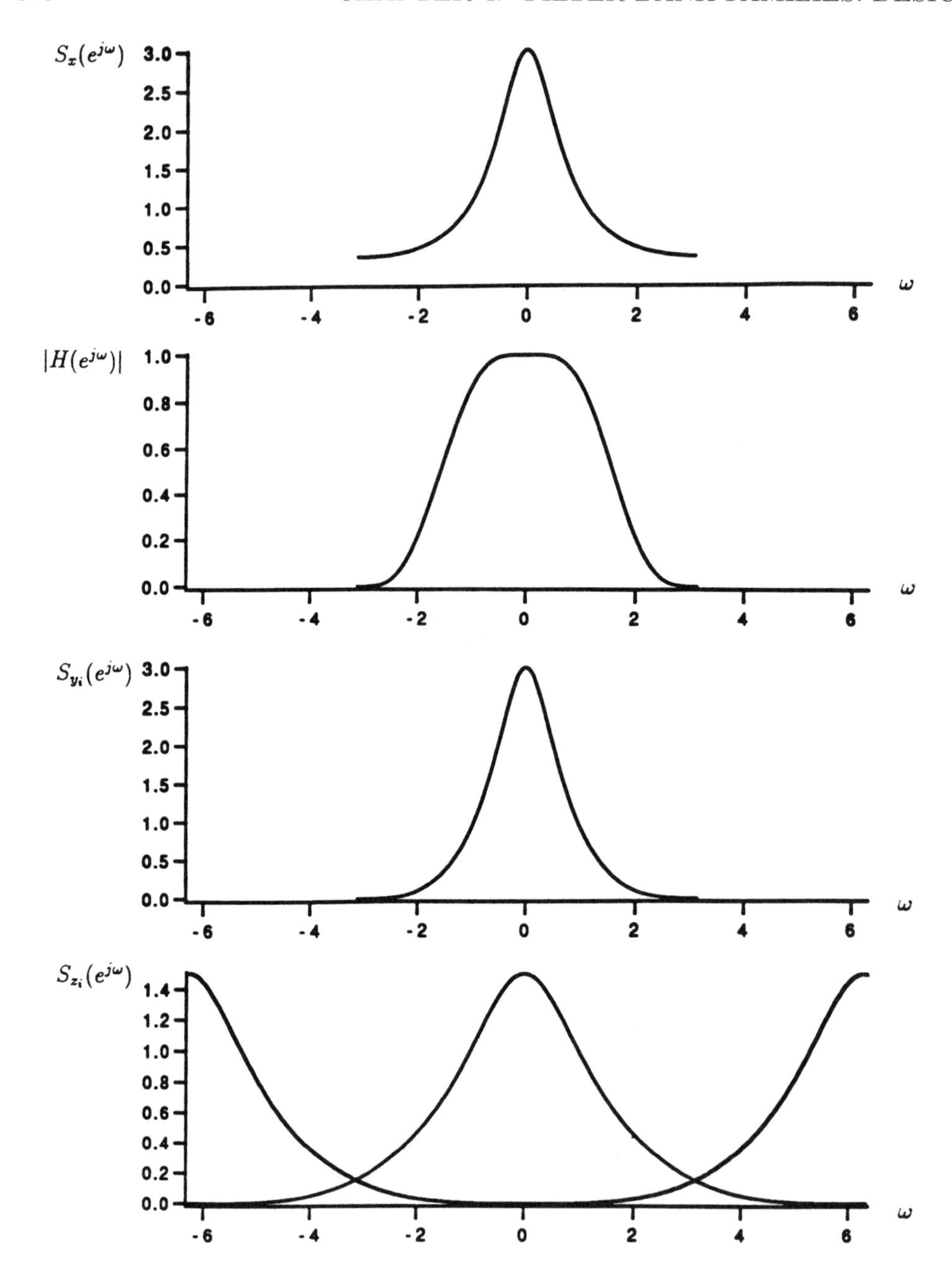

Figure 4.10: The signal spectra of different points in the decimation and interpolation branch of Fig. 4.9 for AR(1) input with $\rho = 0.5$.

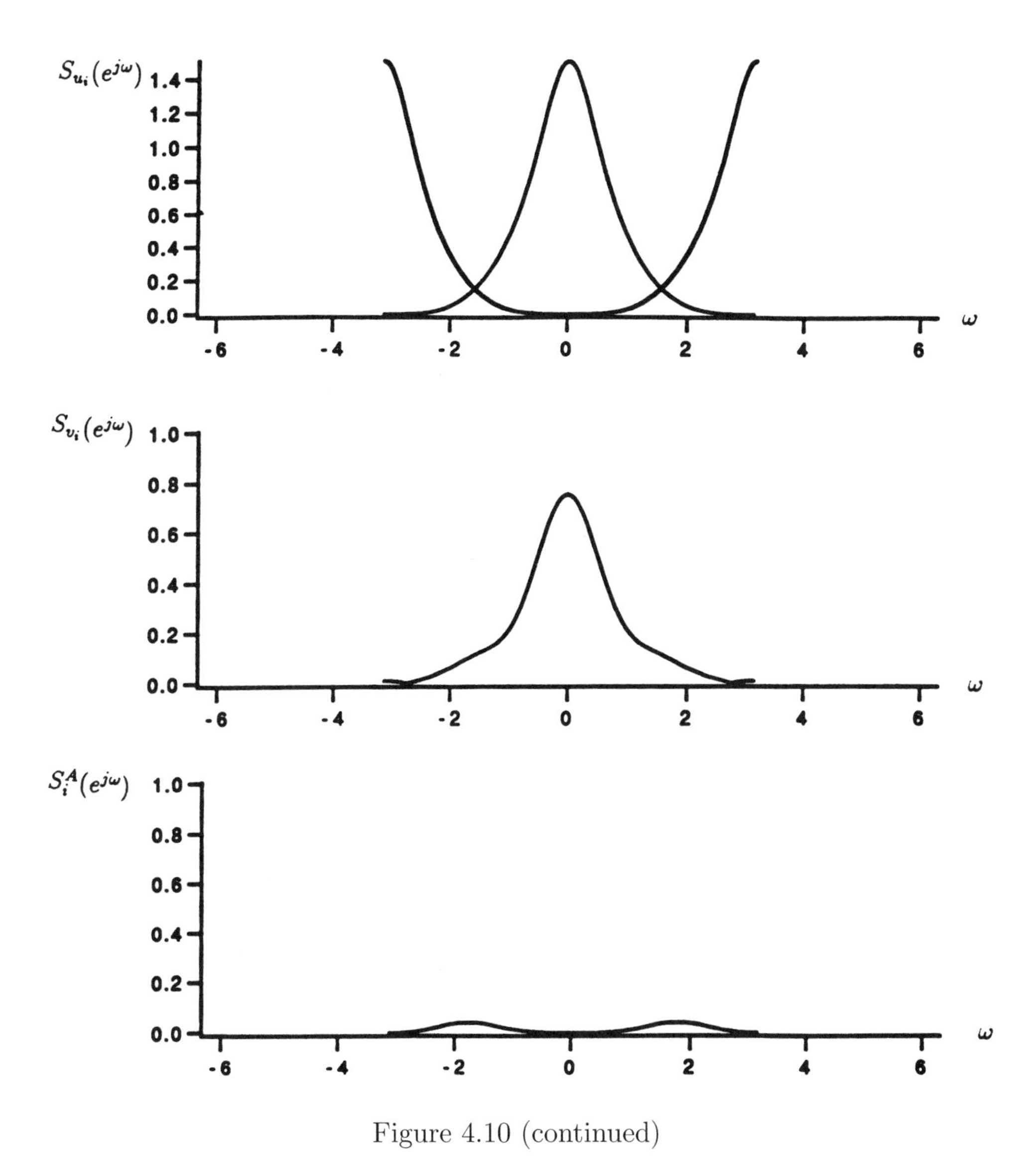

Figure 4.10 (continued)

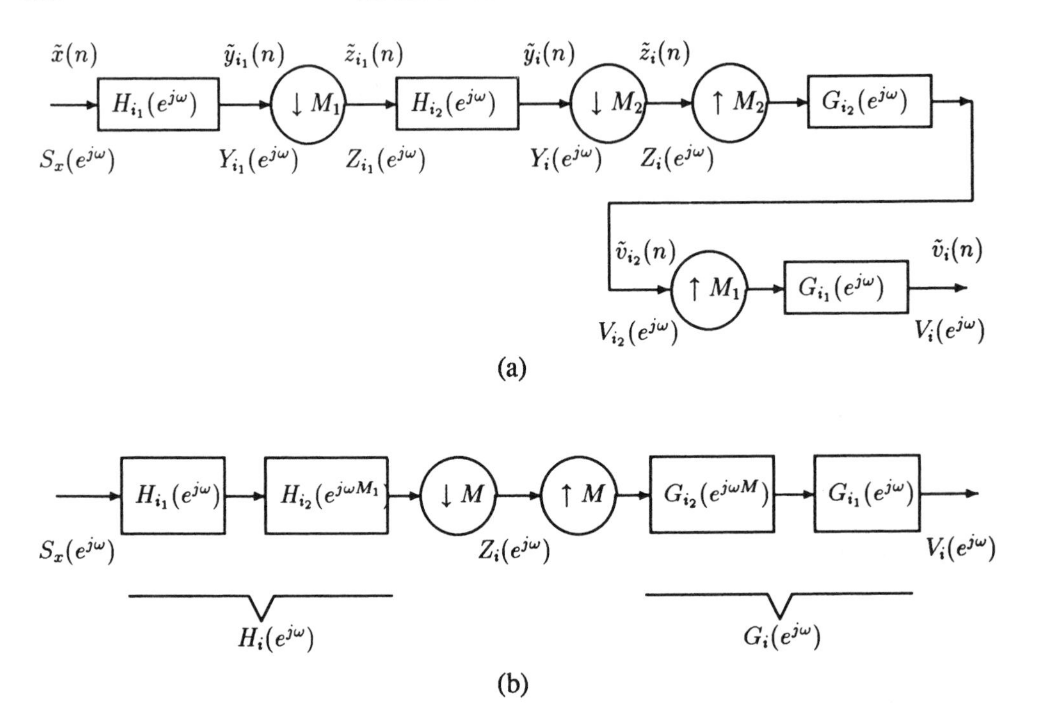

Figure 4.11: (a) Hierarchical decimation/interpolation branch and (b) its equivalent.

The advantage of this analysis in a lossless M-band filter bank structure is its ability to decompose the signal energy into a kind of time-frequency plane. We can express the decomposed signal energy of branches or subbands in the form of an energy matrix defined as (Akansu and Caglar, 1992)

$$[E(i,k)] = \sigma^2_{(i,mod(k+i)_{M-1})} \qquad i,k = 0,1,...,M-1. \tag{4.80}$$

Each row of the matrix E represents one of the bands or channels in the filter bank and the columns correspond to the distributions of subband energies in frequency. The energy matrices of the 8-band DCT, 8-band (3-level) hierarchical filter banks with a 6-tap Binomial-QMF (BQMF), and the most regular wavelet filter (MRWF) (Daubechies) for an AR(1) source with $\rho = 0.95$ follow:

$$E_{DCT} = \begin{bmatrix} 6.6824 & 0.1211 & 0.0280 & 0.0157 & 0.0132 & 0.0157 & 0.0280 & 0.1211 \\ 0.1511 & 0.1881 & 0.1511 & 0.0265 & 0.0113 & 0.0091 & 0.0113 & 0.0265 \\ 0.0345 & 0.0136 & 0.0569 & 0.0136 & 0.0345 & 0.0078 & 0.0046 & 0.0078 \\ 0.0158 & 0.0032 & 0.0050 & 0.0279 & 0.0051 & 0.0032 & 0.0158 & 0.0061 \\ 0.0176 & 0.0032 & 0.0016 & 0.0032 & 0.0176 & 0.0032 & 0.0016 & 0.0032 \\ 0.0065 & 0.0012 & 0.0065 & 0.0022 & 0.0026 & 0.0132 & 0.0026 & 0.0022 \\ 0.0053 & 0.0004 & 0.0001 & 0.0004 & 0.0053 & 0.0033 & 0.0118 & 0.0033 \\ 0.0053 & 0.0002 & 0.0000 & 0.0000 & 0.0000 & 0.0002 & 0.0053 & 0.0155 \end{bmatrix}$$

$$E_{BQMF} = \begin{bmatrix} 7.1720 & 0.0567 & 0.0014 & 0.0005 & 0.0001 & 0.0005 & 0.0014 & 0.0567 \\ 0.0567 & 0.1987 & 0.0567 & 0.0258 & 0.0005 & 0.0014 & 0.0005 & 0.0258 \\ 0.0258 & 0.0025 & 0.0640 & 0.0025 & 0.0258 & 0.0042 & 0.0061 & 0.0042 \\ 0.0042 & 0.0014 & 0.0025 & 0.0295 & 0.0025 & 0.0014 & 0.0042 & 0.0196 \\ 0.0196 & 0.0019 & 0.0001 & 0.0013 & 0.0223 & 0.0013 & 0.0001 & 0.0019 \\ 0.0019 & 0.0061 & 0.0019 & 0.0046 & 0.0013 & 0.0167 & 0.0013 & 0.0045 \\ 0.0045 & 0.0001 & 0.0014 & 0.0001 & 0.0045 & 0.0020 & 0.0162 & 0.0020 \\ 0.0020 & 0.0001 & 0.0001 & 0.0001 & 0.0001 & 0.0001 & 0.0020 & 0.0220 \end{bmatrix}$$

$$E_{MRWF} = \begin{bmatrix} 7.1611 & 0.0589 & 0.0018 & 0.0006 & 0.0001 & 0.0006 & 0.0018 & 0.0589 \\ 0.0589 & 0.1956 & 0.0589 & 0.0262 & 0.0006 & 0.0017 & 0.0006 & 0.0262 \\ 0.0262 & 0.0028 & 0.0628 & 0.0028 & 0.0262 & 0.0043 & 0.0064 & 0.0043 \\ 0.0043 & 0.0018 & 0.0028 & 0.0291 & 0.0028 & 0.0018 & 0.0043 & 0.0196 \\ 0.0196 & 0.0020 & 0.0001 & 0.0014 & 0.0221 & 0.0014 & 0.0001 & 0.0020 \\ 0.0020 & 0.0064 & 0.0020 & 0.0047 & 0.0014 & 0.0164 & 0.0014 & 0.0047 \\ 0.0047 & 0.0001 & 0.0017 & 0.0001 & 0.0047 & 0.0020 & 0.0160 & 0.0020 \\ 0.0020 & 0.0001 & 0.0001 & 0.0001 & 0.0001 & 0.0001 & 0.0020 & 0.0218 \end{bmatrix}$$

We can easily extend this analysis to any branch in a tree structure, as shown in Fig. 4.11(a). We can obtain an equivalent structure by shifting the antialiasing filters to the left of the decimator and the interpolating filter to the right of the up-sampler as shown in Fig. 4.11(b). The extension is now obvious.

4.10.2 Nonaliasing Energy Ratio

The energy compaction measure G_{TC} does not consider the distribution of the band energies in frequency. Therefore the aliasing portion of the band energy is treated no differently than the nonaliasing component. This fact becomes important particularly when all the analysis subband signals are not used for the reconstruction or whenever the aliasing cancellation in the reconstructed signal is not perfectly performed because of the available bits for coding.

From Eqs. (4.78) and (4.79), we define the nonaliasing energy ratio (NER) of an M-band orthonormal decomposition technique as

$$NER = \frac{\sum_{i=0}^{M-1} \sigma_{i_N}^2}{\sum_{i=0}^{M-1} \sigma_i^2} \tag{4.81}$$

where the numerator term is the sum of the nonaliasing terms of the band energies. The ideal filter bank yields NER=1 for any M as the upper bound of this measure for any arbitrary input signal.

4.11 G_{TC} and NER Performance

We consider 4-, 6-, 8-tap Binomial-QMFs in a hierarchical filter bank structure as well as the 8-tap Smith-Barnwell and 6-tap most regular orthonormal wavelet filters, and the 4-, 6-, 8-tap optimal PR-QMFs along with the ideal filter banks for performance comparison. Additionally, 2×2, 4×4, and 8×8 discrete cosine, discrete sine, Walsh-Hadamard, and modified Hermite transforms are considered for comparison purposes. The G_{TC} and NER performance of these different decomposition tools are calculated by computer simulations for an AR(1) source model. Table 4.15 displays G_{TC} and NER performance of the techniques considered with $M = 2, 4, 8$.

It is well known that the aliasing energies become annoying, particularly at low bit rate image coding applications. The analysis provided in this section explains objectively some of the reasons behind this observation. Although the ratio of the aliasing energies over the whole signal energy may appear negligible, the misplaced aliasing energy components of bands may be locally significant in frequency and cause subjective performance degradation.

While larger M indicates better coding performance by the G_{TC} measure, it is known that larger size transforms do not provide better subjective image coding performance. The causes of this undesired behavior have been mentioned in the literature as intercoefficient or interband energy leakages, bad time localization, etc.. The NER measure indicates that the larger M values yield degraded performance for the finite duration transform bases and the source models considered. This trend is consistent with those experimental performance results reported in the literature. This measure is therefore complementary to G_{TC}, which does not consider aliasing.

	M=2 G_{TC} (NER)	M=4 G_{TC} (NER)	M=8 G_{TC} (NER)
DCT	3.2026 (0.9756)	5.7151 (0.9372)	7.6316 (0.8767)
DST	3.2026 (0.9756)	3.9106 (0.8532)	4.8774 (0.7298)
MHT	3.2026 (0.9756)	3.7577 (0.8311)	4.4121 (0.5953)
WHT	3.2026 (0.9756)	5.2173 (0.9356)	6.2319 (0.8687)
Binomial-QMF (4 tap)	3.6426 (0.9880)	6.4322 (0.9663)	8.0149 (0.9260)
Binomial-QMF (6 tap)	3.7588 (0.9911)	6.7665 (0.9744)	8.5293 (0.9427)
Binomial-QMF (8 tap)	3.8109 (0.9927)	6.9076 (0.9784)	8.7431 (0.9513)
Smith-Barnwell (8tap)	3.8391 (0.9937)	6.9786 (0.9813)	8.8489 (0.9577)
Most regular (6 tap)	3.7447 (0.9908)	6.7255 (0.9734)	8.4652 (0.9406)
Optimal QMF (8 tap)*	3.8566 (0.9943)	7.0111 (0.9831)	8.8863 (0.9615)
Optimal QMF (8 tap)**	3.8530 (0.9944)	6.9899 (0.9834)	8.8454 (0.9623)
Optimal QMF (6 tap)*	3.7962 (0.9923)	6.8624 (0.9776)	8.6721 (0.9497)
Optimal QMF (6 tap)**	3.7936 (0.9924)	6.8471 (0.9777)	8.6438 (0.9503)
Optimal QMF (4 tap)*	3.6527 (0.9883)	6.4659 (0.9671)	8.0693 (0.9278)
Optimal QMF (4 tap)**	3.6525 (0.9883)	6.4662 (0.9672)	8.0700 (0.9280)
Ideal filter bank	3.946 (1.000)	7.230 (1.000)	9.160 (1.000)

*This optimal QMF is based on energy compaction.
**This optimal QMF is based on minimized aliasing energy.

Table 4.15: Performance of several orthonormal signal decomposition techniques for AR(1), $\rho = 0.95$ source.

4.12 Quantization Effects in Filter Banks

A prime purpose of subband filter banks is the attainment of data rate compression through the use of pdf-optimized quantizers and optimum bit allocation for each subband signal. Yet scant consideration had been given to the effect of coding errors due to quantization. Early studies by Westerink et al. (1992) and Vandendorpe (1991) were followed by a series of papers by Haddad and his colleagues, Kovacevic (1993), Gosse and Duhamel (1997), and others. This section provides a direct focus on modeling, analysis, and optimum design of quantized filter banks. It is abstracted from Haddad and Park (1995).

We review the gain-plus-additive noise model for the pdf-optimized quantizer advanced by Jayant and Noll (1984). Then we embed this model in the time-domain filter bank representation of Section 3.5.5 to provide an M-band quantization model amenable to analysis. This is followed by a description of an optimum two-band filter design which incorporates quantization error effects in the design methodology.

4.12.1 Equivalent Noise Model

The quantizer studied in Section 2.2.2 is shown in Fig. 4.12(a). We assume that the random variable input x has a known probability density function (pdf) with zero mean. If this quantizer is pdf-optimized, the quantization error $\tilde{x}$ is zero mean and orthogonal to the quantizer output $\hat{x}$ (Prob.2.9), i.e.,

$$E\{\tilde{x}\} = 0, \quad E\{\hat{x}\tilde{x}\} = 0. \tag{4.82}$$

But the quantization error $\tilde{x}$ is correlated with the input so that the variance of the quantization is (Prob. 4.24)

$$\sigma_{\hat{x}}^2 = \sigma_x^2 - \sigma_{\tilde{x}}^2 \tag{4.83}$$

where σ^2 refers to the variance of the respective zero mean signals. Note that for the optimum quantizer, the output signal variance is less than that of the input. Hence the simple input-independent additive noise model is only an approximation to the noise in the pdf-optimized quantizer.

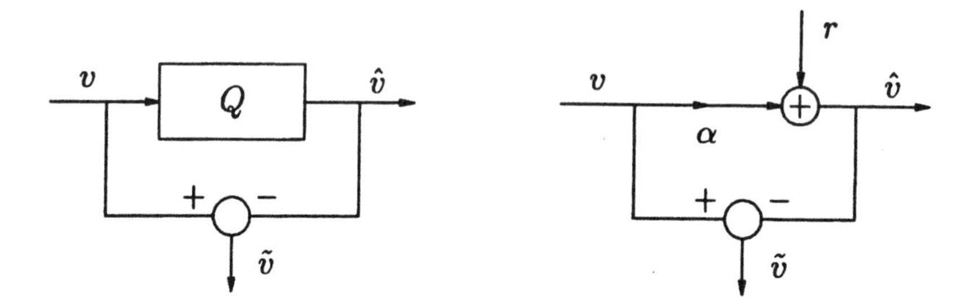

Figure 4.12: (a) pdf-optimized quantizer; (b) equivalent noise model.

Figure 4.12(b) shows a gain-plus-additive noise representation which is to model the quantizer. In this model, we can impose the conditions in Eq. (4.82) and force the input x and additive noise r to be uncorrelated. The model parameters are gain α and variance σ_r^2. With $\hat{x} = \alpha x + r$, the uncorrelated requirement becomes

$$\sigma_{\hat{x}}^2 = \alpha^2\sigma_x^2 + \sigma_r^2. \tag{4.84}$$

Equating $\sigma_{\hat{x}}^2$ in these last two equations gives one condition. Next, we equate $E\{x\tilde{x}\}$ for model and quantizer. From the model,

$$E\{x\tilde{x}\} = E\{x(x - \alpha x - r)\} = (1-\alpha)\sigma_x^2, \tag{4.85}$$

and for the quantizer,

$$E\{x\tilde{x}\} = E\{(\hat{x} + \tilde{x})\tilde{x}\} = E\{\hat{x}\tilde{x}\} + E\{(\tilde{x})^2\} = \sigma_{\tilde{x}}^2. \tag{4.86}$$

These last two equations provide the second constraint. Solving all these gives

$$\alpha = \quad 1 - \frac{\sigma_{\tilde{x}}^2}{\sigma_x^2} \tag{4.87}$$

$$\sigma_r^2 = \quad \alpha(1-\alpha)\sigma_x^2 = \alpha\sigma_{\tilde{x}}^2. \tag{4.88}$$

For the model, r and x are uncorrelated and the gain α and variance σ_r^2 are input-signal dependent.

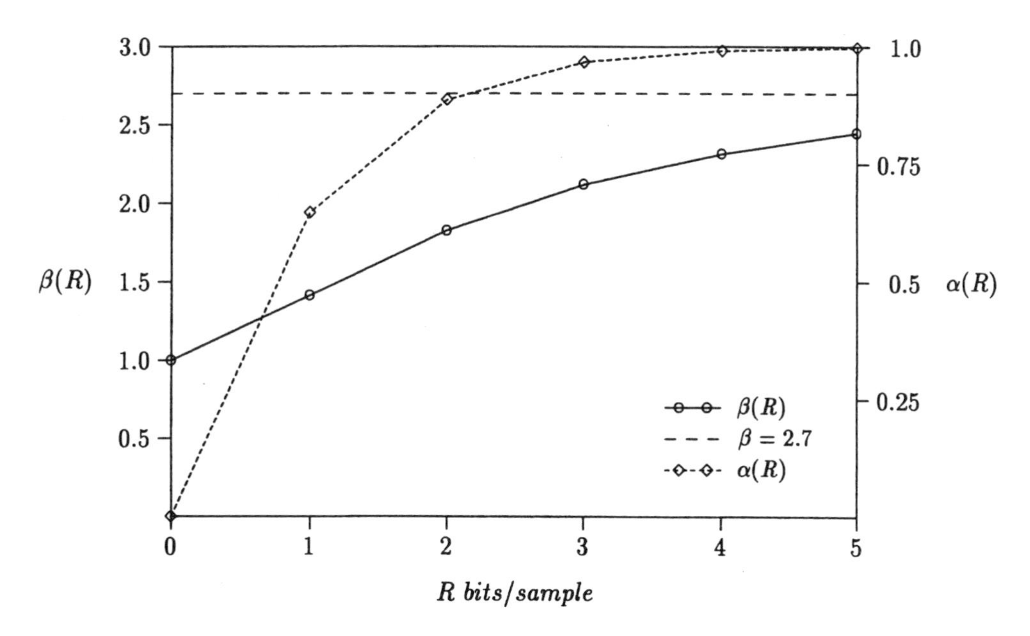

Figure 4.13: $\beta(R)$, $\alpha(R)$ versus R for AR(1) Gaussian input at ρ=0.95.

From rate distortion theory (Berger 1971), the quantization error variance $\sigma_{\tilde{x}}^2$ for the pdf-optimized quantizer is

$$\sigma_{\tilde{x}}^2 = \beta(R)2^{-2R}\sigma_x^2. \tag{4.89}$$

The parameter $\beta(R)$ in Eq. (4.89) depends only on the pdf of the unit variance signal being quantized and on R, the number of bits assigned to the quantizer. It does not depend on the autocorrelation of the input signal. Earlier approaches treated $\beta(R)$ as a constant for a particular pdf. We show the plot of β versus R for a Gaussian input in Fig. 4.13. Jayant and Noll reported β=2.7 for a Gaussian input, the asymptotic value indicated by the dashed line in Fig. 4.13. From Eqs. (4.88) and (4.89) the nonlinear gain α can be evaluated as

$$\alpha = 1 - \beta(R)2^{-2R}. \tag{4.90}$$

Figure 4.13 also shows α vs R using Eq. (4.90). As R gets large, β approaches its asymptotic value, and α approaches unity. Thus, the gain-plus additive noise model parameters α and σ_r^2 are determined once R and the signal pdf are specified. Note that a different plot and different asymptotic value result for differing signal pdfs.

4.12.2 Quantization Model for M-Band Codec

The maximally decimated M-band filter bank with the bank of pdf-optimized quantizers and a bank of scalar compensators (dotted lines) are shown in Fig. 4.14(a). Each quantizer is represented by its equivalent noise model, and the analysis and synthesis banks by the equivalent polyphase structures. This gives the equivalent representation of Fig. 4.14(b), which, in turn, is depicted by the vector-matrix equivalent structure of Fig. 4.14(c). Thus, by moving the samplers to the left and right of the filter banks, and focusing on the slow-clock-rate signals, the system to be analyzed is time-invariant, but nonlinear because of the presence of the signal dependent gain matrix A.

By construction the vectors $\underline{v}[n]$ and $\underline{r}[n]$ are uncorrelated, and A, S are diagonal gain and compensation matrices, respectively, where

$$\begin{aligned} S &= \text{diag}\{s_0, s_1, \cdots s_{M-1}\} \\ A &= \text{diag}\{\alpha_0, \alpha_1, \cdots \alpha_{M-1}\} \qquad (4.91) \\ \alpha_i &= 1 - \frac{\sigma^2_{\tilde{v}_i}}{\sigma^2_{v_i}} \\ \sigma^2_{r_i} &= \alpha_i(1-\alpha_i)\sigma^2_{v_i}. \qquad (4.92) \end{aligned}$$

This representation well now permits us to calculate explicitly the total mean square quantization error in the reconstructed output in terms of analysis and synthesis filter coefficients, the input signal autocorrelation, the scalar compensators, and implicitly in terms of the bit allocation for each band.

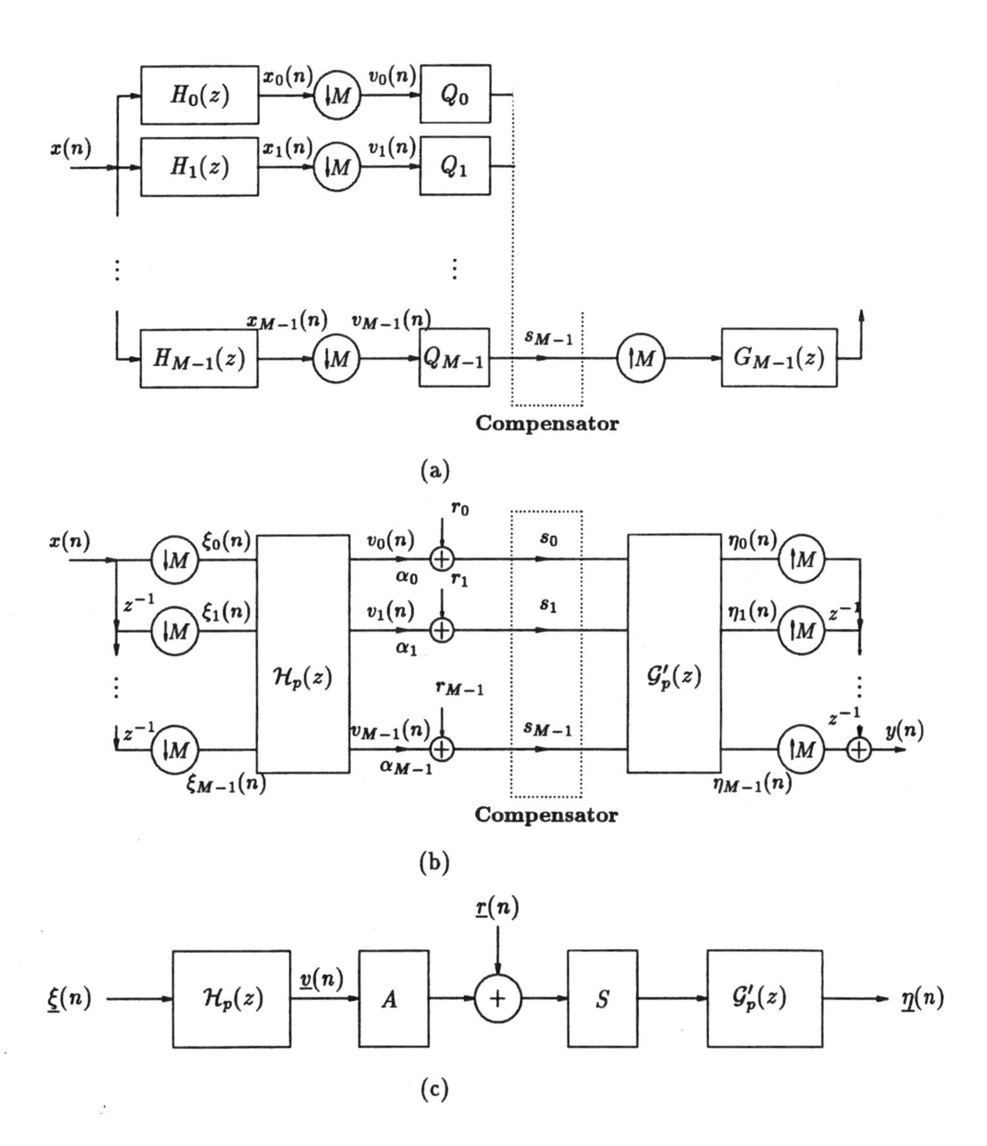

Figure 4.14: (a) M-band filter bank structure with compensators, (b) polyphase equivalent structure, (c) vector-matrix equivalent structure.

We define the total quantization error as the difference

$$\underline{\eta}_q(z) = \underline{\eta}(z) - \underline{\eta}_o(z) \tag{4.93}$$

where the subscript "o" implies the system without quantizers and compensators.

From Fig. 4.14(c) we see that

$$\begin{aligned}\underline{\eta}_q(z) &= \mathcal{G}_p'(z)[SA - I]\mathcal{H}_p(z)\underline{\xi}(z) + \mathcal{G}_p'(z)S\underline{R}(z) & (4.94)\\ &= \mathcal{G}_p'(z)B\underline{V}(z) + \mathcal{G}_p'(z)S\underline{R}(z) \\ &\triangleq \mathcal{C}(z)\underline{V}(z) + \mathcal{D}(z)\underline{R}(z), & (4.95)\end{aligned}$$

where $B = S - I$, and $\underline{V}(z) = \mathcal{H}_p(z)\underline{\xi}(z)$ and $\mathcal{C}(z) = \mathcal{G}_p'(z)B$, $\mathcal{D}(z) = \mathcal{G}_p'(z)S$. We note that $\underline{v}(n)$ and $\underline{r}(n)$ are uncorrelated by construction.

For a time-invariant system with $M \times 1$ input vector $\underline{x}$ and output vector $\underline{y}$, we define $M \times M$ power spectral density (PSD) and correlation matrices as

$$\begin{aligned}R_{yx}[m] &= E[\underline{y}(n)\underline{x}^T(n+m)], \\ S_{yx}(z) &= Z\{R_{yx}\} = \sum_m R_{yx}[m]z^{-m}. & (4.96)\end{aligned}$$

Using these definitions and the fact that $\underline{v}(n)$ and $\underline{r}(n)$ are uncorrelated, we can calculate the PSD $S_{n_q n_q}(z)$ and covariance $R_{n_q n_q}[m]$ for the quantization error $\eta_q(n)$.

It is straightforward to show (Prob. 4.24) that

$$\begin{aligned}S_{n_q n_q}(z) &= \mathcal{C}(z^{-1})S_{vv}(z)\mathcal{C}^T(z) + \mathcal{D}(z^{-1})S_{rr}(z)\mathcal{D}^T(z) & (4.97)\\ R_{n_q n_q}[k] &= C_{-k} * R_{vv}[k] * C_k^T + D_{-k} * R_{rr}(k) * D_k^T & (4.98)\end{aligned}$$

where $\mathcal{C}(z) \leftrightarrow C_k$ and $\mathcal{D}(z) \leftrightarrow D_k$ are Z transform pairs.

At k=0, this becomes

$$R_{n_q n_q}[0] = \sum_j \sum_k C_k R_{vv}[j-k] C_k^T + \sum_j \sum_k D_k R_{rr}[j-k] D_k^T. \tag{4.99}$$

From Fig. 4.14(b), we can demonstrate that $R_{\underline{\eta\eta}}(o)$ is the covariance of the Mth block output vector

$$\begin{aligned}\underline{\eta}^T(n) &= [\eta_0(n), \eta_1(n), \ldots, \eta_{M-1}(n)] \\ &= [y(Mn), y(Mn+1), \ldots, y(Mn+M-1)]. & (4.100)\end{aligned}$$

Consequently,

$$R_{\underline{\eta\eta}}[0] = \begin{bmatrix} R_{yy}(Mn, Mn) & \cdots & R_{yy}(Mn, Mn + Mn + M - 1) \\ R_{yy}(Mn+1, Mn) & \cdots & R_{yy}(Mn+1, Mn + M - 1) \\ \vdots & \ddots & \vdots \\ R_{yy}(Mn + M - 1, Mn) & \cdots & R_{yy}(Mn + M - 1, Mn + M - 1) \end{bmatrix} \quad (4.101)$$

where $R_{yy}(Mn + k, Mn + j) = E[y(Mn + k)y(Mn + j)]$ for k,j=0,1,. . .,$M - 1$. Note that this is cyclostationary; the covariance matrix of the next block of M outputs will also equal $R_{\underline{\eta\eta}}[0]$. Each block of M output samples will thus have same sum of variances. We take the MS value of the output as the average of the diagonal elements of Eq. (4.101),

$$\begin{aligned} \sigma_y^2 &= \overline{E[y^2(n)]} \\ &= \frac{1}{M}\sum_{j=0}^{M-1} R_{yy}(Mn + j, Mn + j) \\ &= \frac{1}{M}\sum_{j=0}^{M-1} E[y^2(Mn + j)] \\ &= \frac{1}{M} trace\{R_{\underline{\eta\eta}}[0]\}. \end{aligned} \quad (4.102)$$

Similarly, if we define $y_q(n)$ as the quantization error in the reconstructed output

$$y_q(n) \stackrel{\triangle}{=} y(n) - y_0(n) \quad (4.103)$$

then the total mean square quantization error (MSE) at the system output is

$$\begin{aligned} \sigma_{y_q}^2 &\stackrel{\triangle}{=} \overline{E[y_q^2(n)]} \\ & \frac{1}{M} trace\{R_{\eta_q\eta_q}[0]\}. \end{aligned} \quad (4.104)$$

Next, by substituting Eq. (4.99) into Eq. (4.104), we obtain

$$\begin{aligned} \sigma_{y_q}^2 &= \frac{1}{M} trace\{\sum_{j=0}^{N-1}\sum_{k=0}^{N-1} C_j R_{vv}[j-k] C_k^T\} + \frac{1}{M} trace\{\sum_{j=0}^{N-1}\sum_{k=0}^{N-1} D_j R_{rr}[j-k] D_k^T\} \\ &= \sigma_d^2 + \sigma_n^2. \end{aligned} \quad (4.105)$$

The first term, σ_d^2, of Eq. (4.105) is the component of the MSE due to the nonlinear gain matrix A and compensation matrix S. The second term σ_n^2 accounts for the additive fictitious random noise $\underline{r}(n)$. These terms σ_d^2, σ_n^2 are called the signal distortion and random noise components of the MSE, respectively. Under PR constraints, σ_d^2 measures the deviation from perfect reconstruction due to the quantizer and compensator. This decomposition of the total MSE enables us to analyze each component error separately. This is the main theoretical consequence of the gain-plus-additive noise quantizer model where the signals $\underline{v}(n)$ and random noise $\underline{r}(n)$ are uncorrelated.

The MSE in Eq. (4.105) can be written in an explicit closed form time-domain expression in terms of the analysis and synthesis filter coefficients. This is achieved by expanding the polyphase coefficient matrices in terms of the synthesis filter coefficients via

$$\mathcal{G}_{p,j} = \begin{bmatrix} g_0(Mj) & g_0(Mj+1) & \cdots & g_0(Mj+M-1) \\ g_1(Mj) & g_1(Mj+1) & \cdots & g_1(Mj+M-1) \\ \vdots & \vdots & \ddots & \vdots \\ g_{M-1}(Mj) & g_{M-1}(Mj+1) & \cdots & g_{M-1}(Mj+M-1) \end{bmatrix} \tag{4.106}$$

and substituting into Eq. (4.105). The results are rather messy and are not presented here. The interested reader can refer to the reference for details. The last step in our formulation requires a further breakdown of $R_{vv}[m]$ in Eq. (4.105). From Fig. 4.14(a) $R_{v_iv_j}[m]$ can be represented as

$$R_{v_iv_j}[m] = E[v_i(n)v_j(n+m)] = R_{x_ix_j}[Mm]. \tag{4.107}$$

By defining the correlation function $\rho_{ji}(m) \stackrel{\triangle}{=} h_i(m) * h_j(-m)$, we have

$$\begin{aligned} R_{x_ix_j}[m] &= R_{xx}[m] * \rho_{ji}(m) \\ &= \sum_k \rho_{ji}(k) R_{xx}[m-k] \\ &= \sum_k [\sum_l h_i(k+l)h_j(l)] R_{xx}[m-k]. \end{aligned} \tag{4.108}$$

This concludes the formulation of the output MSE in terms of the analysis/synthesis filter coefficients $h_i(n)$, $g_i(n)$, the input autocorrelation function $R_{xx}[m]$, the nonlinear gain α_i, and compensator s_i.

Some simplifying assumptions on $R_{\underline{rr}}(k)$ can be argued. First, we note that the decimated signals $\{v_i(n)\}$ occupy frequency bands that can be made to overlap slightly. Hence, $\{v_i(n)\}$ and $\{v_j(n+m)\}$ tend to be weakly correlated. The random errors $\{r_i(n)\}$ due to each quantizer are, by design, uncorrelated with the respective $\{v_i(n)\}$. Therefore, as a simplifying assumption we can say that

$E[r_i(n)r_j(n+m)] \simeq 0$. This makes $R_{\underline{rr}}[n]$ a diagonal matrix. Next, it is often true that the quantization error for a given signal swing (as measured by $\sigma_{v_i}^2$) sweeps over several quantization levels. When this is true, $E[r_i(n)r_i(n+m)] = \sigma_{r_i}^2\delta(m)$. Then, the random component σ_n^2 reduces to a simpler form

$$\sigma_n^2 = \frac{1}{M}\sum_{i=0}^{M-1} s_i^2\sigma_{r_i}^2 \sum_{l=0}^{M-1} g_i^2(l) \tag{4.109}$$

but σ_d^2 remains messy.

From the foregoing, several observations regarding compensators can be noted:

(i) By setting s_i=1, we have no compensation and σ_d^2 in Eq. (4.105), and σ_n^2 in Eq. (4.109) constitute the MSE in the uncompensated structure. As we shall see in the next section, s_i=1 is the optimized selection when paraunitary PR constraints are imposed on the non-quantized system.

(ii) By choosing $s_i = 1/\alpha_i$, the "null compensation," we can eliminate completely the signal distortion term σ_d^2, leaving only the noise term

$$\begin{aligned} \sigma_{y_q}^2 &= \sigma_n^2 \\ &= \frac{1}{M}\sum_{i=0}^{M-1}(1/\alpha_i^2)\sigma_{r_i}^2 \sum_{l=0}^{MN-1} g_i^2(l). \end{aligned} \tag{4.110}$$

(iii) However, this solution is *not optimal* at the stated operating conditions. The quantizer gain $\alpha_i < 1$ and Eq. (4.110) show that we can expect a larger random component than that of the uncompensated structure. In fact, for the uncompensated structure, this random component is dominant. Increasing this component by the null condition is decidedly not optimal.

(iv) However, when the input statistics change from nominal values, the null compensation is found to be superior to the "optimal" one, which is, in fact, optimal only at the nominal values of ρ. In this account, we minimize the total MSE by minimizing jointly the sum of σ_d^2 and σ_n^2 subject to defined PR constraints.

4.12.3 Optimal Design of Bit-Constrained, pdf-Optimized Filter Banks

The design problem is the determination of the optimal FIR filter coefficients, compensators, and integer bit allocation that minimize the MSE subject to constraints of filter length, average bit rate, and PR in the absence of quantizers, for an input signal with a given autocorrelation function.

For the paraunitary case, the orthogonality properties eliminate the cross-correlation between analysis channels, which is implicit in the σ_d^2 component of Eq. (4.105). The MSE in this case reduces to

$$\sigma_d^2 = \frac{1}{M}\sum_{i=0}^{M-1}(\alpha_i s_{i-1})^2, \quad \sigma_n^2 = \frac{1}{M}\sum_{i=0}^{M-1} s_i^2 \sigma_{r_i}^2. \tag{4.111}$$

It is now easy to show that the optimized compensator for this paraunitary condition is $s_i^* = 1$. Then the uncompensated system is optimal for the pdf-optimized paraunitary FB. (On the other hand, $s_i = 1$ is not optimal for the biorthogonal structure because of the cross-correlation between analysis channels.)

Sample designs and simulations for a six-coefficient paraunitary two-band structure for an AR(1) input with $\rho = 0.95$ are shown in Table 4.13. MSE refers to the theoretical calculations and MSE_{sim}, the simulation results. Table 4.13 demonstrates that the optimal filter coefficients are quite insensitive to changes in the average bit rate R and in input correlation ρ. Figure 4.15(a) shows explicitly the distortion and random components of the total MSE. The simulation results closely match the theoretical ones. The random noise σ_n^2 is clearly the dominant component of the MSE. Figure 4.15(b) compares the optimally compensated with the null compensated ($s_i = 1/\alpha_i$) paraunitary systems designed for $\rho = 0.95$. The null compensated is more robust for changing input statistics and performs better than the fixed optimally compensated one when ρ changes from its design value of $\rho = 0.95$.

Similar designs and simulations were executed for the biorthogonal two-band case with equal length (6 taps) analysis and synthesis filters. For the same operating conditions, the biorthogonal structure is superior to the paraunitary in terms of the output MSE. However, the biorthogonal filter coefficients are very sensitive to R, the average number of bits, and to the value of ρ. The paraunitary design is far more robust and emerges as the preferred design when ρ is uncertain.

4.13 Summary

This chapter is dedicated to the description, evaluation, and design of practical QMFs. We described and compared the performance of several known paraunitary two-band PR-QMF families. These were shown to be special cases of a filter design philosophy based on Bernstein polynomials.

We described a new approach to the optimal design of filters using extended performance criteria. This route provides new directions for filter bank designs with particular applications in visual signal processing.

	ρ=0.95			
R	R_0	R_1	MSE	MSE$_{sim}$
1	1	1	0.3533	0.3522
1.5	2	1	0.1182	0.1183
2	3	1	0.0387	0.0391
2.5	4	1	0.0151	0.0154
3	5	1	0.0086	0.0087

(a)

R	$h_0(0)$	$h_0(1)$	$h_0(2)$	$h_0(3)$	$h_0(4)$	$h_0(5)$
1	0.359783	0.806318	0.434517	-0.122522	-0.117625	5.2485e-2
1.5	0.385663	0.796281	0.428142	-0.140852	-0.106698	5.1677e-2
2	0.385662	0.796281	0.428143	-0.140852	-0.106698	5.1677e-2
2.5	0.385659	0.796281	0.428146	-0.140851	-0.106696	5.1677e-2
3	0.385659	0.796281	0.428146	-0.140851	-0.106699	5.1677e-2

(b)

Table 4.13: Optimum designs for the paraunitary FB at $\rho = 0.95$, (a) optimum bits and MSE; (b) optimum filter coefficients

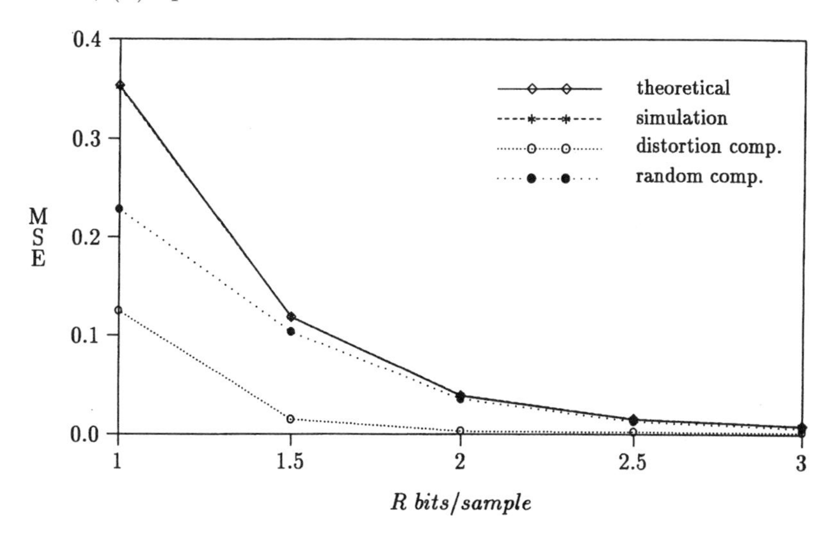

Figure 4.15(a): Theoretical and simulation results of the total output MSE with distortion and random components for the paraunitary FB at ρ=0.95 (b) MSE of optimally compensated, s_i=1, and null compensated, $s_i = 1/\alpha_i$ structures (designed for ρ=0.95) versus ρ for paraunitary FB with AR(1) signal input, σ_x^2=1, R_0=3, R_1=1.

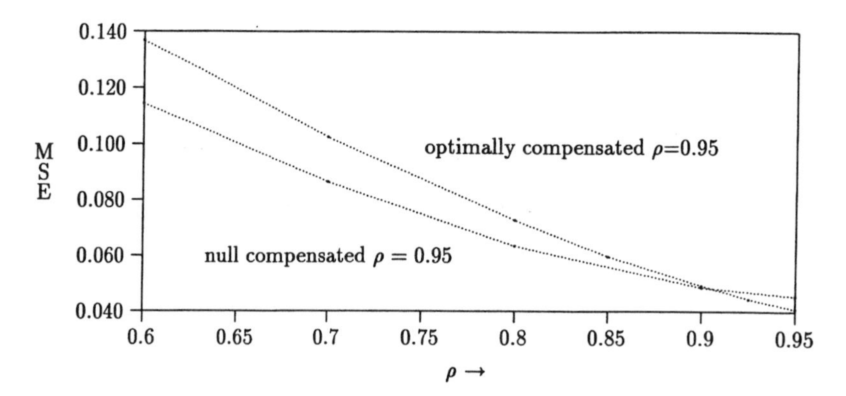

Figure 4.15(b): Theoretical and simulation results of the total output MSE with distortion and random components for the paraunitary FB at ρ=0.95 (b) MSE of optimally compensated, s_i=1, and null compensated, $s_i = 1/\alpha_i$ structures (designed for ρ=0.95) versus ρ for paraunitary FB with AR(1) signal input, σ_x^2=1, R_0=3, R_1=1.

Aliasing energy in a subband tree structure was defined and analyzed along with a new performance measure, the nonaliasing energy ratio (NER). These measures demonstrate that filter banks outperform block transforms for the examples and signal sources under consideration. On the other hand, the time and frequency characteristics of functions or filters are examined and comparisons made between block transforms, hierarchical subband trees, and direct M-band paraunitary filter banks.

We presented a methodology for rigorous modeling and optimal compensation for quantization effects in M-band codecs, and showed how an MSE metric can be minimized subject to paraunitary constraints.

We will present the theory of wavelet transforms in Chapter 6. There we will see that the two-band paraunitary PR-QMF is the basic ingredient in the design of the orthonormal wavelet kernel, and that the dyadic subband tree can provide the fast algorithm for wavelet transform with *proper initialization.* The Binomial-QMF developed in this chapter is the unique maximally flat magnitude square two-band unitary filter. In Chapter 6, it will be identified as a wavelet filter and thus provides a specific example linking subbands and orthonormal wavelets.

References

A. N. Akansu, "Multiplierless Suboptimal PR-QMF Design," Proc. SPIE Visual Communication and Image Processing, Vol. 1818, pp. 723–734, Nov. 1992.

A. N. Akansu, "Some Aspects of Optimal Filter Bank Design for Image-Video Coding," 2nd NJIT Symp. on Multiresolution Image and Video Processing: Subbands and Wavelets, March 1992.

A. N. Akansu and H. Caglar, "A Measure of Aliasing Energy in Multiresolution Signal Decomposition," Proc. IEEE ICASSP, pp. IV 621–624, 1992.

A. N. Akansu, and Y. Liu, "On Signal Decomposition Techniques," Optical Engineering, pp. 912–920, July 1991.

A. N. Akansu, R. A. Haddad, and H. Caglar, "Perfect Reconstruction Binomial QMF-Wavelet Transform," Proc. SPIE Visual Communication and Image Processing, Vol. 1360, pp. 609–618, Oct. 1990.

A. N. Akansu, R. A. Haddad, and H. Caglar, "The Binomial QMF-Wavelet Transform for Multiresolution Signal Decomposition," IEEE Trans. on Signal Processing, Vol. 41, No. 3, pp. 13–20, Jan. 1993.

R. Ansari, C. Guillemot, and J. F. Kaiser, "Wavelet Construction Using Lagrange Halfband Filters," IEEE Trans. Circuits and Systems, Vol. CAS-38, pp. 1116–1118, Sept. 1991.

M. Antonini, M. Barlaud, P. Mathieu, I. Daubechies, "Image Coding Using Vector Quantization in the Wavelet Transform Domain," Proc. ICASSP, pp. 2297–2300, 1990.

T. Berger, *Rate Distortion Theory.* Prentice-Hall, Englewood Cliffs NJ, 1971.

H. Caglar and A. N. Akansu, "PR-QMF Design with Bernstein Polynomials," Proc. IEEE ISCAS, pp. 999–1002, 1992.

H. Caglar, Y. Liu, and A. N. Akansu, "Statistically Optimized PR-QMF Design," Proc. SPIE Visual Communication and Image Processing, pp. 86–94, Nov. 1991.

E. W. Cheney, *Introduction to Approximation Theory,* 2nd edition. Chelsea, New York, 1981.

R. J. Clarke, *Transform Coding of Images.* Academic Press, New York, 1985.

I. Daubechies, "Orthonormal Bases of Compactly Supported Wavelets," Communications on Pure and Applied Math., Vol. XLI, pp. 909–996, 1988.

I. Daubechies, "Orthonormal Bases of Compactly Supported Wavelets. II. Variations on a Theme," Technical Memo #11217-891116-17, AT&T Bell Labs., Murray Hill, 1988.

P. J. Davis, *Interpolation and Approximation.* Ginn-Blaisdell, 1963.

D. E. Dudgeon and R. M. Mersereau, *Multidimensional Digital Signal Processing.* Prentice-Hall, 1984.

D. Esteban and C. Galand, "Application of Quadrature Mirror Filters to Split-band Voice Coding Schemes," Proc. ICASSP, pp. 191–195, 1977.

H. Gharavi and A. Tabatabai, "Sub-band Coding of Monochrome and Color Images," IEEE Trans. on Circuits and Systems, Vol. CAS-35, pp. 207–214, Feb. 1988.

C. Gonzales, E. Viscito, T. McCarthy, D. Ramm, and L. Allman, "Scalable Motion-Compensated Transform Coding of Motion Video: A Proposal for the ISO/MPEG-2 Standard," IBM Research Report, RC 17473, Dec. 9, 1991.

K. Gosse and P. Duhamel, "Perfect Reconstruction vs. MMSE Filter Banks in Source Coding," IEEE Trans. Signal Processing, Vol. 45, No. 9, pp. 2188–2202, Sept. 1997.

R. A. Haddad, "A Class of Orthogonal Nonrecursive Binomial Filters," IEEE Trans. Audio and Electroacoustics, pp. 296–304, Dec. 1971.

R. A. Haddad and A. N. Akansu, "A Class of Fast Gaussian Binomial Filters for Speech and Image Processing," IEEE Trans. on Signal Processing, Vol. 39, pp. 723–727, March 1991.

R. A. Haddad, and B. Nichol, "Efficient Filtering of Images Using Binomial Sequences," Proc. IEEE ICASSP, pp. 1590–1593, 1989.

R. A. Haddad and K. Park, "Modeling, Analysis, and Optimum Design of Quantized M-Band Filter Banks, IEEE Trans. on Signal Processing, Vol. 43, No. 11, pp. 2540–2549, Nov. 1995.

R. A. Haddad and N. Uzun, "Modeling, Analysis and Compensation of Quantization Effects in M-band Subband Codecs", in IEEE Proc. ICASSP, Vol. 3, pp. 173–176, May 1993.

O. Herrmann, "On the Approximation Problem in Nonrecursive Digital Filter Design," IEEE Trans. Circuit Theory, Vol. CT-18, No. 3, pp. 411–413, May 1971.

J. J. Y. Huang and P. M.Schultheiss, "Block Quantization of Correlated Gaussian Random Variables," IEEE Trans. Comm., pp. 289–296, Sept. 1963.

N. S. Jayant and P. Noll, *Digital Coding of Waveforms.* Prentice-Hall Inc., 1984.

J. D. Johnston, "A Filter Family Designed for Use in Quadrature Mirror Filter Banks," Proc. ICASSP, pp. 291–294, 1980.

J. Katto and Y. Yasuda, "Performance Evaluation of Subband Coding and Optimization of its Filter Coefficients," Proc. SPIE Visual Communication and Image Processing, pp. 95–106, Nov. 1991.

J. Kovacevic, "Eliminating Correlated Errors in Subband and Wavelet Coding System With Quantization," Asilomar Conf. Signals, Syst., Comput., pp. 881–885, Nov. 1993.

D. LeGall and A. Tabatabai, "Sub-band Coding of Digital Images Using Symmetric Short Kernel Filters and Arithmetic Coding Techniques," Proc. IEEE ICASSP, pp. 761–764, 1988.

S. P. Lloyd, "Least Squares Quantization in PCM," Inst. Mathematical Sciences Meeting, Atlantic City, NJ, Sept. 1957.

G. G. Lorentz, *Bernstein Polynomials.* University of Toronto Press, 1953.

J. Max, "Quantization for Minimum Distortion," IRE Trans. Information Theory, Vol. IT-6, pp. 7–12, Mar. 1960.

J. A. Miller, "Maximally Flat Nonrecursive Digital Filters," Electronics Letters, Vol. 8, No. 6, pp. 157–158, March 1972.

F. Mintzer, "Filters for Distortion-Free Two-Band Multirate Filter Banks," IEEE Trans. ASSP, Vol. ASSP-33, pp. 626–630, June 1985.

F. Mintzer and B. Liu, "Aliasing Error in the Design of Multirate Filters," IEEE Trans. ASSP, Vol. ASSP-26, pp. 76–88, Feb. 1978.

A. Papoulis, *Probability, Random Variables and Stochastic Processes.* 3rd edition, McGraw-Hill, 1991.

K. Park, *Modeling, Analysis and Optimum Design of Quantized M-channel Subband Codecs.* Ph.D. Thesis, Polytechnic Univ., Brooklyn, NY, Dec. 1993.

K. Park and R. A. Haddad, "Optimum Subband Filter Bank Design and Compensation in Presence of Quantizers," Proc. 27th Asilomar Conf. Sign. Syst. Comput. Pacific Grove, CA, Nov. 1993.

K. Park and R. A. Haddad, "Modeling and Optimal Compensation of Quantization in Multidimensional M-band Filter Bank", in Proc. ICASSP, Vol. 3, pp. 145–148, April 1994.

J. P. Princen and A. B. Bradley, "Analysis/Synthesis Filter Bank Design Based on Time Domain Aliasing Cancellation," IEEE Trans. ASSP, Vol. ASSP-34, pp. 1153–1161, Oct. 1986.

J. P. Princen, A. W. Johnson, and A. B. Bradley, "Subband/Transform Coding Using Filter Bank Designs Based on Time Domain Aliasing Cancellation," Proc. IEEE ICASSP, pp. 2161–2164, April 1987.

L. R. Rajagopaland, S. C. Dutta Roy, "Design of Maximally Flat FIR Filters Using the Bernstein Polynomial," IEEE Trans. Circuits and Systems, Vol. CAS-34, No. 12, pp. 1587–1590, Dec. 1987.

M. J. T. Smith and T. P. Barnwell, "A Procedure for Designing Exact Reconstruction Filter Banks for Tree-Structured Subband Coders," Proc. IEEE ICASSP, pp. 27.1.1–27.1.4, 1984.

M. J. T. Smith and T. P. Barnwell, "Exact Reconstruction Techniques for Tree-Structured Subband Coders," IEEE Trans. ASSP, pp. 434–441, 1986.

A. Tabatabai, "Optimum Analysis/Synthesis Filter Bank Structures with Application to Subband Coding Systems", Proc. IEEE ISCAS, pp. 823–826, 1988.

P. P. Vaidyanathan, and P. Q. Hoang, "Lattice Structures for Optimal Design and Robust Implementation of Two-band Perfect Reconstruction QMF Banks," IEEE Trans. ASSP, Vol. ASSP-36, No.1, pp. 81–94, Jan. 1988.

N. Uzun and R. A. Haddad, "Modeling and Analysis of Quantization Errors in Two Channel Subband Filter Structures," Proc. SPIE Conf. on Visual Comm. and Image Proc., pp. 1446–1457, Nov. 1992.

N. Uzun and R.A Haddad, "Modeling and Analysis of Floating Point Quantization Errors in Subband Filter Structures," Proc. SPIE Conf. on Visual Comm. and Image Proc., pp. 647–653, Nov. 1993.

N. Uzun and R.A Haddad, "Cyclostationary Modeling, Analysis and Optimal Compensation of Quantization Errors in Subband Codecs," IEEE Trans. Signal Processing, Vol. 43, pp. 2109–2119, Sept. 1995.

L. Vandendorpe "Optimized Quantization for Image Subband Coding," Signal Processing, Image Communication, Vol. 4, No. 1, pp. 65–80, Nov. 1991.

E. Viscito and J. Allebach, "The Design of Equal Complexity FIR Perfect Reconstruction Filter Banks Incorporating Symmetries," Tech. Rep., TR-EE 89–27, Purdue Univ., May 1989.

P. H. Westerink, J. Biemond, and D. E. Boekee, "Scalar Quantization Error Analysis for Image Subband Coding Using QMF's," IEEE Transactions on Signal Processing, Vol. 40, No. 2, pp. 421–428, Feb. 1992.

J. W. Woods, Ed., *Subband Image Coding.* Kluwer, 1991.

J. W. Woods and T. Naveen, "A Filter Based Bit Allocation Scheme for Subband Compression of HDTV," IEEE Trans. Image Processing, Vol. 1, No. 3, pp. 436–440, July 1992.

J. W. Woods and S. D. O'Neil, "Subband Coding of Images," IEEE Trans. ASSP, Vol. ASSP-34, No. 5, Oct. 1986.

Chapter 5

Time-Frequency Representations

5.1 Introduction

Time- and frequency-domain characterizations of a signal are not only of classical interest in filter design (Papoulis, 1977) but often dictate the nature of the processing in contemporary signal processing (speech, image, video, etc.). Often signal operations can be performed more efficiently in one domain than the other. By this we imply operations such as compression, excision, modulation, and feature extraction.

Of special interest are nonstationary signals, that is, signals whose salient features change with time. For such signals, we will demonstrate that classical Fourier analysis is inadequate in highlighting local features of a signal.

What is needed is a kernel capable of concentrating its strength over segments in time and segments in frequency so as to allow localized feature extraction. The short-time Fourier (or Gabor) transform and the wavelet transform have this capability for continuous-time signals.

In this chapter, we focus on the description and evaluation of techniques for achieving time-frequency localization on discrete-time signals. We hope to provide the reader with an exposure to current literature on the subject and to serve as a prelude to the wavelet and applications chapters which follow.

First we review the classical analog uncertainty principle and the short-time Fourier transform. Then we develop the discrete-time counterparts to these and show how the binomial sequences emulate the continuous-time Gaussian functions. Following this introduction, we define, calculate, and compare localization features of filter banks and standard block transforms and explore the role of tree-structured filter banks in achieving desired time-frequency resolution. Then we conclude with a section on achieving arbitrary "tiling" of the time-frequency plane using block transforms and demonstrate the utility of this approach with applications to signal compaction and to interference excision in spread spectrum communications systems.

A word on the notation used in this chapter is in order. The terms Z, R, and R^+ denote the set of integers, real numbers, and positive real numbers, respectively; $L^2(R)$ denotes the Hilbert space of measurable, square-integrable functions, i.e., the space of what are termed finite energy signals $f(t)$, or sequences $f(n)$ satisfying

$$\int_{-\infty}^{\infty} |f(t)|^2 dt < \infty, \quad \sum_k |f(k)|^2 < \infty.$$

All one-dimensional functions dealt with in this chapter are assumed to have finite energy. Also, the inner product of two functions is denoted by

$$< f, g > = \int_{-\infty}^{\infty} f(t) g^*(t) dt$$

$$< f, g > = \sum_k f(k) g^*(k).$$

5.2 Analog Background— Time Frequency Resolution

A basic objective in signal analysis is to devise an operator capable of extracting local features of a signal in both time- and frequency-domains. This requires a kernel whose extent or spread is simultaneously narrow in both domains. That is, the transformation kernel $\phi(t)$ and its Fourier transform $\Phi(\Omega)$ should have narrow spreads about selected points t_k, Ω_k in the time-frequency plane. However, the uncertainty principle described below bounds the simultaneous realization of these desiderata. Narrowness in one domain necessarily implies a wide spread in the other.

Standard Fourier analysis decomposes a signal into frequency components and determines the relative strength of each component. It does not tell us *when* the signal exhibited the particular frequency characteristic, since the Fourier kernel $e^{j\Omega t}$ is spread out evenly in time. It is not time-limited.

If the frequency content of the signal were to vary substantially from interval to interval as in a musical scale, the standard Fourier transform

$$F(\Omega) = \int_{-\infty}^{\infty} f(t)\, e^{-j\Omega t} dt \quad \leftrightarrow \quad f(t) = \frac{1}{2\pi} \int_{-\infty}^{\infty} F(\Omega) e^{j\Omega t} d\Omega \qquad (5.1)$$

would sweep evenly over the entire time axis and wash out any local anomalies of the signal (e.g., short duration bursts of high-frequency energy). It is clearly not suitable for nonstationary signals.

Confronted with this challenge, Gabor (1946) resorted to the windowed, short-time Fourier transform (STFT), which moves a fixed-duration window over the time function and extracts the frequency content of the signal within that interval. This would be suitable, for example, for speech signals which generally are locally stationary but globally nonstationary.

The STFT positions a window $g(t)$ at some point τ on the time axis and calculates the Fourier transform of the signal contained within the spread of that window, to wit,

$$F(\Omega, \tau) = \int_{-\infty}^{\infty} f(t) g^*(t - \tau) e^{-j\Omega t} dt. \qquad (5.2)$$

When the window $g(t)$ is Gaussian, the STFT is called the Gabor transform (Gabor, 1946). The STFT basis functions are generated by *modulation* and *translation* of the window function $g(t)$ by parameters Ω and τ, respectively. Typical Gabor basis functions and their associated transforms are shown in Fig. 5.1.

The window function is also called a *prototype* function, or sometimes, a *mother* function. As τ increases this mother function simply translates in time keeping the time-spread of the function constant. Similarly, as seen in Fig. 5.1, as the modulation parameter Ω_k increases, the transform of the mother function also, simply, translates in frequency, keeping a constant bandwidth.

The difficulty with the STFT is that the fixed-duration window $g(t)$ is accompanied by a fixed frequency resolution and thus allows only a fixed time-frequency resolution. This is a consequence of the classical uncertainty principle (Papoulis, 1977). This theorem asserts that for any function $\phi(t)$ with Fourier transform $\Phi(\Omega)$, (and with $\sqrt{t}\phi(t) \to 0$, as $t \to \mp\infty$) it can be shown that

$$\sigma_T \sigma_\Omega \geq 1/2 \qquad (5.3)$$

where σ_T and σ_Ω are, respectively, the RMS spreads of $\phi(t)$ and $\Phi(\Omega)$ around the center values. That is,

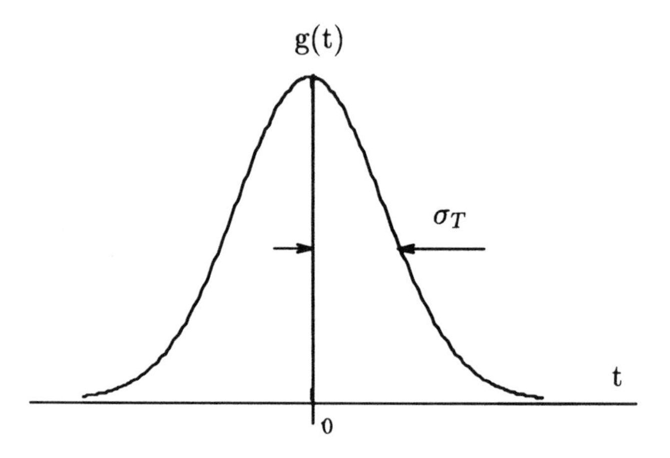

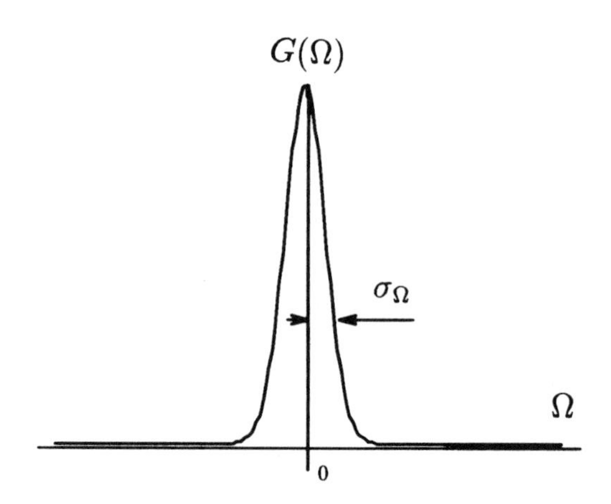

Figure 5.1: Typical basis functions for STFTs and their Fourier transforms.

$$\sigma_T^2 = \frac{\int_{-\infty}^{\infty}(t-\bar{t})^2|\phi(t)|^2 dt}{E}, \tag{5.4}$$

$$\sigma_\Omega^2 = \frac{\frac{1}{2\pi}\int_{-\infty}^{\infty}(\Omega-\bar{\Omega})^2|\Phi(\Omega)|^2 d\Omega}{E}, \tag{5.5}$$

where E is the energy in the signal,

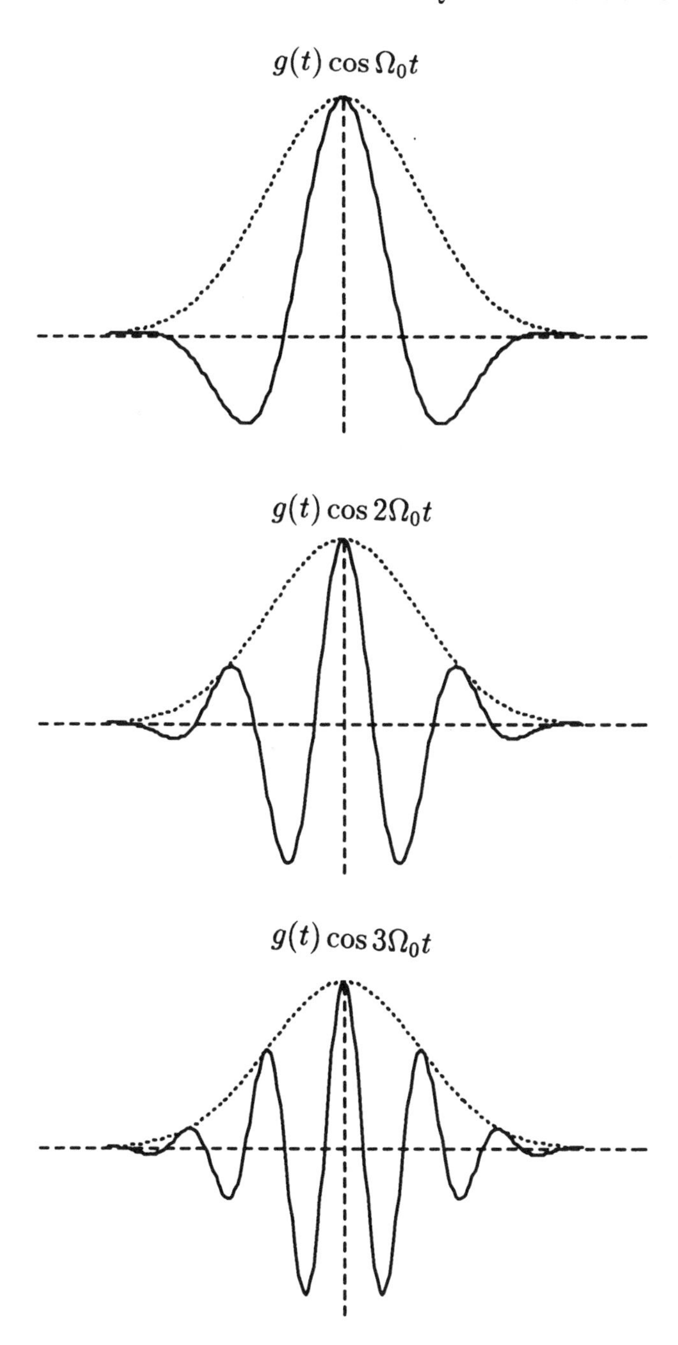

Figure 5.1: (continued)

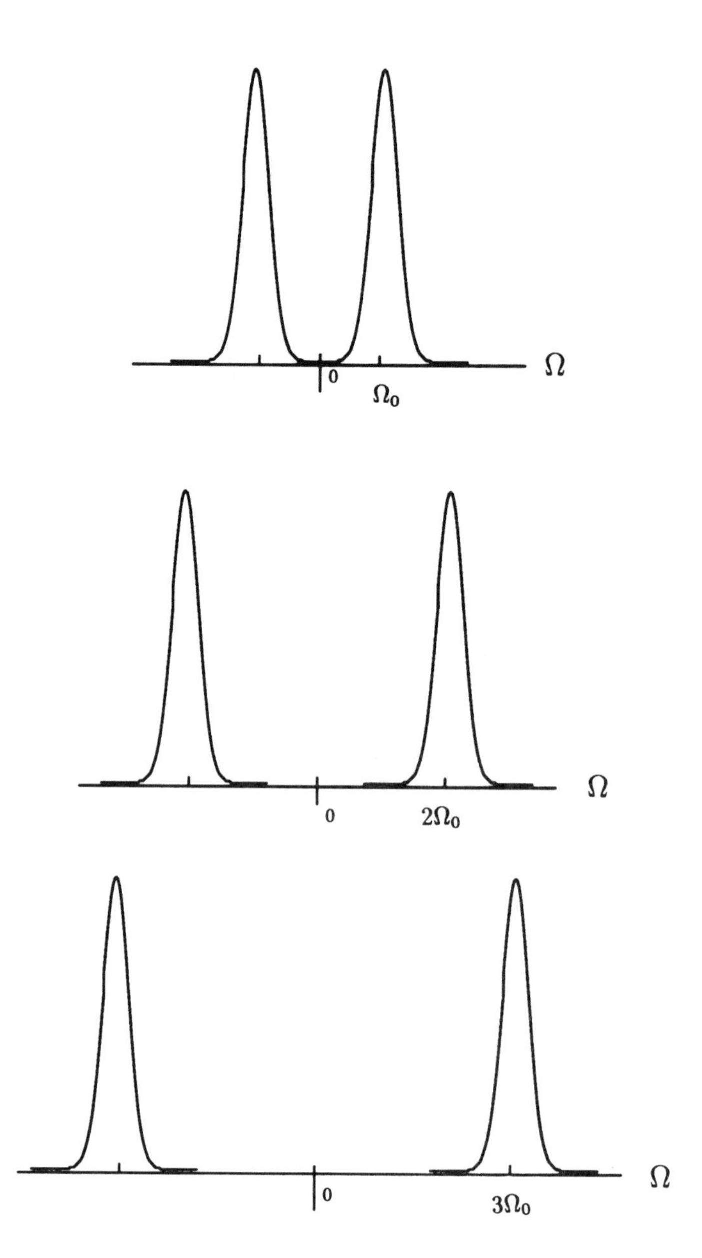

Figure 5.1: (continued)

$$E = \int_{-\infty}^{\infty} |\phi(t)|^2 dt = \frac{1}{2\pi} \int_{-\infty}^{\infty} |\Phi(\Omega)|^2 d\Omega \tag{5.6}$$

and $\bar{t}$ and $\bar{\Omega}$ refer to the center of mass of these kernels,

$$\bar{t} = \frac{\int_{-\infty}^{\infty} t|\phi(t)|^2 dt}{E}, \quad \bar{\Omega} = \frac{\frac{1}{2\pi}\int_{-\infty}^{\infty} \Omega|\Phi(\Omega)|^2 d\Omega}{E}. \tag{5.7}$$

The equal sign holds in Eq. (5.3) if and only if $\phi(t)$ (and consequently its Fourier transform $\Phi(\Omega)$) is Gaussian, of the form $exp(-\alpha t^2)$. The product $\sigma_T \sigma_\Omega$ is called the resolution cell and is a characteristic of the kernel, $\phi(t)$. Let $g(t) \leftrightarrow G(\Omega)$ be a Fourier transform pair and, for convenience, assume that $\bar{t} = 0$ and $\bar{\Omega} = 0$. Then the translated, modulated kernel pair are given by

$$g_{\tau,\beta}(t) = g(t-\tau)e^{j\beta t} \leftrightarrow G_{\tau,\beta}(\Omega) = e^{-j(\Omega-\beta)\tau} G(\Omega - \beta). \tag{5.8}$$

This two-parameter family is centered at (τ, β) in the time-frequency plane, i.e.,

$$\bar{t}_{\tau,\beta} = \tau \quad and \quad \bar{\Omega}_{\tau,\beta} = \beta. \tag{5.9}$$

Now, it is readily shown that the spread of this shifted, modulated kernel is constant in both domains, i.e.,

$$\sigma^2_{\Omega(\tau,\beta)} = \frac{\frac{1}{2\pi}\int_{-\infty}^{\infty} (\Omega-\beta)^2 |G_{\tau,\beta}(\Omega)|^2 d\Omega}{E} = \sigma^2_\Omega \tag{5.10}$$

$$\sigma^2_{T(\tau,\beta)} = \frac{\frac{1}{2\pi}\int_{-\infty}^{\infty} (t-\tau)^2 |g_{\tau,\beta}(t)|^2 dt}{E} = \sigma^2_T \tag{5.11}$$

where σ_Ω, σ_T are the RMS spreads of the unmodulated, untranslated kernels, $g(t) \leftrightarrow G(\Omega)$.

Each element σ_Ω and σ_T of the resolution cell $\sigma_\Omega \sigma_T$ is constant for any frequency Ω and time shift τ as indicated by the rectangles of fixed area and shape in the "tiling" pattern of Fig. 5.2. Any trade-off between time and frequency must be accepted for the entire (Ω, τ) plane.

The wavelet transform, on the other hand, is founded on basis functions formed by *dilation* and *translation* of a prototype function $\psi(t)$. These basis functions are short-duration, high-frequency and long-duration, low-frequency functions. They are much better suited for representing short bursts of high-frequency signals or long-duration, slowly varying signals than the STFT.

This concept is suggested by the *scaling* property of Fourier transforms. If

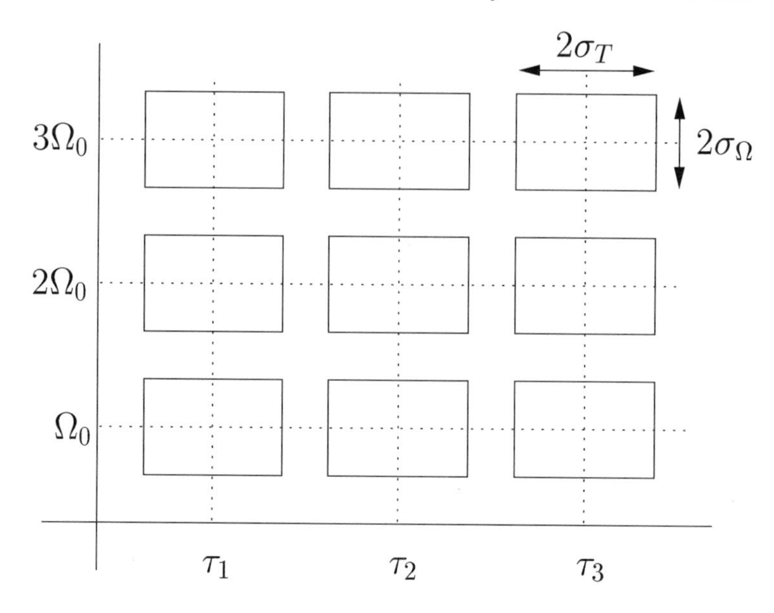

Figure 5.2: Time-frequency plane showing tiling pattern of resolution cells for STFT at times τ_1, τ_2, τ_3.

$$\psi(t) \leftrightarrow \Psi(\Omega)$$

constitute a Fourier transform pair, then

$$\frac{1}{\sqrt{a}}\psi(\frac{t}{a}) \leftrightarrow \sqrt{a}\Psi(a\Omega) \tag{5.12}$$

where $a > 0$ is a continuous variable. Thus a contraction in one domain is accompanied by an expansion in the other, but in a nonuniform way over the time-frequency plane. A typical wavelet and its dilations are shown in Fig. 5.3, along with the corresponding Fourier transforms.

The frequency responses are shown in Fig. 5.4 on a *logarithmic* frequency scale. These are known in the electrical engineering community as *constant* Q resonant circuits, which means that the ratio of RMS bandwidth to center frequency is constant. Alternatively, the RMS bandwidth is constant on the *logarithmic* scale. This may be contrasted with the STFT where the RMS bandwidth is constant on a *linear* scale.

The wavelet family is thus defined by scale and shift parameters a, b as

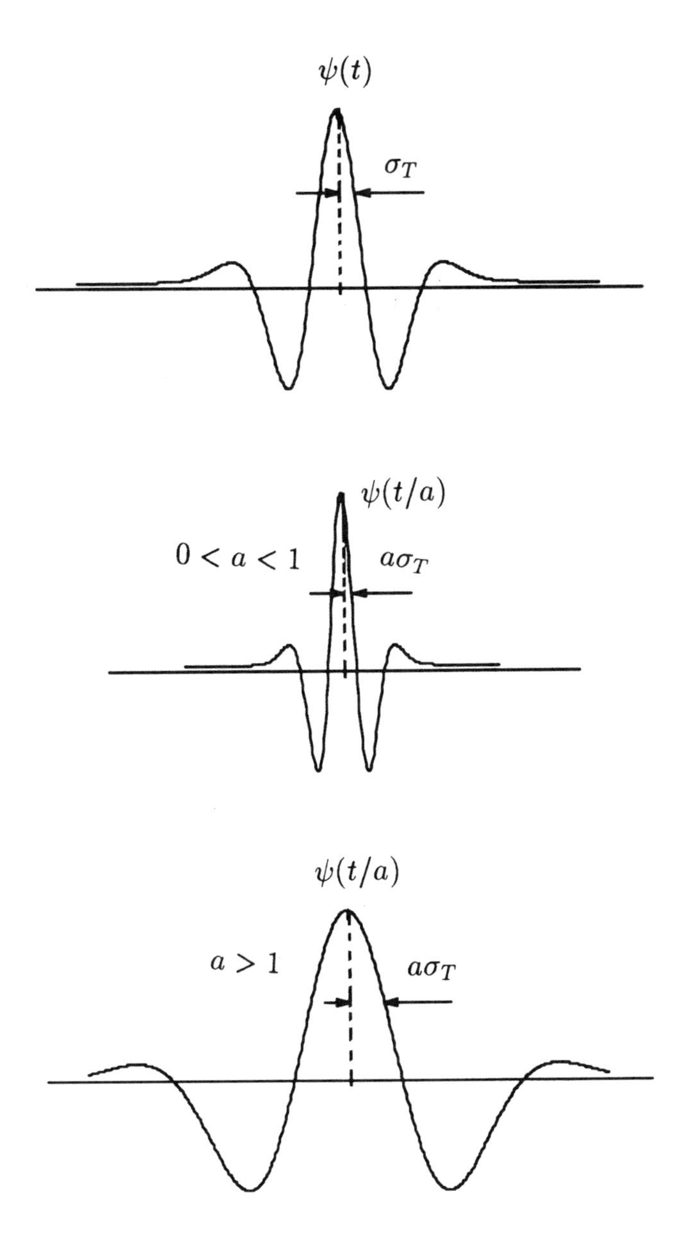

Figure 5.3: Typical wavelet family in time and frequency domains.

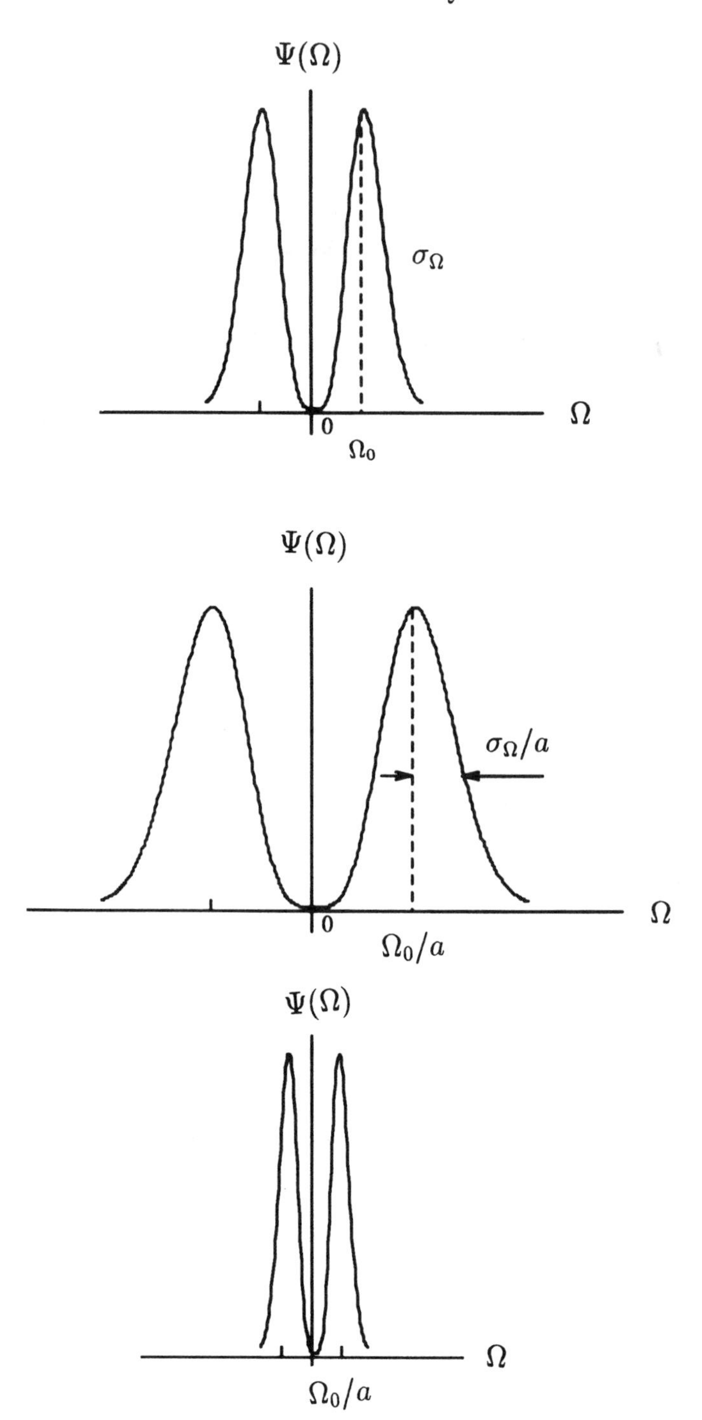

Figure 5.3: (continued)

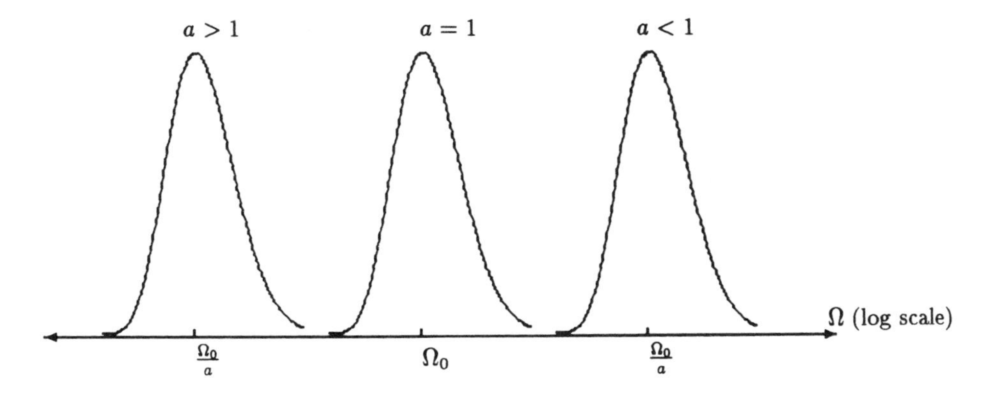

Figure 5.4: Wavelet band-pass filters on a logarithmic frequency scale.

$$\psi_{ab}(t) = \frac{1}{\sqrt{a}}\psi(\frac{t-b}{a}) \tag{5.13}$$

and the wavelet transform is the inner product

$$W(a,b) = \int_{-\infty}^{\infty} \psi_{ab}(t) f^*(t) dt \ = < \psi_{ab}, f > \tag{5.14}$$

where $a \in R^+, \ \ b \in R$.

For large a, the basis function becomes a stretched version of the prototype wavelet, that is, a low-frequency function, while for small a, this basis function is a contracted version of the wavelet function, which is a short time duration, high-frequency function. Depending on the scaling parameter a, the wavelet function $\psi(t)$ dilates or contracts in time, causing the corresponding contraction or dilation in the frequency domain. Thus, the wavelet transform provides a flexible time-frequency resolution. Figure 5.5 displays the time-frequency plane showing resolution cells for the wavelet transform.

5.3 The Short-Time Fourier Transform

In this section we continue with the (continuous) windowed Fourier transform and show how sampling the two-dimensional surface of $F(\Omega, \tau)$ on a rectangular grid yields the discrete STFT of a continuous time function, $f(t)$. Finally, when the time function $f(t)$ is sampled and then transformed, the resulting discrete-time STFT is seen to be the familiar DFT.

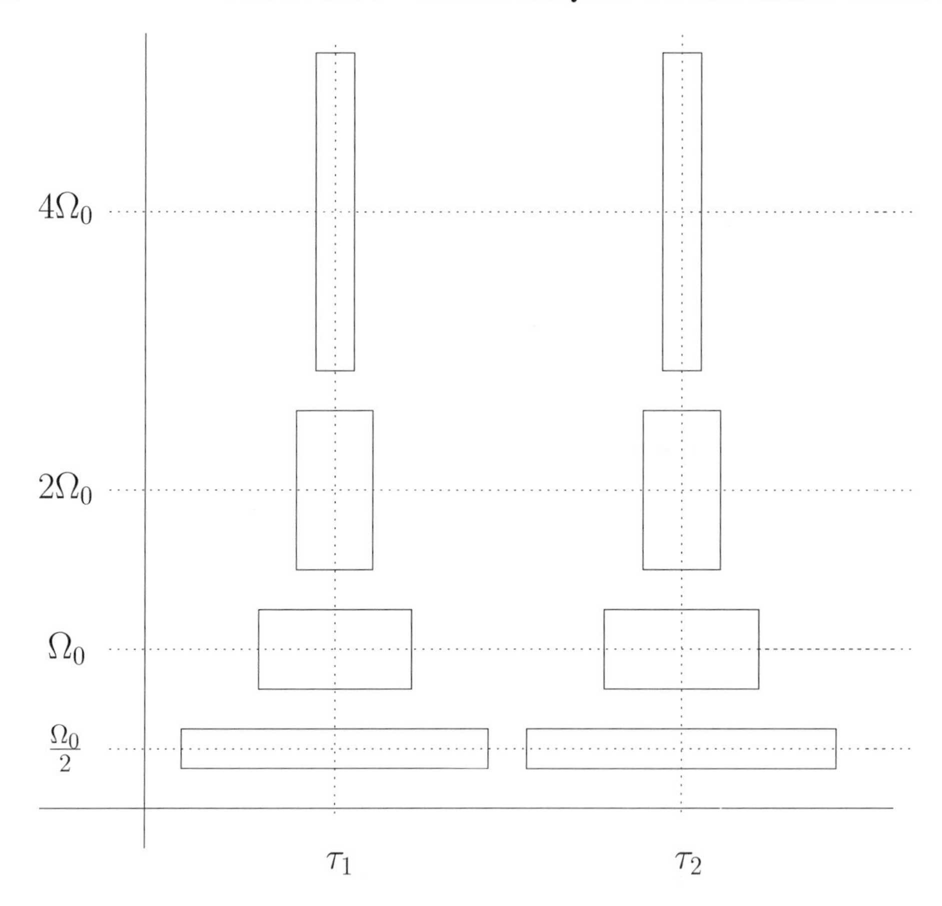

Figure 5.5: Time-frequency plane showing resolution cells for wavelet transform.

5.3.1 The Continuous STFT

We defined the short-time Fourier transform, or STFT, in Eq. (5.2) as a mapping of a function of one continuous variable $f(t)$ into a function $F(\Omega, \tau)$ of two continuous variables. Consequently, the STFT is a highly redundant mapping of $L^2(R) \rightarrow L^2(R^2)$. This transform, $F(\Omega, \tau)$, is a surface in the (Ω, τ) time-frequency plane. If we hold Ω constant and vary τ, we are examining the strength of the signal in a band of frequencies centered at Ω as a function of time. This corresponds to taking a slice of the $F(\Omega, \tau)$ surface parallel to the τ axis. Holding τ fixed and varying Ω gives the windowed (or short-time) Fourier transform, at that instant.

The contour of peak magnitude of $F(\Omega, \tau)$ tracks the frequencies of highest energy as a function of time. This corresponds to tracking the resolution cell with maximum energy in the time frequency plane. The Doppler shift in signals from sources in motion can be tracked in this manner.

The STFT can be regarded as an inner product, or equivalently as a convolution. To summarize the formulations, for $f(t) \in L^2(R)$ the inner product form is

$$F(\Omega,\tau) = < f(t), g(t-\tau)e^{-j\Omega t} > = \int_{-\infty}^{\infty} f(t)g^*(t-\tau)e^{-j\Omega t}dt. \tag{5.15}$$

Equivalently, the STFT can be represented as the convolution of $\tilde{g}(t) \triangleq g^*(-t)$ with the modulated signal $e^{-j\Omega t}f(t)$,

$$F(\Omega,t) = \tilde{g}(t) * e^{-j\Omega t}f(t). \tag{5.16}$$

This view is represented by Fig. 5.6. In this case, Ω is fixed, and the filter output tracks a particular frequency in time.

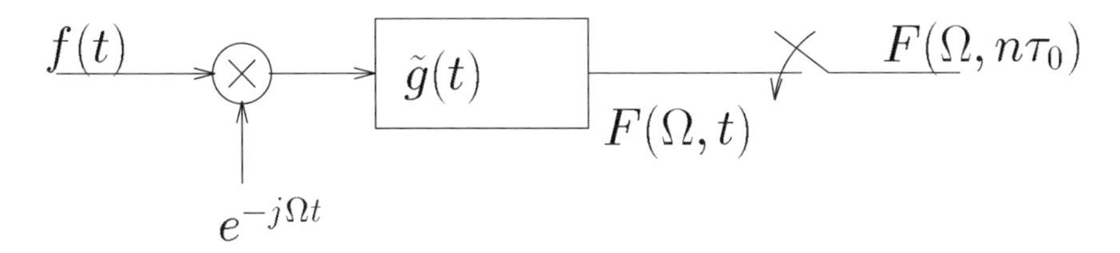

Figure 5.6: The STFT as a convolution.

It can be shown that (Prob.s 5.2, 5.3) the reconstruction formula and Parseval's relation are, respectively,

$$f(t) = \frac{1}{2\pi}\int_{-\infty}^{\infty}\int_{-\infty}^{\infty} F(\Omega,\tau)g(\tau-t)e^{j\Omega t}d\Omega d\tau \tag{5.17}$$

$$\int_{-\infty}^{\infty} |f(t)|^2\,dt = \frac{1}{2\pi}\int_{-\infty}^{\infty}\int_{-\infty}^{\infty} |F(\Omega,\tau)|^2 d\Omega d\tau. \tag{5.18}$$

5.3.2 The Discrete STFT

We can develop an appreciation of the time-frequency localization properties of the short-time Fourier transform by discretizing (Ω,τ) or, in effect, sampling the continuous STFT, Eq. (5.2), on a uniform grid to obtain the discrete STFT. Thus, with

$$\Omega = m\,\Omega_0, \quad \tau = n\,\tau_0 \tag{5.19}$$

we obtain

$$F(m,n) = \int_{-\infty}^{\infty} f(t) g^*(t - n\tau_0) e^{-jm\Omega_0 t}\, dt = < f(t),\, e^{jm\Omega_0 t} g(t - n\tau_0) > . \tag{5.20}$$

This is also the inner product between $f(t)$ and the windowed sinusoid $e^{jm\Omega_0 t}$ over the spread of $g(t - n\tau_0)$. If $f(t)$ has sinusoids at or near the frequency $m\Omega_0$ within the window, the inner product is large. If it has sinusoids different from $m\Omega_0$, the inner product or $F(m,n)$ centered at the resolution cell $m\Omega_0$, $n\tau_0$ will be small. For each window location $n\tau_0$, we can calculate $F(m,n)$ at different frequencies $m\Omega_0$. Then we shift the window location to $(n+1)\tau_0$ and repeat the scanning with frequency. This process generates a two-parameter family $F(m,n)$ that can be plotted on a time-frequency grid as shown in Fig. 5.7. We can then imagine interpolating all these grid points (or in effect letting $m\Omega_0 \rightarrow \Omega$ and $n\tau_0 \rightarrow \tau$) to obtain the two-dimensional surface $F(\Omega,\tau)$ in the (Ω,τ) plane.

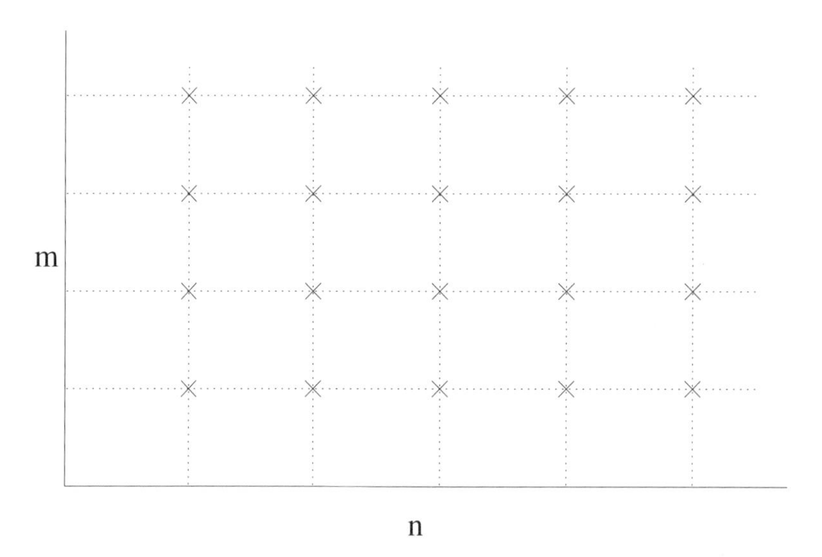

Figure 5.7: Sampling grid for $F(m,n)$, the discrete STFT.

The discrete STFT can be conceptually measured in Fig. 5.6 if Ω is discretized to $m\Omega_0$ and the output sampled at $t = nT_0$. The resulting output is $F(m,n)$ for

m fixed and variable n. That is, we sample $F(\Omega, \tau)$ at grid points indicated in Fig. 5.7.

The coefficients $F(m, n)$ constitute the discrete STFT of the given signal $f(t)$. But how coarsely can we sample $F(\Omega, \tau)$ and still retain enough features of the signal to enable a reconstruction? The answer is rather simple. If $\Omega_0 \tau_0 < 2\pi$, and $g(t) \in L^2(R)$ (Daubechies, 1990), we can (Prob. 5.4) reconstruct $f(t)$ via

$$f(t) = (\frac{\Omega_0 \tau_0}{2\pi}) \sum_m \sum_n F(m, n) e^{jm\Omega_0 t} g(t - n\tau_0). \tag{5.21}$$

The set of functions $g_{m,n}(t) = g(t - n\,\tau_0)\, e^{jm\Omega_0 t}$ is therefore complete, but not linearly independent. The latter implies that there is redundancy in the transform $F(m, n)$. As we try to reduce the redundancy in the transform in $F(m, n)$ by enlarging the sampling grid to force $\Omega_0 \tau_0 \to 2\pi$, the set of functions $g_{m,n}(t)$ approaches an orthonormal basis. But, it is found that such an orthonormal basis is badly localized in either time or frequency. Consequently, an *oversampled* STFT is preferred. As we shall see in Chapter 6 the wavelet transform suffers from no such handicap.

5.3.3 The Discrete-Time STFT, or DFT

We can carry the discretization even further and suppose that the time function in question itself is sampled, $f(n)$. The discrete version of the STFT is then

$$F(k, n) = \sum_{m=-\infty}^{\infty} g^*(m - n) f(m)\, e^{-j\frac{2\pi}{M}km}, \quad k = 0, 1, ..., M - 1 \tag{5.22}$$

where $g(m)$ is a sampled window function of finite extent (or *compact support* M).

This version permits the filter bank interpretation shown in Fig. 5.8. The output of each filter is subsampled by decimation parameter M to produce a windowed discrete-time Fourier transform. If the window is rectangular on $[0, M - 1]$, the modulated filter bank produces the DFT of the input sequence. The output of this bank can then be plotted on the time-frequency grid of Fig. 5.9. The entries in any column represent the DFT for the corresponding batch of data. The entries along any row show how that particular harmonic varies from batch to batch. The reconstruction formula is a special case of Eq. (5.21).

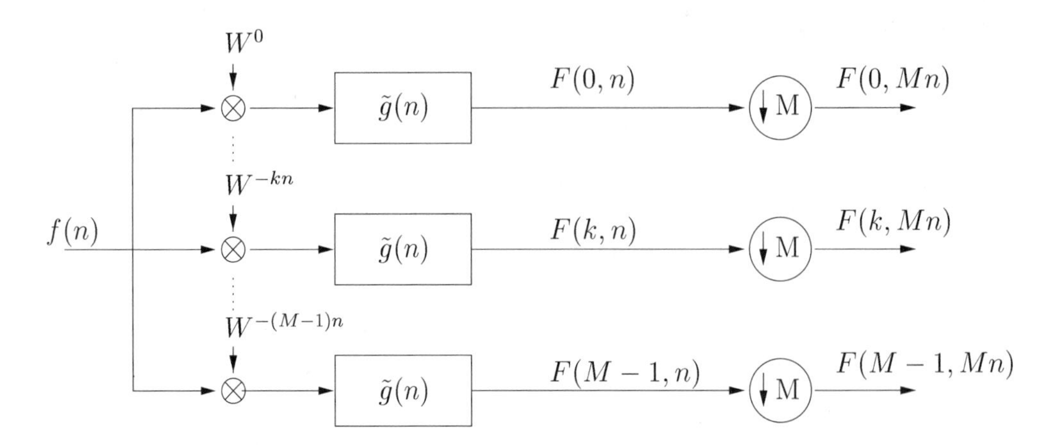

Figure 5.8: The discrete-time STFT as a modulated filter bank.

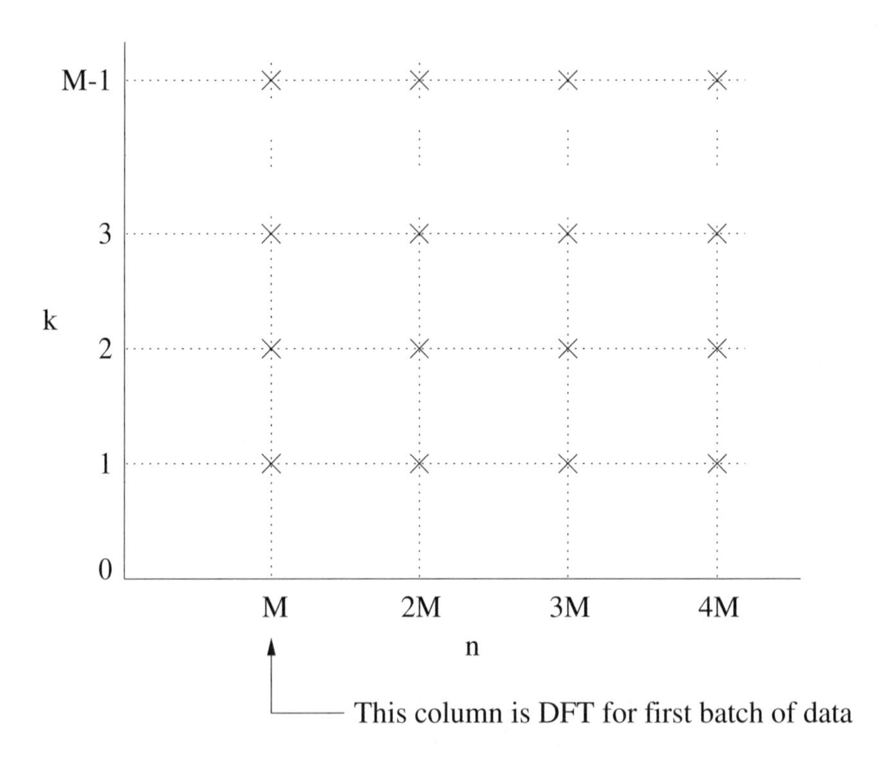

Figure 5.9: The DFT displayed in the time-frequency plane.

5.4 Discrete-Time Uncertainty and Binomial Sequences

In this section, we discuss the discrete-time version of the classical analog uncertainty principle. Then we demonstrate that the binomial sequences are the discrete counterparts to the Gaussian functions which define the optimal resolution cell in the continuous-time case.

5.4.1 Discrete-Time Uncertainty

The discrete-time version of the uncertainty principle is as follows: Let $f(n) \leftrightarrow F(e^{j\omega})$ be a discrete-time Fourier transform pair,

$$F(e^{j\omega}) = \textstyle\sum_{n=-\infty}^{\infty} f(n)e^{-jn\omega}$$

$$f(n) = \frac{1}{2\pi}\int_{-\pi}^{\pi} F(e^{j\omega})e^{jn\omega}\,d\omega. \tag{5.23}$$

By the Parseval theorem, the energy is

$$E = \sum_{n=-\infty}^{\infty} |f(n)|^2 = \frac{1}{2\pi}\int_{-\pi}^{\pi} |F(e^{j\omega}|^2\,d\omega. \tag{5.24}$$

We define the mean (analogous to the center of mass of a distribution) by

$$\bar{\omega} = \frac{\frac{1}{2\pi}\int_{-\pi}^{\pi} \omega|F(e^{j\omega})|^2\,d\omega}{E} \tag{5.25}$$

$$\bar{n} = \frac{\sum_{n=-\infty}^{\infty} n|f(n)|^2}{E}. \tag{5.26}$$

The spread of a function in time and in frequency is

$$\sigma_n^2 = \frac{\sum_{n=-\infty}^{\infty}(n-\bar{n})^2|f(n)|^2}{E}$$

$$\sigma_\omega^2 = \frac{\frac{1}{2\pi}\int_{-\pi}^{\pi}(\omega-\bar{\omega})^2|F(e^{j\omega})|^2\,d\omega}{E}. \tag{5.27}$$

For any real signal $\bar{\omega} = 0$, and without loss of generality, we can also shift the time origin to make $\bar{n} = 0$. For this case, it can be shown that (Haddad, Akansu, and Benyassine, 1993) the time-frequency product $\sigma_n\sigma_\omega$ or resolution cell is given by

$$\sigma_n \sigma_\omega \geq \tfrac{|1-\mu|}{2}$$

$$\mu \triangleq \frac{|F(e^{j\omega})_{\omega=\pi}|^2}{E} = \frac{|F(-1)|^2}{E}. \tag{5.28}$$

In the analog version, $F(\pm\infty) = 0$ and the lower limit is simply 1/2. In the discrete-time case $F(-1)$ need not be zero. Note that in our notation, $F(e^{j\omega})$ at $\omega = 0$ and $\omega = \pi$ are denoted by $F(1)$ and $F(-1)$, respectively.

Remark. The frequency measure in Eq. (5.27) is not suitable for band-pass signals with peak frequency responses centered at $\pm\hat{\omega}$. To obtain a measure of the spread about $\hat{\omega}$, we need to define σ_ω^2 on the interval $[0, \pi]$, rather than $[-\pi, \pi]$. In this case we use

$$\hat{\omega} = \frac{\frac{1}{\pi}\int_0^\pi \omega |F(e^{j\omega})|^2 \, d\omega}{\frac{1}{\pi}\int_0^\pi |F(e^{j\omega})|^2 \, d\omega} \tag{5.29}$$

$$\hat{\sigma}_\omega^2 = \frac{\frac{1}{\pi}\int_0^\pi (\omega - \hat{\omega})^2 |F(e^{j\omega})|^2 \, d\omega}{\frac{1}{\pi}\int_0^\pi |F(e^{j\omega})|^2 \, d\omega} \tag{5.30}$$

and σ_n^2 remains unchanged. It easily follows that

$$\hat{\sigma}_\omega^2 = \sigma_\omega^2 - (\hat{\omega})^2$$

and

$$(\hat{\sigma}_\omega \sigma_n)^2 \geq \frac{1}{4}(1-\mu)^2 - (\hat{\omega})^2 \sigma_n^2. \tag{5.31}$$

Equation (5.31) demonstrates the reduction in the time-frequency product when using the $[0, \pi]$ interval for band-pass signals. An alternative derivation (Haddad, et al., 1993) shows that this product can be expressed as

$$\hat{\sigma}_\omega \sigma_n \geq \tfrac{1}{2}|1 - \mu'|$$

$$\mu' = \frac{\hat{\omega}}{\pi}\frac{|F(1)|^2}{E} + (1 - \frac{\hat{\omega}}{\pi})\frac{|F(-1)|^2}{E}. \tag{5.32}$$

For band-pass signals with zero DC gain, $F(1) = 0$. Equation (5.32) reduces to

$$\mu' = (1 - \frac{\hat{\omega}}{\pi})\frac{|F(-1)|^2}{E}.$$

Additionally, if we have $F(-1) = 0$, then $\mu' = 0$ and

$$\hat{\sigma}_\omega \sigma_n \geq \frac{1}{2}.$$

In the sequel, we concentrate on low-pass filters such that $|F(e^{j\omega})|_{max}$ occurs at $\omega = 0$, and Eqs. (5.27) and (5.28) are used. In this case, there are two classes of filters or signals:

$$Class\, I: \quad F(-1) = 0 \rightarrow \sigma_n \sigma_\omega \geq \frac{1}{2}, \tag{5.33}$$

$$Class\, II: \quad F(-1) \neq 0 \rightarrow \sigma_n \sigma_\omega \geq \frac{|1-\mu|}{2}. \tag{5.34}$$

The bound on the time-frequency product in the first case is the same as that for the continuous-time case (in which $F(\pm\infty) = 0$). In the analog case, we know that the equality in the lower bound is achieved when $\Omega F(\Omega)$ is proportional to $dF/d\Omega$, or $F(\Omega) = K\, exp(-b\Omega^2/2)$, a Gaussian. In the discrete-time formulation (Haddad, et.al., 1993), we have the same form of integral resulting in the differential equation

$$\frac{dF}{d\omega} = -K\omega F^*(e^{j\omega}),$$

whose solution is a Gaussian $exp(-K\omega^2/2)$. This Gaussian function satisfies the differential equation but cannot satisfy the class I boundary condition $F(e^{j\pi}) = 0$. In this case, we conclude that the lower bound cannot be attained and the strict inequality holds, $\sigma_n \sigma_\omega > 1/2$.

We show that the binomial function is a finite impulse response (FIR) approximation to the Gaussian that matches the zero boundary condition at $\omega = \pi$.

For the class II set of functions, the Gaussian can satisfy both the differential equation and the boundary condition, resulting in the equality $\sigma_n \sigma_\omega = \frac{1}{2}|1 - \mu|$.

5.4.2 Gaussian and Binomial Distributions

For the class II signals, the Gaussian is

$$|F(e^{j\omega})|^2 = K\, e^{-\omega^2/2\sigma^2}, \qquad |\omega| < \pi \tag{5.35}$$

$$K = \frac{\sqrt{\pi/2}}{\sigma\ erf(\pi/\sigma)},$$

where *erf* is the error function.

The constant K is chosen to normalize the energy E to unity over $[-\pi, \pi]$. In (Haddad, et al.) (1993), we show that

$$\sigma_\omega = \sigma(1-\mu)^{1/2} \tag{5.36}$$

$$\mu = \frac{|F(\pi)|^2}{E} = K\, e^{\pi^2/2\sigma^2},$$

and, hence,

$$\sigma_\omega \sigma_n = \frac{|1-\mu|}{2}, \rightarrow \sigma_n = \frac{(1-\mu)^{1/2}}{2\sigma}. \tag{5.37}$$

For the narrow-band case, $\sigma \leq \pi/4$, $\mu \leq 10^{-3}$, and $F(-1) \approx 0$, resulting in $\sigma_\omega^2 \cong \sigma^2$, $\sigma_\omega \sigma_n \cong 1/2$. The corresponding time function is found to be approximately Gaussian:

$$f(n) = \sigma \sqrt{\frac{K}{\pi}} e^{-\sigma^2 n^2}. \tag{5.38}$$

Examples of these narrow-band Gaussian functions are shown in Fig. 5.10. Again note that the time-frequency product is very close to 1/2 in these cases.

For the wide-band case, with $\sigma \geq 3\pi/8$, we must use the more exact expansions in Eqs. (5.36) and (5.37). For example, for $\sigma = \pi/2$, we calculate $\mu = 0.22625$, $\sigma_\omega = 1.382$, $\sigma_n \sigma_\omega = 0.3869$ and $\sigma_n = 0.28$ time samples. In this case, there is no simple approximation; $f(n)$ must be computed numerically from the inversion formula, Eq. (5.23). These are shown in Fig. 5.11, in which the very short duration of $f(n)$ is duly noted.

The binomial sequences introduced in Section 2.3.2 are a family generated by successive differences of the binomial kernel, as summarized in Table 5.1.

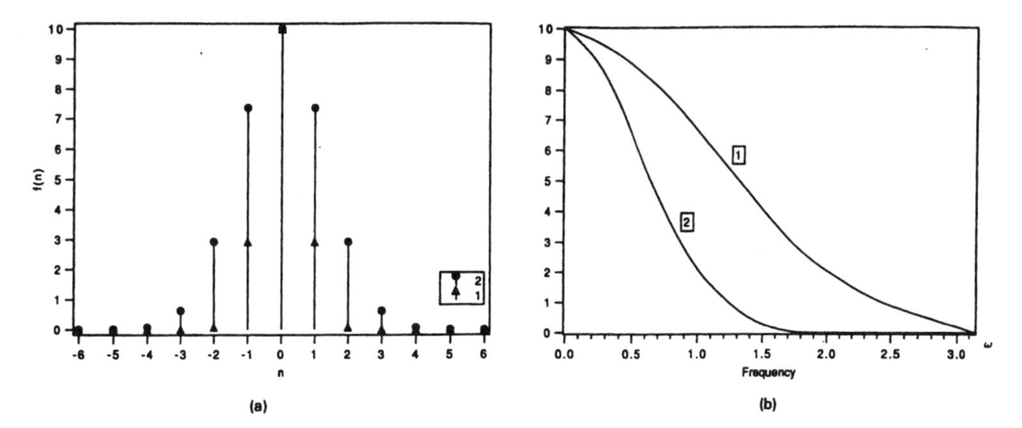

Figure 5.10: (a) Time and (b) frequency plots for narrow-band Gaussian functions: (1) $\sigma_\omega = \pi/4$, $\sigma_n = 0.637$ samples and (2) $\sigma_\omega = \pi/8$, $\sigma_n = 1.274$ samples.

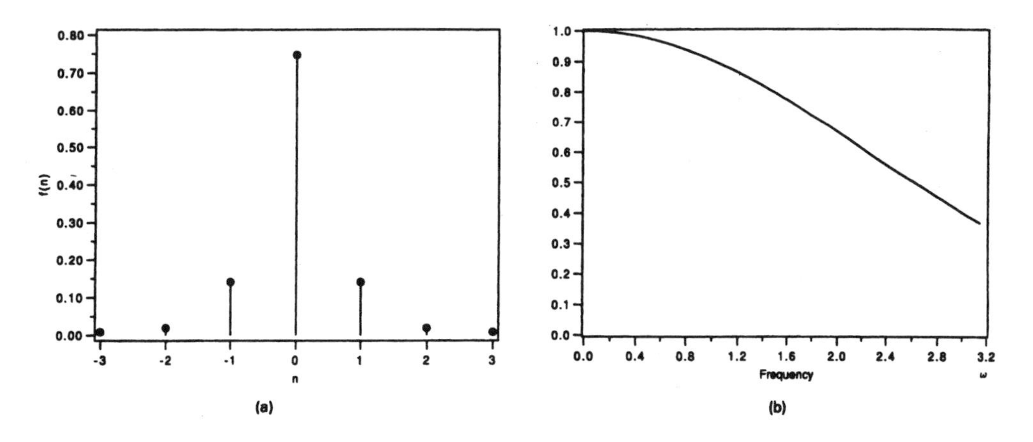

Figure 5.11: (a) Time and (b) frequency plots for the wide-band Gaussian case: $\sigma_\omega = \pi/2$ and $\sigma_\omega \sigma_n = 0.3869$.

The kernel is the binomial sequence $\begin{pmatrix} N \\ k \end{pmatrix}$, which resembles a sampled truncated Gaussian in time, and the frequency response looks like the Gaussian $e^{-\omega^2/2\sigma^2}$. To demonstrate this, let us compare $f(\omega) = [\cos(\omega/2)]^N$, with $g(\omega) = e^{-\omega^2/4\sigma^2}$.

Time Function	Transform
$x_0(k) = \binom{N}{k}, 0 \le k \le N$	$X_0(z) = (1+z^{-1})^N$ $\lvert X_0(e^{j\omega})\rvert = 2^N(\cos\frac{\omega}{2})^N$
$x_r(k) = \nabla^r \binom{N-r}{k}$ $x_r(k) = H_r(k)\binom{N}{k}$	$X_r(z) = (1+z^{-1})^{N-r}(1-z^{-1})^r$ $\lvert X_r(e^{j\omega})\rvert = 2^N(\cos\frac{\omega}{2})^{N-r}(\sin\frac{\omega}{2})^r$

Table 5.1: The binomial family: $H_r(k)$ are discrete Hermite polynomials.

Taking logarithms,

$$\ln[g(\omega)] = -\frac{\omega^2}{4\sigma^2}$$

$$\ln[f(\omega)] = -N\log\cos\frac{\omega}{2} = N[-\frac{1}{2}(\frac{\omega}{2})^2 - \frac{1}{12}(\frac{\omega}{2})^4 \cdots].$$

Matching quadratic terms, $N = 2/\sigma^2$ in these expressions results in a normalized error that is

$$\frac{g(\omega)-f(\omega)}{g(\omega)} \cong \frac{(\omega/\sigma)^4}{48N}, \quad \frac{\omega}{\sigma} < 2.$$

The localization features of the binomial kernel are as follows: $X(1) = 0$, so that $\sigma_n\sigma_\omega > 1/2$. From Haddad et al. (1993),

$$E = \sum_{k=0}^{N}\binom{N}{k}^2 = 2^{2N}\frac{[1\cdot 3\cdot 5\cdots(2N-1)]}{(2\cdot 4\cdot 6\cdots 2N)} \tag{5.39}$$

$$\bar{n} = N/2 \tag{5.40}$$

$$\sigma_n = \frac{N}{2(2N-1)^{1/2}}. \tag{5.41}$$

It can also be shown that

$$\sigma_\omega^2 = \frac{4(A+B)}{E}$$

$$A = \sum_{k=0}^{N-1} \binom{2N}{k} \frac{(-1)^{N-k}}{(N-k)^2}$$

$$B = \binom{2N}{N} \frac{\pi^2}{12}.$$

Sample binomial time-frequency responses are displayed in Fig. 5.12. Note that these approximate the Gaussian very well and the time-frequency products are, respectively, 0.50274 and 0.5002, for $N = 4.12$.

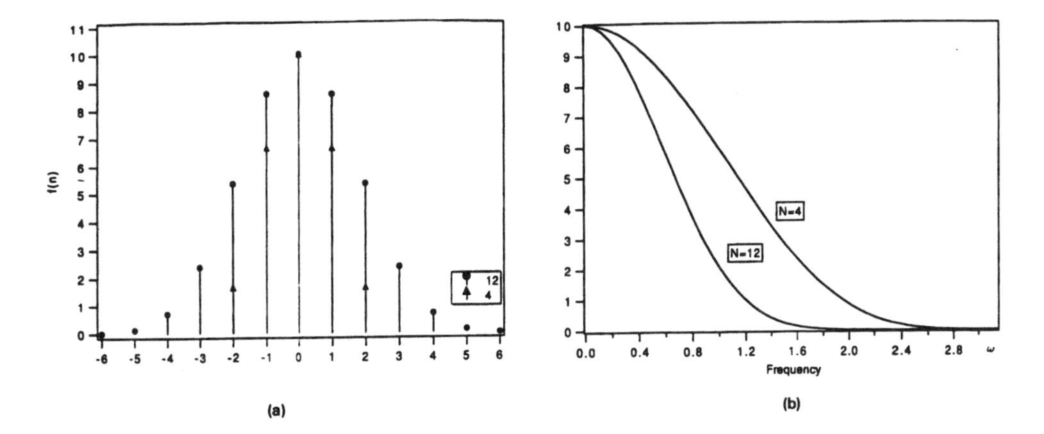

Figure 5.12: Time-frequency plots of binomial sequences: (a) $N = 4$ ($\sigma_\omega = 0.665$, $\sigma_n = 0.756$, and $\sigma_\omega \sigma_n = 0.50274$) and (b) $N = 12$ ($\sigma_\omega = 0.4$, $\sigma_n = 1.25$, and $\sigma_n \sigma_\omega = 0.5002$).

Figure 5.13 shows binomial and Gaussian responses on the same axes for $N = 12$ and $\sigma = \sqrt{2/N} = 1/\sqrt{6}$. Both time and frequency plots are almost indistinguishable. We conclude that the binomial filter provides a simple yet excellent FIR approximation to the optimum Gaussian wave form in the time- and frequency-domains.

5.4.3 Band-Pass Filters

The binomial family can also provide good approximations to Gaussian band-pass filters. The rth member $X_r(z)$ has a magnitude square response of the form

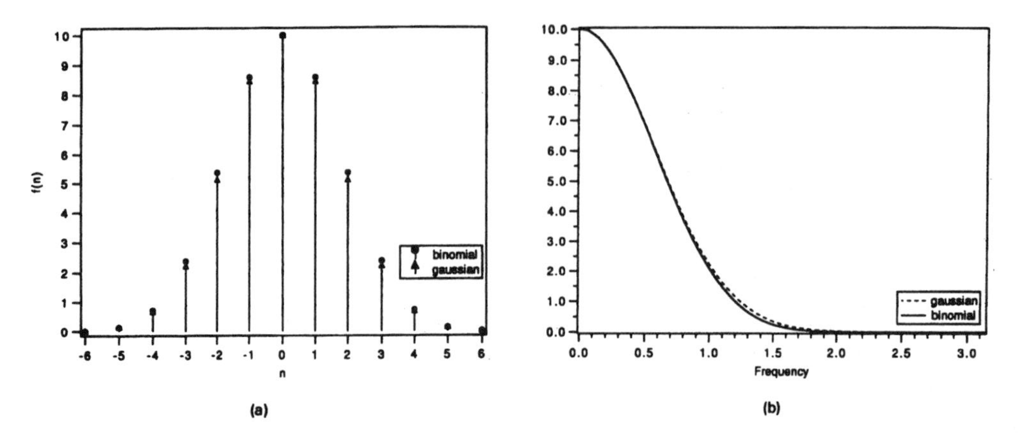

Figure 5.13: Binomial and Gaussian (a) time and (b) frequency plots for $N = 12$ and $\sigma = 1/\sqrt{6}$, respectively.

$$(\sin\frac{\omega}{2})^{2r}(\cos\frac{\omega}{2})^{2(N-r)}.$$

For N large, this response is approximately even symmetric around $\hat{\omega} = 2\ \sin^{-1}(r/N)^{1/2}$, and approximately Gaussian as shown in Fig. 2.8(b). A major advantage of these binomial filters is that they can be synthesized using only add, subtract, and delay operators as indicated in Fig. 2.8(a).

A second class of band-pass filters can be obtained by modulating the low-pass binomial. Let $h(n) \stackrel{\triangle}{=} x_0(n)$, and let

$$g(n) = \frac{1}{\sqrt{2}}h(n)\cos\omega_0 n \leftrightarrow G(e^{j\omega}) = \frac{1}{\sqrt{2}}(H(e^{j(\omega-\omega_0)}) + H(e^{j(\omega+\omega_0)})). \tag{5.42}$$

Now if $\hat{\omega} = \omega_0 >> \sigma = 2/\sqrt{N}$, the leakage of the tail of $H(e^{j(\omega+\omega_0)})$ into the frequency band near $\omega = \omega_0$ is negligible, and over $[0, \pi]$,

$$G(e^{j\omega}) = \frac{1}{\sqrt{2}}H(e^{j(\omega-\omega_0)}) = \frac{1}{\sqrt{2}}X_0(e^{j(\omega-\omega_0)}). \tag{5.43}$$

In this case, it is easy to show that the spread in time- and frequency-domains is the same as for the low-pass binomial prototype window.

5.5 Time-Frequency Localization

In Chapters 2 and 3, we observed that the $M \times M$ block transform is a special case of an M-band orthonormal filter bank wherein the filter length equals the number of channels. In this section we compare the localization properties of established block transforms and certain orthonormal filter families.

5.5.1 Localization in Traditional Block Transforms

The time and frequency responses of several standard block transforms were displayed in Fig. 2.7. These plots demonstrate that these transforms look like a bank of band-pass filters, with time and frequency spreads that are crudely comparable for the same size transform. The frequency responses are not very sharp and the impulse (or basis sequences) responses are widely distributed over the entire interval. The time-frequency spreads for two of these, the DCT and the WHT, are given in Table 5.2. The trade-offs in σ_n and σ_ω as a function of filter length (or transform size) are obvious in this table. To sharpen the frequency resolution, the transform size or filter length is increased. The frequency spread σ_ω is decreased significantly, but at an appreciable increase in σ_n. The time-frequency products or resolution cell areas also increase with filter length.

To obtain narrower frequency bands σ_ω as needed for compression, we can use orthonormal M-band filter banks where the length of each filter $N \geq M$ can be freely chosen. As a case in point, Table 5.3 displays the time frequency localization for an 8×16 DCT LOT (here $N = 2M$) that can be contrasted with that for 8×8 DCT in Table 5.2. The LOT frequency resolution is much sharper and the resolution cell products are smaller.

5.5.2 Localization in Uniform M-Band Filter Banks

Uniform M-band filter banks can be realized either by multilevel binary hierarchical tree as indicated in Fig. 3.21 and 3.24, or by the one-level M-band unitary filter bank shown in Fig. 3.32. These structures allow filters of arbitrary length and hence can provide sharp frequency responses.

For our present purposes, we want to evaluate the time-frequency localization properties of some known filter banks. Table 5.4 lists these characteristics for three different eight-tap two-band filter banks: the binomial QMF (Akansu, Haddad and Caglar, 1990), the Smith and Barnwell conjugate quadrature filter (CQF) (Smith and Barnwell, 1986), and the multiplierless PR QMF (Akansu, 1992). Tables 5.5 and 5.6 continue this comparison for hierarchical structure four-band (22-tap product filters) and eight-band (50-tap product filters) configurations. In all

	$\bar{\omega}$	$\bar{n}$	σ_ω^2	σ_n^2	$\sigma_\omega^2 \times \sigma_n^2$
2x2 DCT	0	0.50	1.2899	0.2500	0.3225
and WHT	π	0.50	1.2899	0.2500	0.3225
4x4 DCT	0	1.50	0.6787	1.2500	1.2234
	1.27	1.50	0.3809	1.9570	0.7454
	1.85	1.50	0.2424	1.2500	0.3030
	π	1.50	0.4896	0.5428	0.2657
4x4 WHT	0	1.50	0.6787	1.2500	0.8484
	1.29	1.50	0.2424	1.2500	0.3030
	1.85	1.50	0.2424	1.2500	0.3030
	π	1.50	0.6787	1.2500	0.8484
8x8 DCT	0	3.50	0.3447	5.2500	1.8097
	0.74	3.50	0.3021	8.4054	2.5393
	1.02	3.50	0.2413	5.9572	1.4375
	1.36	3.50	0.1957	5.4736	1.0712
	1.71	3.50	0.1488	5.2500	0.7812
	2.08	3.50	0.1206	5.0263	0.6062
	2.45	3.50	0.0797	4.5428	0.3621
	π	3.50	0.1388	2.0955	0.2908
8x8 WHT	0	3.50	0.3447	5.2500	1.8097
	0.82	3.50	0.3485	5.2500	1.8296
	1.15	3.50	0.2977	5.2500	1.5629
	1.43	3.50	0.1488	5.2500	0.7812
	1.72	3.50	0.1488	5.2500	0.7812
	1.99	3.50	0.2977	5.2500	1.5629
	2.33	3.50	0.3485	5.2500	1.8296
	π	3.50	0.3447	5.2500	1.8097

Table 5.2: Time-frequency localizations of DCT and WHT bases for two- and eight-band cases.

	$\bar{\omega}$	$\bar{n}$	σ_ω^2	σ_n^2	$\sigma_\omega^2 \times \sigma_n^2$
8x16	0	7.50	0.0917	4.654	0.4269
DCT-LOT	0.59	7.50	0.0549	7.615	0.418
	0.98	7.50	0.0345	8.387	0.2898
	1.37	7.50	0.0523	8.645	0.4523
	1.76	7.50	0.0367	8.35	0.3070
	2.16	7.50	0.0608	7.549	0.4596
	2.55	7.50	0.0389	7.778	0.3026
	π	7.50	0.0119	5.360	0.6419

Table 5.3: Time-frequency localizations of $8x16$ DCT LOT.

	$\bar{\omega}$	$\bar{n}$	σ_ω^2	σ_n^2	$\sigma_\omega^2 \times \sigma_n^2$
B-QMF (8-tap)	0	1.46	0.9468	0.6025	0.5704
	π	5.54	0.9468	0.6025	0.5704
Multiplierless (8-tap)	0	2.50	0.9743	0.3750	0.3654
	π	4.50	0.9743	0.3750	0.3654
Smith-Barnwell (8-tap)	0	4.17	0.9174	0.5099	0.4678
	π	2.83	0.9174	0.5099	0.4678

Table 5.4: Time-frequency localizations of three different eight-tap, two-band PR-QMF banks.

	$\bar{\omega}$	$\bar{n}$	σ_ω^2	σ_n^2	$\sigma_\omega^2 \times \sigma_n^2$
B-QMF Hierarchical 4 Band Tree (22-tap product filters)	0	4.05	0.2526	2.7261	0.6886
	1.23	12.88	0.1222	3.8269	0.4676
	1.91	16.28	0.1222	2.7757	0.3392
	π	8.80	0.2526	2.2622	0.5714
Multiplierless (22-tap product filters)	0	7.50	0.2747	1.5817	0.4345
	1.24	11.50	0.1346	2.1683	0.2918
	1.90	13.49	0.1346	2.1675	0.2918
	π	9.50	0.2747	1.5818	0.4245
Smith-Barnwell (22-tap product filters)	0	12.45	0.2339	2.1458	0.5019
	1.22	9.88	0.1077	2.9463	0.3173
	1.92	8.45	0.1077	3.0185	0.3251
	π	11.22	0.2339	2.0772	0.4859

Table 5.5: Time-frequency localizations of hierarchical subband trees for two-level (four bands) cases.

these cases, the multiplierless structure has the best time-frequency product $\sigma_n \sigma_\omega$, followed by the Smith and Barnwell CQF and the binomial QMF. As expected, longer duration filters have a narrower σ_ω and a wider σ_n. Again, as expected, the eight-band, eight-tap block transforms (Table 5.2) have a much narrower σ_n than any of the eight-band tree-structured filter banks, but very poor frequency localization.

Figure 5.14 displays the impulse responses of the product filters of the two-band binomial QMF-based hierarchical tree for the two-, four-, and eight-band cases. Figure 5.15 shows the corresponding frequency responses. These demonstrate the drawbacks of blindly repeating a two-band PR-QMF module in a hierarchical subband tree. The time spread increases considerably while the time-frequency product degrades. This suggests two possibilities: Either design the M-band,

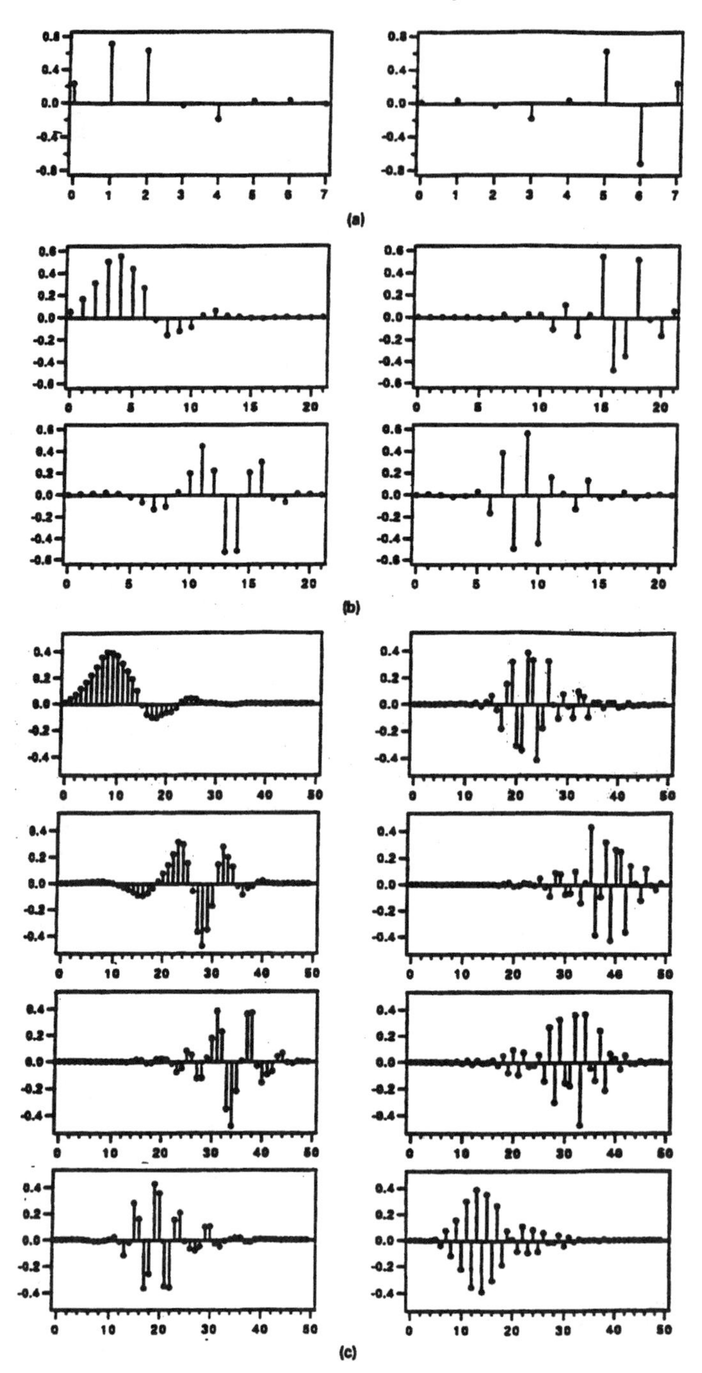

Figure 5.14: Impulse responses of the product filters of the two-band binomial QMF-based hierarchical tree for the (a) two-band, (b) four-band, and (c) eight-band case.

	$\bar{\omega}$	$\bar{n}$	σ_ω^2	σ_n^2	$\sigma_\omega^2 \times \sigma_n^2$
B-QMF Hierarchical 8-Band Tree (50-tap product filters)	0	9.12	0.0644	11.726	0.7552
	0.63	26.96	0.0490	15.953	0.7817
	1.01	34.11	0.0961	11.326	1.0884
	1.45	19.65	0.0496	9.7846	0.4853
	1.68	22.56	0.0496	10.510	0.5213
	2.13	37.99	0.0961	12.013	1.1544
	2.52	31.54	0.0490	14.950	0.7326
	π	14.36	0.0644	10.777	0.6940
Multiplierless (50-tap product filters)	0	17.53	0.0724	6.3415	0.4591
	0.64	25.46	0.0688	8.8171	0.6066
	1.02	29.46	0.1193	9.1282	1.0890
	1.45	21.54	0.0558	7.2005	0.4018
	1.68	23.47	0.0558	7.2099	0.4023
	2.11	31.53	0.1193	9.1234	1.0884
	2.50	27.54	0.0688	8.8269	0.6073
	π	19.47	0.0724	6.3371	0.4588
Smith-Barnwell (50-tap product filters)	0	28.86	0.0591	8.6494	0.5112
	0.6137	24.03	0.0321	11.837	0.3800
	0.9951	21.22	0.0688	12.623	0.8685
	1.4488	26.55	0.0436	9.5939	0.4183
	1.6927	25.32	0.0436	9.6769	0.4219
	2.1465	19.57	0.0688	12.599	0.8668
	2.5279	22.51	0.0321	11.912	0.3824
	π	27.93	0.0591	8.5379	0.5046

Table 5.6: Time-frequency localizations of hierarchical subband trees for three-level (eight bands) cases.

single-level structure directly or, in using the hierarchical tree structure, monitor the PR-QMF module from level to level.

The "best" filter bank from a localization standpoint depends on the application at hand. For example, time (or spatial) localization is known to be more important in visual signal processing and coding applications for human perception, while frequency localization is the predominant concern for certain compression considerations. In any case, the joint time-frequency characteristics of the basis functions must be carefully monitored for subjective performance improvement.

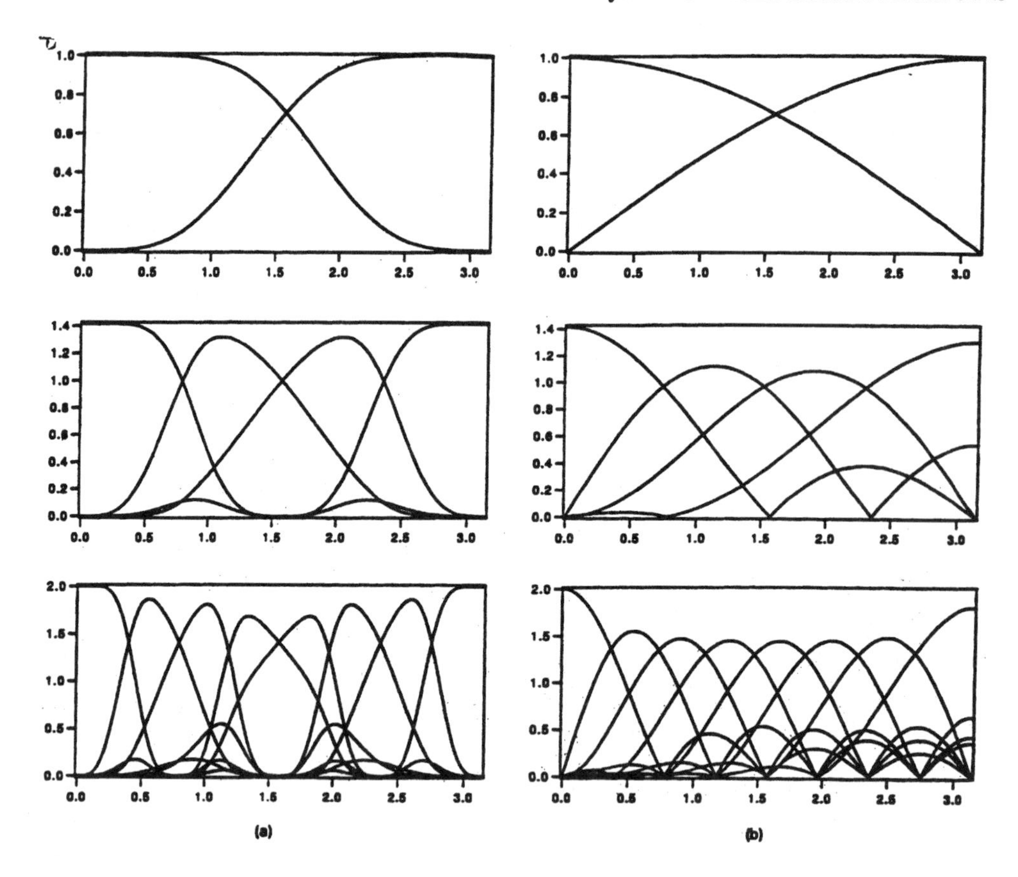

Figure 5.15: Frequency responses of (a) 2×2, 4×4 and 8×8 DCT, and (b) 2-band, 4-band, and 8-band binomial QMF-based hierarchical structure.

5.5.3 Localization in Dyadic and Irregular Trees

Filter banks based on dyadic tree structures as indicated in Fig. 3.27 [and repeated here as Fig. 5.16(a)] realize octave band frequency splits or concentrations. These lead to the time-frequency tiling patterns of Fig. 5.16(b), which is similar to that for the wavelet transform. The LLL frequency band is concentrated over 1/8 of the frequency range but requires three levels of the tree, resulting in a product filter with length equal to 50. The single-level H-band filter occupying one-half the frequency range is realized with one 8-tap filter.

An irregular tree structure results in irregular frequency band split which can be related to a corresponding tiling pattern as shown in Fig. 5.17. Note the trade-

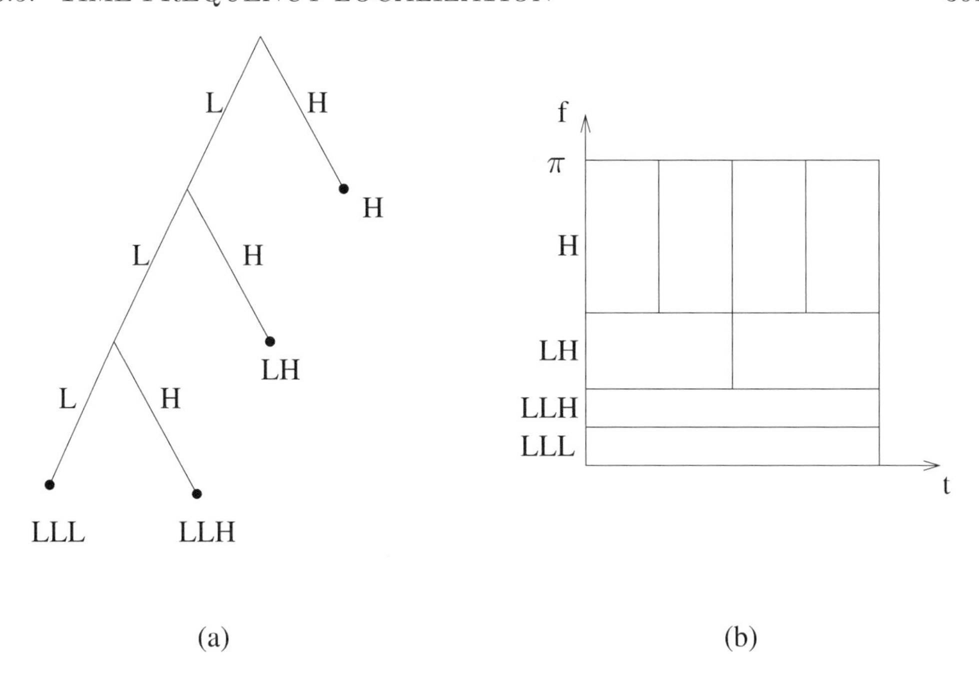

Figure 5.16: (a) Dyadic tree. (b) Time-frequency tiling pattern.

offs in time and frequency resolutions suggested by the patterns in Figs. 5.16 and 5.17.

These examples are based on a binary split at each node or level. Figure 5.18 illustrates a mixed tertiary and binary split and the resulting tiling pattern.

As we will see in Chapter 6, the orthonormal wavelet filters are constructed from the binary dyadic tree using the two-band paraunitary filters $H_0(z)$ and $H_1(z)$ of Eq. (3.50) with the added constraint that $H(-1) = 0$. The Daubechies wavelet filters (Daubechies, 1988) are identical to the maximally flat binomial QMF filters of Chapter 4.

Other wavelet families (e.g., the most regular, Coiflets) are devised by imposing other requirements on $H_0(z)$ as presented in Chapter 6.

Table 5.7 compares the time-frequency resolutions of scaling and wavelet functions for three wavelet families generated by six-tap paraunitary filters—the Daubechies, most regular, and Coiflet—along with the localization properties of the progenitor discrete-time filters. Table 5.7 demonstrates that the time-frequency localizations are important measures in the evaluation of a wavelet family as an analog filter bank. In particular, the role of regularity in wavelet transforms should

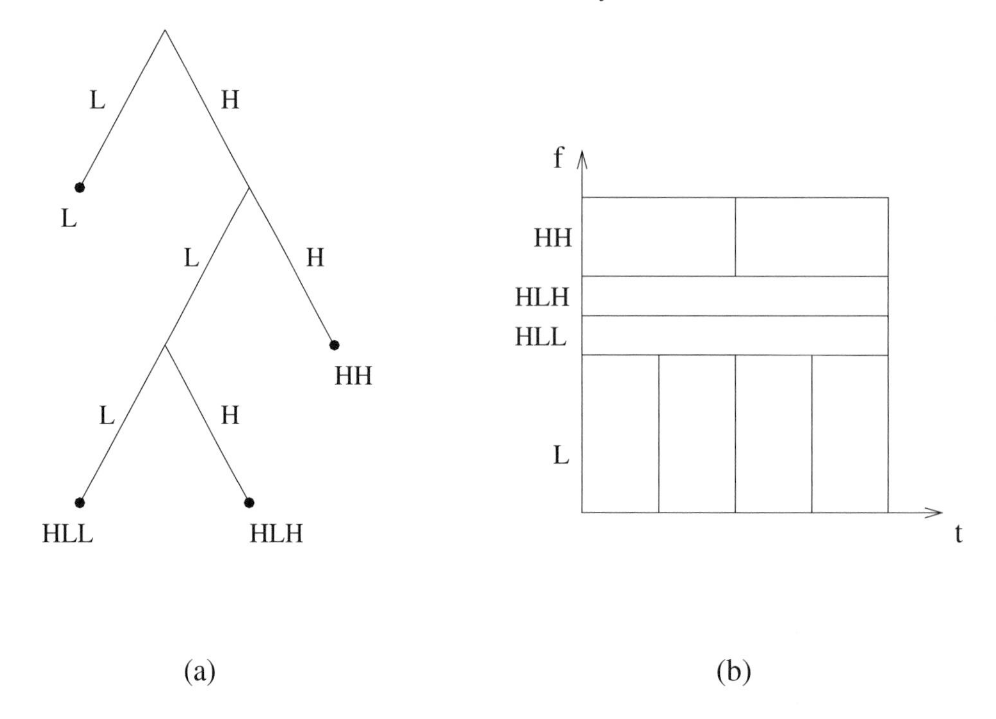

Figure 5.17: (a) Irregular tree. (b) Corresponding tiling pattern.

be evaluated for signal processing applications. We will return to these aspects of wavelets in Chapter 6.

5.6 Block Transform Packets

A block transform packet (BTP) is an orthonormal block transform that is synthesized from conventional block transforms so as to realize an arbitrary tiling of the time-frequency plane. The BTP has time localizability and can be adapted to deal with nonstationary signals. The BTP also preserves the computational efficiency of the progenitor block transform.

In this section, we show how to generate a BTP from a specified tiling pattern, and hence we show how to specify a desirable or optimum tiling pattern for a given signal. This leads to interesting applications in signal compaction and interference excision.

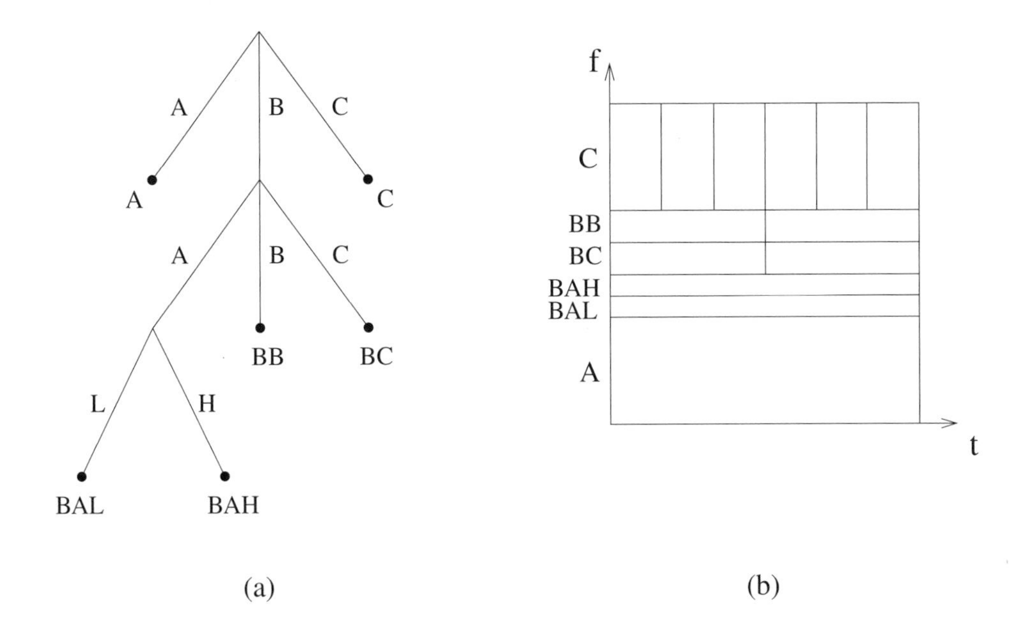

Figure 5.18: (a) Mixed tertiary/binary tree. (b) Tiling pattern.

		Daubechies	Mostregular	Coiflet
Scaling Function	σ_T^2	0.314	0.143	0.086
	σ_Ω^2	5.22	5.77	11.86
	$\sigma_T^2\sigma_\Omega^2$	0.699	0.825	1.02
Wavelet Function	σ_T^2	0.178	0.188	0.108
	σ_Ω^2	8.97	11.70	39.36
	$\sigma_T^2\sigma_\Omega^2$	1.596	2.199	4.25
Lowpass PR-QMF	σ_n^2	0.453	0.470	0.305
	σ_ω^2	0.987	0.996	1.059
High-Pass PR-QMF	σ_n^2	0.453	0.470	0.305
	σ_ω^2	0.987	0.996	1.059

Table 5.7: Time-frequency localizations of six-tap wavelet filters and corresponding scaling and wavelet functions.

5.6.1 From Tiling Pattern to Block Transform Packets

The plots of several orthonormal 8×8 block transforms are shown in Fig. 2.7. These depict the basis sequences of each transform in both time- and frequency-domains. These plots demonstrate that the basis sequences are spread over all eight time slots whereas the frequency plots are concentrated over eight separate frequency bands. The variation in these from transform to transform is simply a matter of degree rather than of kind. The resulting time-frequency tiling then would have the general pattern shown in Fig. 5.19(a). There are eight time slots and eight frequency slots, and the energy concentration is in the frequency bands. The basis functions of these transforms are clearly frequency-selective and can be regarded as FIR approximations to a "brick wall" (i.e., ideal rectangular band-pass filter) frequency pattern which of course would necessitate infinite sinc function time responses.

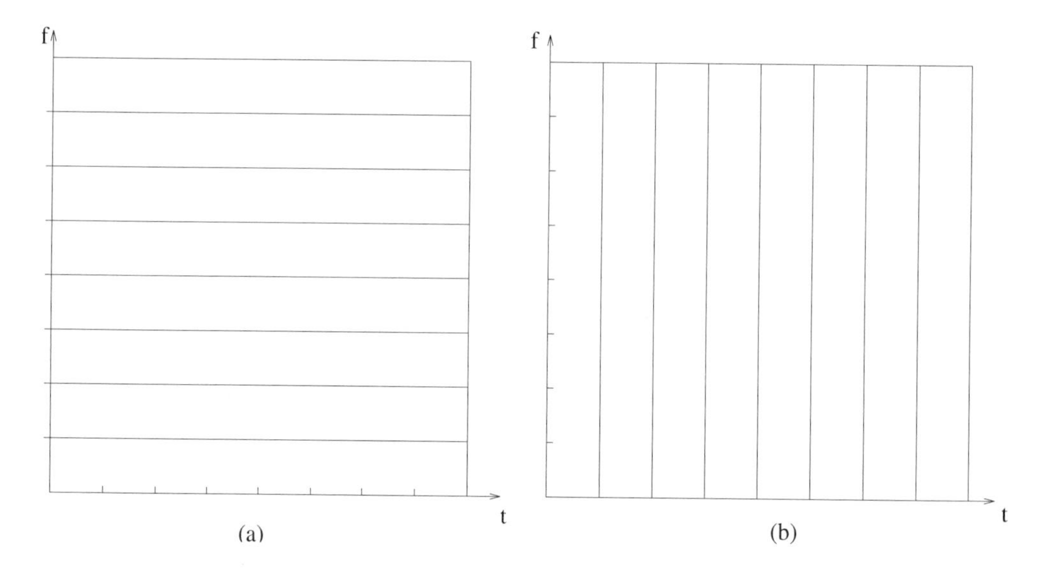

Figure 5.19: (a) Tiling pattern for frequency-selective transform, (b) Tiling pattern for time-selective transform

The other extreme is that of the shifted Kronecker delta sequences as basis functions as mentioned in Section 2.1. The time- and frequency-domain plots are shown in Fig. 5.20. This realizable block transform (i.e., the identity matrix) has perfect resolution in time but no resolution in frequency. Its tiling pattern is shown in Fig. 5.19(b) and can be regarded as a realizable brick-wall-in-time pattern, the dual of the nonrealizable brick-wall-in-frequency.

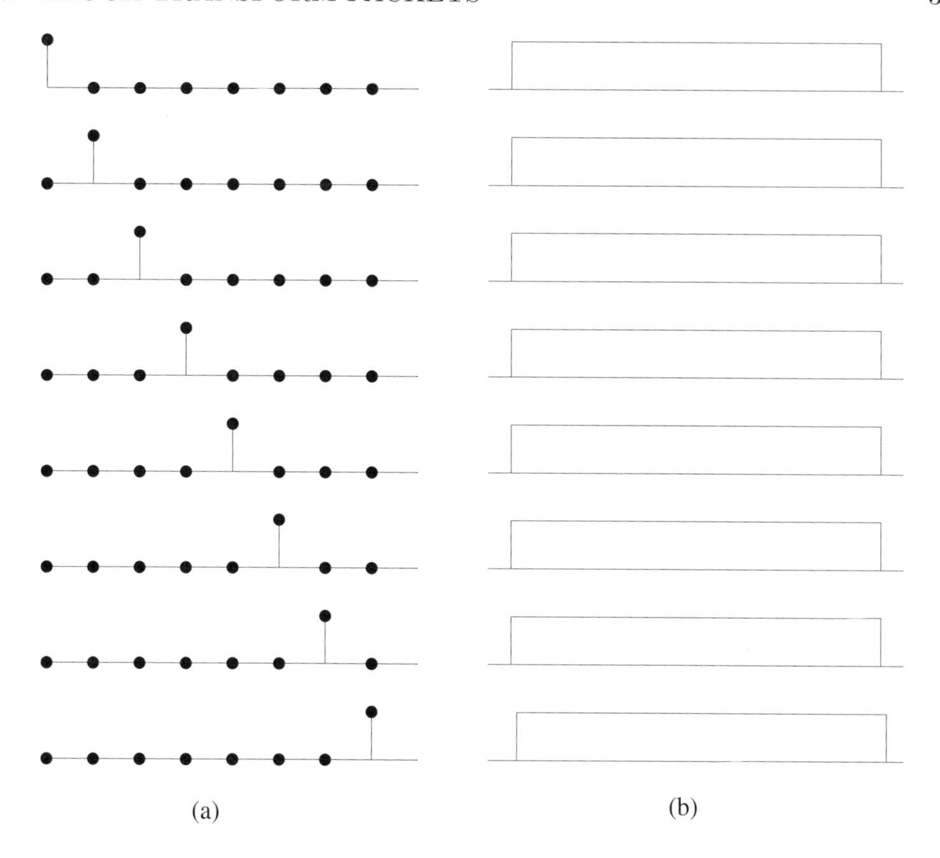

Figure 5.20: (a) Basis functions in time; (b) magnitude of Fourier transform of basis functions.

The challenge here is to construct a specified but arbitrary tiling pattern while retaining the computational efficiencies inherent in certain block transforms, using the DFT, DCT, MHT, and WH. Our objective then is to develop desired time-localized patterns starting from the frequency-selective pattern of Fig. 5.19(a), and conversely, to create frequency-localized tiling from the time-localized Kronecker delta pattern of Fig. 5.19(b).

The first case is the time-localizable block transform, or TLBT (Kwak and Haddad, 1994), (Horng and Haddad, 1996). This is a unitary block transform which can concentrate the energy of its basis functions in desired time intervals—hence, time-localizable.

We start with a frequency-selective block transform, *viz.*, the DCT, DFT, WH, whose basis functions behave as band-pass sequences, from low-pass to high-pass,

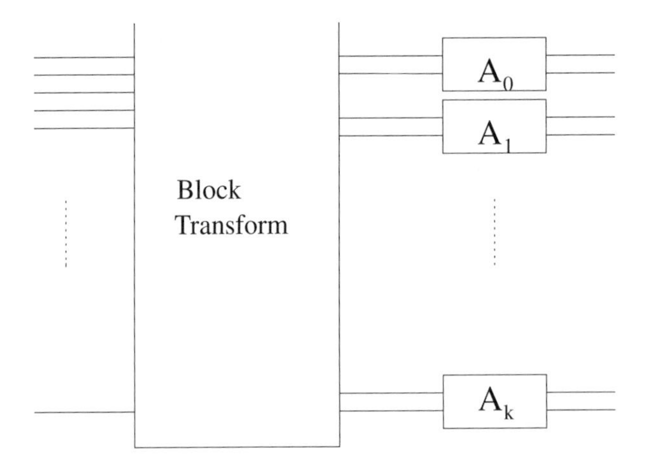

Figure 5.21: Structure of the block transform packet.

as in Fig. 2.7. Consider a subset of M_k basis functions with contiguous frequency bands. We then construct a new set of M_k time-localized basis functions as a linear combination of the original M_k (frequency-selective) basis functions in such a way that each of the new M_k basis functions is concentrated over a desired time interval but distributed over M_k frequency bands. Hence we can swap frequency resolution for time resolution in any desired pattern. The construction of the TLBT system is shown in Fig. 5.21.

Let $\phi_i(n),\ i,n = 0,1,...,N-1$, and $\psi_i(n),\ i,n = 0,1,...,N-1$, be the original set of orthonormal basis sequences, and the TLBT basis functions, respectively. These are partitioned into subsets by the $N \times N$ diagonal block matrix,

$$\begin{bmatrix} \psi_0(n) \\ \psi_1(n) \\ \vdots \\ \psi_{N-1}(n) \end{bmatrix} = \begin{bmatrix} A_0 & & & 0 \\ & A_1 & & \\ & & \ddots & \\ 0 & & & A_L \end{bmatrix} \begin{bmatrix} \phi_0(n) \\ \phi_1(n) \\ \vdots \\ \phi_{N-1}(n) \end{bmatrix}, \tag{5.44}$$

where each coefficient matrix A_k is an M_kxM_k unitary matrix, and $\sum_{k=0}^{L} M_k = N$. Consider the kth partition indicated by

$$\begin{bmatrix} \psi_{k,0}(n) \\ \psi_{k,1}(n) \\ \vdots \\ \psi_{k,M_{k-1}}(n) \end{bmatrix} = A_k \begin{bmatrix} \phi_{k,0}(n) \\ \phi_{k,1}(n) \\ \vdots \\ \phi_{k,M_{k-1}}(n) \end{bmatrix} \; ; \; or \;\; \underline{\Psi}_k(n) = A_k\, \underline{\Phi}_k(n) \tag{5.45}$$

and let $\underline{\alpha}_i^*$ be the ith row of A_k, such that $\psi_{k,i}(n)$ is the inner product $\psi_{k,i}(n) = \underline{\alpha}_i^* \, \underline{\Phi}_k(n)$. We want to find the coefficient vector $\underline{\alpha}_i$ such that the TLBT basis sequence $\psi_{k,i}(n)$ maximally concentrates its energy in the interval $I_i : [i(\frac{N}{M_k}), (i+1)(\frac{N}{M_k}) - 1]$. We choose to *minimize* the energy of $\psi_i(n)$ *outside* the desired I_i, i.e., to minimize

$$J_i' = \sum_{n \notin I_i} |\psi_{k,i}(n)|^2 \tag{5.46}$$

subject to orthonormality constraint on the rows of A_k, $\underline{\alpha}_i^* \underline{\alpha}_j = \delta_{i-j}$. Hence, we minimize the objective function

$$J = \sum_{i=0}^{M_k-1} [J_i' - \lambda_i(\underline{\alpha}_i^* \underline{\alpha}_i - 1)]. \tag{5.47}$$

It can be shown (Prob. 5.5) that the optimal coefficient vector, $\underline{\alpha}_i$, is the eigenvector of a matrix E_i^k which depends only on $\underline{\Phi}_k$,

$$E_i^k = \sum_{n \notin I_i} \underline{\Phi}_k(n) \underline{\Phi}_k^\dagger(n), \tag{5.48}$$

where $\underline{v}^\dagger$ indicates the conjugate transpose.

We now have a procedure for retiling the time-frequency plane so as to meet any set of requirements.

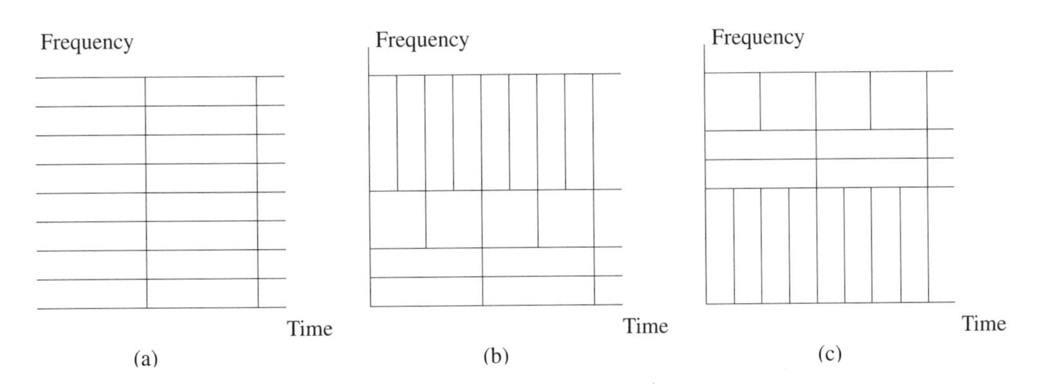

Figure 5.22: (a) Original tiling pattern for frequency concentrated 8×8 transform. (b) Tiling pattern for $M_k = 1, 1, 2, 4$. (c) Tiling pattern for $M_k = 4, 1, 1, 2$.

It is noted that the selection of the M_k values determines the time-frequency tiling patterns. The larger the value of M_k, the more time resolution can be obtained at the cost of sacrificing the resolution in frequency. For example, if we use an 8×8 block transform to construct the time-frequency tiling pattern in Fig. 5.22(b), the entire set of basis functions should be partitioned into four subsets with sizes $M_k = 1, 1, 2, 4$. For the time-frequency tiling in Fig. 5.22(c), the values of M_k are 4,1,1,2.

Figure 5.23(a) shows a portion of a time-frequency tiling based on the 64×64 DCT transform. According to Fig. 5.23(a) the entire set of DCT basis functions is partitioned into several subsets: $\{\phi_0(n), ..., \phi_3(n)\}$ for the subspace S_0 with $M_0 = 4$, $\{\phi_4(n)\}$ for S_1 with $M_1 = 1$, $\{\Phi_5(n)\}$ for S_2 with $M_2 = 1$, $\{\Phi_6(n), ...\Phi_{13}(n)\}$ for S_4 with $M_4 = 8, \cdots$. For S_0, the subset of the TLBT basis functions is $\Psi_0(n) = \{\psi_0(n), ..., \psi_3(n)\}$. In Fig. 5.23(a), the number on each cell represents the order of the TLBT basis functions. Cells Z_0 and Z_2 are the regions where $\psi_0(n)$ and $\psi_2(n)$ will concentrate their energies. The energy distributions in both the time- and frequency-domains for $\psi_0(n)$ and $\psi_2(n)$ are illustrated in Figs. 5.23(b) and 5.23(c), respectively. These figures demonstrate that $\psi_0(n)$ and $\psi_2(n)$ concentrate their energies both around $(1.5)(\frac{2\pi}{64})$ in the frequency-domain, but in the different time intervals $[0-15]$ and $[32-47]$, respectively. From these figures, we see that the TLBT basis functions concentrate most of their energies in the desired time interval and frequency band, specified in Fig. 5.23(a).

The dual case is that of constructing a frequency-localized transform FLBT from the time-localized Kronecker delta sequences. Here, we select M of these to be transformed into $\psi_{k,0}, ..., \psi_{k,M-1}$ via the unitary $M \times M$ matrix B, where

$$\psi_{k,i}(n) = \sum_{j=0}^{M-1} b_{ij}\phi_{k,j}(n). \tag{5.49}$$

We define J_i as the energy concentration in the frequency domain in the ith frequency band,

$$J_i = \int_{I_{\omega_i}} |\Psi_{k,i}(\omega)|^2 \, d\omega, \tag{5.50}$$

where $\Psi_{k,i}(\omega)$ is the Fourier transform of the basis function $\psi_{k,i}(n)$, and the given frequency band is $I_{\omega_i} = \{2\pi i(N/M) \leq \omega \leq [2\pi(i+1)(N/M)-1]\}$. J_i is maximized if we choose B to be the DFT matrix, i.e.,

$$b_{mn} = \frac{1}{\sqrt{M}} e^{j\frac{2\pi}{M}(m+a)n}, \quad a = \frac{1}{2}[1 - \frac{M}{N}]. \tag{5.51}$$

In other words, the DFT is the optimum sequence to transform the information from time-domain to the frequency-domain. Horng and Haddad (1998) describe a

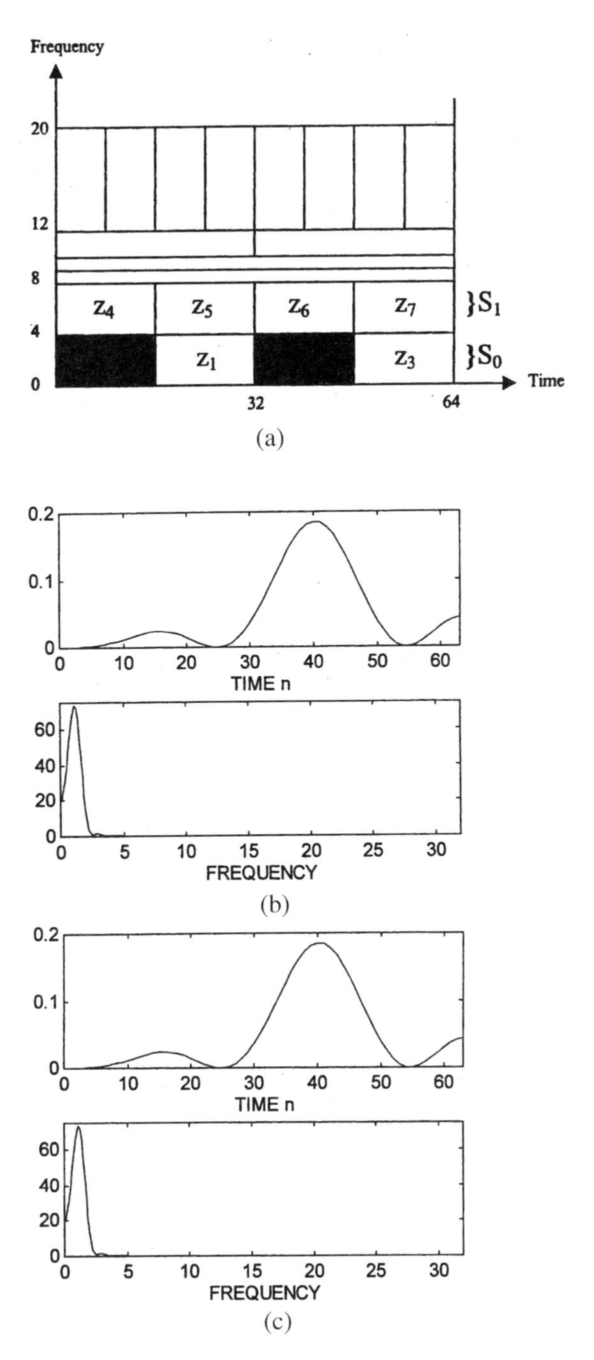

Figure 5.23: (a) Portion of desired tiling pattern; (b) energy distribution of $\psi_0(n)$ in both time- and frequency-domain associated with cell 1 in (a); (c) energy distribution of $\psi_1(n)$ in both time- and frequency-domain for cell 3 in (a).

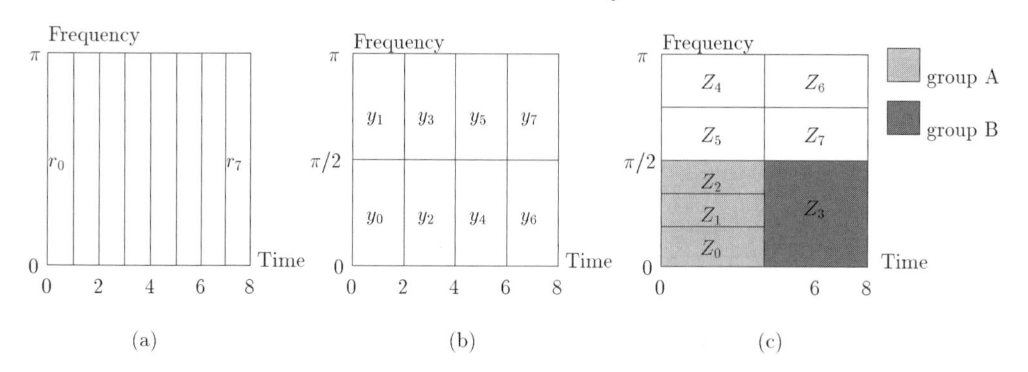

Figure 5.24: (a) Time localized Kronecker delta tiling pattern; (b) Intermediate pattern; (c) Desired tiling pattern.

procedure for constructing a FLBT that matches a desired tiling pattern starting from the delta sequences. The process involves a succession of diagonal block DFT matrices separated by permutation matrices. The final transform can be expressed as the product of a sequence of matrices with DFT blocks along the diagonal and permutation matrices. In this procedure, no tree formulations are needed, and we are able to build a tiling pattern that cannot be realized by pruning a regular or, for that matter, an irregular tree in the manner suggested earlier, as in Figs. 5.16 and 5.17.

The procedure is illustrated by the following example. Figure 5.24(c) is the desired pattern for an 8×8 transform. Note that this pattern is not realizable by pruning a binary tree, nor any uniformly structured tree as reported in the papers by Herley et. al. (1993).

(1) We note that Fig. 5.24(c) is divided into two broad frequency bands, $[0, \pi/2]$ and $[\pi/2, \pi]$; therefore, we split the tiling pattern of Fig. 5.24(a) into two bands using a 2×2 DFT transform matrix. This results in the pattern of Fig. 5.24(b). The output coefficient vector $y = [y_0, y_1, ..y_7]$ corresponding to the input data vector $\underline{f}$ is given by

$$\underline{y} = A_2 \, \underline{f} \tag{5.52}$$

where $A_2 = diag[\Phi_2, \Phi_2, \Phi_2, \Phi_2]$ and Φ_k is a $k \times k$ DFT matrix.

The 2×2 DFT matrix takes two successive time samples and transforms them into two frequency-domain coefficients. Thus Φ_2 operating on the first two time-domain samples, f_0 and f_1, generates transform coefficients y_0 and y_1, which represent the frequency concentration over $[0, \pi/2]$ and $[\pi/2, \pi]$, respectively. These are represented by cells y_0, y_1 in Fig. 5.24(b).

(2) We apply a permutation matrix P to regroup the coefficients y_i into same frequency bands. In this case, $P^T = [\delta_0^T, \delta_2^T, \delta_4^T, \delta_6^T, \delta_1^T, \delta_3^T, \delta_5^T, \delta_7^T]$, where $\delta_k = [0, ..., 0, 1, 0, ..., 0]$ is the Kronecker delta.

(3) Next we observe that the lower frequency band in Fig. 5.24(c) consists of two groups: Group A has 3 narrow bands of width $(\pi/6)$ each and time duration 6 (from 0 to 5), and group B has one broad band of width $(\pi/2)$ and duration 2 (from 6 to 7). This is achieved by transformation matrix $= diag[\Phi_3 \Phi_1]$ applied to the lower half of Fig. 5.24(b). The top half of Fig. 5.24(c) is obtained by splitting the high-frequency band of Fig. 5.24(b) from $\pi/2$ to π into two bands of width $\pi/4$ and time duration 4. This is achieved by transformation matrix $= diag[\Phi_2 \Phi_2]$ applied to the top half of Fig. 5.24(b). Thus,

$$\underline{z} = A_3 \, P \underline{y} \tag{5.53}$$

where $A_3 = diag[\Phi_3, \Phi_1, \Phi_2, \Phi_2]$.

The final block transform is then

$$\underline{z} = A \, \underline{f} \tag{5.54}$$

where $A = A_3 \, P \, A_2$ as given in Eq. (5.55) and $C = e^{j2\pi/3}$.

$$A = \begin{bmatrix} 1 & 1 & 1 & 1 & 1 & 1 & 0 & 0 \\ 1 & 1 & c & c & c^* & c^* & 0 & 0 \\ 1 & 1 & c^* & c^* & c & c & 0 & 0 \\ 0 & 0 & 0 & 0 & 0 & 0 & 1 & 1 \\ 1 & -1 & 1 & -1 & 0 & 0 & 0 & 0 \\ 1 & -1 & -1 & 1 & 0 & 0 & 0 & 0 \\ 0 & 0 & 0 & 0 & 1 & -1 & 1 & -1 \\ 0 & 0 & 0 & 0 & 1 & -1 & -1 & 1 \end{bmatrix}. \tag{5.55}$$

The basis sequences corresponding to cells z_1,z_6 are the corresponding rows of the A matrix. Concentration of these sequences in both time- and frequency-domains is shown in Figs. 5.25 and 5.26. $z_1(n)$ is concentrated over first 6 time slots as shown in the plot of $|z_1(n)|^2$. The associated frequency response $|Z_1(e^{j\omega})|^2$ is shown in Fig. 5.25(b), which is concentrated in the frequency band $(\pi/6, 2\pi/6)$. $z_6(n)$ is concentrated over last four time slots in Fig. 5.26(a) and $|Z_6(e^{j\omega})|^2 = |\sin(\omega/2) - \sin(2\omega/2)|^2$ concentrates over $(3\pi/4, \pi)$ as in Fig. 5.26(b).

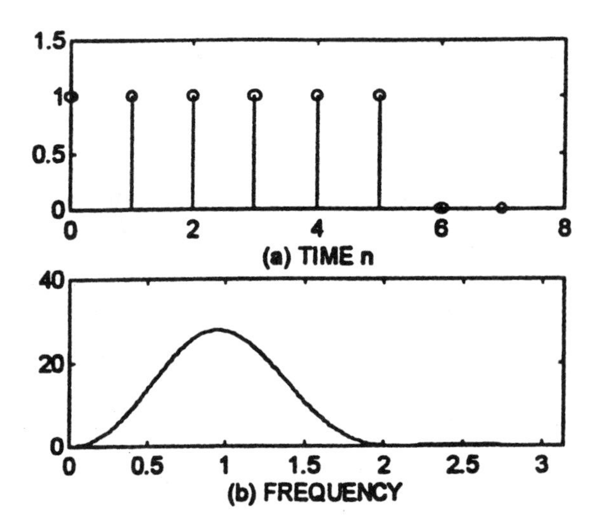

Figure 5.25: Energy distribution for cell Z_1: (a) time-domain; (b) frequency-domain.

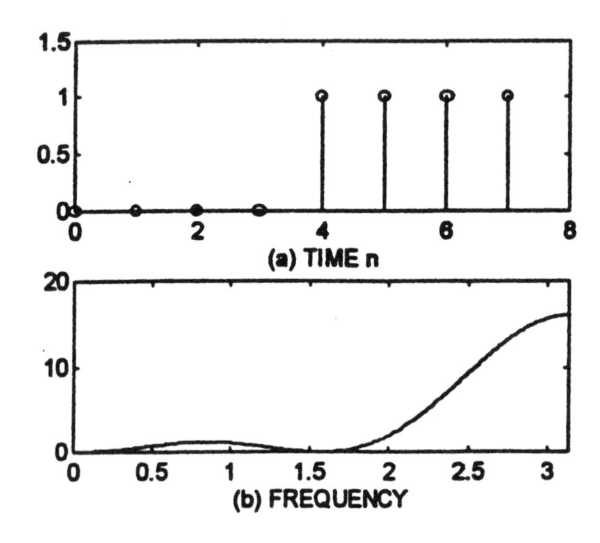

Figure 5.26: Energy distribution for cell Z_6: (a) time-domain; (b) frequency-domain.

5.6.2 Signal Decomposition in Time-Frequency Plane

We have seen how to synthesize block transform packets with specified time-frequency localization while maintaining the computational efficiency of the progenitor transform. The next and perhaps more challenging problem is the determination of the tiling pattern that "best" portrays the time-frequency energy properties of a signal. To achieve this goal, we will first review the differing ways of representing continuous time signals, and then work these into useful tiling patterns for discrete-time signals. Our description of classical time-frequency distributions is necessarily brief, and the reader is encouraged to read some of the cited literature for a more rigorous and detailed treatment (Cohen, 1989, Hlawatsch and Bartels, 1992).

The short-time Fourier transform (STFT) and the wavelet transform are examples of two-dimensional representations of the time-frequency and time-scale characteristics of a signal. Accordingly, these are often called spectrograms, and scalograms, respectively. The classical time-frequency distribution tries to describe how the energy in a signal is distributed in the time-frequency plane. These distributions $P(t, \Omega)$, then, are functions structured to represent the energy variation over the time-frequency plane.

The most famous of these is the Wigner (or Wigner-Ville) (Wigner, 1932; Ville, 1948) distribution for continuous-time signals. This distribution, $W(t, \Omega)$ represents the energy density at time t and frequency Ω, and $W(t, \Omega)\Delta t \Delta\Omega$ is the fractional energy in the time-frequency cell $\Delta t \Delta \Omega$, at the point (t, Ω). It is defined as

$$W(t, \Omega) = \int_{-\infty}^{\infty} x^*\left(t + \frac{\tau}{2}\right) x\left(t - \frac{\tau}{2}\right) e^{-jw\tau} d\tau \tag{5.56}$$

and the total energy

$$E = \int\int W(t, \Omega) dt d\Omega = 1. \tag{5.57}$$

Using $x(t) \leftrightarrow X(\Omega)$ as a Fourier transform pair, this distibution can also be expressed as

$$W(t, \Omega) = \frac{1}{2\pi}\int_{-\infty}^{\infty} X\left(\Omega - \frac{\nu}{2}\right) X^*\left(\Omega + \frac{\nu}{2}\right) e^{-j\nu t} d\nu. \tag{5.58}$$

This classical function has the following properties:

(1) It satisfies the marginals, i.e.,

$$\begin{aligned} \int W(t, \Omega) dt &= |X(\Omega)|^2 \\ \frac{1}{2\pi}\int W(t, \Omega) d\Omega &= |x(t)|^2 \end{aligned} \tag{5.59}$$

where $|x(t)|^2$ and $|X(\Omega)|^2$ are instantaneous energy per unit time and per unit frequency, respectively.
(2) $W(t, \Omega) = W^*(t, \Omega)$, i.e., it is real.
(3) Support properties: If $x(t)$ is strictly time-limited to $[t_1, t_2]$, then $W(t, \Omega)$ is also time-limited to $[t_1, t_2]$. By duality, a similar statement holds in the frequency domain.
(4) Inversion formula states that $W(t, \Omega)$ determines $x(t)$ within a multiplicative constant,

$$x(t)x^*(0) = \int W\left(\frac{t}{2}, \Omega\right) e^{jwt} dw. \tag{5.60}$$

(5) Time and frequency shifts: If $x(t) \leftrightarrow W_{xx}(t, \Omega)$, then

$$\overline{x}(t) = x(t - t_0)e^{j\Omega_0 t} \leftrightarrow W_{\overline{x}\overline{x}}(t, \Omega) = W_{xx}(t - t_0, \Omega - \Omega_0). \tag{5.61}$$

While the foregoing distribution has a very nice Fourier-like properties, it nevertheless suffers from the following disadvantages:
(1) It is computationally burdensome for the discrete-time case to be considered subsequently.
(2) $W(t, \Omega)$ can take on negative values, which is inappropriate for an "energy" function.
(3) The Wigner distribution is not zero at intervals where the time function $x(t)$ (or frequency function $X(\Omega)$ is zero.
(4) It has spurious terms or artifacts. For example, the Wigner distribution for the sum of two *sine* waves at frequencies Ω_1, Ω_2 has sharp peaks at these frequencies, but also a spurious term at $(\Omega_1 + \Omega_2/2)$. The Choi-Williams (1989) distribution ameliorates such artifacts by modifying the kernel of the Wigner distribution.

Still other time-frequency energy distributions have been proposed, two examples of which are the positive density function,

$$P(t, \Omega) = |X(\Omega)|^2 |x(t)|^2 \Phi(u, v), \tag{5.62}$$

where

$$\int \Phi(u, v)du = \int \Phi(u, v)dv = 1$$

and

$$u(t) = \int_{-\infty}^{t} |x(t)|^2 dt$$

$$v(\Omega) = \int_{-\infty}^{\Omega} |X(\nu)|^2 d\nu;$$

and the complex-valued Kirkwood (1933) density

$$P(t,\Omega) = x(t)X^*(\Omega)e^{-j\Omega t}. \tag{5.63}$$

Each of these satisfies the marginals and has some advantages over the Wigner form, primarily ease of computation.

The discrete time-frequency version of these distributions has been examined in the literature (Peyrin and Prost, 1986), and various forms have been advanced. The simplest Wigner form is

$$W(n,k) = \sum_m x^*(n-m)x(n+m)e^{-j4\pi km/L}, \tag{5.64}$$

where L is the DFT length. When derived from continuous time signals, sampling and aliasing considerations come into play. Details can be found in the literature. In particular, see Peyrin and Prost (1986).

A discrete version of the Kirkwood distribution is the real part of

$$P(n,k) = x^*(n)X(k)\Phi(k,n) \tag{5.65}$$

where $x(n) \leftrightarrow X(k)$ are a DFT pair and $[\Phi(k,n)] = [e^{j2\pi kn/N}]$ is the DFT matrix.

This last $P(n,k)$ satisfies the marginals

$$\sum_{n=0}^{N-1} P(n,k) = |X(k)|^2$$

and

$$\frac{1}{N}\sum_{n=0}^{N-1} P(n,k) = |x(n)|^2. \tag{5.66}$$

For the tiling study, and for ease of computation, we define the quasi-distribution

$$P(n,k) = \frac{1}{N}\frac{|x(n)|^2|X(k)|^2}{E} \tag{5.67}$$

with the properties

$$\begin{aligned} \sum_n P(n,k) &= \frac{1}{N}|X(k)|^2 \\ \sum_k P(n,k) &= |x(n)|^2 \\ \sum_n\sum_k P(n,k) &= E. \end{aligned} \tag{5.68}$$

For our computational purposes, we choose the $X(k)$ to be the DCT, rather than the DFT, as our tiling measure. Therefore, for our purposes, we define the time-frequency energy metric as the normalized product of instantaneous energy in each domain,

$$P(n,k) = C|x(n)|^2|X(k)|^2, \tag{5.69}$$

where C normalizes $\sum_n \sum_k P(n,k) = 1$. In the next section, the $P(n,k)$ of Eq. (5.69) is called a "microcell," and the distribution of these microcells defines the energy distribution in the time-frequency plane. The tiling pattern as discussed in the next section consists of the rank-ordered partitioning of the plane into clusters of microcells, each of which constitutes a resolution cell as described in Section 5.6.1.

5.6.3 From Signal to Optimum Tiling Pattern

We have seen how to construct block transform packets with specified time-frequency localization while maintaining the computational efficiency of the progenitor transform.

The next question is what kind of tiling pattern should we use to fit the signal characteristics? A resolution cell is a rectangle of constant area and a given location in the time-frequency plane. The tiling pattern is the partitioning of the time-frequency plane into contiguous resolution cells. This is a feasible partitioning. Associated with each resolution cell is a basis function or "atom." Each coefficient in the expansion of the signal in question using the new transform basis function represents the signal strength associated with that resolution cell. We want to find the tiling pattern corresponding to the maximum energy concentration for that particular signal. From an energy compaction point of view, the tiling pattern should be chosen such that the energies concentrate in as few coefficients as possible.

In order to answer this question, we need to define an appropriate time-frequency energy distribution which can be rapidly computed from the given signal.

Microcell Approach

The Kronecker delta sequence resolves the time-domain information, and the frequency-selective block transforms provide the frequency information. Combining these two characterizations together gives the energy sampling grid in the time-frequency plane. Let $x_i = |f(i)|^2$, the amplitude square of the function $f(i)$,

$0 \leq i \leq N-1$, at time t_i and $y_j = |F(j)|^2$ the magnitude square of the coefficient of the frequency-selective block transform (e.g., DCT) at frequency slot f_i. Take outer product of these two groups of samples to obtain the quasi distribution, Eq. (5.69), $P_{i,j} = x_i y_j$, $i, j = 0, 1, \ldots, N-1$. Each $P(i,j)$ represents the energy strength in the corresponding area in the time-frequency plane. The area corresponding to each $P(i,j)$ is called a microcell. $P = P(i,j)$ is the microcell energy pattern or distribution for a given signal. Totally we have N^2 microcells and each resolution cell is composed of N microcells. Take $N = 8$ as an example: We have 64 microcells and each resolution cell consists of 8 microcells arranged in a rectangular pattern 1×8, 2×4, 4×2, and 8×1. Therefore, our task here is to group the microcells such that the tiling pattern has the maximum energy concentration.

Search for the Most Energetic Resolution Cell

The most energetic resolution cell in P is the rectangular region which is composed of N microcells and has the maximum energy strength. Our objective is to search $P = \{P(i,j)\}$, the pattern of N^2 energy microcells in the T-F plane, to find the feasible pattern of N resolution cells Z_i, $0 \leq i \leq N-1$, such that the signal energy is optimally concentrated in as few cells as possible. We can perform an exhaustive search of P using rectangular windows of size N to find the most energetic resolution cell, and then the second most energetic resolution cell, and so on. With some assumptions, we can improve the search efficiency as follows. Assume that the most energetic microcell $P^*(i,j)$ is included in the most energetic resolution cell Z_i^*. We search the neighborhood of $P^*(i,j)$ to find the rectangular cluster of N microcells with the most energy. That cluster defines the most energetic resolution cell Z_i^*. Therefore, starting from the most energetic microcell, we group the microcells to find the most energetic resolution cell. The procedure is as follows:

(1) Rank order $P(i,j)$, and P_i, $i = 1, \ldots, N^2$ are the rank-ordered microcells.

(2) Form the smallest rectangle A_1 specified by P_1 and P_2. We test if these can be included in one resolution cell by simply calculating the area of A_1, $||A_1||$, $||A_1|| \leq N$. If not, test P_1 and P_3.

(3) Form the smallest rectangle A_2 specified by A_1 and next available P_i, and repeat the test.

(4) Repeat forming rectangles and tests until $||A_{last}|| = N$. A_{last} is the most energetic resolution cell Z_i^*.

Repeat this search for the next most energetic resolution cell. Eventually, a complete T-F tiling pattern can be obtained. This procedure is tedious and not practical for large transforms. In the following, we describe a more efficient way, a sequentially adaptive approach.

Adaptive Approach

The objective of the proposed method is to expand our signal in terms of BTP basis functions in a sequential fashion, i.e., find one resolution cell from a succession of N T-F tiling patterns rather than N cells from one T-F pattern. The concept of matching pursuit (Mallat and Zhang, 1993), as embodied in Fig. 5.27, suggests the following adaptive scheme:

(1) Start at stage $q = 1$. We construct P_1 from $f(n)$ and use the microcell and search algorithm to find the most energetic resolution cell Z_1 with its associated basis function $\psi_1(n)$ and block transform packet T_1. The projection of $f(n)$ onto $\psi_1(n)$ gives the coefficient β_1 and our first approximation

$$\hat{f}_1(n) = \beta_1 \psi_1(n) \quad and \quad \beta_1 = < f, \psi_1 > . \tag{5.70}$$

(2) Take the residual $\tilde{f}_1(n)$ as the input to the next stage where

$$\tilde{f}_1(n) = f(n) - \hat{f}_1(n) = f(n) - \beta_1 \psi_1(n). \tag{5.71}$$

(3) Repeat (1) and (2) for $q \geq 1$ where the residual signal $\tilde{f}_i(n)$ at ith stage is

$$\tilde{f}_i(n) = \tilde{f}_{i-1}(n) - \hat{f}_i(n) = \tilde{f}_{i-1}(n) - \beta_i \psi_i(n), \tag{5.72}$$

and $\psi_i(n)$ is the most energetic basis function corresponding to tiling pattern P_i and BTP T_i.

In general, the basis functions $\psi_i(n)$ need not be orthonormal to each other. However, at each stage the BTP is a unitary transform and therefore, $||f(n)|| > ||\tilde{f}_1(n)||$ and $||\tilde{f}_{i-1}(n)|| > ||\tilde{f}_i(n)||$. Thus the norm of the residual $\tilde{f}_i(n)$ monotonically decreases and converges to zero. Similar to the matching pursuit algorithm, this procedure maintains the energy conservation property.

Because this representation is adaptive, it will be generally concentrated in a very small subspace. As a result, we can use a finite summation to approximate the signal with a residual error as small as one wishes. The approximated signal can be expressed as

$$\hat{f}(n) = \sum_{i=1}^{L} \beta_i \psi_i(n). \tag{5.73}$$

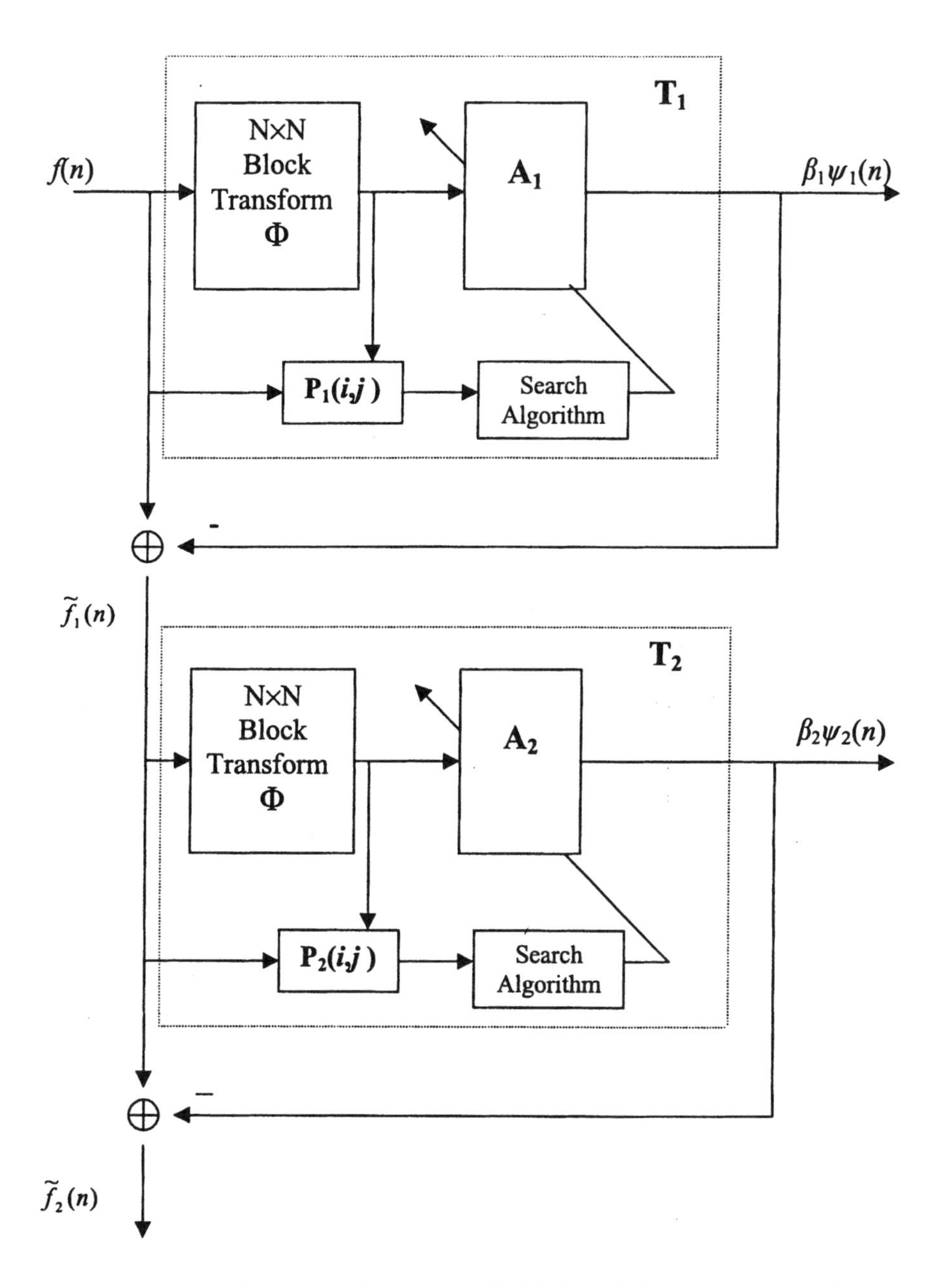

Figure 5.27: Block diagram of adaptive BTP-based decomposition algorithm.

The error energy for that frame using L coefficients is

$$\Omega_L = \sum_{n=1}^{N-1} |\tilde{f}_L(n)|^2. \tag{5.74}$$

For a long-length signal, this scheme can be adapted from frame to frame.

5.6.4 Signal Compaction

In this section, two examples are given to show the energy concentration properties of the adaptive BTP. The BTP is constructed from DCT bases with block size 32. In each example the signal length is 1024 samples. The data sequence is partitioned into 32 frames consisting of 32 samples per frame. For each frame, we compute the residual $\tilde{f}_i(n)$ and the corresponding error energy Ω_i, $1 \leq i \leq 4$. The average of these Ω_i's over 32 frames is then plotted for each example to show the compression efficiency. For comparison purpose, the standard DCT codec is also used.

Figure 5.28(a) shows the energy concentration property in terms of the number of coefficients for BTP and DCT codecs. The testing signal is a narrow band Gaussian signal S_1 with bandwidth$= 0.2$ rad and central frequency $5\pi/6$. Because of the frequency-localized nature of this signal, BTP has only slight compaction improvement over the DCT. The signal used in Fig. 5.28(b) is the narrow-band Gaussian signal S_1 plus time-localized white Gaussian noise S_2 with 10% duty cycle and power ratio $(S_1/S_2) = -8$dB. Basically, it is a combination of frequency-localized and time-localized signals and therefore, it cannot be resolved only in the time- or in the frequency-domain. As expected, BTP shows the compaction superiority over DCT in Fig. 5.28(b). It demonstrates that BTP is a more efficient and robust compaction engine over DCT.

It is noted that the BTP codec, like any adaptive tree codec, needs some side information for decompression. They are the starting point and size of the post matrix A_k in Eq. (5.45), and the location of the most energetic coefficient which defines the location and shape of the most energetic resolution tile. If one uses the adaptive approach, side information is necessary at each stage. Therefore, the compression efficiency will be reduced significantly. One possible solution is to use one tiling pattern for each frame of data. Another possible solution is to use the same BTP basis functions for adjacent frames of data. Both will reduce the side-information effect and improve the compression efficiency.

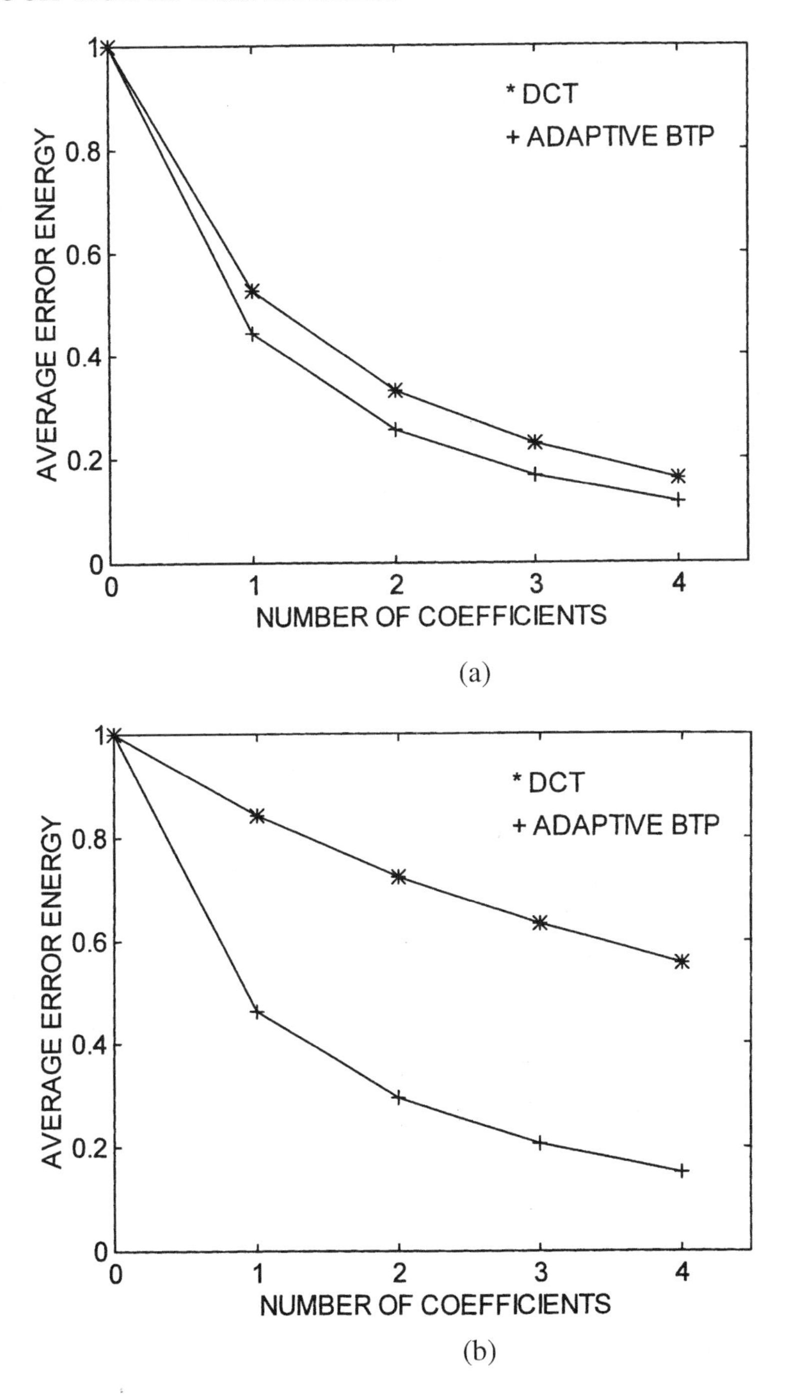

Figure 5.28: Compaction efficiency comparisons for (a) narrow-band Gaussian signal S_1, (b) S_1 plus time localized Gaussian signal S_2 with power ratio $S_1/S_2 = -8$ dB.

5.6.5 Interference Excision

Spread spectrum communication systems provide a degree of interference rejection capability. However, if the level of interference becomes too great, the system will not function properly. In some cases, the interference immunity can be improved significantly using signal processing techniques which complement the spread spectrum modulation (Milstein, 1988).

The most commonly used type of spread spectrum is the direct sequence spread spectrum (DSSS), as shown in Fig. 5.29, in which modulation is achieved by superimposing a pseudo-noise (PN) sequence upon the data bits. During the transmission, the channel adds the noise term $\underline{n}$ and an interference $\underline{j}$. Therefore, the received signal $\underline{f}$ can be written as

$$\underline{f} = \underline{s} + \underline{j} + \underline{n}, \tag{5.75}$$

where the desired signal $\underline{s} = d\underline{c}$ is the product of data bit stream d and the spreading sequence $\underline{c}$. In general, $\underline{n}$ is assumed to be additive white Gaussian noise with parameter N_0 and $\underline{j}$ could be the narrow-band or time-localized Gaussian interference. In the absence of jammers, both the additive white Gaussian noise $\underline{n}$ and PN modulated sequence $\underline{s}$ are uniformly spread out in both time and frequency domains. Because of the presence of the jamming signal, the spectrum of the received signal will not be flat in the time-frequency plane. The conventional fixed transform based excisers map the received signal into frequency bins and reject the terms with power greater than some threshold. This system works well if the jammers are stationary and frequency localized. In most cases jamming signals are time-varying and not frequency concentrated. Furthermore, the discrete wavelet transform bases are not adapted to represent functions whose Fourier transforms have a narrow high frequency support (Medley et al., 1994). Therefore, conventional transform-domain based techniques perform poorly in excising nonstationary interference such as spikes (Tazebay and Akansu, 1995).

Adaptive BTPs provide arbitrary T-F resolutions and are suitable for dealing with such problems. The energetic resolution cells indicate the location of the jamming signal in the T-F plane; this jamming signal can be extracted from the received signal by using adaptive BTP based techniques. Figure 5.27 shows the adaptive scheme for multistage interference excision.

(1) In the first stage, $q = 1$, we construct BTP T_1 for a frame of the received signal $\underline{f}$ by using the microcell and search algorithm and find the basis function $\psi_1(n)$ and coefficient β_1 associated with the most energetic cell.

(2) If the interference is present (the time-frequency spectrum is not flat as determined by comparison of the most energetic cell with a threshold based on

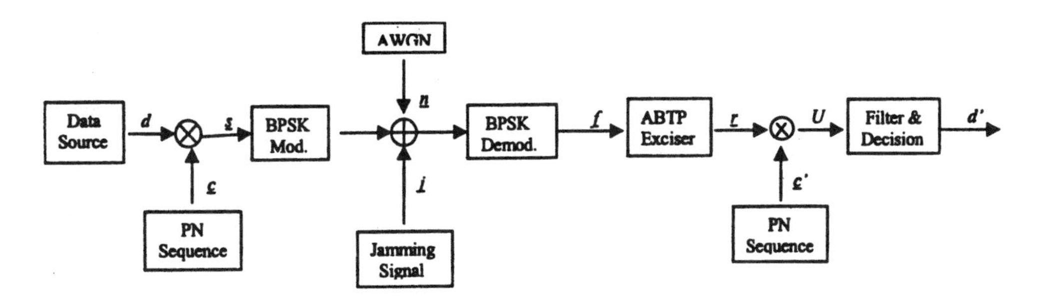

Figure 5.29: Block diagram of a DSSS communication system.

average of energy in all other cells), take the residual $\tilde{f}_1(n)$ as the input to the next stage as in Eq. (5.71).

(3) Repeat (1) and (2) for $q > 1$ where the residual signal $\tilde{f}_i(n)$ at ith stage as in Eq. (5.72) and $\psi_i(n)$ is the most energetic basis function corresponding to BTP T_i.

(4) Stop this process at any stage where the spectrum of the residual signal at that stage is flat.

The performance of the proposed ABTP exciser is compared with DFT and DCT excisers. A 32-chip PN code is used to spread the input bit stream. The resulting DSSS signal is transmitted over an AWGN channel. Two types of interference are considered: a narrow-band jammer with uniformly distributed random phase ($\theta \in [0, 2\pi]$), and a pulsed (time-localized) wide-band Gaussian jammer. Figure 5.30(a) displays the bit error rate (BER) performance of the ABTP exciser along with DFT and DCT based excisers for the narrow-band jammer case where the signal to interference power ratio (SIR) is -15 dB. The jamming signal j can be expressed as

$$j(t) = A\sin(\omega_0 t + \theta), \tag{5.76}$$

where $\omega_0 = \pi/2$ and $\theta \in [0, 2\pi]$. Three largest bins are removed for DFT and DCT based excisers. Because of the frequency concentrated nature of the jamming signal, all systems perform comparably. Figure 5.30(b) shows the results for a time-localized wideband Gaussian jammer. The jammer is an on/off type that is randomly switched with a 10% duty cycle. In this scenario, as expected, none of the fixed-transform-based excisers is effective for interference suppression. However, the ABTP exciser has significant improvement over the fixed transform based exciser. The ABTP exciser also has consistent performance at several other SIR values.

It should be noted that neither the duty cycle nor the switching time of the interference is known a priori in this scheme. The ABTP exciser performs slightly better than the DCT exciser for the single tone interference, but is far superior to the DCT and DFT for any combination of time-localized wide-band Gaussian jammers or time-localized single-tone interference.

The excision problem is revisited in Section 7.2.2 from the standpoint of adaptive pruning of a subband tree structure. A smart time-frequency exciser (STFE) that is domain-switchable is presented. Its superior performance over existing techniques is presented and interpreted from the time-frequency perspective.

5.6.6 Summary

Traditional Fourier analysis views the signal over its entire extent in time or in frequency. It is clearly inadequate for dealing with signals with nonstationary characteristics. The STFT, the wavelet transform, and the block transform packet are analysis techniques which can extract signal features in the time-frequency plane.

In this chapter, we compared the localization properties of standard block transforms and filter banks from this vantage point.

The time-frequency approach described in this chapter sets the stage for innovative and adaptive methods to deal with "problem" signals, some of which are described here, and others outlined in Chapter 7.

(Figure on facing page) Bit error rate (BER) performance for adaptive BTP exciser: (a) BER for narrow-band interference, $SIR = -15$ dB, (b) BER for time-localized wide-band Gaussian jammer, 10% duty cycle, and $SIR = -15$ dB.

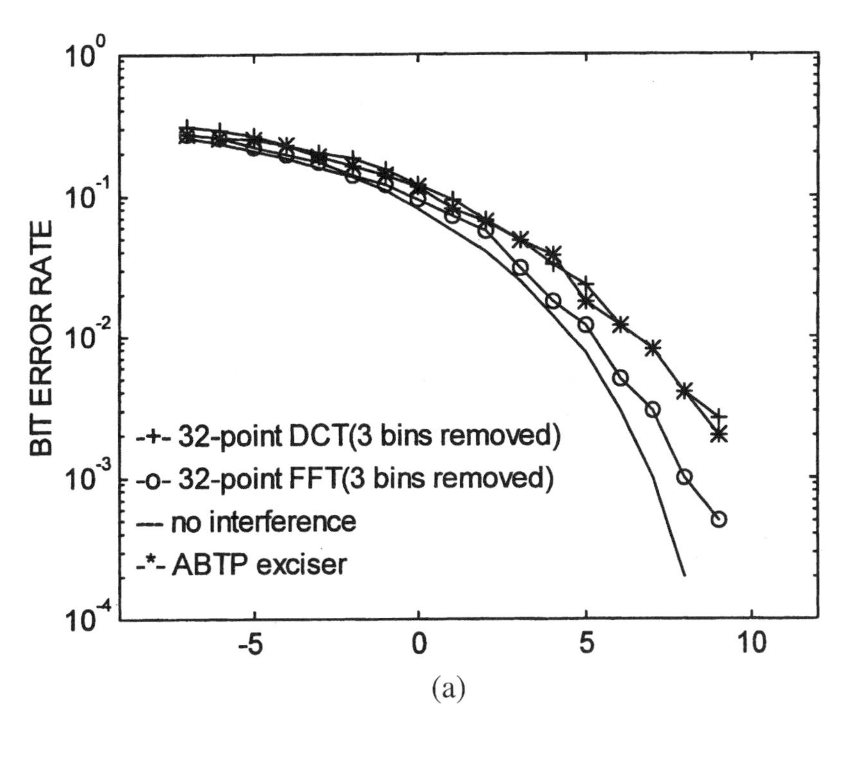

(a)

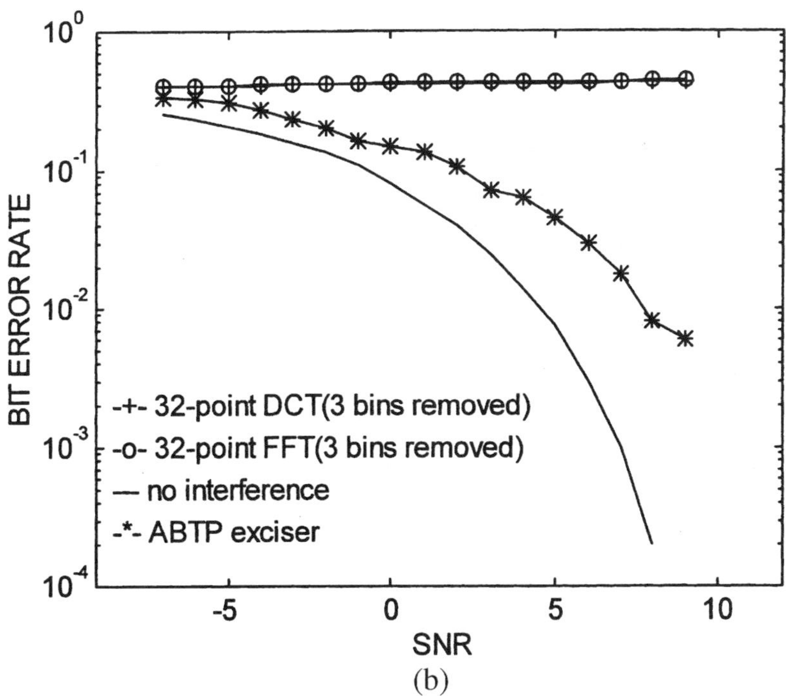

(b)

Figure 5.30

References

A. N. Akansu, "Multiplierless Suboptimal PR-QMF Design," Proc. SPIE Visual Communication and Image Processing, pp. 723–734, Nov. 1992.

A. N. Akansu, R. A. Haddad and H. Caglar, "The Binomial QMF-Wavelet Transform for Multiresolution Signal Decomposition," IEEE Trans. Signal Processing, Vol. 41, No. 1, pp. 13–19, Jan. 1993.

A. N. Akansu and Y. Liu, "On Signal Decomposition Techniques," Optical Engineering, pp. 912–920, July 1991.

A. N. Akansu, R. A. Haddad and H. Caglar, "Perfect Reconstruction Binomial QMF-Wavelet Transform," Proc. SPIE Visual Communication and Image Processing, Vol. 1360, pp. 609–618, Oct. 1990.

J. B. Allen and L. R. Rabiner, "A Unified Approach to Short-Time Fourier Analysis and Synthesis," Proc. IEEE, Vol. 65, pp. 1558–1564, 1977.

M. J. Bastiaans, "Gabor's Signal Expansion and Degrees of Freedom of a Signal," Proc. IEEE, Vol. 68, pp. 538–539, 1980.

R. B. Blackman and J. W. Tukey, *The Measurement of Power Spectra.* Dover, 1958.

B. Boashash and A. Riley, "Algorithms for Time-Frequency Signal Analysis," in B. Boashash, Ed., *Time-Frequency Signal Analysis.* Longman, 1992.

G. F. Boudreaux-Bartels and T. W. Parks, "Time-Varying Filtering and Signal Estimation Using Wigner Distribution Synthesis Techniques," IEEE Trans. ASSP, Vol. ASSP-34, pp. 442–451, 1986.

H. Caglar, Y. Liu and A. N. Akansu, "Statistically Optimized PR-QMF Design," Proc. SPIE Visual Communication and Image Processing, Vol. 1605, pp. 86–94, 1991.

L. C. Calvez and P. Vilbe, "On the Uncertainty Principle in Discrete Signals," IEEE Trans. Circ. Syst.-II: Analog and Digital Signal Processing, Vol. 39, No. 6, pp. 394–395, June 1992.

D. C. Champeney, *A Handbook of Fourier Transforms.* Cambridge University Press, 1987.

H. I. Choi and W. J. Williams, "Improved Time-Frequency Representation of Multicomponent Signals Using Exponential Kernels," IEEE Trans. ASSP, Vol. ASSP-37, 1989.

T. A. C. M. Claasen and W. F. G. Mecklenbrauker, "The Wigner Distribution-A Tool for Time-Frequency Signal Analysis; Part I: Continuous-Time Signals," Philips J. Res., Vol. 35, pp. 217–250, 1980.

T. A. C. M. Claasen and W. F. G. Mecklenbrauker, "The Wigner Distribution- A Tool for Time-Frequency Signal Analysis; Part II: Discrete-Time Signals," Philips J. Res., Vol. 35, pp. 276–300, 1980.

L. Cohen, "Generalized Phase-Spaced Distribution Functions," J. Math. Phys., Vol. 7, pp. 781–806, 1966.

L. Cohen, "Time Frequency Distributions- A Review," Proc. of IEEE, Vol. 77, No. 7, pp. 941–981, July 1989.

R. Coifman and Y. Meyer, "The Discrete Wavelet Transform," Technical Report, Dept. of Math., Yale Univ., 1990.

R. Coifman and Y. Meyer, "Orthonormal Wave Packet Bases," Technical Report, Dept. of Math., Yale Univ., 1990.

J. M. Combes, A. Grossman, and P. H. Tchamitchian, Eds., *Time-Frequency Methods and Phase Space.* Springer, 1989.

R. E. Crochiere and L. R. Rabiner, *Multirate Digital Signal Processing.* Prentice-Hall, 1983.

I. Daubechies, "Orthonormal Bases of Compactly Supported Wavelets," Communications in Pure and Applied Math., Vol. 41, pp. 909–996, 1988.

I. Daubechies, "The Wavelet Transform, Time-Frequency Localization and Signal Analysis," IEEE Trans. Information Theory, Vol. 36, No. 5, pp. 961–1005, Sept. 1990.

I. Daubechies, A. Grossmann, and Y. Meyer, "Painless Nonorthogonal Expansions," J. Math. Phys., Vol. 27, No. 5, pp. 1271–1283, 1986.

J. B. J. Fourier, "Theorie Analytique de la Chaleur," in *Oeuvres de Fourier*, tome premier, G. Darboux, ed. Gauthiers-Villars, 1888.

D. Gabor, "Theory of Communications," J. of the IEE (London), Vol. 93, pp. 429–457, 1946.

J. Gevargiz, P. K. Das, and L. B. Milstein, "Adaptive Narrowband Interference Rejection in a DS Spread Spectrum Intercept Receiver Using Transform Signal Processing Techniques," IEEE Trans. Communications, Vol. 37, No. 12, pp. 1359–1366, Dec. 1989.

R. A. Haddad, A. N. Akansu and A. Benyassine, "Time-Frequency Localization in Transforms, Subbands and Wavelets: A Critical review," Optical Engineering, Vol. 32, No. 7, pp. 1411–1429, July 1993.

C. Herley, J. Kovacevic, K. Ramchandran, and M. Vetterli, "Tilings of the Time-Frequency Plane: Construction of Arbitrary Orthogonal Bases and Fast

Tiling Algorithm," IEEE Trans. Signal Processing, Vol. 41, No. 12, pp. 3341–3359, Dec. 1993.

C. Herley, Z. Xiong, K. Ramchandran, and M. T. Orchard, "An Efficient Algorithm to Find a Jointly Optimal Time-Frequency Segmentation Using Time-Varying Filter Banks," Proc. IEEE ICASSP, Vol. 2, pp. 1516–1519, 1995.

F. Hlawatsch and G. F. Boudreaux-Bartels, "Linear and Quadratic Time-Frequency Signal Representations," IEEE Signal Processing Magazine, pp. 21–67, April 1992.

J. Horng and R. A. Haddad, "Variable Time-Frequency Tiling Using Block Transforms," Proc. IEEE DSP Workshop, pp. 25–28, Norway, Sept. 1996.

J. Horng and R. A. Haddad, "Signal Decomposition Using Adaptive Block Transform Packets", Proceedings of ICASSP'98, Int'l Conf. on Acoustics, Speech, and Signal Processing, pp. 1629–1632, May 1998.

J. Horng and R. A. Haddad, "Interference Excision in DSSS Communication System Using Time-Frequency Adaptive Block Transform," Proc. IEEE Symp. on Time-Frequency and Time-Scale Analysis, pp. 385–388, Oct. 1998.

J. Horng and R. A. Haddad, "Block Transform Packets - An Efficient Approach to Time-Frequency Decomposition", IEEE Int'l Symp. on Time-Frequency and Time-Scale Analysis, Oct. 1998.

A. J. E. M. Janssen, "Gabor Representation of Generalized Functions," J. Math. Appl., Vol. 80, pp. 377–394, 1981.

W. W. Jones and K. R. Jones, "Narrowband Interference Suppression Using Filter Bank Analysis/Synthesis Techniques," Proc. IEEE MILCOM, pp. 38.1.1–38.1.5, 1992.

S. M. Kay, *Modern Spectral Estimation.* Prentice-Hall, 1987.

J. G. Kirkwood, "Quantum Statistics of Almost Classical Ensembles," Phys. Rev., Vol. 44, pp. 31–37, 1933.

K. Y. Kwak and R. A. Haddad, "A New Family of Orthonormal Transforms with Time Localizability Based on DFT," in Proc. SPIE Visual Comm. and Image Processing, Vol. 2308, Oct. 1994.

K. Y. Kwak and R. A. Haddad, "Time-Frequency Analysis of Time Localizable Linear Transform based on DFT," Proc. 28th Asilomar Conf. on SSC, pp. 1095–1099, 1994.

H. J. Landau and H. O. Pollak, "Prolate Spheroidal Wave Functions, Fourier Analysis and Uncertainty, II," Bell Syst. Tech. J., Vol. 40, pp. 65–84, 1961.

M. J. Levin, "Instantaneous Spectra and Ambiguity Functions," IEEE Trans. Information Theory, Vol. IT-13, pp. 95–97, 1967.

S. G. Mallat and Z. Zhang, "Matching Pursuit with Time-Frequency Dictionaries," IEEE Trans. Signal Processing, Vol. 41, No. 12, pp. 3397–3415, Dec. 1993.

H. Margenau and R. N. Hill, "Correlation Between Measurements in Quantum Theory," Prog. Theor. Phys., Vol. 26, pp. 722–738, 1961.

S. L. Marple, *Digital Spectral Analysis with Applications.* Prentice-Hall, 1987.

M. Medley, G. Saulnier, and P. Das, "Applications of Wavelet Transform in Spread Spectrum Communications Systems," Proc. SPIE, Vol. 2242, pp. 54–68, 1994.

A. Mertins, *Signal Analysis.* John Wiley, 1999.

L. B. Milstein, "Interference Rejection Techniques in Spread Spectrum Communications," Proc. IEEE, Vol. 76, No. 6, pp. 657–671, June 1988.

S. H. Nawab and T. Quartieri, "Short Time Fourier Transforms," A Chapter in J. S. Lim and A. V. Oppenheim, Eds., *Advanced Topics in Signal Processing.* Prentice-Hall, 1988.

C. H. Page, "Instantaneous Power Spectra," J. Appl. Phys., Vol. 23, pp. 103–106, 1952.

A. Papoulis, *Signal Analysis.* McGraw-Hill, 1977.

F. Peyrin and R. Prost, "A Unified Definition for the Discrete-Time Discrete-Frequency, and Discrete-Time/Frequency Wigner Distributions," IEEE Trans. ASSP, Vol. ASSP-34, pp. 858–867, 1986.

M. R. Portnoff, "Time-Frequency Representation of Digital Signals and Systems Based on the Short-Time Fourier Analysis," IEEE Trans. ASSP, Vol. 28, pp. 55–69, Feb. 1980.

S. Qian and D. Chen, "Discrete Gabor Transform," IEEE Trans. Signal Processing, Vol. 41, No. 7, pp. 2429–2439, July 1993.

S. Qian, D. Chen and K. Chen, "Signal Approximation via Data-Adaptive Normalized Gaussian Functions and its Applications for Speech Processing," Proc. IEEE ICASSP, Vol. 1, pp. 141–144, March 1992.

K. Ramchandran and M. Vetterli, "Best Wavelet Packet Bases in a Rate-Distortion Sense," IEEE Trans. Image Processing, Vol. 2, No. 2, pp. 160–175, April 1993.

W. Rihaczek, "Signal Energy Distribution in Time and Frequency," IEEE Trans. Information Theory, Vol. IT-14, pp. 369–374, 1968.

D. Slepian and H. O. Pollak, "Prolate Spheroidal Wave Functions, Fourier Analysis and Uncertainty, I," Bell Syst. Tech. J., Vol. 40, pp. 43–64, 1961.

M. J. T. Smith and T. Barnwell, "The Design of Digital Filters for Exact Reconstruction in Subband Coding," IEEE Trans. ASSP, Vol. ASSP-34, No. 6, pp. 434–441, June 1986.

J. M. Speiser, "Wide-Band Ambiguity Functions," IEEE Trans. Info. Theory, pp. 122–123, 1967.

M. V. Tazebay and A. N. Akansu, "Adaptive Subband Transforms in Time-Frequency Excisers for DSSS Communications Systems," IEEE Trans. Signal Processing, Vol. 43, No. 11, Nov. 1995.

L. C. Trintinalia and H. Ling, "Time-Frequency Representation of Wideband Radar Echo Using Adaptive Normalized Gaussian Functions," Proc. IEEE Antennas and Propagation Soc. Int'l Symp., Vol. 1, pp. 324–327, June 1995.

J. Ville, "Theorie et Applications de la Notion de Signal Analytique," Cables et Transmission, Vol. 2A, pp. 61–74, 1948.

E. P. Wigner, "On the Quantum Correction for Thermodynamic Equilibrium," Phys. Rev., Vol. 40, pp. 749–759, 1932.

J. E. Younberg and S. F. Boll, "Constant-Q Signal Analysis and Synthesis," Proc. IEEE ICASSP, pp. 375–378, 1978.

Chapter 6

Wavelet Transform

The wavelet transforms, particularly those for orthonormal and biorthogonal wavelets with finite support, have emerged as a new mathematical tool for multiresolution decomposition of continuous-time signals with potential applications in computer vision, signal coding, and others. Interestingly enough, these wavelet bases are closely linked with the unitary two-band perfect reconstruction quadrature mirror filter (PR-QMF) banks and biorthogonal filter banks developed in Chapter 3.

In this chapter, we present the continuous wavelet transform as a signal analysis tool with the capability of variable time-frequency localization and compare it with the fixed localization of the short-time Fourier transform. Our prime interest, however, is in the multiresolution aspect of the compactly supported wavelet and its connection with subband techniques. The two-band PR filter banks will play a major role in this linkage. The multiresolution aspect becomes clear upon developing the one-to-one correspondence between the coefficients in the fast wavelet transform and the dyadic or octave-band subband tree.

The Haar wavelets, which are discontinuous in time, and the Shannon wavelets, discontinuous in frequency, are introduced to provide simple and easily understood examples of multiresolution wavelet concepts.

Upon establishing the link between wavelets and subbands, we then focus on the properties of wavelet filters and compare these with other multiresolution decomposition techniques developed in this book. In particular the Daubechies filters (1988) are shown to be identical to the Binomial QMFs of Chapter 4; other wavelet filters are seen to be special cases of the parameterized PR-QMF designs of Chapter 4. Finally, we comment on the concept of wavelet "regularity" from a subband coding perspective.

We strive to explain concepts and some of the practical consequences of wavelets rather than to focus on mathematical rigor. The interested reader is provided with list of references for more detailed studies.

We conclude this chapter with some discussions and suggest avenues for future studies in this active research topic.

6.1 The Wavelet Transform

The wavelet transform (WT) is another mapping from $L^2(R) \longrightarrow L^2(R^2)$, but one with superior time-frequency localization as compared with the STFT. In this section, we define the continuous wavelet transform and develop an *admissibility* condition on the wavelet needed to ensure the invertibility of the transform. The discrete wavelet transform (DWT) is then generated by sampling the wavelet parameters (a, b) on a grid or lattice. The question of reconstruction of the signal from its transform values naturally depends on the *coarseness* of the sampling grid. A fine grid mesh would permit easy reconstruction, but with evident redundancy, i.e., *oversampling*. A too-coarse grid could result in loss of information. The concept of *frames* is introduced to address these issues.

6.1.1 The Continuous Wavelet Transform

The continuous wavelet transform (CWT) is defined by Eq. (6.1) in terms of dilations and translations of a prototype or *mother* function $\psi(t)$. In time and Fourier transform domains, the wavelet is

$$\psi_{ab}(t) = \frac{1}{\sqrt{a}}\psi(\frac{t-b}{a}) \leftrightarrow \Psi_{ab}(\Omega) = \sqrt{a}\Psi(a\Omega)e^{-jb\Omega}. \tag{6.1}$$

The CWT maps a function $f(t)$ onto time-scale space by[1]

$$W_f(a,b) = \int_{-\infty}^{\infty} \psi_{ab}(t)f(t)dt =< \psi_{ab}(t), f(t) > . \tag{6.2}$$

The transform is invertible if and only if the *resolution of identity* holds (Klauder and Sudarshan, 1968) and is given by the superposition

$$f(t) = \underbrace{\frac{1}{C_\psi}\int_{-\infty}^{\infty}\int_0^{\infty}\frac{dadb}{a^2}}_{summation} \quad \underbrace{W_f(a,b)}_{\substack{Wavelet \\ coefficients}} \quad \underbrace{\psi_{ab}(t)}_{Wavelet} \tag{6.3}$$

[1] For the remainder of this chapter, all *time* functions (including wavelets) are assumed to be real.

where

$$C_\psi = \int_0^\infty \frac{|\Psi(\Omega)|^2}{\Omega} d\Omega, \tag{6.4}$$

provided a real $\psi(t)$ satisfies the *admissibility* condition. The wavelet is called *admissible* if $C_\psi < \infty$ (Grossmann, Morlet, and Paul, 1985-86) (see Appendix A).

This in turn implies that the *DC gain* $\Psi(0) = 0$,

$$\Psi(0) = \int_{-\infty}^{\infty} \psi(t)dt = 0. \tag{6.5}$$

Thus, $\psi(t)$ behaves as the impulse response of a band-pass filter that decays at least as fast as $|t|^{1-\epsilon}$. In practice, $\psi(t)$ should decay much faster to provide good time-localization.

Another way of stating admissibility is to require that for any two functions, $f, g \in L^2(R)$,

$$\int\int W_f(a,b) W_g^*(a,b) \frac{dadb}{a^2} = C_\psi \int f(t) g^*(t) dt. \tag{6.6}$$

In particular, the Parseval relation becomes

$$C_\psi \int |f(t)|^2 dt = \int\int |W_f(a,b)|^2 \frac{dadb}{a^2}. \tag{6.7}$$

The latter states that within a scale factor, the wavelet transform is an *isometry* from $L^2(R)$ into $L^2(R^2)$.

To prove this, note that $W_f(a,b)$ is the convolution of $f(\tau)$ with the time-reversed wavelet $\psi_{ab}(-\tau)$ evaluated at $t = b$.

Hence,

$$W_f(a,b) = f(t) * \psi_{ab}(-t)|_{t=b} \Longleftrightarrow F(\Omega)|a|^{1/2}\Psi^*(a\Omega)e^{j\Omega b}.$$

As derived in Appendix A, this leads directly to

$$\begin{aligned} \int_{-\infty}^{\infty}\int_{-\infty}^{\infty} |W_f(a,b)|^2 \frac{dadb}{a^2} &= \frac{1}{2\pi}\int_{-\infty}^{\infty} |F(\Omega)|^2 \underbrace{\int_{-\infty}^{\infty} \frac{|\Psi(a\Omega)|^2}{|a|} da}_{C_\psi} \, d\Omega \\ &= C_\psi \int_{-\infty}^{\infty} |f(t)|^2 dt. \end{aligned} \tag{6.8}$$

This is another interpretation of the *admissibility* condition introduced earlier.

It is worth noting that the orthonormal wavelet transform also preserves energy between the different scales such that[2]

$$\int_{-\infty}^{\infty} |\psi(t)|^2 dt = \int_{-\infty}^{\infty} \frac{1}{|a|} |\psi_{ab}(t)|^2 dt. \tag{6.9}$$

An often-quoted example of a wavelet is the second derivative of a Gaussian,

$$\psi(t) = (1 - t^2)e^{-t^2/2} \longleftrightarrow \Psi(\Omega) = \sqrt{2\pi\Omega^2}e^{-\Omega^2/2}. \tag{6.10}$$

This pair is sketched in Fig. 6.1. This function has excellent localization in time and frequency and clearly satisfies the admissibility condition.

The *admissibility condition* ensures that the continuous wavelet transform is complete if $W_f(a, b)$ is known for all a, b. Figure 5.3 displays a typical wavelet and its dilations. It shows the band-pass nature of $\psi(t)$ and the time-frequency resolution of the wavelet transform.

We have seen in Chapter 5 that the STFT yields the decomposition of a signal into a set of equal bandwidth functions sweeping the frequency spectrum. On the other hand the wavelet transform provides the decomposition of a signal by a set of constant Q (or equal bandwidth on a logarithmic scale) band-pass functions. The constant bandwidth condition on a logarithmic scale is implicit in Eq. (6.1). The roles played by the transform parameters are also different for STFT and wavelet transforms. The time parameter τ in the STFT refers to the actual time instant in the signal, while the parameter b in the continuous wavelet transform refers to the time instant $a^{-1}b$. In other words, the parameter in the wavelet representation indicates time by using a yardstick that is local in scale. The time scale thus adapts to the frequency scale under examination. The coarse or fine frequency scale is accompanied by a time-scale that is accordingly long or short. This is the primary reason for the efficiency of the wavelet transform inherent in a multiresolution environment over the STFT. This fact also suggests that the wavelet transform can represent a multirate system.

There is a time-frequency resolution trade-off in the wavelet transform. To quantify how the continuous wavelet transform spans the time-frequency plane, the measures of time and frequency resolutions must be defined. Let σ_t and σ_Ω be the RMS extent of the mother wavelet function $\psi(t)$ in time and frequency domains defined as

$$\sigma_t^2 = \int (t - t_0)^2 |\psi(t)|^2 dt \tag{6.11}$$

[2]This is the normalization of the wavelet at different scales. The wavelet transform of a function will have different energies in different scales.

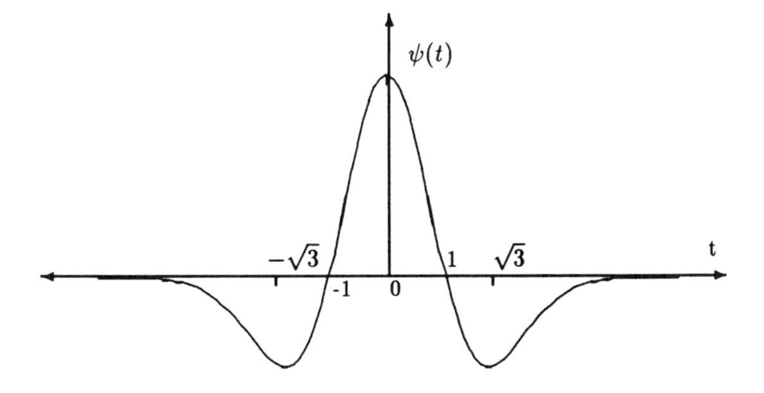

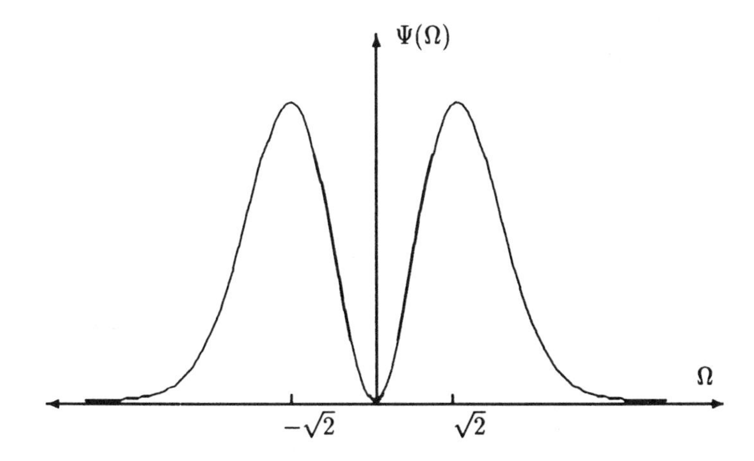

Figure 6.1: Gaussian based (second derivative) wavelet function and its Fourier transform.

$$\sigma_{\Omega}^2 = \int (\Omega - \Omega_0)^2 |\Psi(\Omega)|^2 d\Omega \tag{6.12}$$

where the wavelet function $\psi(t)$ is centered at (t_0, Ω_0) in the time-frequency plane. Hence, $\psi(\frac{t-b}{a})$ is centered at $(t_0, \Omega_0/a)$. At this point the RMS extent σ_{ab_t} and σ_{ab_Ω} are

$$\sigma_{ab_t}^2 = \int_{-\infty}^{\infty} (t - t_0)^2 \, |\psi_{ab}(t)|^2 \, dt$$

$$= a^2\sigma_t^2 \tag{6.13}$$

and

$$\begin{aligned}\sigma_{ab\Omega}^2 &= \int_0^\infty (\Omega - \frac{\Omega_0}{a})^2 |\Psi_{ab}(\Omega)|^2 d\Omega \\ &= \frac{1}{a^2}\sigma_\Omega^2. \end{aligned} \tag{6.14}$$

These results explain the role of the scaling parameter a in the wavelet transform. Figure 6.2(b) and (c) depict time-frequency resolutions of the Daubechies wavelet (6-tap) and scaling functions, which will be introduced in Section 6.3.2, for different values of the dilation parameter a.

6.1.2 The Discrete Wavelet Transform

The continuous wavelet transform suffers from two drawbacks: redundancy and impracticality. The first is obvious from the nature of the wavelet transform and the second from the fact that both transform parameters are continuous. We can try to solve both problems by sampling the parameters (a, b) to obtain a set of wavelet functions in discretized parameters. The questions that arise are:

- Is the set of discrete wavelets complete in $L^2(R)$?
- If complete, is the set redundant?
- If complete, how coarse can the sampling grid be such that the set is minimal, i.e., nonredundant?

We will address these questions and show that the *tightest* set is the *orthonormal wavelet* set, which can be synthesized through a multiresolution framework, which is the focus of our efforts in this text. The reader interested in a deeper treatment on the choice of sampling grids may consult Duffin and Schaeffer (1952) and Young (1980).

Let the sampling lattice be

$$a = a_0^m \qquad b = nb_0a_0^m$$

so that

$$\psi_{mn}(t) = a_0^{-m/2}\psi(a_0^{-m}t - nb_0) \tag{6.15}$$

where $m, n \in Z$. If this set is complete in $L^2(R)$ for some choice of $\psi(t)$, a, b, then the $\{\psi_{mn}\}$ are called *affine* wavelets. Then we can express any $f(t) \in L^2(R)$ as the superposition

$$f(t) = \sum_m \sum_n d_{m,n}\psi_{mn}(t) \tag{6.16}$$

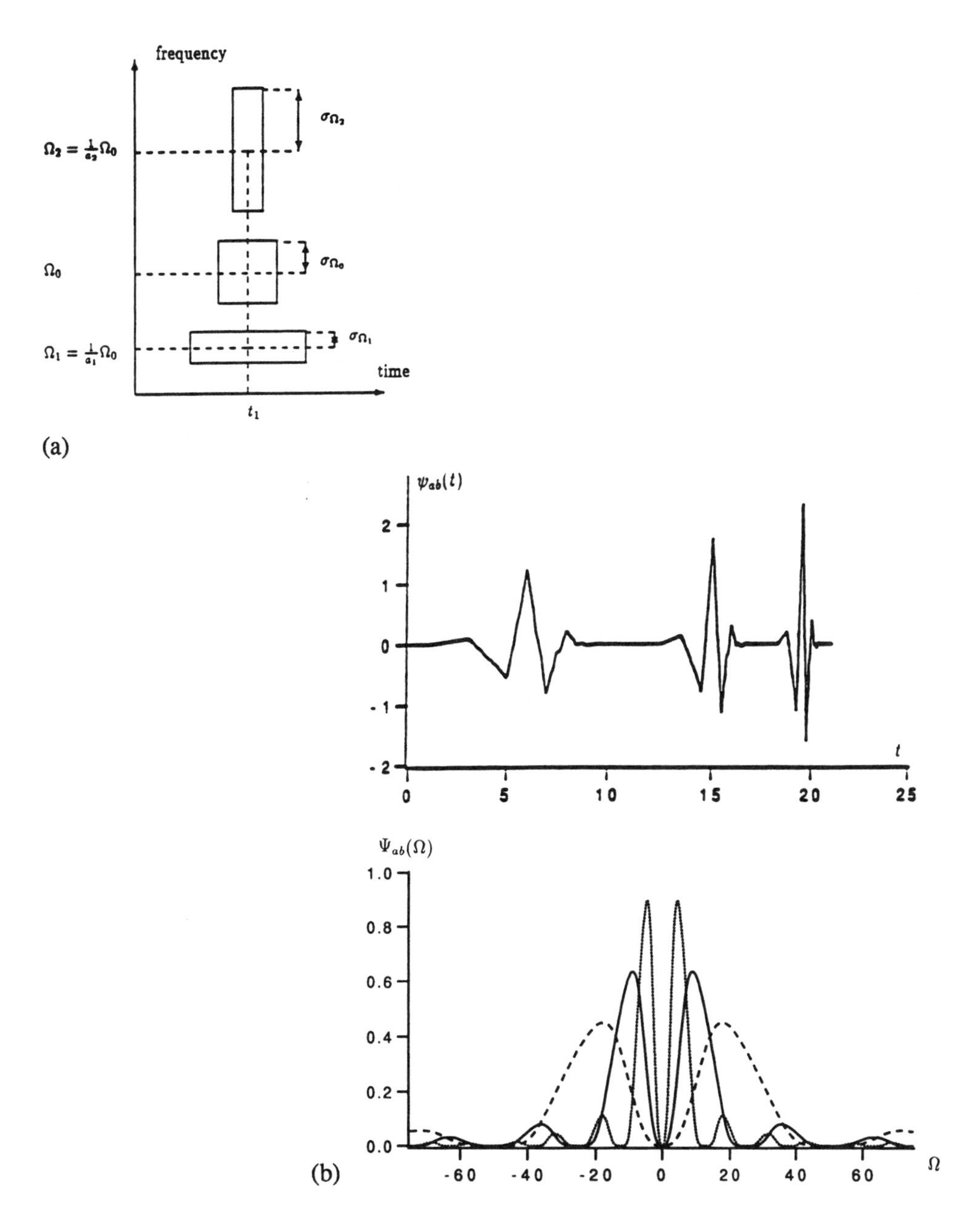

Figure 6.2: (a) Display of time-frequency cells of a wavelet function and its Fourier transform. (b) Daubechies (6-tap) wavelet function and its dilations, for $a = 2$ and $1/2$, along with their frequency responses.

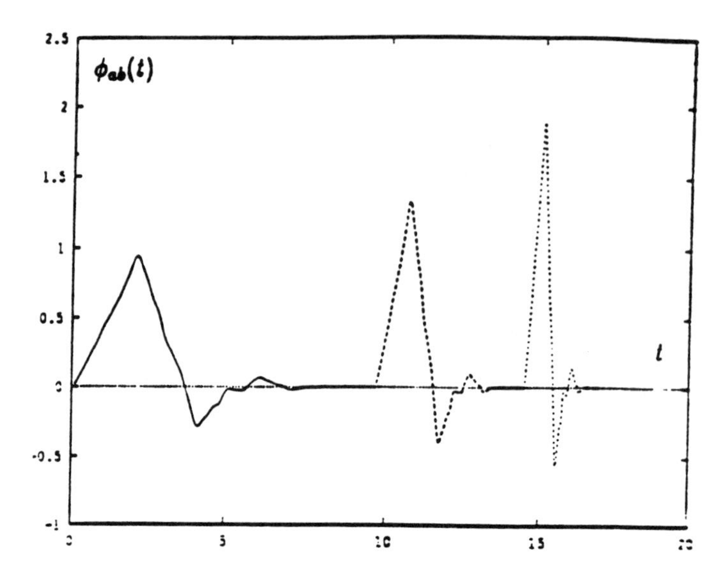

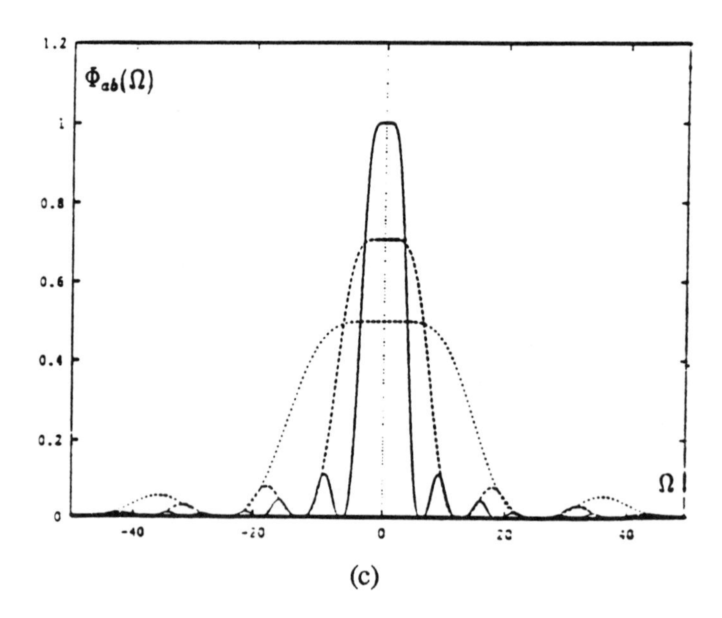

(c)

Figure 6.2: (c) Daubechies (6-tap) scaling function and its dilations, for $a = 2$ and $1/2$, along with their frequency responses.

where the wavelet coefficient $d_{m,n}$ is the inner product

$$\begin{aligned} d_{m,n} &= < f(t), \psi_{mn}(t) > \\ &= \frac{1}{a_0^{m/2}} \int f(t)\psi(a_0^{-m}t - nb_0)dt. \end{aligned} \tag{6.17}$$

Such complete sets are called *frames*. They are not yet a basis. Frames do not satisfy the Parseval theorem, and also the expansion using frames is not unique. In fact, it can be shown that (Daubechies, 1990)

$$A\|f\|^2 \leq \sum_m \sum_n | < f, \psi_{mn} > |^2 \leq B\|f\|^2 \tag{6.18}$$

where

$$\|f\|^2 \triangleq \int |f(t)|^2 dt.$$

The family $\psi_{mn}(t)$ constitutes a frame if $\psi(t)$ satisfies *admissibility*, and $0 < A < B < \infty$. Then the frame bounds are constrained by the inequalities

$$A \leq \frac{\pi}{b_0 log a_0} \int \frac{|\Psi(\Omega)|^2}{|\Omega|} d\Omega \leq B. \tag{6.19}$$

These inequalities hold for any choices of a_0 and b_0. These bounds diverge for nonadmissible wavelet functions (Daubechies, 1990).

Next, the frame is *tight* if $A = B = 1$. But, $\{\psi_{mn}(t)\}$ is still not necessarily linearly independent. There still can be redundancy in this frame. A frame is *exact* if removal of a function leaves the frame incomplete. Finally, a *tight, exact* frame with $A = B = 1$ constitutes an *orthonormal basis* for $L^2(R)$. This implies that the Parseval energy relation holds. The orthonormal wavelet expansion can be thought of as the wavelet counterpart to the critically sampled subband filter bank of Chapter 3.

The orthonormal wavelets $\{\psi_{mn}(t)\}$ satisfy

$$\int \psi_{mn}(t)\psi_{m'n'}(t)dt = \begin{cases} 1, & m = m', n = n' \\ 0, & otherwise \end{cases} \tag{6.20}$$

and are orthonormal in *both indices*. This means that for the same scale m they are orthonormal in time, and they are also orthonormal across the scales. We will elaborate further on this point in the multiresolution expansion developed in the following section for the octave or dyadic grid, where $a_0 = 2$, $b_0 = 1$.

Similarly, the scaling functions (to be defined in Section 6.2) satisfy an orthonormality condition, but only *within the same scale*, i.e.,

$$\int \phi_{mn}(t)\phi_{ml}(t)dt = \delta_{n-l}. \tag{6.21}$$

It will be seen that the scaling function is a low-pass filter that complements the wavelet function in representing a signal at the same scale. A signal $f \in L^2(R)$ can be approximated at scale m by its projection onto scale space. Section 6.2.1 develops this view.

We can imagine the wavelet coefficients as being generated by the wavelet filter bank of Fig. 6.3. The convolution of $f(t)$ with $\psi_m(-t)$ is

$$y_m(t) = \int f(\tau)\psi_m(\tau - t)d\tau$$

where

$$\psi_m(t) = 2^{-m/2}\psi(2^{-m}t).$$

Sampling $y_m(t)$ at $n2^m$ gives

$$y_m(n2^m) = 2^{-m/2}\int f(\tau)\psi(2^{-m}\tau - n)d\tau = d_{m,n}.$$

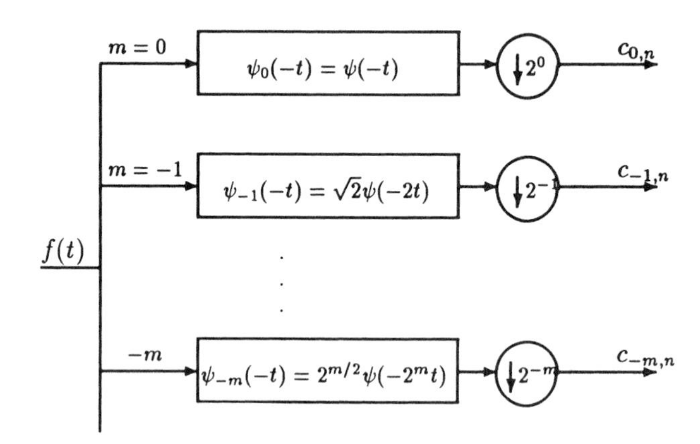

Figure 6.3: Discrete wavelet transform filter bank.

This filter bank could be contrasted with the STFT bank of Fig. 5.6. Note that the wavelet down-sampler varies with position or scale in the bank; in the STFT the down-sampler is the same for every branch.

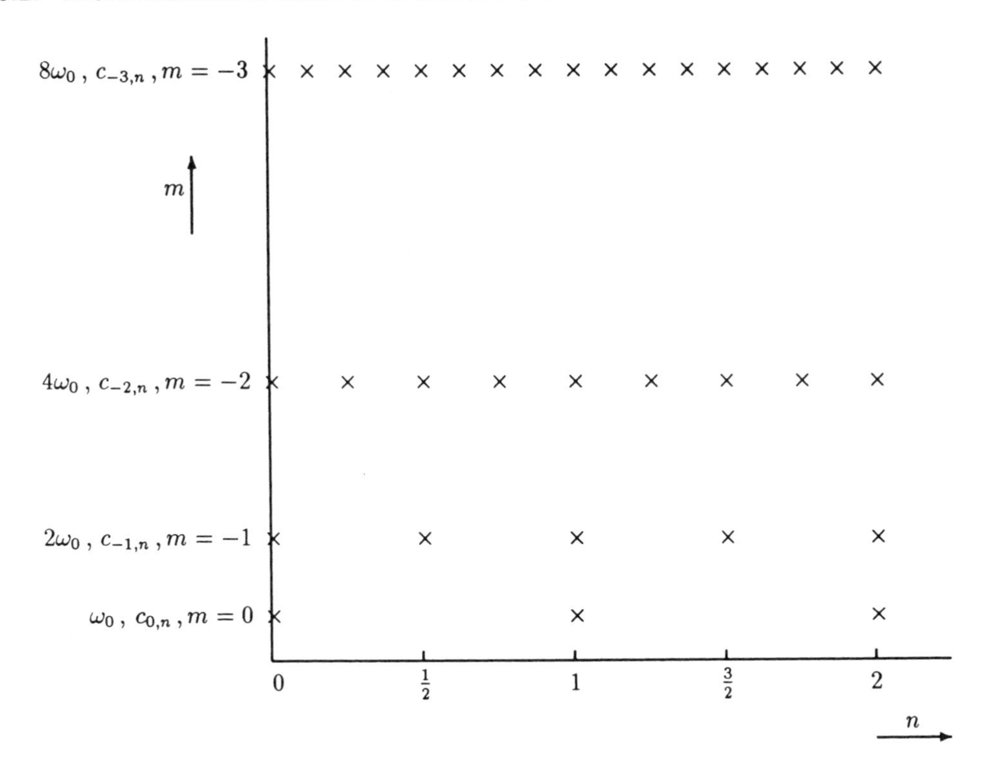

Figure 6.4: Sampling grid (dyadic) for discrete wavelet transform.

This effect is further illustrated in the DWT dyadic sampling grid shown in Fig. 6.4. Each node corresponds to a wavelet basis function $\psi_{mn}(t)$ with scale 2^{-m} and shift $n2^{-m}$. This wavelet grid can be contrasted with the uniform STFT grid of Fig. 5.7.

In the following sections, we will show how *compactly supported* wavelet bases can be constructed from a multiresolution signal analysis, and we will link these to the dyadic tree PR-QMF structure of Chapter 3.

6.2 Multiresolution Signal Decomposition

Here we describe the approach taken by Mallat and Meyer for constructing orthonormal wavelets of compact support. Our intent is to show the link between these wavelet families and the pyramid-dyadic tree expansions of a signal. Finally we will show that FIR PR-QMF with a special property called *regularity* provides a procedure for generating compactly supported orthonormal wavelet bases.

6.2.1 Multiresolution Analysis Spaces

Multiresolution signal analysis provides the vehicle for these links. In this representation, we express a function $f \in L^2$ as a limit of successive approximations, each of which is a smoothed version of $f(t)$. These successive approximations correspond to different resolutions — much like a pyramid. This smoothing is accomplished by convolution with a low-pass kernel called the *scaling function* $\phi(t)$.

A multiresolution analysis consists of a sequence of closed subspaces $\{V_m | m \in Z\}$ of $L^2(R)$ which have the following properties:

• Containment:

$$...V_2 \subset V_1 \subset V_0 \subset V_{-1} \subset V_{-2}...$$
$$\leftarrow Coarser \quad Finer \rightarrow$$

• Completeness:

$$\bigcap_{m \in Z} V_m = \{0\} \qquad \bigcup_{m \in Z} V_m = L^2(R).$$

• Scaling property:

$$f(x) \in V_m \Longleftrightarrow f(2x) \in V_{m-1} \text{ for any function } f \in L^2(R).$$

• The Basis/Frame property: There exists a scaling function $\phi(t) \in V_0$ such that $\forall m \in Z$, the set

$$\{\phi_{mn}(t) = 2^{-m/2}\phi(2^{-m}t - n)\}$$

is an orthonormal basis for V_m, i.e.,

$$\int \phi_{mn}(t)\phi_{mn'}(t)dt = \delta_{n-n'}.$$

Let W_m be the orthogonal complement of V_m in V_{m-1}, i.e.,

$$\begin{aligned} V_{m-1} &= V_m \oplus W_m \\ V_m &\perp W_m. \end{aligned} \tag{6.22}$$

Furthermore, let the direct sum of the possibly infinite spaces W_m span $L^2(R)$:

$$.. \oplus W_j \oplus W_{j-1} ... \oplus W_0 ... \oplus W_{-j+1} \oplus W_{-j+2} ... = L^2(R). \tag{6.23}$$

We will associate the scaling function $\phi(t)$ with the space V_0, and the wavelet function with W_0. Next, we introduce projection operators P_m, Q_m from $L^2(R)$ onto V_m and W_m, respectively. The completeness property ensures that $\lim_{m \to -\infty} P_m f = f$, for any $f \in L^2(R)$. The containment property implies that as m decreases $P_m f$ leads to successively better approximations of f.

Any function f can be approximated by $P_{m-1}f$, its projection onto V_{m-1}. From Eq. (6.22) this can be expressed as a sum of projections onto V_m and W_m:

$$P_{m-1}f = P_m f + Q_m f. \tag{6.24}$$

$P_m f$ is the low-pass part of f in V_m and $Q_m f$ is the high-frequency detail or difference, i.e., the increment in information in going from V_m to V_{m-1}.

Equation (6.24) can be expressed as $Q_m f = P_{m-1}f - P_m f$, where $Q_m f \in W_m$. Hence, we can say that the orthogonal or complementary space W_m is given by the difference $V_{m-1} \ominus V_m$. Now $\phi(t-n) \in V_0$ and $\phi(2t-n) \in V_{-1}$. Since $V_0 = span\{\phi(t-n)\}$ and $V_{-1} = span\{\phi(2t-n)\}$, it is reasonable to expect the existence of a function $\psi(t) \in W_0$, such that $W_0 = span\{\psi(t-n)\}$. This function $\psi(t)$ is the wavelet function associated with the multiscale analysis. Clearly, by the scaling property, $W_m = span\{\psi(2^{-m}t - n)\}$. The term W_m is also generated by the translates and dilations $\{\psi_{mn}(t)\}$ of a single wavelet or kernel function $\psi(t)$. The containment and completeness properties, together with $W_m \perp V_m$ and $V_{m-1} = V_m \oplus W_m$, imply that the spaces W_m are all mutually orthogonal and that their direct sum is $L^2(R)$. Since for each m, the set $\{\psi_{mn}(t)\ ;\ n \in Z\}$ constitutes an orthonormal basis for W_m, it follows that the whole collection $\{\psi_{mn}(t)\ ;\ m, n \in Z\}$ is an orthonormal wavelet basis for $L^2(R)$. The set $\{\psi_{mn}(t) = 2^{-m/2}\psi(2^{-m}t - n)\}$ is the wavelet basis associated with the multiscale analysis, with property

$$\int \psi_{mn}(t)\psi_{m'n'}(t)dt = \delta_{m-m'}\delta_{n-n'}.$$

Since

$$V_{i-1} = W_i \oplus V_i = W_i \oplus W_{i+1} \oplus W_{i+2} \oplus \ldots\ , \tag{6.25}$$

the function $P_{m-1}f$ at a given resolution can be represented as a sum of added details at different scales.

Suppose we start with a scaling function $\phi(t)$, such that its translates $\{\phi(t-n)\}$ span V_0. Then V_{-1} is spanned by $\phi(2t-n)$, dilates of the function in V_0. The basis functions in V_{-1} are then

$$\phi(2(t-n)), \qquad \text{n even}$$

$$\phi(2(t-n)-1), \qquad \text{n odd}\ .$$

Thus V_{-1} is generated by integer translates of two functions, and $\phi(t)$ can be expressed as a linear combination of even and odd translates of $\phi(2t)$, or

$$\phi(t) = 2\sum_n h_0(n)\phi(2t-n). \tag{6.26}$$

This last equation is an inherent consequence of the containment property. The coefficient set $\{h_0(n)\}$ are the *interscale basis coefficients.* We shall see that this is the low-pass unit sample response of the two-band paraunitary filter bank of Chapter 3.

Similarly, the band-pass wavelet function can be expressed as a linear combination of translates of $\phi(2t)$:

$$\psi(t) = 2\sum_n h_1(n)\phi(2t-n). \tag{6.27}$$

This is the *fundamental* wavelet equation. The coefficients $\{h_1(n)\}$ will be identified with the high-pass branch in the two-band PR filter bank structure.

6.2.2 The Haar Wavelet

To fix ideas, we pause to consider the case of Haar functions. These functions are sufficiently simple yet of great instructional value in illustrating multiresolution concepts. Let V_m be the space of piecewise constant functions,

$$V_m = \{f(t) \in L^2(R);\ f \text{ is constant on } [2^m n, 2^m(n+1)]\ \ \forall n \in Z\}. \tag{6.28}$$

Sample functions in these spaces are shown in Fig. 6.5.

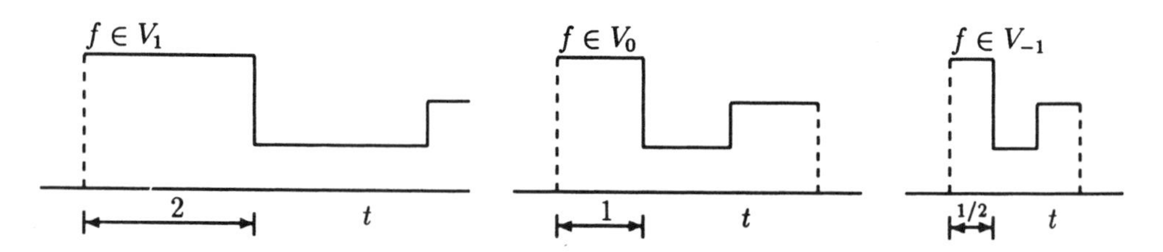

Figure 6.5: Piecewise constant functions in V_1, V_2, and V_{-1}.

First we observe that $..V_1 \subset V_0 \subset V_{-1}..$ and that $f(t) \in V_0 \longleftrightarrow f(2t) \in V_{-1}$ so that the containment property is satisfied; for example, functions in V_0 that are constant over integers $(n, n+1)$ are also constant over the half integers of V_{-1}.

The completeness and scaling property are also obvious. The integer translates of the scaling function

$$\phi(t) = \begin{cases} 1, & 0 \leq t \leq 1 \\ 0, & \text{otherwise} \end{cases} \tag{6.29}$$

with transform

$$\Phi(\Omega) = e^{-j\frac{\Omega}{2}} \frac{\sin \Omega/2}{\Omega/2} \tag{6.30}$$

constitute an orthonormal basis for V_0. This is obvious since for $n \neq m$, $\phi(t-n)$ and $\phi(t-m)$ do not overlap, and

$$\int \phi(t-n)\phi(t-m)dt = \delta_{n-m}.$$

The waveforms in Fig. 6.6 show that $\phi(t)$ is a linear combination of the even and odd translates of $\phi(2t)$:

$$\phi(t) = \phi(2t) + \phi(2t-1).$$

Next, since $V_{-1} = V_0 \oplus W_0$ and $Q_0 f = (P_{-1}f - P_0 f) \in W_0$ represents the detail in going from scale 0 to scale -1, then W_0 must be spanned by $\psi(t-n)$, where

$$\psi(t) = \phi(2t) - \phi(2t-1) = \begin{cases} 1, & 0 \leq t < 1/2 \\ -1, & 1/2 \leq t < 1 \\ 0, & \text{otherwise.} \end{cases}$$

The Fourier transform is that of a band-pass analog filter, shown in Fig. 6.6(e):

$$\Psi(\Omega) = je^{-j\frac{\Omega}{2}} \frac{\sin^2 \Omega/4}{\Omega/4}. \tag{6.31}$$

This $\psi(t)$ is then the mother wavelet function from which all the babies spring. The dilations and translates of $\psi(t)$ are the *Haar functions* $\{\psi_{mn}(t)\}$, which are known basis functions for $L^2(R)$. The Haar wavelets are shown in Fig. 6.7. Those wavelets $\psi_{mn}(t) = 2^{-\frac{m}{2}}\psi(2^{-m}t - n)$ are seen to be orthonormal at the *same scale*

$$\int \psi_{mn}(t)\psi_{mn'}(t)dt = \delta_{n-n'} \tag{6.32}$$

and orthonormal *across* the scales

$$\int \psi_{mn}(t)\psi_{m'n'}(t)dt = \delta_{m-m'}\delta_{n-n'}. \tag{6.33}$$

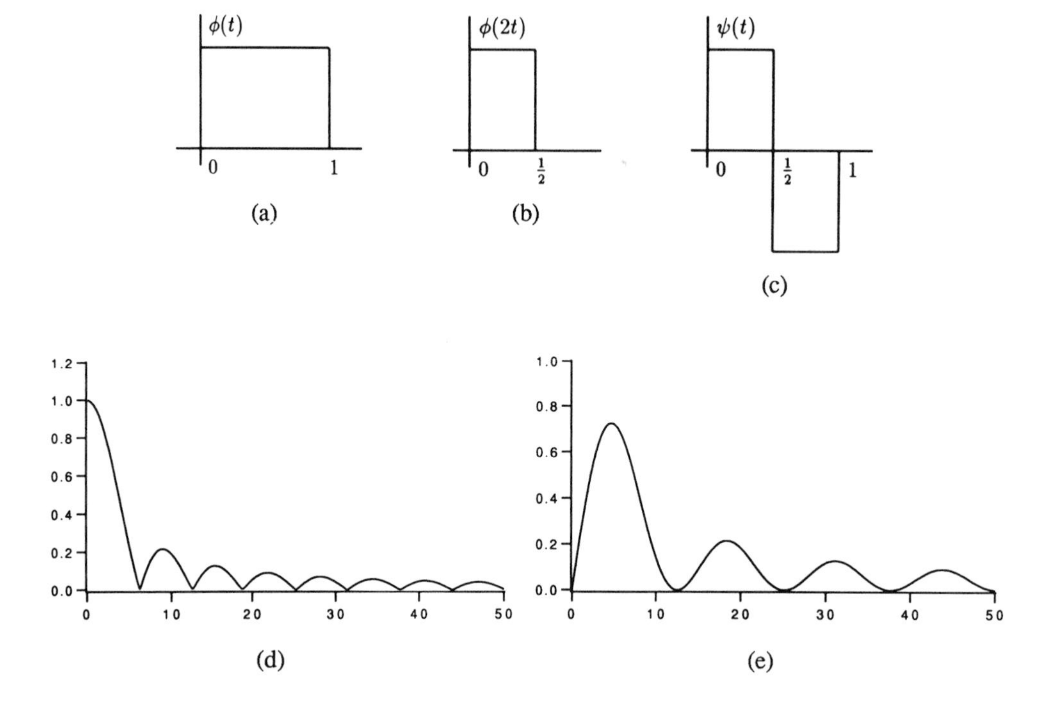

Figure 6.6: (a) and (b) Haar scaling basis functions; (c) Haar wavelet; (d) Fourier transform of Haar scaling function; and (e) Haar wavelet function.

It is easily verified that[3]

$$\phi_{m+1,n} = \frac{1}{\sqrt{2}}[\phi_{m,2n} + \phi_{m,2n+1}]$$

and

$$\psi_{m+1,n} = \frac{1}{\sqrt{2}}[\phi_{m,2n} - \phi_{m,2n+1}]. \tag{6.34}$$

The approximations $P_0 f$, $P_{-1} f$ and the detail $Q_0 f$ for a sample function $f(t)$ are displayed in Fig. 6.8. Note how the detail $Q_0 f$ adds to the coarse approximation $P_0 f$ to provide the next finer approximation $P_{-1} f$.

At scale m, the scaling function coefficient is

$$c_{m,n} = < f, \phi_{mn} > = 2^{-m/2} \int_{2^m n}^{2^m (n+1)} f(t) dt$$

[3]For ease of notation, the explicit time dependency in $\phi_{m,n}$ and $\psi_{m,n}$ is not shown, but implied.

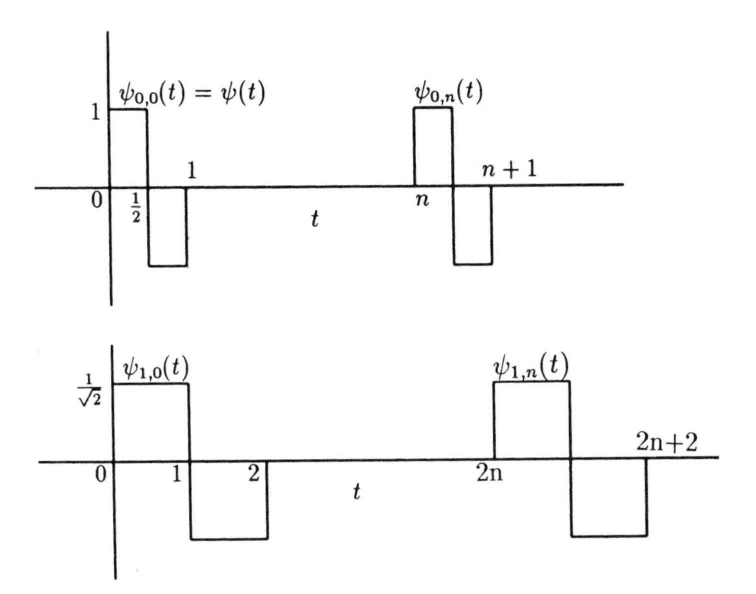

Figure 6.7: Typical Haar wavelets at scales 0,1.

and the approximation of f at scale m is

$$P_m f = \sum_n c_{m,n}\phi_{mn}(t) = \sum_n c_{m,n}2^{-m/2}\phi(2^{-m}t - n).$$

Since $V_{m+1} \subset V_m$, the next coarser approximation is $P_{m+1}f$, and

$$\phi_{m+1,n} = \frac{1}{\sqrt{2}}(\phi_{m,2n} + \phi_{m,2n+1}).$$

The orthogonal complement of $P_{m+1}f$ is then

$$\begin{aligned} Q_{m+1}f &= P_m f - P_{m+1}f \\ &= 2^{-1}\sum_n (c_{m,2n} - c_{m,2n+1})(\phi_{m,2n} - \phi_{m,2n+1}) \\ &= \sum_n \gamma_{m+1,n}\psi_{m+1,n}. \end{aligned}$$

Consequently, $\psi_{mn}(t) = 2^{-m/2}\psi(2^{-m}t - n)$ and for fixed m, $\{\psi_{mn}(t)\}$ spans W_m; for $m, n \in Z$, $\{\psi_{mn}(t)\}$ spans $L^2(R)$.

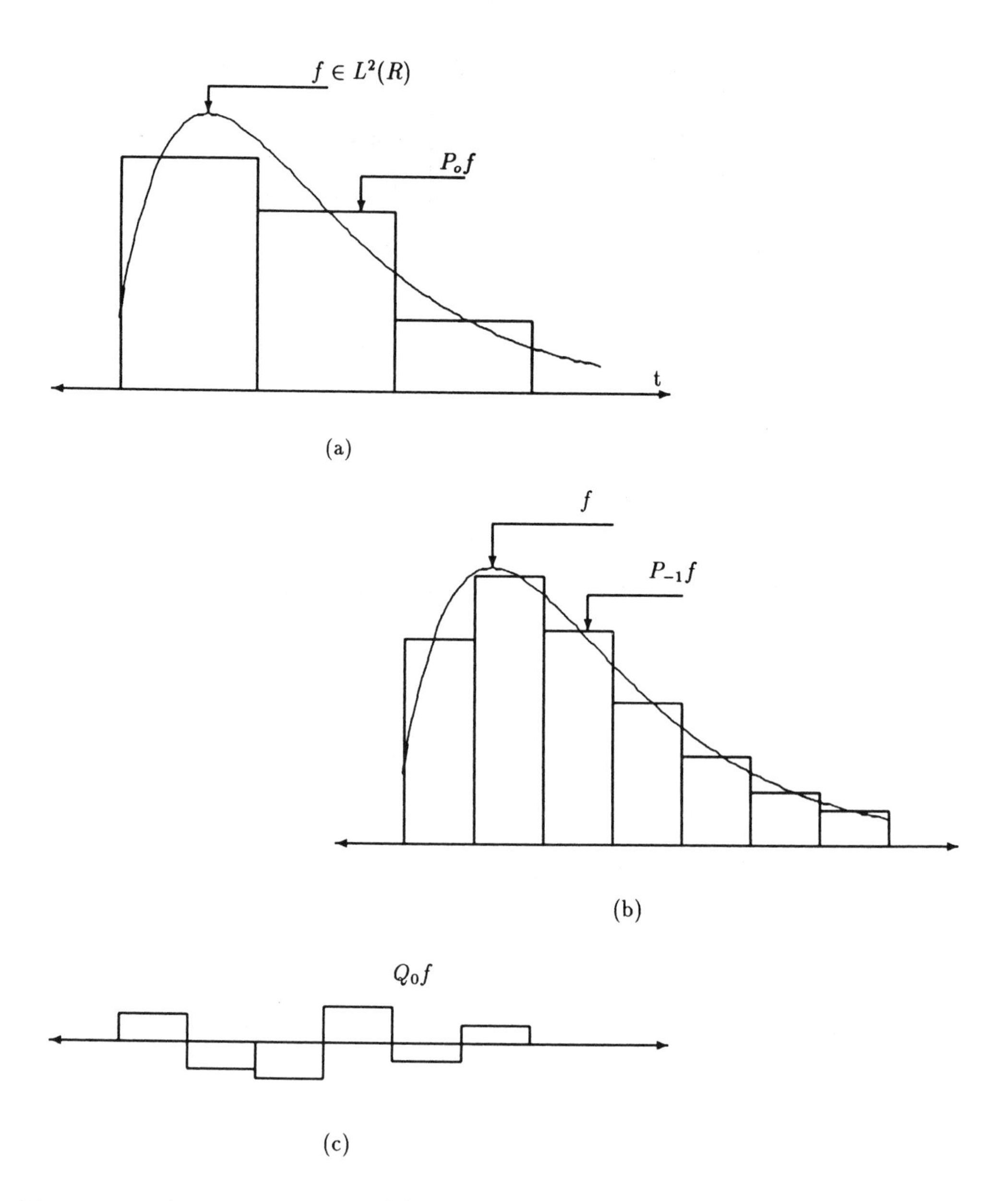

Figure 6.8: Approximations of (a) $P_0 f$ and (b) $P_{-1} f$; (c) detail $Q_0 f$, where $P_0 f + Q_0 f = P_{-1} f$.

The Fourier transforms of these Haar scaling and wavelet functions

$$\begin{aligned}\Phi(\Omega) &= e^{-j\Omega/2}\frac{\sin\Omega/2}{\Omega/2}\\ \Psi(\Omega) &= je^{-j\Omega/2}\frac{\sin^2\Omega/4}{\Omega/4}\end{aligned} \tag{6.35}$$

are sketched in Fig. 6.6(d) and (e).

These Haar functions are very well localized in *time*, but the frequency localization is seen to be poor owing to the discontinuities in the time-domain approximation.

For the Haar basis, we see that the interscale coefficients and their system functions are

$$h_0(n) = \begin{cases} 1/\sqrt{2}, & n = 0, 1 \\ 0, & \text{otherwise} \end{cases} \longleftrightarrow H_0(e^{j\omega}) = \sqrt{2}e^{-j\omega/2}\cos\omega/2$$

$$h_1(n) = \begin{cases} 1/\sqrt{2}, & n = 0 \\ -1/\sqrt{2}, & n = 1 \\ 0, & \text{otherwise} \end{cases} \longleftrightarrow H_1(e^{j\omega}) = j\sqrt{2}e^{-j\omega/2}\sin\omega/2. \tag{6.36}$$

These are *paraunitary* filters, albeit very simple ones. Examples of smoother time-frequency wavelet representations are developed subsequently. Since this last result is not obvious, we will indicate the details in the derivation.

Let $P_m f$ be decomposed into even and odd indices; then

$$\begin{aligned} P_m f &= \sum_n c_{m,n}\phi_{m,n} \\ &= \sum_{n\ even} + \sum_{n\ odd} \\ &= \sum_n c_{m,2n}\phi_{m,2n} + \sum_n c_{m,2n+1}\phi_{m,2n+1}. \end{aligned} \tag{6.37}$$

The coefficient at scale $m+1$, can be expressed as a smoothing of two finer scale coefficients via

$$\begin{aligned} c_{m+1,n} &= < f, \phi_{m+1,n} > = < f, \frac{1}{\sqrt{2}}(\phi_{m,2n} + \phi_{m,2n+1}) > \\ &= \frac{1}{\sqrt{2}} < f, \phi_{m,2n} > + \frac{1}{\sqrt{2}} < f, \phi_{m,2n+1} > \\ &= \frac{1}{\sqrt{2}}[c_{m,2n} + c_{m,2n+1}] \end{aligned} \tag{6.38}$$

Equations (6.38) and (6.34) allow us to write

$$P_{m+1}f = \sum_n \frac{1}{\sqrt{2}}(c_{m,2n} + c_{m,2n+1})\frac{1}{\sqrt{2}}(\phi_{m,2n} + \phi_{m,2n+1}). \tag{6.39}$$

Subtracting Eq. (6.39) from Eq. (6.37) and rearranging gives

$$Q_{m+1}f = \sum_n d_{m+1,n}\psi_{m+1,n} \tag{6.40}$$

where

$$d_{m+1,n} = \frac{1}{\sqrt{2}}(c_{m,2n} - c_{m,2n+1})$$

$$\psi_{m+1,n} = \frac{1}{\sqrt{2}}(\phi_{m,2n} - \phi_{m,2n+1}).$$

Thus, the projection of f onto W_{m+1} is representable as a linear combination of translates and dilates of the mother function $\psi(t)$.

Another important observation is the relationships between the wavelet and scaling coefficients at scale $m+1$ and the scaling coefficient at the finer scale m. We have seen that

$$h_0(n) = \frac{1}{2}(\delta_n + \delta_{n-1}),$$

and

$$h_1(n) = \frac{1}{2}(\delta_n - \delta_{n-1}).$$

From Eqs. (6.38) and (6.40) we conclude that $c_{m+1,n}$ and $d_{m+1,n}$ can be obtained by convolving $c_{m,n}$ with $\sqrt{2}h_0(n)$ and $\sqrt{2}h_1(n)$, respectively, followed by a 2-fold down-sampling as shown in Fig. 6.9. Hence the interscale coefficients can be represented by a decimated two-band filter bank. The output of the upper decimator represents the coefficients in the approximation of the signal at scale $m+1$, while the lower decimator output represents the detail coefficients at that scale.

In the next section, we show that any orthonormal wavelet of compact support can be representable in the form of the two-band unitary filter bank developed here. More interesting wavelets with smoother time-frequency representation are also developed in the sequel.

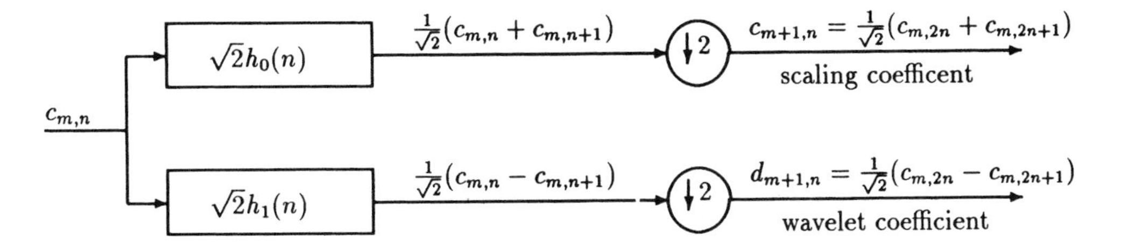

Figure 6.9: Interscale coefficients as a two-band filter bank.

6.2.3 Two-Band Unitary PR-QMF and Wavelet Bases

Here we resume the discussion of the interscale basis coefficients in Eq. (6.26). But first, we must account for the time normalization implicit in translation. Hence, with $\phi(t) \longleftrightarrow \Phi(\Omega)$ as a Fourier Transform pair, we then have

$$\phi(2t) \longleftrightarrow \frac{1}{2}\Phi(\frac{\Omega}{2})$$

and

$$\phi(2(t-\frac{nT_0}{2})) = \phi(2t-nT_0) \longleftrightarrow \frac{1}{2}e^{-jn\Omega\frac{T_0}{2}}\Phi(\frac{\Omega}{2}).$$

Taking the Fourier transform of both sides of Eq. (6.26) gives

$$\Phi(\Omega) = \sum_n h_0(n)e^{-j\frac{\Omega}{2}nT_0}\Phi(\frac{\Omega}{2}). \tag{6.41}$$

Now with $\omega = \Omega T_0$ as a normalized frequency and $H_0(e^{j\omega})$ as the transform of the sequence $\{h_0(n)\}$,

$$H_0(e^{j\omega}) = \sum_n h_0(n)e^{-j\omega n} \tag{6.42}$$

we obtain[4]

$$\Phi(\Omega) = H_0(e^{j\frac{\omega}{2}})\Phi(\frac{\Omega}{2}), \quad \omega = \Omega T_0. \tag{6.43}$$

The variables Ω and ω in this equation run from $-\infty$ to ∞. In addition, $H_0(e^{j\omega})$ is periodic with period 2π. Similarly, for the next two adjacent resolutions,

$$\phi(2t) = 2\sum_n h_0(n)\phi(4t-n)$$

[4]We will use Ω as the frequency variable in a continuous-time signal, and ω for discrete-time signals, even though $\Omega = \omega$ for $T_0 = 1$.

and

$$\Phi(\frac{\Omega}{2}) = H_0(e^{j\frac{\omega}{4}})\Phi(\frac{\Omega}{4}).$$

Therefore, $\Phi(\Omega)$ of Eq. (6.43) becomes

$$\Phi(\Omega) = H_0(e^{j\frac{\omega}{2}})H_0(e^{j\frac{\omega}{4}})\Phi(\frac{\Omega}{4}).$$

Note that $H_0(e^{j\frac{\omega}{4}})$ has a period of 8π. If we repeat this procedure infinitely many times, and using $\lim_{n\to\infty}\Omega/2^n = 0$, we get $\Phi(\Omega)$ as the iterated product

$$\Phi(\Omega) = \Phi(0)\prod_{k=1}^{\infty} H_0(e^{j\frac{\omega}{2^k}}). \tag{6.44}$$

We can show that the completeness property of a multiresolution approximation implies that any scaling function satisfies a nonzero mean constraint (Prob. 6.1)

$$\Phi(0) = \int_{-\infty}^{\infty} \phi(t)dt \neq 0.$$

If $\phi(t)$ is real, it is determined uniquely, up to a sign, by the requirement that $\phi_{0n}(t)$ be orthonormal. Therefore,

$$\int_{-\infty}^{\infty} \phi(t)dt = \pm 1 \tag{6.45}$$

which is equivalent to

$$\begin{aligned} |\Phi(0)| &= |\int_{-\infty}^{\infty} \phi(t)dt| \\ &= |H_0(e^{j\omega})|_{\omega=0} = 1. \end{aligned} \tag{6.46}$$

Hence the Fourier transform of the continuous-time scaling function is obtained by the infinite resolution product of the discrete-time Fourier transform of the interscale coefficients $\{h_0(n)\}$. If the duration of the interscale coefficients $\{h_0(n)\}$ is finite, the scaling function $\phi(t)$ is said to be *compactly supported.* Furthermore, if $h_0(n)$ has a duration $0 \leq n \leq N-1$, then $\phi(t)$ is also supported within $0 \leq t \leq (N-1)T_0$. (Prob. 6.2) For convenience, we take $T_0 = 1$ in the sequel (Daubechies, 1988).

In the Haar example, we had $N = 2$, and the duration of $h_0(n)$ was $0 \leq n \leq 1$; accordingly, the support for $\phi(t)$ is $0 < t \leq 1$.

Next we want to find the constraints $H_0(e^{j\omega})$ must satisfy so that $\phi(t)$ is a scaling function, and for any *given scale* m, the set $\{\phi_{mn}(t)\}$ is *orthonormal*. In particular, if $\{\phi(t-n)\}$ spans V_0, then we show in Appendix B that the corresponding $\Phi(\Omega)$ must satisfy the *unitary* condition in frequency

$$\sum_k |\Phi(\Omega + 2\pi k)|^2 = 1. \tag{6.47}$$

Next, after substituting

$$\Phi(2\Omega) = H_0(e^{j\omega})\Phi(\Omega)$$

into the preceding orthonormality condition and after some manipulations (Prob. 6.4), we obtain

$$\sum_k |H_0(e^{j(\omega+k\pi)})|^2 |\Phi(\Omega + k\pi)|^2 = 1.$$

This can be rewritten as an even and odd indexed sum,

$$|H_0(e^{j(\omega+\pi)})|^2 \sum_{2k+1} |\Phi(\Omega + (2k+1)\pi)|^2 + |H_0(e^{j\omega})|^2 |\sum_{2k} \Phi(\Omega + 2k\pi)|^2 = 1. \tag{6.48}$$

This last equation yields the magnitude square condition of the interscale coefficient sequence $\{h_0(n)\}$,

$$|H_0(e^{j\omega})|^2 + |H_0(e^{j(\omega+\pi)})|^2 = 1. \tag{6.49}$$

This is recognized as the low-pass filter requirement in a maximally decimated unitary PR-QMF of Eq. (3.129). We proceed in a similar manner to obtain filter requirements for the orthonormal wavelet bases.

First, it is observed that if the scaling function $\phi(t)$ is compactly supported on $[0, N-1]$, the corresponding wavelet $\psi(t)$ generated by Eq. (6.27) is compactly supported on $[1-\frac{N}{2}, \frac{N}{2}]$, Again, for the Haar wavelet, we had $N = 2$. In that case the duration of $h_1(n)$ is $0 \leq n \leq 1$, as is the support for $\psi(t)$.

Letting $h_1(n) \longleftrightarrow H_1(e^{j\omega})$ and transforming Eq. (6.27) gives the transform of the wavelet as

$$\Psi(\Omega) = H_1(e^{j\frac{\omega}{2}})\Phi(\frac{\Omega}{2}).$$

If we replace the second term on the right-hand side of this equation with the infinite product derived earlier in Eq. (6.44),

$$\Psi(\Omega) = H_1(e^{j\frac{\omega}{2}}) \prod_{k=2}^{\infty} H_0(e^{j\frac{\omega}{2^k}}). \tag{6.50}$$

The orthonormal wavelet bases are complementary to the scaling bases. These satisfy the *intra-* and *interscale* orthonormalities

$$< \psi_{mn}(t), \psi_{kl}(t) >= \delta_{m-k}\delta_{n-l},$$

where m and k are the scale and n and l the translation parameters. Notice that the orthonormality conditions of wavelets hold for different scales, in addition to the same scale, which is the case for scaling functions. Since $\{\psi(t-n)\}$ forms an orthonormal basis for W_0, their Fourier transforms must satisfy the unitary condition

$$\sum_k |\Psi(\Omega + 2\pi k)|^2 = 1. \tag{6.51}$$

As before, using

$$\Psi(2\Omega) = H_1(e^{j\omega})\Phi(\Omega) \tag{6.52}$$

in this last equation leads to the expected result

$$|H_1(e^{j\omega})|^2 + |H_1(e^{j(\omega+\pi)})|^2 = 1$$

or

$$H_1(z)H_1(z^{-1}) + H_1(-z)H_1(-z^{-1}) = 1, \tag{6.53}$$

which corresponds to the high-pass requirement in the two-channel unitary PR-QMF, Eq. (3.129).

Finally, these scaling and wavelet functions also satisfy the orthonormality condition between themselves,

$$< \phi_{mn}(t), \psi_{kl}(t) >= 0.$$

Note that the orthonormality of wavelet and scaling functions is satisfied at different scales, as well as at the same scale. This time-domain condition implies its counterpart in the frequency-domain as

$$\sum_k \Phi(\Omega - 2k\pi)\Psi^*(\Omega - 2\pi k) = 0. \tag{6.54}$$

Now, if we use Eqs. (6.44), (6.50), and (6.54) we can obtain the frequency-domain condition for alias cancellation [see Eq. (3.130)]:

$$H_0(e^{j\omega})H_1(e^{-j\omega}) + H_0(e^{j(\omega+\pi)})H_1(e^{-j(\omega+\pi)}) = 0 \tag{6.55}$$

or

$$H_0(z)H_1(z^{-1}) + H_0(-z)H_1(-z^{-1}) = 0.$$

The three conditions required of the transforms of the interscale coefficients, $\{h_0(n)\}$ and $\{h_1(n)\}$ in Eqs. (6.49), (6.53), and (6.55) in the design of compactly supported orthonormal wavelet and scaling functions are then equivalent to the requirement that the alias component (AC) matrix $H_{AC}(e^{j\omega})$ of Chapter 3 for the two-band filter bank case,

$$H_{AC}(e^{j\omega}) = \begin{bmatrix} H_0(e^{j\omega}) & H_1(e^{j\omega}) \\ H_0(e^{j(\omega+\pi)}) & H_1(e^{j(\omega+\pi)}) \end{bmatrix}, \tag{6.56}$$

be *paraunitary* for all ω.

In particular, the cross-filter orthonormality, Eq. (6.55), is satisfied by the choice

$$H_1(z) = z^{-(N-1)} H_0(-z^{-1}), \qquad \text{N even}$$

or in the time-domain,

$$h_1(n) = (-1)^{n+1} h_0(N-1-n) \tag{6.57}$$

In addition, since

$$|H_0(e^{j\omega})|^2 + |H_1(e^{j\omega})|^2 = 1$$

and we have already argued that

$$H_0(e^{j\omega})|_{\omega=0} = 1,$$

then $H_1(e^{j\omega})$ must be a high-pass filter with

$$H_1(e^{j\omega})|_{\omega=0} = \sum_k h_1(k) = 0. \tag{6.58}$$

Thus the wavelet must be a band-pass function, satisfying the admissibility condition

$$\Psi(0) = \int \psi(t)dt = 0. \tag{6.59}$$

Therefore, $H_0(z)$ and $H_1(z)$ must each have at least one zero at $z = -1$ and $z = 1$, respectively. It is also clear from Eq. (6.57) that if $h_0(n)$ is FIR, then so is $h_1(n)$. Hence the wavelet function is of *compact support* if the scaling function is.

In summary, compactly supported orthonormal wavelet bases imply a paraunitary, 2-band FIR PR-QMF bank; conversely, a paraunitary FIR PR-QMF filter pair with the constraint that $H_0(z)$ have at least one zero at $z = -1$ imply a compactly supported orthonormal wavelet basis (summarized in Table 6.1). This is needed to ensure that $\Psi(0) = 0$. Orthonormal wavelet bases can be constructed by multiresolution analysis, as described next.

Paraunitary 2-band FIR PR-QMF	$\longleftrightarrow$	Orthonormal Wavelets of Compact Support
$\tilde{H}_{AC}(z)H_{AC}(z) = I$		$\Phi(\Omega) = \prod_{k=1}^{\infty} H_0(e^{j\omega/2k})$
$\sum_k h_r(k)h_s(k+2n) = \begin{cases} 0, & r \neq s \\ \delta(n), & r = s \end{cases}$		$\Psi(\Omega) = H_1(e^{j\omega/2}) \prod_{k=2}^{\infty} H_0(e^{j\omega/2^k})$
		$H_1(z) = z^{-(N-1)} H_0(-z^{-1})$
$H_r(z)H_s(z^{-1}) + H_r(-z)H_s(-z^{-1}) = \begin{cases} 2, & r = s = 0,1 \\ 0, & r \neq s \end{cases}$		$\phi(t) = \sum h_0(n)\phi(2t-n)$ $\psi(t) = \sum h_1(n)\psi(2t-n)$
$H_0(e^{j\omega})\vert_{\omega=\pi} = 0$		$< \psi_{m,n}(t), \psi_{m',n'}(t) >= \delta_{m-m'}\delta_{n-n'}$
		$< \phi_{m,n}(t), \phi_{m,n'}(t) >= \delta_{n-n'}$
		$< \psi_{m,n}(t), \phi_{m',n'}(t) >= 0$

Table 6.1: Summary of relationships between paraunitary 2-band FIR PR-QMF's and compactly supported orthonormal wavelets.

6.2.4 Multiresolution Pyramid Decomposition

The multiresolution analysis presented in the previous section is now used to decompose the signal into successive layers at coarser resolutions plus detail signals, also at coarser resolution. The structure of this multiscale decomposition is the same as the pyramid decomposition of a signal, described in Chapter 3.

Suppose we have a function $f \in V_0$. Then, since $\{\phi(t-n)\}$ spans V_0, f can be represented as a superposition of translated scaling functions:

$$f(t) = \sum_n c_{0,n}\phi(t-n) = \sum_n c_{0,n}\phi_{0n}(t) \tag{6.60}$$

where

$$c_{0,n} =< f, \phi_{0n} >= \int f(t)\phi(t-n)dt. \tag{6.61}$$

Next, since $V_0 = V_1 \oplus W_1$, we can express f as the sum of two functions, one lying entirely in V_1 and the other in the orthogonal complement W_1:

$$f(t) = P_1 f + Q_1 f = \underbrace{\sum_n c_{1,n}\phi_{1n}(t)}_{f_v^1(t)} + \underbrace{\sum_n d_{1,n}\psi_{1n}(t)}_{f_w^1(t)}. \tag{6.62}$$

Here, the scaling coefficients $c_{1,n}$ and the wavelet coefficients $d_{1,n}$ are given by

$$c_{1,n} = < f_v^1, \phi_{1n} > = \frac{1}{\sqrt{2}} \int f_v^1(t) \phi(\frac{t}{2} - n) dt$$

$$d_{1,n} = < f_w^1, \psi_{1n} > = \frac{1}{\sqrt{2}} \int f_w^1(t) \phi(\frac{t}{2} - n) dt. \tag{6.63}$$

In the example using Haar functions, we saw that for a given starting sequence $\{c_{0,n}\}$, the coefficients in the next resolution $\{c_{1,n}\}$ and $\{d_{1,n}\}$ can be represented, respectively, as the convolution of $c_{0,n}$ with $\tilde{h}_0 = h_0(-n)$ and of $c_{0,n}$ with $\tilde{h}_1(n) = h_1(-n)$, followed by down-sampling by 2. Our contention is that this is generally true. To appreciate this, multiply both sides of Eq. (6.62) by $\phi_{1n}(t)$ and integrate

$$< f, \phi_{1n} > = < f_v^1, \phi_{1n} > + < f_w^1, \phi_{1n} > . \tag{6.64}$$

But $f_w^1(t)$ is a linear combination of $\{\psi_{1k}(t)\}$, each component of which is orthogonal to $\phi_{1n}(t)$. Therefore, the second inner product in Eq. (6.64) is zero, leaving us with

$$< f, \phi_{1n} > = < f_v^1, \phi_{1n} > = c_{1,n}. \tag{6.65}$$

(Another way of showing this explicitly is to write

$$\begin{aligned} < f_w^1, \phi_{1n} > &= \int \phi_{1n}(t) f_w^1(t) dt = \int \phi_{1n}(t) \sum_k d_{1,k} \psi_{1k}(t) dt \\ &= \sum_k d_{1,k} \int \phi_{1n}(t) \psi_{1k}(t) dt = 0. \end{aligned}$$

This last integral is zero by orthogonality.)

Therefore,

$$c_{1,n} = \frac{1}{\sqrt{2}} \int f(t) \phi(\frac{t}{2} - n) dt.$$

But from Eq. (6.26)

$$\phi(\frac{t}{2} - n) = 2 \sum_k h_0(k) \phi(t - 2n - k).$$

Therefore,

$$\begin{aligned} c_{1,n} &= \sqrt{2} \int f(t) \sum_k h_0(k) \phi(t - 2n - k) dt \\ &= \sqrt{2} \sum h_0(k) \int f(t) \phi(t - 2n - k) dt \\ &= \sqrt{2} \sum h_0(k) c_{0,2n+k} = \sqrt{2} \sum h_0(k - 2n) c_{0,k}. \end{aligned} \tag{6.66}$$

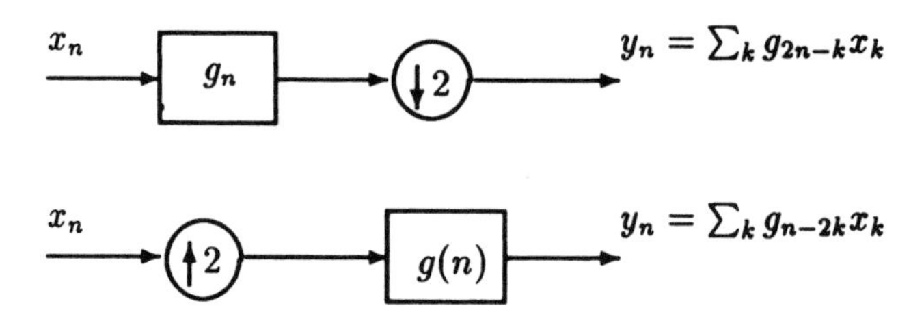

Figure 6.10: Scaling and filtering operators.

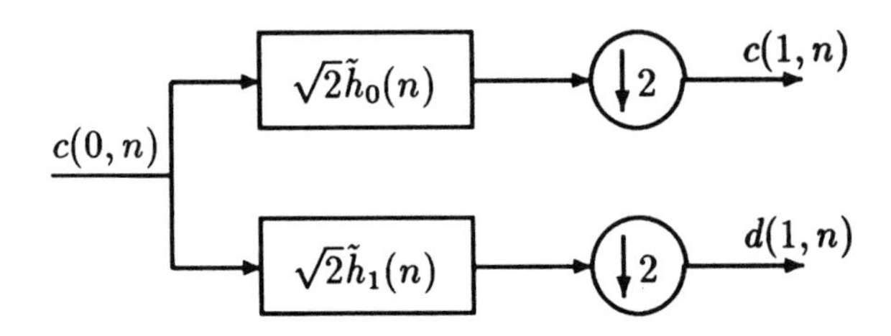

Figure 6.11: First stage of multiresolution signal decomposition.

In a similar way, we can arrive at

$$d_{1,n} = \sqrt{2} \sum h_1(k-2n) c_{0,k}. \tag{6.67}$$

Figure 6.10 shows twofold decimation and interpolation operators. So our last two equations define convolution followed by subsampling as shown in Fig. 6.11. This is recognized as the first stage of a subband tree where $\{\tilde{h}_0(n), \tilde{h}_1(n)\}$ constitute a *paraunitary* FIR pair of filters. The discrete signal $d_{1,n}$ is just the discrete wavelet transform coefficient at resolution 1/2. It represents the detail or difference information between the original signal $c_{0,n}$ and its smoothed down-sampled approximation $c_{1,n}$. These signals $c_{1,n}$ and $d_{1,n}$ are said to have a resolution of 1/2, if $c_{0,n}$ has unity resolution. Every down-sampling by 2 reduces the resolution by that factor.

The next stage of decomposition is now easily obtained. We take $f_v^1 \in V_1 = V_2 \oplus W_2$ and represent it by a component in V_2 and another in W_2:

$$\begin{aligned} f_v^1(t) &= f_v^2(t) + f_w^2(t) \\ f_v^2(t) &= \sum c_{2,n} \phi_{2n}(t) \\ f_w^2(t) &= \sum d_{2,n} \psi_{2n}(t). \end{aligned} \tag{6.68}$$

Following the procedure outlined, we can obtain the coefficients of the smoothed signal (approximation) and of the detail signal (approximation error) at resolution 1/4:

$$c_{2,n} = \sqrt{2} \sum c_{1,k} h_0(k-2n)$$

$$d_{2,n} = \sqrt{2} \sum c_{1,k} h_1(k-2n).$$

These relations are shown in the two-stage multiresolution pyramid displayed in Fig. 6.12. The decomposition into coarser, smoothed approximation and detail can be continued as far as we please.

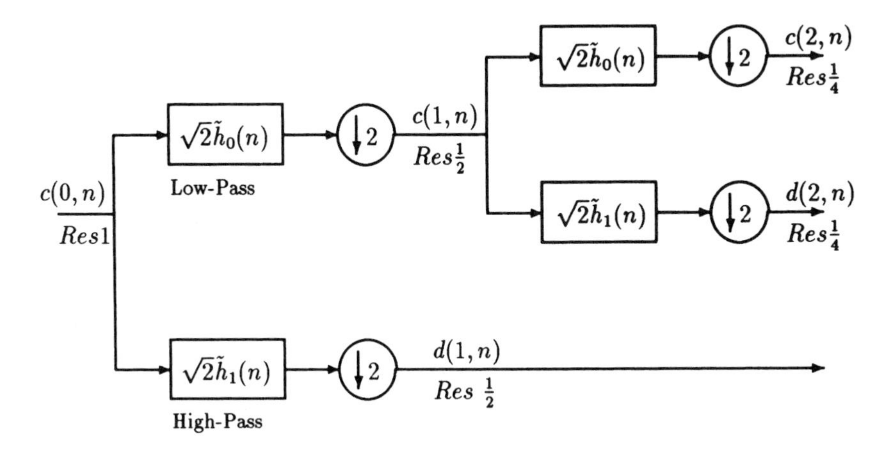

Figure 6.12: Multiresolution pyramid decomposition.

To close the circle we can now reassemble the signal from its pyramid decomposition. This reconstruction of $c_{0,n}$ from its decomposition $c_{1,n}$, and $d_{1,n}$ can be achieved by up-sampling and convolution with the filters $h_0(n)$, and $h_1(n)$ as in Fig. 6.13. This is as expected, since the front end of the one-stage pyramid is simply the analysis section of a two-band, PR-QMF bank. The reconstruction therefore must correspond to the synthesis bank. To prove this, we need to represent $h_0(n)$ and $h_1(n)$ in terms of the scaling and wavelet functions. Note that analysis filters $\tilde{h}_i(n) \stackrel{\triangle}{=} h_i(-n)$ as shown are anticausal when synthesis filters are causal.

Recall that $\{h_0(n)\}$, $\{h_1(n)\}$, are the interscale coefficients

$$\phi(t) = 2\sum_n h_0(n)\phi(2t-n), \qquad \psi(t) = 2\sum h_1(n)\psi(2t-n).$$

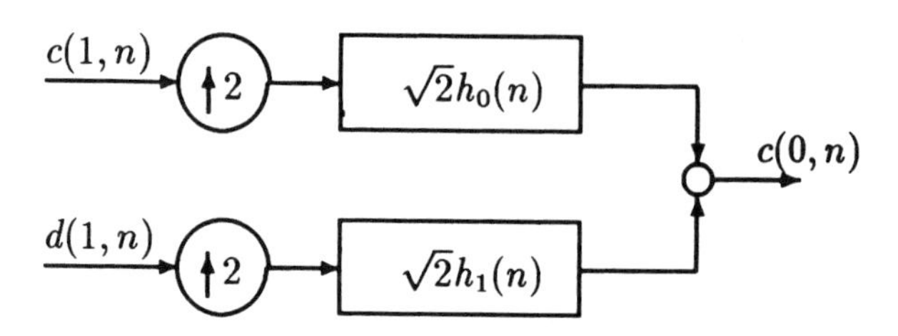

Figure 6.13: Reconstruction of a one-stage multiresolution decomposition.

Then

$$\phi(\frac{t}{2}) = 2\sum_k h_0(k)\phi(t-k)$$

and

$$\begin{aligned}\int dt\phi(\frac{t}{2})\phi(t-n) &= 2\int dt \sum h_0(k)\phi(t-k)\phi(t-n)dt \\ &= \sum_k 2h_0(k)\underbrace{\int \phi(t-k)\phi(t-n)dt}_{\delta_{k-n}} \\ &= 2h_0(n).\end{aligned}$$

Hence,

$$h_0(n) = \frac{1}{2}\int \phi(\frac{t}{2})\phi(t-n)dt. \tag{6.69}$$

Similarly

$$h_1(n) = \frac{1}{2}\int \psi(\frac{t}{2})\phi(t-n)dt.$$

The coefficient $c_{0,n}$ can be written as the sum of inner products

$$\begin{aligned}c_{0,n} &= < f, \phi_{0n} >=< f_v^1, \phi_{0n} > + < f_w^1, \phi_{0n} > \\ &= y_v(n) + y_w(n)\end{aligned}$$

where the interpolated low-pass signal is

$$\begin{aligned}y_v(n) &= \int f_v^1(t)\phi_{0n}(t)dt = \int \phi_{0n}(t)\sum_k c_{1,k}\phi_{1k}(t)dt \\ &= \sum_k c_{1,k}\frac{1}{\sqrt{2}}\int \phi(t-n)\phi(\frac{t}{2}-k)dt\end{aligned}$$

Equation (6.69) reveals that this inner integral is $2h_0(n-2k)$. Hence,

$$y_v(n) = \sqrt{2}\sum_k c_{1,k} h_0(n-2k). \tag{6.70}$$

Similarly, we can easily show

$$y_w(n) = \sqrt{2}\sum_k d_{1,k} h_1(n-2k). \tag{6.71}$$

These last two synthesis equations are depicted in Fig. 6.13.

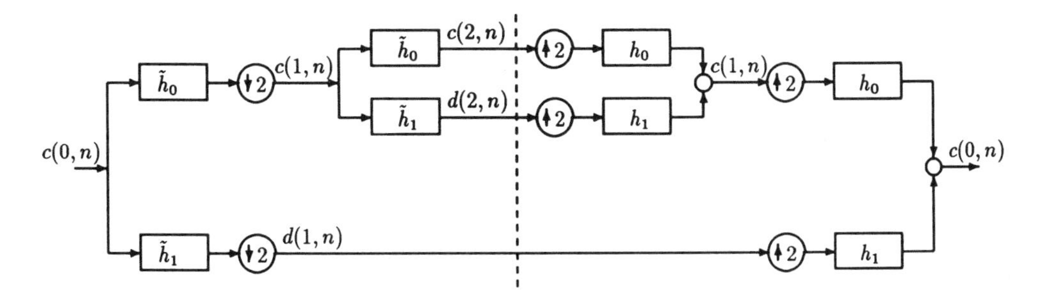

Figure 6.14: Multiresolution (pyramid) decomposition and reconstitution structure for a two-level dyadic subband tree; $\tilde{h}_i(n) = h_i(-n)$.

We can extrapolate these results for the multiscale decomposition and reconstitution for the dyadic subband tree as shown in Fig. 6.14. The gain of $\sqrt{2}$ associated with each filter is not shown explicitly. We have therefore shown that orthonormal wavelets of compact support imply FIR PR-QMF filter banks. But the converse does not follow unless we impose a regularity requirement, as discussed in the next section. Thus, if one can find a paraunitary filter $H_0(e^{j\omega})$ with regularity, then the mother wavelet can be generated by the infinite product in Eq. (6.50).

This regularity condition imposes a smoothness on H_0. Successive iteration of this operation, as required by the infinite product form, should lead to a *nicely behaved* function. This behavior is assured if $H_0(z)$ has one or more zeros at $z=-1$, a condition naturally satisfied by Binomial filters.

6.2.5 Finite Resolution Wavelet Decomposition

We have seen that a function $f \in V_0$ can be represented as

$$f(t) = \sum c_{0,n}\phi(t-n)$$

and decomposed into the sum of a lower-resolution signal (approximation) plus detail (approximation error)

$$\begin{aligned} f(t) &= f_v^1(t) + f_w^1(t) \\ &= \sum c_{1,n} 2^{-\frac{1}{2}} \phi(\tfrac{t}{2} - n) + \sum d_{1,n} 2^{-\frac{1}{2}} \psi(\tfrac{t}{2} - n). \end{aligned}$$

The coarse approximation $f_v^1(t)$ in turn can be decomposed into

$$f_v^1(t) = f_v^2(t) + f_w^2(t)$$

so that

$$f(t) = f_v^2(t) + f_w^2(t) + f_w^1(t).$$

Continuing up to $f_v^L(t)$, we have

$$f(t) = f_v^L(t) + f_w^L(t) + f_w^{L-1}(t) + ... + f_w^1$$

or

$$f(t) = \sum_{n=-\infty}^{\infty} c(L,n) 2^{-L/2} \phi(\frac{t}{2^L} - n) + \sum_{m=1}^{L} \sum_{n=-\infty}^{\infty} d(m,n) 2^{-m/2} \psi(\frac{t}{2^m} - n). \quad (6.72)$$

The purely wavelet expansion of Eq. (6.16) requires an infinite number of resolutions for the complete representation of the signal. On the other hand, Eq. (6.72) shows that $f(t)$ can be represented as a low-pass approximation at scale L plus the sum of L detail (wavelet) components at different resolutions. This latter form clearly is the more practical representation and points out the complementary role of the scaling basis in such representations.

6.2.6 The Shannon Wavelets

The Haar functions are the simplest example of orthonormal wavelet families. The orthonormality of the scaling functions in the time-domain is obvious — the translates do not overlap. These functions which are discontinuous in time are associated with a very simple 2-tap discrete filter pair. But the discontinuity in time makes the frequency resolution poor. The Shannon wavelets are at the other extreme — discontinuous in frequency and hence spread out in time. These are interesting examples of multiresolution analysis and provide an alternative basis connecting multiresolution concepts and filter banks in the frequency domain. However, it should be said that these are *not* of compact support.

Let V_0 be the space of bandlimited functions with support $(-\pi, \pi)$. Then from the Shannon sampling theorem, the functions

$$\phi(t-k) = \frac{\sin \pi(t-k)}{\pi(t-k)}, \qquad k \in Z,$$

constitute an orthonormal basis for V_0. Then any function $f(t) \in V_0$ can be expressed as

$$f(t) = \sum_k f_k \frac{\sin \pi(t-k)}{\pi(t-k)} = \sum_k f_k \phi(t-k). \tag{6.73}$$

The orthonormality can be easily demonstrated in the frequency domain. With

$$\phi(t) = \frac{\sin \pi t}{\pi t} \longleftrightarrow \Phi(\Omega) = \begin{cases} 1, & |\Omega| < \pi \\ 0, & \text{otherwise} \end{cases} \tag{6.74}$$

the inner product $< \phi_{0,k}, \phi_{0,l} >$ is just

$$\begin{aligned} \int \phi(t-k)\phi^*(t-l)dt &= \frac{1}{2\pi} \int \Phi_{0,k}(\Omega)\Phi^*_{0,l}(\Omega)d\Omega \\ &= \frac{1}{2\pi} \int_{-\pi}^{\pi} e^{-j(k-l)\Omega} d\Omega = \delta_{k-l}. \end{aligned} \tag{6.75}$$

Next, let V_{-1} be the space of functions band limited to $[-2\pi, 2\pi]$, and W_0 the space of band-pass signals with support $(-2\pi, -\pi) \bigcup (\pi, 2\pi)$. The succession of multiresolution subspaces is shown in Fig. 6.15. By construction, we have $V_{-1} = V_0 \oplus W_0$, where W_0 is the orthogonal complement of V_0 in V_{-1}. It is immediately evident that $< \phi_{-1,l}, \phi^*_{-1,k} >= \delta_{k-l}$. Furthermore, any band-pass signal in W_0 can be represented in terms of the translated Shannon wavelet, $\psi(t-k)$, where

$$\begin{aligned} \psi(t) &= \left(\frac{\sin \frac{\pi}{2} t}{\frac{\pi}{2} t} \right) \cos \frac{3\pi}{2} t \\ \Psi(\Omega) &= \begin{cases} 1, & \pi < |\Omega| < 2\pi \\ 0, & \text{otherwise} \end{cases} . \end{aligned} \tag{6.76}$$

This Shannon wavelet is drawn in Fig. 6.16. The orthogonality of the wavelets at the same scale is easily shown by calculating the inner product $< \psi(t-k), \psi(t-l) >$ in the frequency-domain. The wavelet orthogonality *across* the scales is manifested by the nonoverlap of the frequency-domains of W_k, and W_l as seen in Fig. 6.15. This figure also shows that V_i can be expressed as the infinite direct sum

$$V_i = W_{i+1} \oplus W_{i+2} \oplus W_{i+3} \oplus \ldots \quad . \tag{6.77}$$

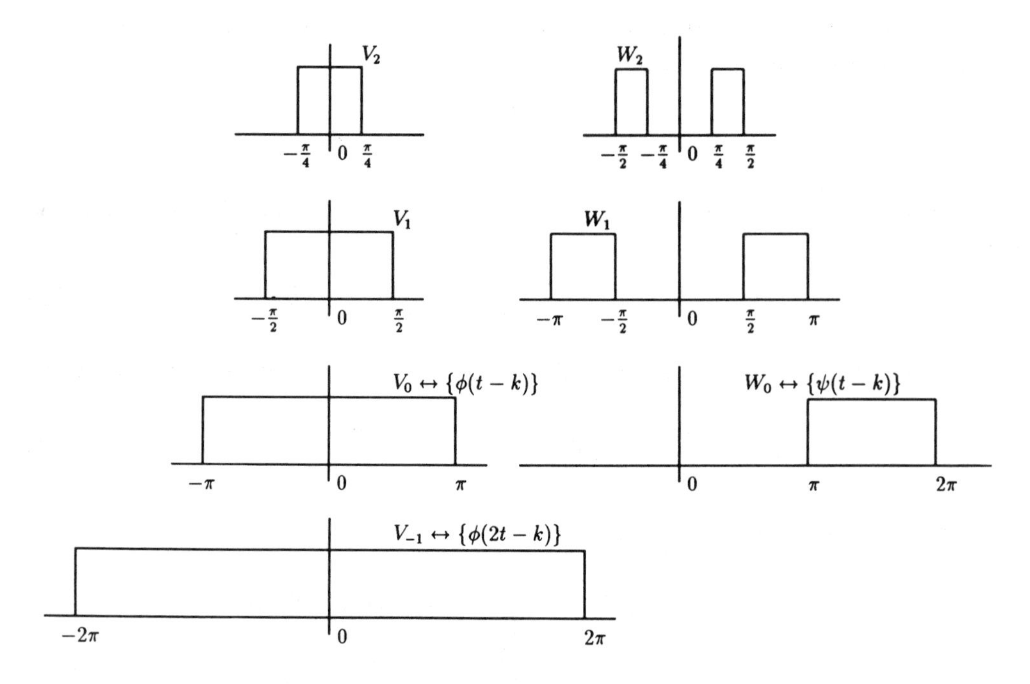

Figure 6.15: Succession of multiresolution subspaces.

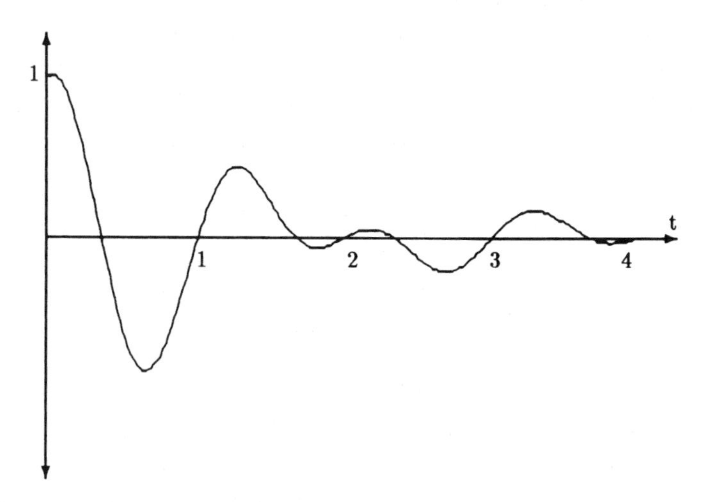

Figure 6.16: Shannon wavelet, $\psi(t) = \frac{\sin \frac{\pi}{2} t}{\frac{\pi}{2} t} \cos \frac{3\pi}{2} t$.

This is the space of L^2 functions band-limited to $-\pi/2^{i-1}, \pi/2^{i-1}$, *excluding* $\omega = 0$ (The latter exception results since a DC signal is not square integrable).

Since $\phi(t)$ and $\psi(t)$ are of infinite support, we expect the interscale coefficients to have the same property. Usually, we use a pair of appropriate PR-QMF filters to generate $\Phi(\Omega)$, $\Psi(\Omega)$ via the infinite product representation of Eqs.(6.44) and (6.50). In the present context, we reverse this process for illustrative purposes and compute $h_0(n)$ and $h_1(n)$ from $\phi(t)$ and $\psi(t)$, respectively. From Eq. (6.43) the product of $\Phi(\Omega/2)$ band-limited to $\pm 2\pi$ and $H_0(e^{j\omega/2})$ must yield $\Phi(\Omega)$ band-limited to $\pm\pi$:

$$\Phi(\Omega) = \Phi(\frac{\Omega}{2})H_0(e^{j\omega/2}).$$

Therefore, the transform of the discrete filter $H_0(e^{j\omega/2})$ with period 4π must itself be band-limited to $\pm\pi$. Hence, $H_0(e^{j\omega})$ must be the ideal half-band filter

$$H_0(e^{j\omega}) = \begin{cases} 1, & |\omega| < \frac{\pi}{2} \\ 0, & \text{otherwise} \end{cases} \quad \text{on } |\omega| < \pi \tag{6.78}$$

and correspondingly,

$$h_0(n) = \frac{1}{2}\frac{\sin(n\pi/2)}{(n\pi/2)}. \tag{6.79}$$

From Eq. (6.57) with $N = 2$ the high-frequency half-bandwidth filter is then

$$\begin{aligned} H_1(e^{j\omega}) &= e^{-j\omega}H_0(e^{-j(\omega+\pi)}) \\ h_1(n) &= (-1)^{n+1}h_0(1-n) \end{aligned} \tag{6.80}$$

The frequency and time responses of these discrete filters are displayed in Fig. 6.17. These Shannon wavelets are clearly not well localized in time — decaying only as fast as $1/t$. In the following sections, we investigate wavelets that lie somewhere between the two extremes of the Haar and Shannon wavelets. These will be smooth functions of both time and frequency, as determined by a property called *regularity*.

6.2.7 Initialization and the Fast Wavelet Transform

The major conclusion from multiresolution pyramid decomposition is that a continuous time function, $f(t)$, can be decomposed into a low-pass approximation at the 1/2 resolution plus a sum of L detail wavelet (band-pass) components at successively finer resolutions. This decomposition can be continued indefinitely. The coefficients in this pyramid expansion are simply the outputs of the paraunitary subband tree. Hence the terminology *fast wavelet transform.*

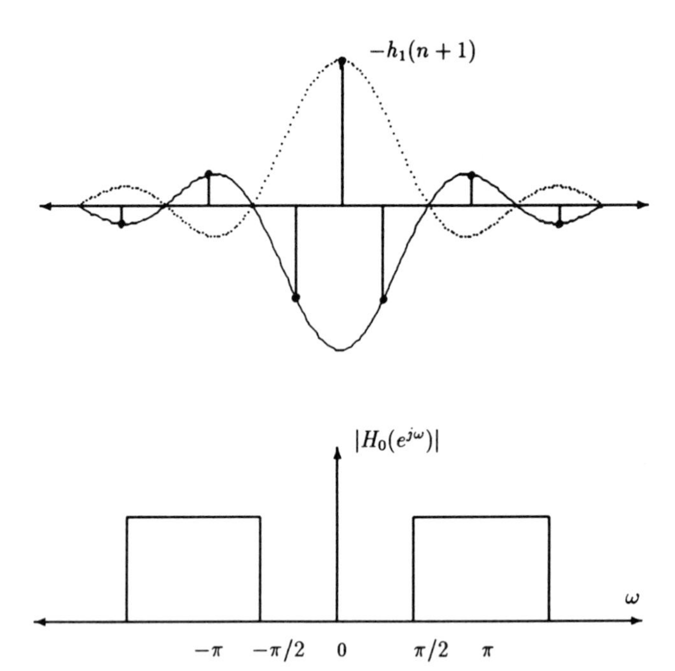

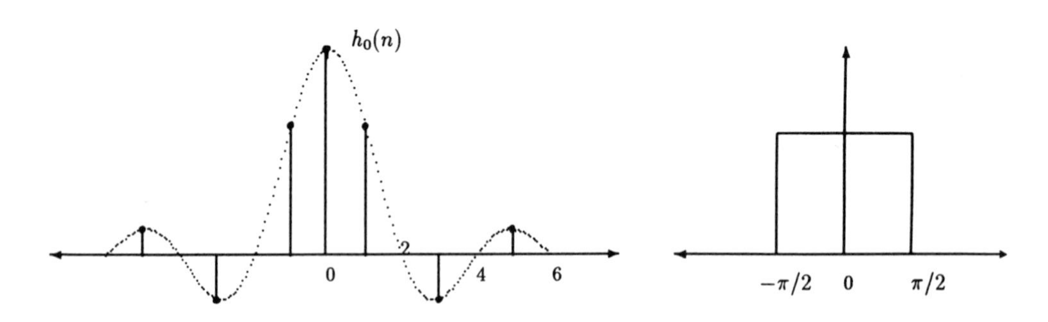

Figure 6.17: Ideal half-band filters for Shannon wavelets.

The fly in the ointment here is the initialization of the subband tree by $\{c_{0,n}\}$. If this starting point Eq. (6.61) is only an approximation, then the expansion that follows is itself only an approximation. As a case in point, suppose $f(t)$ is a band-limited signal. Then

$$\phi_{0,n} = Sin(\pi(t-n))/\pi(t-n)$$

is an orthonormal basis and

$$f(t) = \sum f(n)Sin(\pi(t-n))/\pi(t-n).$$

In this case, $h_0(n)$ and $h_1(n)$ must be ideal "brick wall" low-pass and high-pass filters. If $f(n) = c_{0,n}$ is inputted to the dyadic tree with filters that only approximate the ideal filters, then the resulting coefficients, or subband signals $\{d_{m,n}\}$ and $\{c_{m,n}\}$, are themselves only approximations to the exact values.

6.3 Wavelet Regularity and Wavelet Families

The wavelet families, Haar and Shannon, discussed thus far have undesirable properties in either frequency- or time-domains. We therefore need to find a set of interscale coefficients that lead to smooth functions of compact support in time and yet reasonably localized in frequency. In particular we want to specify properties for $H_0(e^{j\omega})$ so that the infinite product $\Phi(\Omega) = \prod_{k=1}^{\infty} H_0(e^{j\omega/2^k})$ converges to a smooth function, rather than breaking up into fractals.

6.3.1 Regularity or Smoothness

The concept of *regularity* (Daubechies, 1988) provides a measure of smoothness for wavelet and scaling functions. The regularity of the scaling function is defined as the maximum value of r such that

$$|\Phi(\Omega)| \leq \frac{c}{(1+|\Omega|)^{r+1}}, \qquad \Omega \in R. \tag{6.81}$$

This in turn implies that $\phi(t)$ is *m-times* continuously *differentiable*, where $r \geq m$. The decay of $\Phi(\Omega)$ determines the regularity, i.e., smoothness, of $\phi(t)$ and $\psi(t)$.

We know that $H_0(z)$ must have at least one zero at $z = -1$. Suppose it has L zeros at that location and that it is FIR of degree $N-1$; then

$$H_0(z) = \left(\frac{1+z^{-1}}{2}\right)^L P(z), \qquad H_0(1) = 1 \tag{6.82}$$

or

$$|H_0(e^{j\omega})| = |\cos\frac{\omega}{2}|^L |P(e^{j\omega})|, \tag{6.83}$$

where $P(z)$ is a polynomial in z^{-1} of degree $N-1-L$ with real coefficients. To see the effect of these L zeros at $z=-1$ on the decay of $\Phi(\Omega)$, substitute Eq. (6.83) into the infinite product form, Eq. (6.44),

$$|\Phi(\Omega)| = \prod_{k=1}^{\infty} \left|(\cos\frac{\omega}{2^{k+1}})\right|^L \prod_{k=1}^{\infty} |P(e^{j\omega/2^k})|. \tag{6.84}$$

But

$$\cos\frac{\omega}{2} = \frac{\sin\omega}{2\sin\omega/2}.$$

The first product term in Eq. (6.84) is therefore

$$\lim_{M\to\infty}\left[\prod_{k=1}^{M}\left|\left(\frac{\sin\omega/2^k}{2\sin\omega/2^{k+1}}\right)\right|\right]^L = \left\{\lim_{M\to\infty}\left[\left|\frac{\sin\omega/2}{2^M\sin\omega/2^{M+1}}\right|\right]\right\}^L = \left|(\frac{\sin\omega/2}{\omega/2})\right|^L$$

$$|\Phi(\Omega)| = \left|\frac{\sin\omega/2}{\omega/2}\right|^L \prod_{k=1}^{\infty}|P(e^{j\omega/2^k})| \tag{6.85}$$

The $(sinc\frac{\omega}{2})^L$ term contributes to the decay of $\Phi(\Omega)$ provided the second term can be bounded. This form has been used to estimate the regularity of $\phi(t)$. One such estimate is as follows. Let $P(e^{j\omega})$ satisfy

$$\underset{\omega\in R}{\text{Max}} \; |\prod_{k=0}^{l} P(e^{j\frac{\omega}{2}k})| \leq 2^{l(N-m-1)} \tag{6.86}$$

for some $l > 1$; then $h_0(n)$ defines a scaling function $\phi(t)$ that is m-times continuously differentiable. Tighter estimates of regularity have been reported in the literature (Daubechies and Lagarias, 1991).

We have seen in Eq. (6.85) the implication of the L zeros of $H_0(z)$ at $z=-1$ on the decay of $\Phi(\Omega)$. These zeros also imply a flatness on the frequency response of $H_0(e^{j\omega})$ at $\omega=\pi$, and consequent vanishing moments of the high-pass filter $h_1(n)$. With

$$H_0(e^{j\omega}) = e^{-j\omega L/2}\left(\cos\frac{\omega}{2}\right)^L P(e^{j\omega}) = \left(\cos\frac{\omega}{2}\right)^L f(\omega) = \sum_n h_0(n)e^{-jn\omega} \tag{6.87}$$

we find that

$$\frac{d^r H_0(e^{j\omega})}{d\omega^r} = \left(\cos\frac{\omega}{2}\right)^{L-r} f_r(\omega) = \sum_n(-jn)^r h_0(n)e^{-jn\omega}. \tag{6.88}$$

The $(\cos\omega/2)^{L-r}$ term makes these derivatives zero at $\omega = \pi$ for $r = 0, 1, 2, ..., L-1$, leaving us with (Prob. 6.6)

$$\left.\frac{d^r H_0(e^{j\omega})}{d\omega^r}\right|_{\omega=\pi} = (-j)^r \sum_n n^r (-1)^n h_0(n) = 0 \qquad r = 0, \cdots, L-1. \tag{6.89}$$

This produces a smooth low-pass filter.

From Eq. (6.57), the high-pass filter $H_1(z)$ has L zeros at $z = 1$. Hence we can write

$$H_1(e^{j\omega}) = \left(\sin\frac{\omega}{2}\right)^L g(\omega).$$

The $(\sin\omega/2)^L$ term ensures the vanishing of the derivatives of $H_1(e^{j\omega})$ at $\omega = 0$ and the associated moments, that is,

$$\left.\frac{d^r H_1(e^{j\omega})}{d\omega^r}\right|_{\omega=0} = 0, \qquad r = 0, 1, ..., L-1,$$

implying

$$\sum_n n^r h_1(n) = 0.$$

Several proposed wavelet solutions are based on Eq. (6.82). To investigate further the choice of L in this equation, we note that since $P(z)$ is a polynomial in z^{-1} with real coefficients, $Q(z) = P(z)P(z^{-1})$ is a symmetric polynomial:

$$Q(z) = \sum_{-M}^{M} q_n z^{-n}, \quad M = N - 1 - L, \text{and } q_n = q_{-n}.$$

Therefore,

$$Q(e^{j\omega}) = |P(e^{j\omega})|^2 = q_0 + 2\sum_{n=1}^{M} q_n \cos(n\omega).$$

But $\cos(n\omega)$ can be expressed as a polynomial in $\cos\omega$, which in turn can be represented in terms of $\sin^2\omega/2$. Therefore, $|P(e^{j\omega})|^2$ is some polynomial $f(.)$, in $(\sin^2\omega/2)$ of degree $N-1-L$:

$$|H_0(e^{j\omega})|^2 = (\cos^2\frac{\omega}{2})^L f(\sin^2\omega/2) = (1-x)^L f(x), \tag{6.90}$$

where

$$x = \sin^2\omega/2$$

Substituting into the *power complementary* equation

$$|H_0(e^{j\omega})|^2 + |H_0(e^{j(\omega+\pi)})|^2 = 1.$$

gives

$$(1-x)^L f(x) + x^L f(1-x) = 1. \tag{6.91}$$

This equation has a solution of the form

$$f(x) = \sum_{k=0}^{L-1} \binom{L-1-k}{k} x^k + x^L R(1-2x) \tag{6.92}$$

where $R(x)$ is an odd polynomial such that

$$R(x) = -R(1-x).$$

Different choices for $R(x)$ and L lead to different wavelet solutions. We will comment on two solutions attributed to Daubechies.

6.3.2 The Daubechies Wavelets

If we choose $R(x) \equiv 0$ in Eq. (6.92), L reaches its maximum value, which is $L = N/2$ for an N-tap filter. This corresponds to the unique *maximally flat* magnitude square response in which the number of vanishing derivatives of $|H_0(e^{j\omega})|^2$ at $\omega = 0$ and $\omega = \pi$ are equal. This interscale coefficient sequence $\{h_0(n)\}$ is identical to the unit sample response of the Binomial-QMF derived in Chapter 4.

The regularity of the Daubechies wavelet function $\psi(t)$ increases linearly with its support width, i.e., on the length of FIR filter. However, Daubechies and Lagarias have proven that the maximally flat solution does not lead to the highest regularity wavelet. They devised counterexamples with higher regularity for the same support width, but with a reduced number of zeros at $z = -1$.

In Chapter 3, we found that paraunitary linear-phase FIR filter bank did not exist for the two-band case (except for the trivial case of a 2-tap filter). It is not surprising then to discover that it is equally *impossible* to obtain an orthonormal compactly supported wavelet $\psi(t)$ that is either symmetric or antisymmetric, except for the trivial Haar case. In order to obtain $h_0(n)$ as close to linear-phase as possible we have to choose the zeros of its magnitude square function $|H_0(e^{j\omega})|^2$ alternatively from inside and outside the unit circle as frequency increases. This leads to nonminimum-phase FIR filter solutions. For N sufficiently large, the unit sample responses of $h_0(n)$ and $h_1(n)$ have more acceptable symmetry or antisymmetry. In the Daubechies wavelet bases there are $2^{[N/4]-1}$ different filter solutions.

However, for $N = 4$ and 6, there is effectively only one pair of $\phi(t)$ and $\psi(t)$. For $N \geq 8$ we could choose the solution that is closest to linear-phase.

If one selects the nonminimum-phase solution for the analysis filters in order to enhance the phase response, perfect reconstruction requires that the associated minimum-phase solution be used in the synthesis stage. The latter has poorer phase response. From a coding standpoint, the more linear-like phase should be used on the analysis side so as to reduce the effects of quantization (Forchheimer and Kronander, 1989).

The Daubechies solution involved solving for $P(e^{j\omega})$ from $|P(e^{j\omega})|^2$ by spectral factorization. The same filters were found from an entirely different starting point using the elegant properties of the Binomial sequences. The latter approach also suggested the efficient realizations of these filters using the Binomial QMF structures of Fig. 4.1. Table 4.2 gives the coefficient values of these filters for tap lengths of 4, 6, and 8.

6.3.3 The Coiflet Bases

For a given support, the Daubechies wavelet $\psi(t)$ has the maximum number of vanishing moments. The scaling function $\phi(t)$ does not satisfy any moment condition, except $\int \phi(t)dt = 1$. For numerical analysis applications, it may be useful to trade off some of the zero moments of the wavelet in order to obtain some zero moments for the scaling function $\phi(t)$ such as

$$\begin{aligned} \int \phi(t)dt &= 1 \\ \int t^{\nu}\phi(t)dt &= 0 \quad \text{for } \nu = 1, 2, ..., L-1 \\ \int t^{\nu}\psi(t)dt &= 0 \quad \text{for } \nu = 0, 1, ..., L-1. \end{aligned} \tag{6.93}$$

It is seen that the wavelet and scaling functions have an equal number of vanishing moments in this case. Imposing such vanishing moments on the scaling function $\phi(t)$ also increases its symmetry.

In the frequency domain, these conditions directly impose *flatness* on the transforms of the scaling and wavelet functions,

$$\begin{aligned} \Phi(0) &= 1 \\ \frac{d^{\nu}}{d\Omega^{\nu}}\Phi(0) &= 0 \quad \text{for} \quad \nu = 1, 2, ..., L-1 \\ \frac{d^{\nu}}{d\Omega^{\nu}}\Psi(0) &= 0 \quad \text{for} \quad \nu = 0, 1, ..., L-1. \end{aligned}$$

In terms of $H_0(e^{j\omega})$, these conditions become

$$\begin{aligned} \frac{d^\nu}{d\omega^\nu} H_0(e^{j\omega})|_{\omega=0} &= 0 \quad \text{for} \quad \nu = 1, 2, ..., L-1 \\ \frac{d^\nu}{d\omega^\nu} H_0(e^{j\omega})|_{\omega=\pi} &= 0 \quad \text{for} \quad \nu = 0, 1, ..., L-1. \end{aligned} \tag{6.94}$$

In order to satisfy those conditions $H_0(e^{j\omega})$ must have the form

$$H_0(e^{j\omega}) = 1 + (1 - e^{-j\omega})^L V(e^{j\omega}) \tag{6.95}$$

From Eq. (6.94), $H_0(e^{j\omega})$ has a zero of order L at $\omega = \pi$. Consequently, $H_0(e^{j\omega})$ must also satisfy

$$H_0(e^{j\omega}) = \left[\frac{1}{2}(1 + e^{-j\omega})\right]^L P(e^{j\omega}) \tag{6.96}$$

where $P(e^{j\omega})$ is as found earlier,

$$|P(e^{j\omega})|^2 = \sum_{k=0}^{L-1} \binom{L-1+k}{k} \sin^{2k}(\frac{\omega}{2}) + \sin^{2L}(\frac{\omega}{2}) R(\cos\omega)$$

where $R(x)$ is an odd polynomial. Equations (6.95) and (6.96) lead to L independent linear constraints on the coefficients of $V(e^{j\omega})$ which are very difficult to solve for $L > 6$.

Daubechies (Tech. Memo) has indicated an indirect approach. For $L = 2K$, she shows that $H_0(e^{j\omega})$ can be expressed as

$$H_0(e^{j\omega}) = 1 + (\sin^2(\frac{\omega}{2}))^K \left[-\sum_{k=0}^{K-1} \binom{K-1+k}{k} (\cos^2(\frac{\omega}{2}))^k + (\cos^2(\frac{\omega}{2}))^k f(e^{j\omega}) \right] \tag{6.97}$$

in order to satisfy both equations simultaneously. Remaining $f(e^{j\omega})$ must then be chosen such that the PR conditions are satisfied. The interscale coefficients or filters of these Coiflet bases are given in Table 6.2 for the lengths $N = 6, 12, 18, 24$. It is noteworthy that samples of Coiflet filters were derived earlier as special cases of the Bernstein polynomial approach to filter bank design of Section 4.3.

6.4 Biorthogonal Wavelets and Filter Banks

In Chapter 3, we described a design procedure for a two-band PR FIR lattice structure with linear phase. This was possible because the paraunitary constraint

n	h(n)	h(n)	h(n)	h(n)
0	0.000630961046	-0.002682418671	0.011587596739	-0.051429728471
1	-0.001152224852	0.005503126709	-0.029320137980	0.238929728471
2	-0.005194524026	0.016583560479	-0.047639590310	0.602859456942
3	0.011362459244	-0.046507764479	0.273021046535	0.272140543058
4	0.018867235378	-0.043220763560	0.574682393857	-0.051429972847
5	-0.057464234429	0.286503335274	0.294867193696	-0.011070271529
6	-0.039652648517	0.561285256870	-0.054085607092	
7	0.293667390895	0.302983571773	-0.042026480461	
8	0.553126452562	-0.050770140755	0.016744410163	
9	0.307157326198	-0.058196250762	0.003967883613	
10	-0.047112738865	0.024434094321	-0.001289203356	
11	-0.068038127051	0.011229240962	-0.000509505539	
12	0.027813640153	-0.006369601011		
13	0.017735837438	-0.001820458916		
14	-0.010756318517	0.000790205101		
15	-0.004001012886	0.000329665174		
16	0.002652665946	-0.000050192775		
17	0.000895594529	-0.000024465734		
18	-0.000416500571			
19	-0.000183829769			
20	0.000044080354			
21	0.000022082857			
22	-0.000002304942			
23	-0.000001262175			

Table 6.2: Coiflet PR-QMF filter coefficients for taps $N = 6, 12, 18, 24$ (Daubechies, 1990).

was removed. Vetterli and Herley (1992) demonstrate that linear phase analysis and synthesis filters can be obtained in the two-band case using the rubric of *biorthogonality*. We will outline their approach and then show its implications for a multiresolution wavelet decomposition. It is also interesting that the biorthogonal filter banks are identical to the modified Laplacian pyramid, which was introduced earlier in Section 3.4.6.

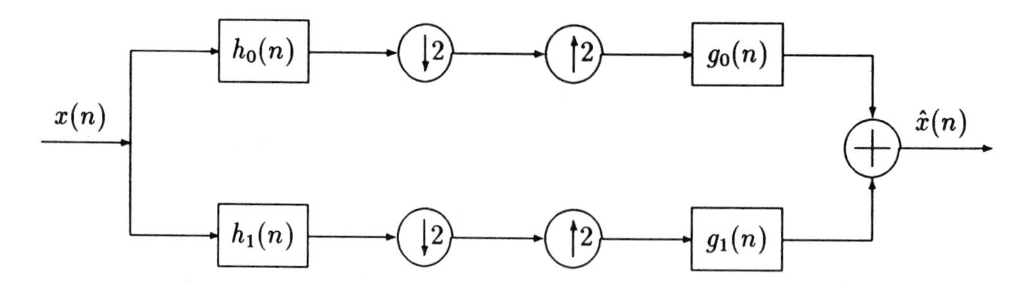

Figure 6.18: Biorthogonal filter bank structure.

The standard two-band filter bank is shown in Fig. 6.18. Our analysis in Chapter 3 showed that

$$\begin{aligned}\hat{X}(z) &= T(z)X(z) + S(z)X(-z) \\ T(z) &= \frac{1}{2}[G_0(z)H_0(z) + G_1(z)H_1(z)] \\ S(z) &= \frac{1}{2}[G_0(z)H_0(-z) + G_1(z)H_1(-z)].\end{aligned} \tag{6.98}$$

For perfect reconstruction, we require

$$\begin{aligned}S(z) &= 0 \\ T(z) &= cz^{-n_0}\end{aligned} \tag{6.99}$$

where alias cancellation is ensured by selecting $G_0(z) = -H_1(-z)$, and $G_1(z) = H_0(-z)$, resulting in

$$T(z) = [G_0(z)H_0(z) + G_0(-z)H_0(-z)] = cz^{-n_0}.$$

The biorthogonal solution is as follows. The PR conditions are satisfied by imposing orthogonality across the analysis and synthesis sections (Vetterli and Herley, 1992),

$$\begin{aligned}&< \tilde{g}_0(n-2k), h_1(n-2l) >= 0 \\ &< \tilde{g}_1(n-2k), h_0(n-2l) >= 0\end{aligned} \tag{6.100}$$

where

$$\tilde{g}_i(n) = g_i(-n)$$

and

$$\begin{aligned}&< \tilde{g}_0(n-2k), h_0(n) >= \delta_k \\ &< \tilde{g}_1(n-2k), h_1(n) >= \delta_k.\end{aligned} \tag{6.101}$$

(The noncausal paraunitary solution was $\tilde{g}_0(n) = h_0(n)$, and $\tilde{g}_1(n) = h_1(n)$). The added flexibility in the biorthogonal case permits use of linear-phase and unequal length filters.

In the parlance of wavelets, we can define two hierarchies of approximations (Cohen, Daubechies, and Feauveau)

$$\begin{array}{c} ...V_2 \subset V_1 \subset V_0 \subset V_{-1} \subset V_{-2}... \\ ...\bar{V}_2 \subset \bar{V}_1 \subset \bar{V}_0 \subset \bar{V}_{-1} \subset \bar{V}_{-2}... \quad . \end{array} \tag{6.102}$$

The subspace W_j is complementary to V_j in V_{j-1}, but it is not the *orthogonal* complement. Instead W_j is the orthogonal complement of $\bar{V}_j$. Similarly, $\bar{W}_j \perp V_j$. Thus,

$$V_{j-1} = V_j \oplus \bar{W}_j \quad \text{and} \quad \bar{V}_{j-1} = \bar{V}_j \oplus W_j. \tag{6.103}$$

The associated scaling and wavelet functions are then

$$\begin{aligned} \phi(t) &= 2\sum h_0(n)\phi(2t-n) \\ \bar{\phi}(t) &= 2\sum \tilde{g}_0(n)\bar{\phi}(2t-n) \end{aligned} \tag{6.104}$$

and

$$\begin{aligned} \psi(t) &= 2\sum h_1(n)\phi(2t-n) \\ \bar{\psi}(t) &= 2\sum \tilde{g}_1(n)\bar{\phi}(2t-n). \end{aligned} \tag{6.105}$$

We should expect, and indeed find, that the following orthogonality among the scaling and wavelet functions

$$\begin{aligned} < \bar{\phi}(t-k), \phi(t-l) > &= \delta_{k-l} \\ < \bar{\psi}(t-k), \psi(t-l) > &= \delta_{k-l} \end{aligned} \tag{6.106}$$

and

$$\begin{array}{c} < \bar{\phi}(t-k), \psi(t-l) >= 0 \\ < \bar{\psi}(t-k), \phi(t-l) >= 0. \end{array} \tag{6.107}$$

These relations confirm that $\{\psi_{mn}(t), \bar{\psi}_{m'n'}\}$ are an orthogonal set across the scales such that

$$< \psi_{mn}, \bar{\psi}_{m'n'} >= \delta_{mm'}\delta_{nn'}. \tag{6.108}$$

This permits us to express any function $f \in L^2(R)$ in the form

$$\begin{aligned} f(t) &= \sum_j \sum_k < f, \bar{\psi}_{jk} > \psi_{jk}(t) \\ &= \sum_j \sum_k < f, \psi_{jk} > \bar{\psi}_{jk}(t). \end{aligned} \tag{6.109}$$

Biorthogonality provides additional degrees of freedom so that both perfect reconstruction and linear-phase filters can be realized simultaneously. For example, the requirement on the low-pass branch of Fig. 6.18 can be stated in time- and transform-domains as

$$\sum h_0(n)\tilde{g}_0(n+2k) = \delta_k \longleftrightarrow H_0(z)G_0(z^{-1}) + H_0(-z)G_0(-z^{-1}) = 2. \tag{6.110}$$

Constraints of this type can be satisfied with linear-phase filters. Note that both low-pass filters $h_0(n)$ and $g_0(n)$ have at least one zero at $\omega = \pi$. Following the design procedures given for compactly supported orthonormal scaling and wavelet functions in Section 6.2.3, one can easily obtain biorthogonal (dual) scaling and wavelet functions from biorthogonal filters in the Fourier domain as

$$\begin{aligned} \Phi(\Omega) &= \Phi(0) \prod_{k=1}^{\infty} H_0(e^{j\omega/2^k}) \\ \overline{\Phi}(\Omega) &= \overline{\Phi}(0) \prod_{k=1}^{\infty} G_0(e^{j\omega/2^k}) \end{aligned} \tag{6.111}$$

and

$$\begin{aligned} \Psi(\Omega) &= H_1(e^{j\omega/2}) \prod_{k=2}^{\infty} H_0(e^{j\omega/2^k}) \\ \overline{\Psi}(\Omega) &= G_1(e^{j\omega/2}) \prod_{k=2}^{\infty} G_0(e^{j\omega/2^k}). \end{aligned} \tag{6.112}$$

Detailed design procedures for linear-phase biorthogonal wavelets and filter banks are described in the references.

There is a caveat to be noted in this structure, however. The biorthogonal nature of the filter bank allows different filter lengths in the analysis section and consequently an unequal split of the signal spectrum into low-band and high-band segments. Since each of these bands is followed by a down-sampler of rate 2, there is an inherent mismatch between the antialiasing filters and the decimation factor. In turn, the synthesis or interpolation stage has the same drawback. In a hierarchical filter bank structure it might become an issue to be monitored carefully. Indeed, Cohen and Daubechies (1993) reported that the extension of wavelet packets to the biorthogonal case might generate an unstable decomposition due to the frequency-domain spreads of basis functions. Quantizing and encoding these decimated subband signals can cause more degradation than would be the case for an almost linear-phase response paraunitary solution. Therefore, one needs to be aware of the potential problem in applications. It is possible to use different filters at different levels of a subband tree as a solution for this concern (Tazebay and Akansu, 1994). Cohen and Sere (1996) independently suggested the same solution for handling possible instabilities of nonstationary wavelet packets.

It should also be noted that the biorthogonal filter bank is a critically sampled solution in the modified Laplacian pyramid of Section 3.4.6.

6.5 Discussions and Conclusion

In this chapter, we established the link between the two-band paraunitary PR-QMF filter bank and the wavelet transform. The former provides an FIR transfer function $H_0(e^{j\omega})$ whose infinite iterated product can be made to converge to a wavelet or mother function. Additionally, with proper initialization, the dyadic tree subband structure provides a vehicle for the fast computation of the coefficients in a wavelet expansion — hence a *fast wavelet transform.*

We also saw that discrete orthonormal wavelet filters are simply the filters of a paraunitary two-band QMF bank with a zero-mean condition on the high-pass filter. This constraint is a desirable feature in any signal decomposition technique since a "DC" component can be represented using *only one* basis function. In wavelet terminology this implies a degree of *regularity.* However, it should be noted that wavelet filters with maximum regularity, while mathematically appealing, do not have any established special properties for signal processing applications. Furthermore, it should be emphasized that the wavelet transform is defined on a *continuous* variable and can therefore serve as a transform tool for analog signals. As indicated previously, wavelet expansions for such signals can be done with the discrete time dyadic subband tree structure, but only if properly initialized.

References

A. N. Akansu and M. J. Medley, Eds., *Wavelet, Subband and Block Transforms in Communications and Multimedia.* Kluwer, 1999.

A. N. Akansu and M. J. T. Smith, Eds., *Subband and Wavelet Transforms: Design and Applications.* Kluwer, 1996.

A. N. Akansu, R. A. Haddad, and H. Caglar, "Perfect Reconstruction Binomial QMF-Wavelet Transform," Proc. SPIE Visual Communication and Image Processing, Vol. 1360, pp. 609–618, Oct. 1990.

A. N. Akansu, R. A. Haddad, and H. Caglar, "The Binomial QMF-Wavelet Transform for Multiresolution Signal Decomposition," IEEE Trans. Signal Processing, Vol. 41, No. 1, pp. 13–19, Jan. 1993.

R. A. Ansari, C. Guillemot, and J. F. Kaiser, "Wavelet Construction Using Lagrange Halfband Filters," IEEE Trans. Circuits and Systems, Vol. CAS-38, pp. 1116–1118, Sept. 1991.

A. Arnedo, G. Grasseau, and M. Holschneider, "Wavelet Transform of Multifractals," Phys. Review Letters, Vol. 61, pp. 2281–2284, 1988.

P. Auscher, *Ondelettes Fractales et Applications.* Ph.D. Thesis, Univ. of Paris IX, 1989.

L. Auslander, T. Kailath, S. Mitter, Eds., *Signal Processing, Part I: Signal Processing Theory.* Inst. for Maths. and its Applications, Vol. 22, Springer-Verlag, 1990.

G. Battle, "A Block Spin Construction of Ondelettes. Part I: Lemarie Functions," Comm. Math. Phys., Vol. 110, pp. 601–615, 1987.

G. Beylkin, R. Coifman, and V. Rokhlin, "Fast Wavelet Transforms and Numerical Algorithms. I," Technical Report, Dept. of Math., Yale Univ., 1991.

C. S. Burrus, R. A. Gopinath and H. Guo, *Introduction to Wavelets and Wavelet Transforms: A Primer.* Prentice-Hall, 1998.

P. J. Burt and E. H. Adelson, "The Laplacian Pyramid as a Compact Image Code," IEEE Trans. on Comm., Vol. 31, pp. 532–540, April 1983.

H. Caglar, *A Generalized, Parametric PR-QMF/Wavelet Transform Design Approach for Multiresolution Signal Decomposition.* Ph.D. Thesis, New Jersey Institute of Technology, Jan. 1992.

A. P. Calderon, "Intermediate Spaces and Interpolation, the Complex Method," Studia Math., Vol. 24, pp. 113–190, 1964.

A. P. Calderon and A. Torchinsky, "Parabolic Maximal Functions Associated to a Distribution, I," Adv. Math., Vol. 16, pp. 1–64, 1974.

C. K. Chui, *An Introduction to Wavelets.* Academic Press, 1992.

C. K. Chui, Ed., *Wavelets: A Tutorial in Theory and Applications.* Academic Press, 1992.

A. Cohen and I. Daubechies, "On the Instability of Arbitrary Biorthogonal Wavelet Packets," SIAM J. Math. Anal., pp. 1340–1354, 1993.

A. Cohen and E. Sere, "Time-Frequency Localization with Non-Stationary Wavelet Packets," in A. N. Akansu and M. J. T. Smith, Eds., *Subband and Wavelet Transforms: Design and Applications.* Kluwer, 1996.

A. Cohen, I. Daubechies, and J. C. Feauveau, "Biorthogonal Bases of Compactly Supported Wavelets," Technical Memo., #11217–900529–07, AT&T Bell Labs., Murray Hill.

R. Coifman and Y. Meyer, "The Discrete Wavelet Transform," Technical Report, Dept. of Math., Yale Univ., 1990.

R. Coifman and Y. Meyer, "Orthonormal Wave Packet Bases," Technical Report, Dept. of Math., Yale Univ., 1990.

J. M. Combes, A. Grossman, and P. Tchamitchian, Eds., *Wavelets, Time-Frequency Methods and Phase Space.* Springer-Verlag, 1989.

R. E. Crochiere and L. R. Rabiner, *Multirate Digital Signal Processing.* Prentice-Hall, 1983.

I. Daubechies, "Orthonormal Bases of Compactly Supported Wavelets," Communications in Pure and Applied Math., Vol. 41, pp. 909–996, 1988.

I. Daubechies, "The Wavelet Transform, Time-Frequency Localization and Signal Analysis," IEEE Trans. Inf. Theo., Vol. 36, pp. 961–1005, Sept. 1990.

I. Daubechies, *Ten Lectures on Wavelets.* SIAM, 1992.

I. Daubechies, "Orthonormal Bases of Compactly Supported Wavelets II. Variations on a Theme," Technical Memo., #11217–891116–17, AT&T Bell Labs., Murray Hill, 1988.

I. Daubechies and J. C. Lagarias, "Two-Scale Difference Equations I. Existence and Global Regularity of Solutions," SIAM J. Math. Anal., Vol. 22, pp. 1388–1410, 1991.

I. Daubechies and J. C. Lagarias, "Two-Scale Difference Equations II. Local Regularity, Infinite Products of Matrices and Fractals," SIAM J. Math. Anal., 24(24): 1031–1079.

I. Daubechies, A. Grossmann, and Y. Meyer, "Painless Non-orthogonal Expansions," J. Math. Phys., Vol. 27, pp. 1271–1283, 1986.

R. J. Duffin and A. C. Schaeffer, "A Class of Nonharmonic Fourier Series," Trans. Am. Math. Soc., Vol. 72, pp. 341–366, 1952.

R. Forchheimer and T. Kronander, "Image Coding - From Waveforms to Animation", IEEE Trans. ASSP, Vol. 37, pp. 2008–2023, Dec. 1989.

J. B. J. Fourier, "Theorie Analytique de la Chaleur," in *Oeuvres de Fourier*, tome premier, G. Darboux. Ed., Gauthiers-Villars, 1888.

D. Gabor, "Theory of Communication," J. of the IEE, Vol. 93, pp. 429–457, 1946.

R. A. Gopinath, *The Wavelet Transforms and Time-Scale Analysis of Signals.* M. S. Thesis, Rice University, 1990.

R. A. Gopinath and C. S. Burrus, "State-Space Approach to Multiplicity M Orthonormal Wavelet Bases," Tech. Rep., CML TR-91–22, Rice Univ., Nov. 1991.

R. A. Gopinath and C. S. Burrus, "On Cosine-Modulated Wavelet Orthonormal Bases," Tech. Rep., CML TR-92–06, Rice Univ., 1992.

R. A. Gopinath, W. M. Lawton, and C. S. Burrus, "Wavelet-Galerkin Approximation of Linear Translation Invariant Operators," Proc. IEEE ICASSP, 1991.

R. A. Gopinath, J. E. Odegard, and C. S. Burrus, "On the Correlation Structure of Multiplicity M Scaling Functions and Wavelets," Proc. IEEE ISCAS, pp. 959–962, 1992.

P. Goupillaud, A. Grossmann, and J. Morlet, "Cycle-Octave and Related Transforms in Seismic Signal Analysis," Geoexploration, Vol. 23, pp. 85–102, Elsevier Science Pub., 1984.

A. Grossmann and J. Morlet, "Decomposition of Hardy Functions into Square Integrable Wavelets of Constant Shape," SIAM J. Math. Anal., Vol. 15, pp. 723–736, July 1984.

A. Grossmann, J. Morlet, and T. Paul, "Transforms Associated to Square Integrable Group Representations," I, J. Math. Phys., Vol. 26, pp. 2473–2479, 1985, II, Ann. Inst. Henri Poincare, Vol. 45, pp. 293–309, 1986.

A. Haar, "Zur Theorie der Orthogonalen Funktionen-systeme," Math. Annal., Vol. 69, pp. 331–371, 1910.

C. E. Heil and D. F. Walnut, "Continuous and Discrete Wavelet Transforms," SIAM Review, Vol. 31, pp. 628–666, Dec. 1989.

IEEE Trans. on Information Theory, Special Issue on Wavelet Transforms and Multiresolution Signal Analysis, March 1992.

IEEE Trans. on Signal Processing, Special Issue on Theory and Application of Filter Banks and Wavelet Transforms, April 1998.

A. J. E. M. Janssen, "Gabor Representation of Generalized Functions," J. Math. Appl., Vol. 80, pp. 377–394, 1981.

B. Jawerth and W. Sweldens, "Biorthogonal Smooth Local Trigonometric Bases," J. Fourier Anal. Appl., Vol. 2, 1996.

J. R. Klaunder and E. Sudarshan, *Fundamentals of Quantum Optics.* Benjamin, 1968.

J. Kovacevic, *Filter Banks and Wavelets: Extensions and Applications.* Ph.D. Thesis, Columbia University, 1991.

R. Kronland-Martinet, J. Morlet, and A. Grossmann, "Analysis of Sound Patterns through Wavelet Transforms," Int. J. Pattern Rec. and Artif. Intell., Vol. 1, pp. 273–301, 1987.

P. G. Lemarie and Y. Meyer, "Ondelettes et bases Hilbertiennes," Rev. Math. Iberoamericana, Vol. 2, pp. 1–18, 1986.

S. Mallat, "A Theory for Multiresolution Signal Decomposition: the Wavelet Representation," IEEE Trans. on Pattern Anal. and Mach. Intell., Vol. 11, pp. 674–693, July 1989.

S. Mallat, "Multifrequency Channel Decompositions of Images and Wavelet Models," IEEE Trans. on ASSP, Vol. 37, pp. 2091–2110, Dec. 1989.

S. Mallat, "Multiresolution Approximations and Wavelet Orthonormal Bases of $L^2(R)$," Trans. Amer. Math. Soc., Vol. 315, pp. 69–87, Sept. 1989.

S. G. Mallat, Ed., *A Wavelet Tour of Signal Processing.* Academic Press, 1998.

A. Mertins, *Signal Analysis.* Wiley, 1999.

Y. Meyer, *Ondelettes et Operateurs*, Tome I. Herrmann Ed., 1990.

Y. Meyer, *Wavelets: Algorithms and Applications.* SIAM, 1993.

S. H. Nawab and T. F. Quatieri, "Short-Time Fourier Transform," Chapter 6 in *Advanced Topics in Signal Processing*, J. S. Lim and A. V. Oppenheim, Eds. Prentice-Hall, 1988.

A. Papoulis, *Signal Analysis.* McGraw-Hill, 1977.

M. R. Portnoff, "Time-Frequency Representation of Digital Signals and Systems Based on the Short-Time Fourier Analysis," IEEE Trans. on ASSP, Vol. 28, pp. 55–69, Feb. 1980.

O. Rioul and M. Vetterli, "Wavelets and Signal Processing," IEEE Signal Processing Magazine, Vol. 8, pp. 14–38, Oct. 1991.

M. B. Ruskai, G. Beylkin, R. Coifman, I. Daubechies, S. Mallat, Y. Meyer, and L. Raphael, Eds., *Wavelets and Their Applications.* Jones and Bartlett, Boston, MA 1992.

M. J. Shensa, "The Discrete Wavelet Transform: Wedding the a trous and Mallat Algorithms," IEEE Trans. Signal Processing, Vol. 40, No. 10, pp. 2464–2482, Oct. 1992.

G. Strang and T. Nguyen, *Wavelets and Filter Banks.* Wellesley-Cambridge Press, 1996.

B. Suter, *Multirate and Wavelet Signal Processing.* Academic Press, 1998.

M. V. Tazebay and A. N. Akansu, "Progressive Optimality in Hierarchical Filter Banks," Proc. IEEE ICIP, pp. 825–829, 1994.

A. H. Tewfik and P. E. Jorgensen, "The Choice of a Wavelet for Signal Coding and Processing," Proc. IEEE ICASSP, 1991.

A. H. Tewfik and M. Y. Kim, "Multiscale Statistical Signal Processing Algorithms," Proc. IEEE ICASSP, pp. IV 373–376, 1992.

M. Vetterli and C. Herley, "Wavelets and Filter Banks: Relationships and New Results," Proc. IEEE ICASSP, pp. 1723–1726, 1990.

M. Vetterli and C. Herley, "Wavelets and Filter Banks: Theory and Design," IEEE Trans. Signal Processing, Vol. 40, No. 9, pp. 2207–2232, Sept. 1992.

M. Vetterli and J. Kovacevic, *Wavelets and Subband Coding.* Prentice-Hall, 1995.

H. Volkmer, "Asymptotic Regularity of Compactly Supported Wavelets," SIAM J. Math. Anal., Vol. 26, pp. 1075–1087, 1995.

J. E. Younberg and S. F. Boll, "Constant-Q Signal Analysis and Synthesis," Proc. IEEE ICASSP, pp. 375–378, 1978.

R. M. Young, *An Introduction to Nonharmonic Fourier Series.* Academic Press Inc., 1980.

S. H. Zou and A. H. Tewfik, "M-band Wavelet Decompositions," Proc. IEEE ICASSP, pp. IV 605–608, 1992.

Chapter 7

Applications

7.1 Introduction

Transform-domain processing of signals has been successfully used for several decades in many engineering fields spanning the application areas from communications to oil exploration. The most popular block transform has been the discrete Fourier transform (DFT), which found a variety of applications due to its performance and low cost of implementation, i.e., fast Fourier transform (FFT). More recently, the discrete cosine transform (DCT) has become the industry standard for still frame image and video compression applications, e.g., JPEG, H.261, H.263, and MPEG compression algorithms. The subband transform with its multiresolution feature has also been forwarded as an alternative to the DCT for low bit rate image and video coding. In addition to these conventional applications of transform-domain signal processing, there has been a rapidly growing activity in new application areas like spread spectrum communication, discrete multitone (DMT) modulation, low probability of intercept (LPI) communication, radar signal processing, biomedical signal processing, and many others. Our intent in this chapter is to demonstrate how the fundamental concepts of linear transforms can lead to meaningful applications in representable areas as subband coding, interference excision in spread spectrum communications, discrete multitone modulation, and orthogonal code division multiple access (CDMA) user codes. We stress the concepts, not the details, which are adequately discussed in the literature (Akansu and Smith, 1996; Akansu and Medley, 1999).

It is now well understood by engineers that the block, subband, and wavelet transforms are subsets of the general linear transform family. Each one of these members has certain types of time- and frequency-domain properties which might be suitable for certain applications. The classical "uncertainty principle" asserts that no function can be optimally localized in both the time- and frequency-domains. This is the fundamental point which led to the different types of linear transforms. For example, block transforms utilize the minimum time duration functions in the set. This implies that the frequency selectivity of the basis functions is limited. Therefore, longer duration time functions in the set are necessary in order to obtain frequency-domain functions with good selectivity, e.g., the infinite duration sinc function provides a brick-wall frequency function. This need was the primary impetus for the graceful move from block to subband transforms. On the other hand, better frequency selectivity demands a better match to the Nyquist criterion. Hence, sampling rate conversion is needed and it provides the theoretical foundation for multiresolution or multirate signal processing.

The wavelet transform has been forwarded for continuous time signal processing. There have been a flurry of wavelet papers in the literature that deal with sampled or discrete-time signals. Following the basics of wavelet transforms discussed in Chapter 6, it is clear that most of these studies represent an approximation of wavelet analysis. In contrast, subband or filter bank theory is complete for discrete-time signal processing. Therefore, the applications presented next are discrete-time in nature; block or subband transforms are utilized.

The intended application of the subband transform determines the configuration to be used, as in the following:

(a) Analysis/synthesis subband transform configuration

(b) Synthesis/analysis subband transform configuration (transmultiplexer)

(c) Analysis or synthesis only subband transform configurations

We present these transform configurations and their applications in the following sections. The philosophy and justifications behind these application areas are discussed in detail. Since the block transform is a subset of subband transform or filter bank, we use the subband transform as the describing example in the sequel.

7.2 Analysis/Synthesis Configuration

Figure 7.1 displays an equal-bandwidth, single-level, maximally decimated M-band FIR QMF, analysis/synthesis subband transform or filter bank configuration. The analysis/synthesis filter bank was extensively treated in Chapter 3. It was shown in the earlier chapters that the forward/inverse block transform structure is a special case of analysis/synthesis filter bank configuration where the time duration of the analysis and synthesis filters is equal to the number of functions in the transform basis. Therefore, the block transform is the minimum

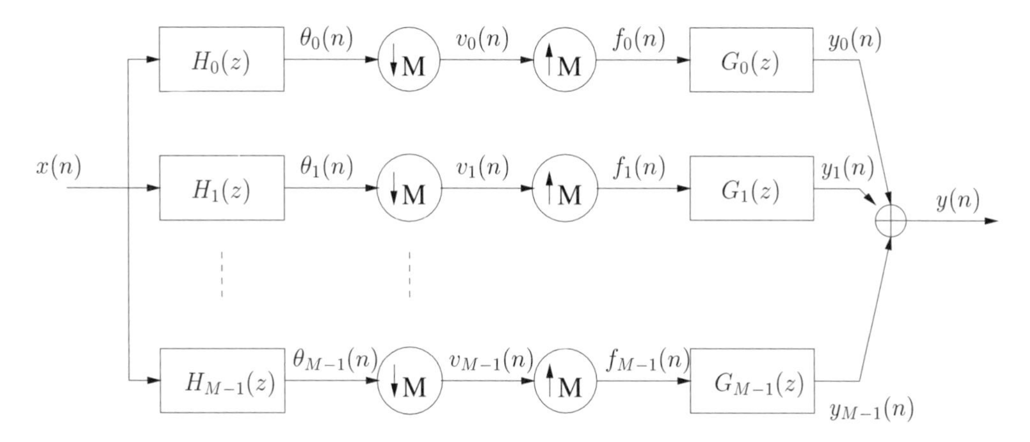

Figure 7.1: An equal-bandwidth, single-level, maximally decimated M-band subband transform analysis/synthesis configuration.

possible time-duration subset of a general M-band, equal-bandwidth, single-level analysis/synthesis subband transform configuration.

The following points are raised from Fig. 7.1, which might be of practical significance in some application areas.

7.2.1 Selection of Analysis and Synthesis Filters

The theory and design methodologies of subband transforms were discussed in detail in Chapters 3 and 4, respectively. The theoretical significance of the analysis and synthesis functions, in order to compensate the effects of down- and up-samplers, was presented. It was emphasized in Chapter 4 that there are available degrees of freedom in the design of analysis and synthesis functions which can be utilized for optimization purposes. The optimization methodologies discussed earlier basically aim to shape the time- and frequency-domain features of the functions in the set.

It was also shown that the analysis, $H_k(z)$, and synthesis, $G_k(z)$, filters need not have identical magnitude functions in frequency. Solutions of this nature are called biorthogonal subband transforms in the literature.

It was shown in Chapter 3 that there cannot be any linear-phase, two-band orthonormal QMF solution. On the other hand, it is possible to design linear-phase M-band filter banks for $M > 2$. Therefore, hierarchical M-band filter banks utilizing two-band PR-QMFs will have nonlinear-phase response product filters. One should carefully monitor the nonlinearities of the phase responses if the application at hand is phase-sensitive. Another approach might be the use of a two-band biorthogonal filter bank as the generating cell of a hierarchical M-band filter bank. The critical issue in this linear-phase solution case is the unequal bandwidth of the low and high-pass filters. Since both of these filters are used with rate 2 down-and upsamplers, they must be equally binded by the Nyquist criterion. Therefore, in the case of a biorthogonal two-band filter bank, one should design the linear phase low- and high-pass filters with almost equal bandwidths. It was reported by some researchers in the literature that the biorthogonal two-band filter bank based hierarchical subband transform performs better than the two-band PR-QMF based case for image and video coding (Bradley, Brislawn, and Hopper, 1993).

It will be seen later that the selection is driven by application-specific considerations. For example, the brick-wall-shaped ideal filter functions will be desirable if the application requires a frequency domain selective or localized signal processing. In contrast, the spread spectrum PR-QMF codes for code division multiple access (CDMA) communication, which will be introduced in Section 7.3.2, are jointly spread in both time- and frequency-domains. Therefore, the very high level of aliasing among the functions is desired for that class of applications. On the other hand, the time-domain autocorrelation and crosscorrelation of the basis functions are minimized.

The design of PR-QMF banks has degrees of freedom. Therefore, the engineering art is to find the best possible analysis and synthesis functions from among infinitely many available solutions based on the measures of interest for a given application.

7.2.2 Spectral Effects of Down- and Up-samplers

It was shown in Chapter 3 that a decimation operator consists of a filtering operation followed by a down-sampler of proper rate. Similarly, an interpolation operator is an up-sampler followed by the interpolation filter. The rate of the up-sampler and the bandwidth of the interpolation filter should be in match according to the Nyquist criterion.

The aliasing and imaging effects of down- and up-sampler operators, respectively, are almost inevitable in real-world applications where finite duration analysis and synthesis functions are used. Therefore, the negative spectral effects of down- and up-samplers should be carefully monitored if the application is sensitive to it. As an example, it was found that these operators degrade the system performance drastically in a transform-domain interference excision scheme used in a direct sequence spread spectrum (DSSS) communication system as discussed in Section 7.2.2. The rate converters, down- and up-samplers, are omitted and oversampled PR-QMF banks were used successfully in that application (Tazebay and Akansu, 1995). In contrast, the down- and up-samplers are critical for image and video processing applications presented in Section 7.2.1 where multirate/multiresolution property is a desirable feature.

7.2.3 Tree Structuring Algorithms for Hierarchical Subband Transforms

Since one of the early applications of subband transforms has been image and video coding, hierarchical filter banks with an inherent multiresolution property have been widely used in the literature. A subband tree consists of repetitive use of two- and three-band generic PR-QMF banks. Such a tree is appealing because of its design and implementation efficiencies although it is a more constrained solution compared with a single stage filter bank.

The fundamental issue in a hierarchical subband transform is how to define the most proper subband tree for given input signal and processing tasks (Akansu and Liu, 1991). Some authors have referred to these subband transform trees as wavelet packets (Coifman and Wickerhauser, 1992). More recently, it was shown that improved product functions in a tree structure can be designed by optimizing the constituent two- and three-band filter banks at different nodes of a subband tree in order to optimize the product functions of hierarchical decomposition. This approach is called "progressive optimality of subband trees" and the reader is referred to Tazebay and Akansu (1994) and Cohen and Sere (1996) for further discussions.

A subband tree structuring algorithm (TSA) based on energy compaction was proposed first in Akansu and Liu (1991) and successfully used for transform domain interference excision in a direct sequence spread spectrum communications system. TSA, by utilizing the energy compaction measure discussed in Chapter 2, can effectively track the spectral variations of the input signal. This in turn suggests the most proper subband tree structure needed to achieve the desired spectral decomposition. The use of TSA for interference excision in DSSS Communications is discussed later in Section 7.2.2. Similarly, TSA has also been utilized for the selection of basis functions (orthogonal carriers) in a synthesis/analysis filter bank configuration (orthogonal transmultiplexer) for a given communications channel. A discrete multitone (DMT) modulation scheme is discussed in Section 7.3.1. Interested readers are referred to the references for the details of subband tree structuring algorithms.

Two distinct application areas of analysis/synthesis subband transform configuration are presented in the following sections:

(1) Subband coding

(2) Interference excision in direct sequence spread spectrum (DSSS) communications

7.2.4 Subband Coding

Introduction

Multirate properties of filterbanks make them attractive signal processing tools for image and video processing and for coding applications where multiresolution representation is natural. Therefore, subband transforms have found popular applications in subband image and video coding.

The first principle of source coding is to minimize redundancy of the information source. The redundancy of the source is directly related to the shape of its spectral density function. Figure 7.2 displays the spectral density function of a LENA image where the energy is concentrated in the low-frequency components. Rate-distortion theory shows that a desirable source encoder decomposes the source into its uneven spectral energy bands, and codes them independently. Hence, the unevenness of the spectrum is the deciding factor for the efficiency of the subband coder along with the subband transform basis utilized to decompose the signal into its spectral bands. Note that any spectrum of flat shape needs no subband decomposition and will be encoded in signal domain.

Recall that block transforms (i.e., DCT) are merely a special class of subband transforms. Their poor frequency selectivity in low bit rate image and video coding generates blockiness artifacts that are perceptually unpleasant. Historically, this was another practical concern that generated significant research around subband image coding as an alternative technique to DCT coding. The longer duration of basis functions of the subband transform can reduce the blockiness artifacts at low bit rates. On the other hand, very long duration subband filters cause ringing effects that are undesirable as well. Therefore, midrange filter durations are used in subband image coding. In contrast, subband audio coding applications require a good spectral selectivity and utilize longer filters for that purpose. In the next section a one-dimensional subband codec will be used as an example to discuss details of the application.

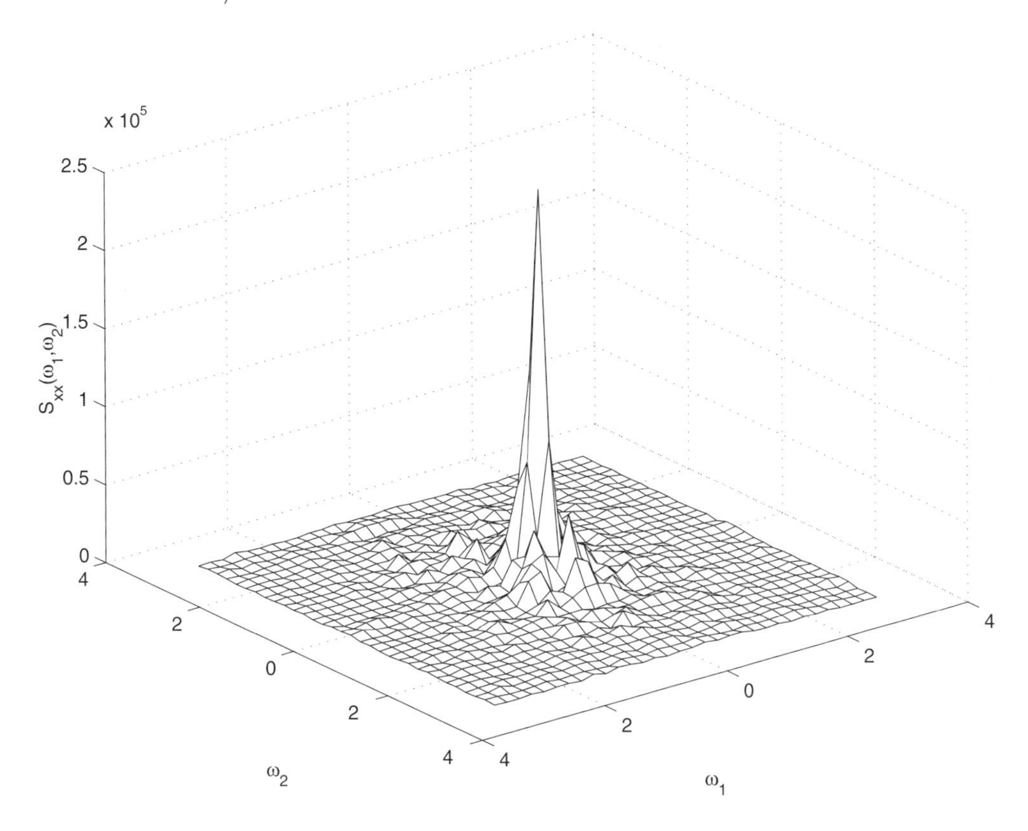

Figure 7.2: Spectral density function of LENA image.

One-Dimensional Subband Codec

Figure 7.3 displays the block diagram of a one-dimensional subband encoder/decoder or codec. The input signal $x(n)$ goes through a spectral decomposition via an analysis filter bank. The subbands of the analysis filter bank should be properly designed to match the shape of the input spectrum. This is a very important point that significantly affects performance of the system. Compression bits are then allocated to the subband signals based on their spectral energies. These allocated bits are used by quantizers. An entropy encoder follows the quantizers to remove any remaining redundancy. The compressed bit stream $\{b_i\}$ is transmitted through a communication channel or stored in a storage medium. We assume an ideal channel or storage medium in this example. Similarly, entropy decoding, inverse quantization, and synthesis filtering operations are performed at the receiver in order to obtain the decompressed signal $\hat{x}(n)$. In reality, a communications channel introduces some bit errors during transmissions that degrade the quality of the synthesized signal at the decoder (receiver).

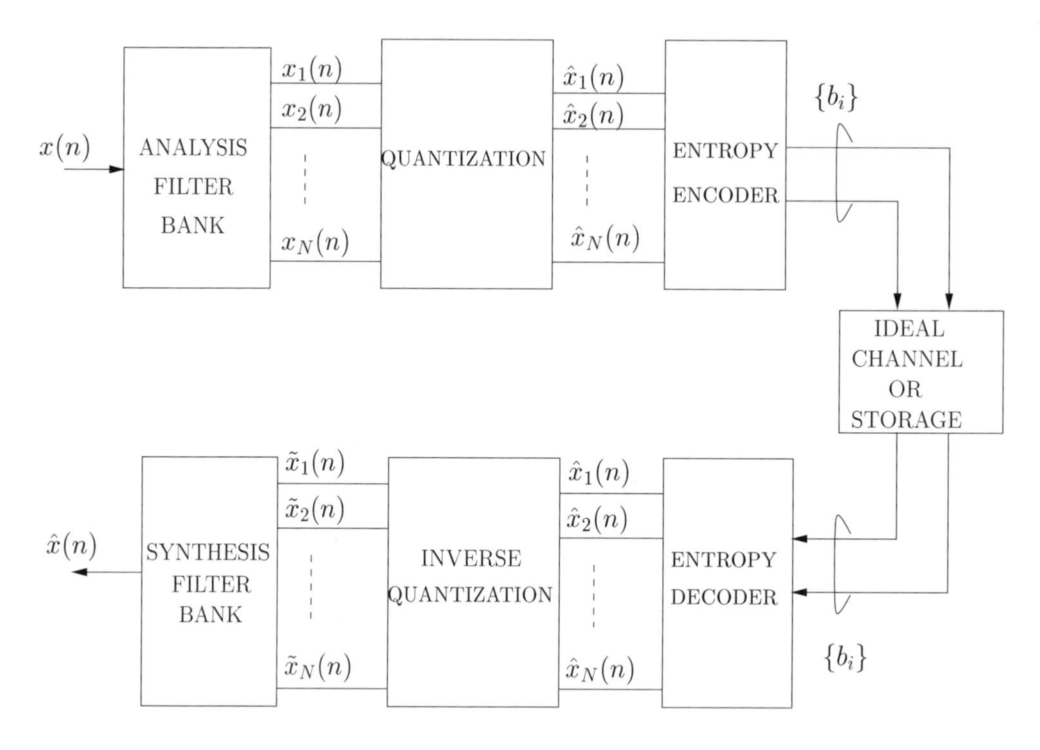

Figure 7.3: The block diagram of a subband codec.

The blocks in the subband codec system (Fig. 7.3) are briefly described as follows:

Analysis Filterbank Hierarchical filter banks are used in most coding applications. The subband tree structure which defines the spectral decomposition of the input signal should match input spectrum. Additionally, several time- and frequency-domain tools were introduced in Chapter 4 for optimal filter bank design. The implementation issues along with the points made here will yield practical solutions.

Quantization Lossy compression techniques require an efficient entropy reduction scheme. A quantizer is basically a bit compressor. It reduces the bit rate and introduces irreversable quantization noise. Hence, it is called lossy compression.

Entropy Encoder The quantizer generates an output with some redundancy. Any entropy encoder, such as the Huffmann coder, exploits this redundancy. Note that the entropy encoder encodes the source in a lossless fashion that is perfectly reversible. The output bit stream of the entropy encoder is compressed and ready for transmission or storage.

Channel or Storage Medium The capacity of a communications channel or storage medium at a given bit error rate is the defining factor. The encoder aims to achieve the necessary compression rate in order to fit the original source data into the available channel or storage capacity. Note that wireline (e.g., telephone lines) and wireless (e.g., cell phones) channels have different physical media and engineering properties that are handled accordingly.

Similarly, entropy decoders, inverse quantizers, and synthesis filter banks perform inverse operations at the receiver or decoder.

Subband Image Codec

The multiresolution or scaleability feature for visual signals is a desirable one that generated significant research and development on subband image coding. Scaleability allows the transmitted bit stream to be decoded in different spatial resolutions for different transmission channel properties or application requirements. Digital image/video libraries, on-demand image/video retrieval over the Internet, and real-time video conferencing are three examples that naturally benefit from a scaleable bit stream.

Figure 7.4: (a) Original input LENA image; (b) L and H subbands (Horizontal); (c) LL, LH, HL and HH decomposed subband images.

Separable 2D subband decomposition basically employs 1D filter bank operations back-to-back, in both horizontal and vertical dimensions. Figure 7.4 displays images of a single-stage (four band) subband image encoder that first decomposes an input image into four image subband signals; $x_{LL}(n)$, $x_{LH}(n)$, $x_{HL}(n)$ and $x_{HH}(n)$. The available bit budget for quantization (entropy reduction) purposes is distributed among these subband images based on their energies. The quantized subband images go through an entropy encoder like a Huffmann coder, and four bit streams, namely $\{b_{LL}\}$, $\{b_{LH}\}$, $\{b_{HL}\}$, $\{b_{HH}\}$ are obtained. Note that all these bit streams are required at the decoder in order to reconstruct the compressed version of the original image. On the other hand, only the $\{b_{LL}\}$ bit stream is needed if one desires to reconstruct only a quarter-size version of the compressed image at the receiver. Hence, the compressed bit stream is scaleable.

A practical subband image encoder repeatedly uses a four-band (single stage) analysis filter bank cell for further spatial resolutions and improved compression efficiency. In most cases, the low-pass band goes through additional decompositions since significant image energy resides in that region of spectrum.

The purpose of this section is to connect subband theory with subband coding applications. We provided only broad discussions of fundamental issues in this application area, without the rigor which is beyond the scope of this book.

The literature is full of excellent books, book chapters, and technical papers on subband image and video coding. The reader is referred to Nosratinia et al. (1999); Girod, Hartung, and Horn (1996); Clarke (1995); and Woods (1991) for further studies.

7.2.5 Interference Excision in Direct Sequence Spread Spectrum Communications

Direct Sequence Spread Spectrum Communications

Spread spectrum modulation techniques generate a transmission signal with a bandwidth that is much wider than the original information bandwidth. In a direct sequence spread spectrum (DSSS) communications system, the spreading of the information bits is performed by their modulation with a pseudo-noise (PN) sequence before transmission. At the receiver, the received spread spectrum signal is "despreaded" by correlating it with a local replica of the PN code. The correlation operation spreads the narrow band interference over the bandwidth of the PN signal, while the desired information component of the received signal shrinks to its original bandwidth (Ziemer and Peterson, 1985).

The DSSS transmitter, shown in Fig. 7.5, spreads the spectrum of incoming data bit stream d_b, where $d_b \in -1, 1$ for all b, by multiplying them individually with the length L spreading binary PN code $\underline{c}$; $c_i \in -1, 1$ for $i = 1, 2, ..., L$. During the transmission, the channel adds white Gaussian noise (AWGN) term $\underline{n_b}$ and

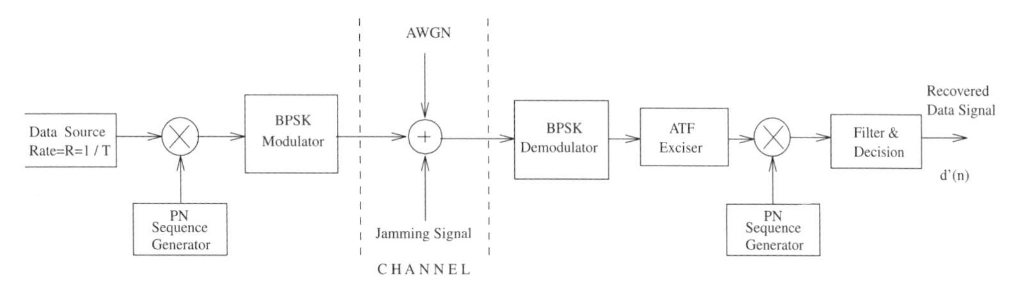

Figure 7.5: Block diagram of a direct sequence spread spectrum communications system.

other undesired interferences $\underline{j_b}$ (e.g., jamming signal). Therefore, the received signal can be expressed as

$$\underline{r_b} = \sqrt{P} d_b\, \underline{c} + \underline{j_b} + \underline{n_b}. \tag{7.1}$$

The transmitted signal power $\sqrt{P}$ can be assumed to be unity. The data bit stream d_b has a time duration of T_d seconds per bit. The PN spreading code has a chipping rate of T_c seconds per chip where $T_d \gg T_c$. Hence, the length of the PN code is expressed as $L = \frac{T_d}{T_c}$ chips per code. The received DSSS signal has a flat and wide spectrum in case of no interference signal $\underline{j_b}$ and no interference exciser. The receiver correlates the receiver signal with a properly synchronized version of the spreading PN code $\underline{c}$ where $\underline{c}\underline{c}^T = \sum_{i=1}^{L} c_i^2 = L$. Therefore, the decision variable at the detector is expressed as

$$\xi = \underline{r_b}\underline{c}^T = d_b\underline{c}\underline{c}^T + \underline{j_b}\underline{c}^T + \underline{n_b}\underline{c}^T = Ld_b + \underline{j_b}\underline{c}^T + \underline{n_b}\underline{c}^T. \tag{7.2}$$

Equation (7.2) shows that the spreading operation emphasizes the desired component of received signal while spreading the interference. The receiver makes a binary decision as to whether $+1$ or -1 was sent depending on the value of the decision variable, $\xi <> 0$. The DSSS receiver fails to operate whenever the interference signal power is greater than the jamming margin of the system. The interference immunity of a DSSS receiver can be further improved by excising the interference component $\underline{j_b}$ of the received signal $\underline{r_b}$.

Interference Excision Techniques in DSSS Communications

It has been shown in the literature that the performance of a conventional DSSS receiver can be substantially improved by eliminating the interference component of the received signal in Eq. (7.1) prior to the correlation as displayed in Fig. 7.5. Previous work in this area primarily involved classes of interference excision schemes which are summarized in this section (Saulnier et al., 1996).

The first class is the parametric modeling and estimation of the interference by means of a linear prediction filter (Ketchum and Proakis, 1982). Since the PN code and white Gaussian noise of the channel have relatively flat spectra, they cannot be properly predicted from their past values. However, the narrow-band or band-pass interference can be accurately predicted. The stationary and narrow-band assumptions of interference are crucial to the performance of this parametric excision technique. Otherwise, the system performance degrades drastically.

The second class is the transform-domain excisers. The discrete Fourier (DFT) has been the most popular transform-domain signal processing method used for narrow-band interference excision (Davidovici and Kanterakis, 1989). The DFT, however, suffers from its fixed frequency resolution and poor side-lobe attenuation. More recently, fixed subband transforms with an improved frequency localization and side-lobe attenuation were forwarded for transform-domain interference excision (Jones and Jones, 1992). The latest contribution in this arena is the time-frequency adaptive block transform excisers described in Chapter 5.

The shortcomings of fixed block and subband transform based excisers are threefold:
(i) They can only handle narrow-band interference
(ii) They have fixed time-frequency resolution
(iii) They have a high level of interband spectral leakage
Narrow-band interference falling into one of the transform bins or subbands can be efficiently suppressed. However, the spectral variations of the interference between transform bins or subbands cause a dynamic contamination in the desired signal. In order to suppress this kind of interference, more transform bins have to be removed, resulting in an additional loss of the desired signal spectrum which causes a performance degradation of the DSSS communications system.

The last two of the three points raised above can be overcome by using the tree structuring algorithm (TSA) discussed in the previous section. For a given input spectrum, TSA recommends the best subband tree, regular or irregular tree (equal or unequal bandwidth subbands), consisting of two-band and/or three-band (equal bandwidth) prototype filter bank cells. The TSA considers both two-band and three-band PR-QMF banks in order to handle the transition band frequency regions around $w = \pi/3$, $\pi/2$, or $2\pi/3$ which might be of practical significance.

The TSA algorithm analyzes the spectra at each node of the tree with the assumption of ideal filters, and either justifies further decomposition or prunes the tree. A subband node is further decomposed if the energy compaction measure at that node exceeds a predefined threshold. Therefore, the best subband tree for the given input spectrum is generated in order to localize the interference. The bins that contain the interference are nullified before the synthesis stage. Hence, the excised version of the received signal is reconstructed and fed to the correlator. Figure 7.6 depicts the flexible spectral resolution achieved in a seven-band unequal bandwidth subband tree. The decision thresholds set in TSA yield the minimum number of functions in the set with the best possible desired frequency selectivity. In real-world applications, the ideal filters are replaced with finite duration functions.

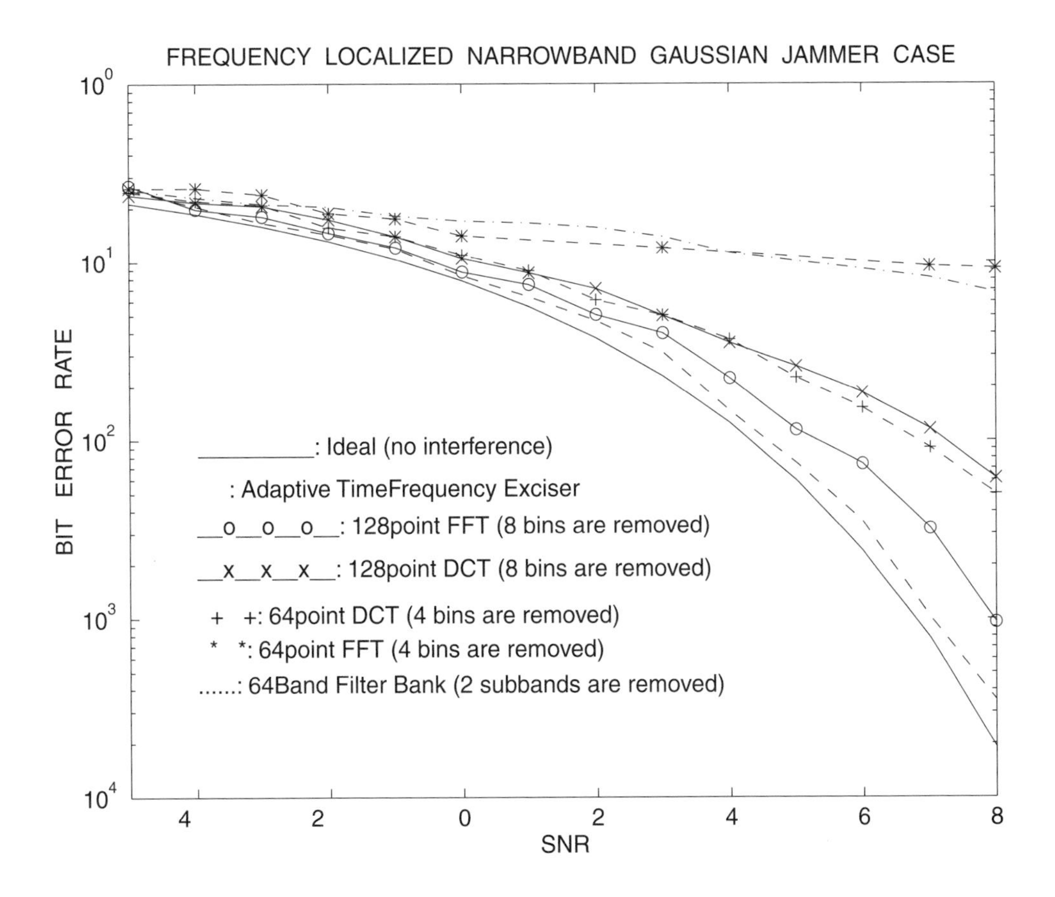

Figure 7.6: Bit error rate curves for frequency localized narrow band Gaussian jammer case (center frequency = $\pi/2$ rad, SIR = -20 dB).

A smart time-frequency exciser (STFE) was devised to answer all of the three points just raised. The STFE first examines the time-domain features of the received signal in order to decide on the domain of excision. If the interference is time localized, a simple time-domain exciser naturally outperforms any transform-domain excision technique. For the case of frequency localized interference, STFE utilizes the TSA discussed earlier. TSA changes the recommended subband tree structure whenever the input spectrum varies. Therefore, the spectral decomposition (subband transform) tracks the variations of the input spectrum. The implementation details and superior performance of STFE over the conventional excision techniques are found in Tazebay and Akansu (1995). The bit error rate (BER) performance of STFE along with the other excision techniques are displayed in Fig. 7.7 . The robustness of STFE performance is clearly observed from Fig. 7.8. The references Tazebay (1996) and Medley (1995) are excellent for the theoretical and implementation issues of the excision techniques discussed in this section.

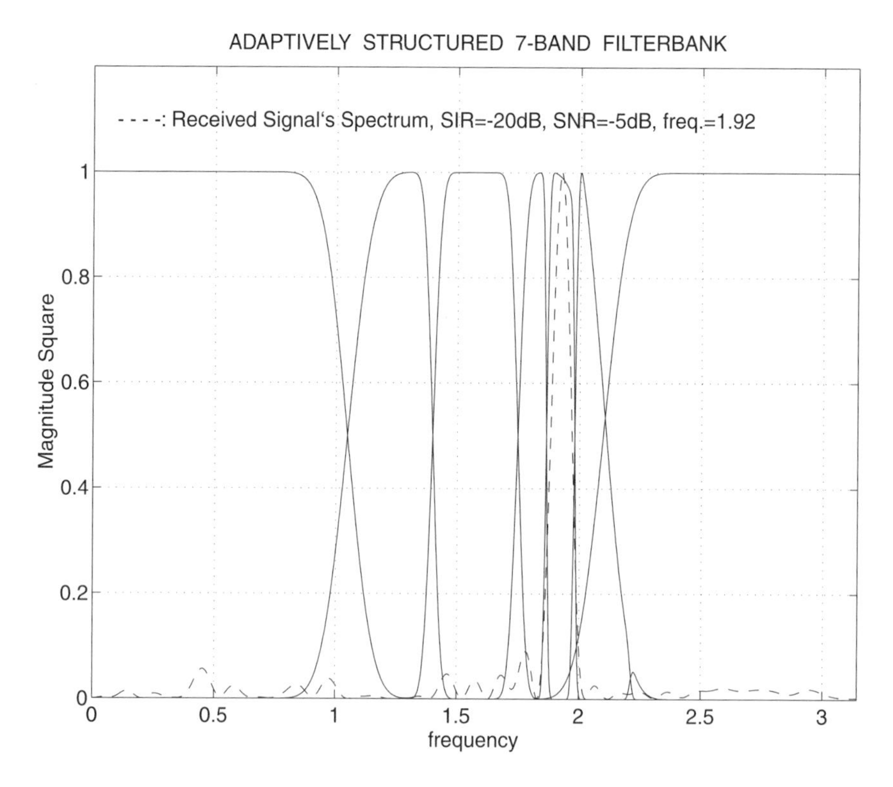

Figure 7.7: Adaptive filter bank structure for single tone jammer case (tone frequency = 1.92 rad, SIR = -20 dB, and SNR = -5 dB).

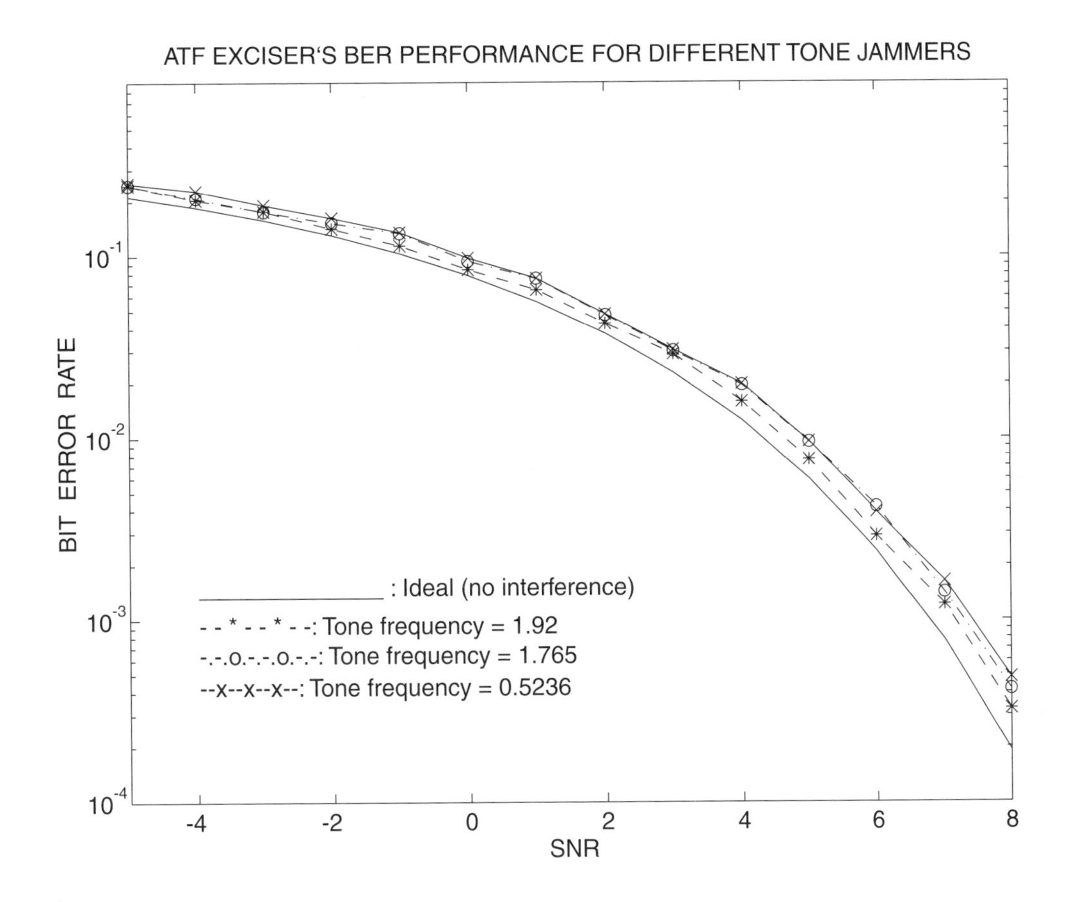

Figure 7.8: Bit error rate curves of STFE for different frequency tone jammers (SIR = -20 dB, $\omega_1 = 0.5236$ rad, $\omega_2 = 1.765$ rad, and $\omega_3 = 1.92$ rad.

7.3 Synthesis/Analysis Configuration

The transmultiplexer has been a very useful spectral processing tool for allocating available channel resources among its multiple users in a communications scenario. Figure 7.9 displays a synthesis/analysis filter bank configuration which serves as an M-band transmultiplexer. The duality between filter banks and multiplexers was discussed in Section 3.8. The most popular version of transmultiplexers is of frequency division multiplexing (FDM) type. In this case, the available channel spectrum is divided into nonoverlapping subspectra and each subspectrum is as-

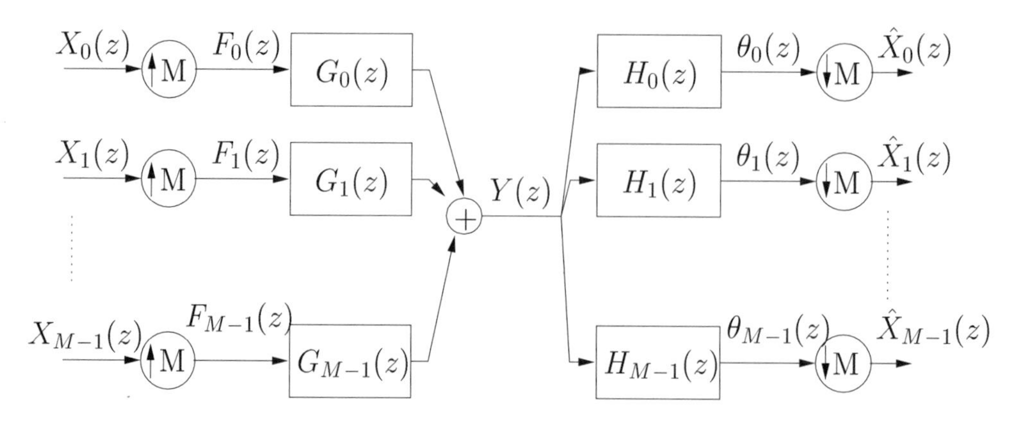

Figure 7.9: M-band transmultiplexer structure (critically sampled synthesis/analysis filter bank configuration).

signed to a specific user. The synthesis filters $G_i(z)$ must have good frequency selectivity in order to achieve FDM. Similarly, the analysis filters at the receiver, $H_i(z)$, must also have good frequency responses. Therefore, the synthesis/analysis filter bank configuration functions as a time division multiplexing TDM-to-FDM (synthesis) and then FDM-to-TDM (analysis) converters. Figure 7.10 displays signal spectra at the different points of an M-band transmultiplexer (Fig. 7.9). There are two important points drawn from Fig. 7.10:

(a) Spectral effects of up- and down-samplers that were treated in Chapter 3;

(b) Significance of synthesis and analysis filters, $\{G_i(z)\}$ and $\{H_i(z)\}$, respectively, on the type of multiplexing. For example, bandlimited ideal filters are used in Fig. 7.9 in order to achieve TDM-to-FDM conversion for channel utilization. As discussed later in Section 7.3.2, spectrally spread $\{G_i(z)\}$ and $\{H_i(z)\}$ filters (code) provide a transmultiplexer configuration for spread spectrum code division multiple access (CDMA) communications. In this case, filter functions are not frequency selective. They are spread spectrum user codes.

In a real world the filter functions $\{G_i(z)\}$ and $\{H_i(z)\}$ are not ideal brick-wall shaped. Then spectral leakage from one subchannel to another, or cross-talk, is of major concern. Therefore, cross-talk cancellation has become a critical measure in the design of multiplexers. It is a mature subject and there are many excellent references in the literature on transmultiplexers (IEEE Trans. Communications, May 1978 and July 1982 special issues; Koilpillai, Nguyen, and Vaidyanathan, 1991).

The analysis of synthesis/analysis filter bank configuration is given in Section 3.8. It is shown that the design problem of an orthogonal transmultiplexer is

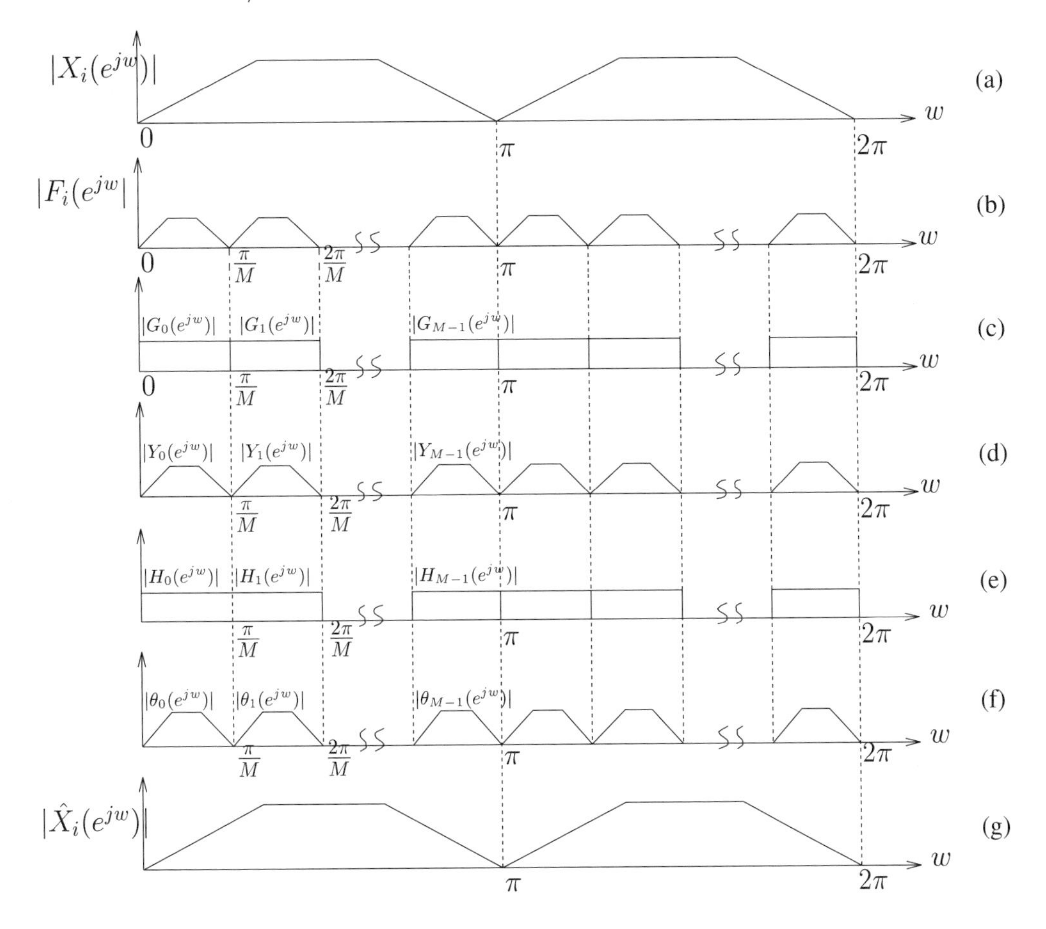

Figure 7.10: Spectra at different points of an M-band transmultiplexer.

a special case of PR-QMF design with certain delay properties. Interested readers are referred to Section 3.8 for detailed treatment of this topic.

There are several popular single and multiuser communications applications that utilize orthogonal transmultiplexers. Some of these applications are presented in the following sections.

7.3.1 Discrete Multitone Modulation for Digital Communications

Discrete multitone (DMT) or orthogonal frequency division multiplexing (OFDM) is a class of frequency division digital modulation. This concept of multicarrier modulation dates back to the mid-1960s (Chang, 1966; Saltzberg, 1967;

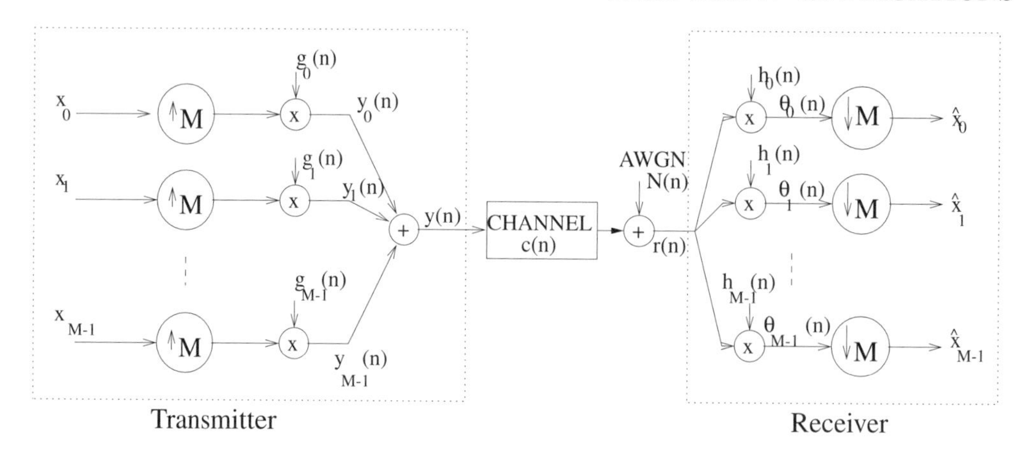

Figure 7.11: Basic structure of a DMT modulation based digital communications system.

Weinstein and Ebert, 1971; Peled and Ruiz, 1980). However, it received more attention recently for digital audio broadcasting (DAB) and asymmetric digital subscriber line (ADSL) communication applications. The synthesis/analysis filter bank configuration discussed in the previous section is used for DMT modulation. Since it is of FDM type, the synthesis and analysis filter functions, $\{G_i(z)\}$ and $\{H_i(z)\}$ in Fig. 7.9, should be frequency selective and cross-talk-free. Figure 7.11 displays the basic structure of a DMT modulation based digital communications system.

It is seen that Fig. 7.11 is similar to the synthesis/analysis filter bank configuration of Fig. 7.9 with the exceptions of channel $c(n)$ and additive white Gaussian noise (AWGN) introduced by the channel between the synthesis and analysis sections. Therefore, the orthogonality properties of the complete system is destroyed due to the non-ideal channel properties in a real-world application. The imperfectness of the channel is compensated by an equalizer in order to improve the communications performance.

The subsymbols $\{x_i\}$ in Fig. 7.11 that are applied to the orthogonal modulating functions $\{g_i(n)\}$ are usually complex for quadrature amplitude modulation (QAM) schemes and real for the pulse amplitude modulation (PAM) case. These subsymbols are formed by grouping blocks of incoming bits in the constellation step. The parsing of the incoming bits to the subsymbols is controlled by the spectral properties of the channel $c(n)$ (channel power levels). Since the transmitted signal $y(n)$ is the composite of M independent subchannels or carriers, each of the orthogonal subchannels will carry more bits of information. This discussion leads to the concept of optimal bit allocation among the subchannels (orthogonal carriers) from the incoming bit stream. This is fundamental in a DMT based system used for ADSL communications. The basics of such a system are introduced in the following section.

Orthogonal Transforms in ADSL and HDSL Communications

DMT or OFDM based digital communication systems have been proposed as a standard for high-speed digital subscriber line (HDSL) and asymmetric digital subscriber line (ADSL) data transmission applications over twisted-pair cable of plain old telephone service (POTS) that will not affect existing telephone service. The distance of the communications link (1.5 to 5 miles) and its data transmission speed are inversely related. The DFT- based DMT communication system has become a reference model recommended by American National Standards Institute (ANSI)'s T1E1.4 Working Group for ADSL data transmission. This standard sets the guidelines for an expanded use of existing copper communication lines. The ADSL communications standard is designed to operate on two-wire twisted metallic cable pairs with mixed gauges. The same technology can also be utilized for high-speed communications over coaxial cable TV channels. The recommended standard handles downstream bit rates of 1.536 to 6.144 Mbits/sec. In contrast, it can provide an upstream channel capacity of 16 to 640 kbits/sec. Therefore, it is called asymmetric communications system (ADSL). The examples of potential ADSL services and applications include movies and music on demand, high-speed Internet access, interactive TV, distant class rooms, video conferencing, telecommuting, telemedicine, and many others. Interested readers are referred to Draft American National Standard for Telecommunications. T1E1.4 (95-007R2) for the details of the ADSL standard.

The fundamentals of a DMT based ADSL system (Fig. 7.11) with transform techniques are summarized in the following.

a. Subchannels and Optimal Bits/Subsymbol (Coefficient) It is assumed that the communications channel virtually consists of subchannels. Therefore, each subchannel will be assumed as an independent transmission medium implying its own noise properties. Since a composite signal generated by contributions of subchannels is transmitted through a physical channel, the orthogonalities of these subchannels are of critical importance.

For that reason, an orthogonal function set is used to represent subchannels. It is seen from Fig. 7.11 that an inverse transform (synthesis operation) is performed on defined transform coefficients x_i (subsymbols or subband signals) to generate the composite signal $y(n)$. This signal is put through the channel $c(n)$.

It is noted that the channel spectrum varies as a function of frequency. Therefore, each subchannel has its own spectral properties (channel noise, attenuation, etc.). It implies an optimal bit allocation procedure among subchannels that results in a uniform bit error rate over all channels. An excellent treatment of this topic is found in Kalet (1996) and Bingham (1990).

The current technology described in Draft American National Standard for Telecommunications. T1E1.4 (95-007R2) uses DFT of size 512 (256 subbands). There have been other studies reported in the literature that use equal or unequal bandwidth orthogonal carriers with frequency responses better than DFT (Tzannes et al., 1993; Benyassine and Akansu, 1995).

b. Effects of Nonideal Channel on Orthogonalities of Carriers Because of the imperfectness of the channel's frequency response and additive channel noise (AWGN), the orthogonality properties of the carriers are lost. This is going to cause a severe intersymbol interference (ISI) problem that degrades the system performance significantly. For the ideal case, the channel impulse response will be equal to the Kronecker delta function, $c(n) = \delta(n)$, where the channel output will be equal to its input $y(n)$ in Fig. 7.11. Therefore, orthogonality properties of subchannel carriers are maintained in the absence of channel noise $N(n)$. The subsymbols will be obtained at the receiver after a forward transform operation on the received signal $r(n)$.

The cyclic prefix method is successfully used in case of DFT-based DMT systems to overcome this problem (Peled and Ruiz, 1980). If one uses a better frequency-selective subband basis instead of DFT, the orthogonal carriers will have longer time durations. Hence, ISI distortion becomes more dominant with the benefit of reduced interchannel interference (ICI). The optimal basis selection and equalization problems for DMT communications have been investigated by some researchers (Lin and Akansu, 1996; de Courville et al., 1996).

Digital Audio Broadcasting (DAB)

One of the earlier applications of DMT (OFDM) modulation is in digital audio broadcasting (DAB). The DAB channel for mobile receivers has a hostile transmission environment with multipaths, interference, and impulsive noise. The impulse response of such a communications channel is over several microseconds. Therefore, high-speed data transmission over DAB channel is not a trivial problem.

A DMT-based DAB system basically splits the available transmission band into many subchannels. More subchannels imply longer duration orthogonal carriers with narrower bandwidths. This helps to reduce the severe ISI problem inherent in a typical DAB channel with long impulse response. A receiver would only like to receive a single radio channel (program), while the available orthogonal carriers (subchannels) are distributed among multiple radio transmitters. The subchannel allocators in a multiple radio transmission scenario are visualized in Fig. 7.12.

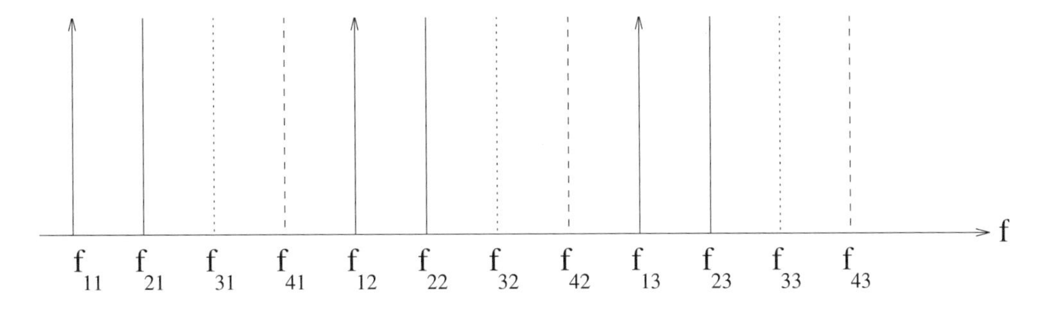

Figure 7.12: Allocation of orthogonal carriers among multiple radio stations.

In this example, each of four radio stations is utilizing four uniformly located subchannels within the available total channel spectrum. Therefore, this application utilizes a DMT structure given in Fig. 7.11 for multiple incoming bit streams. For the scenario of Fig. 7.12, there are four simultaneously transmitting radio stations where each uses three uniformly spaced orthogonal carriers.

The receiver has the ability to pick one of four radio transmissions at a time. It picks a set of subchannels in order to decode the desired radio transmission, e.g., f_{i1}, f_{i2}, f_{i3} for radio stations $i = 1, 2, 3$ in Fig. 7.12. Similar to the DMT-based ADSL technology, the current DAB systems also utilize DFT basis as its orthogonal carriers. Duhamel and de Courville (1999) present a nice discussion on DMT-based DAB technology and its trade-offs from a communications systems engineering point of view. It is reported that although DMT-based modulation overcomes the multipath problem in DAB to mobile receivers, it does not by any means handle the fading problem. Therefore, a channel coding scheme is of a critical importance in a real DAB system (Alard and Lasalle, 1987; Akansu et al., 1998).

7.3.2 Spread Spectrum PR-QMF Codes for CDMA Communications

In the previous section we said that an orthogonal transmultiplexer (synthesis/analysis filter bank configuration) has been successfully utilized for FDM-based multiuser communications. Each user is assigned to a branch of the orthogonal transmultiplexer displayed in Fig. 7.9 with the corresponding subspectrum of the total channel spectrum (see Fig. 7.10). Therefore, a user can only use an allocated subchannel exclusively at any time. This naturally limits the maximum available transmission rate to any user.

The synthesis/analysis filter bank structure (Fig. 7.9) provides a useful theoretical basis for an orthogonal transmultiplexer. It serves as a common communications configuration for all possible popular multiuser techniques such as FDMA, TDMA, and CDMA. The core component of these various multiuser communications types is the synthesis and analysis filter functions, $\{g_i(n)\}$ and $\{h_i(n)\}$, respectively, used in a synthesis/analysis filter bank. Basically, the time-frequency properties of these basis functions or user codes define the type of multiuser communications system, e.g., TDMA, FDMA, or CDMA.

Recent advances in wireless and mobile radio communications suggest CDMA as a potential alternative to the existing TDMA-based systems. All users of a CDMA communications system are equally entitled to use any time and frequency slots. This implies that all the user codes are spread both in the time and frequency domains. Therefore, CDMA is advantageous when compared with the conventional multiplexing techniques such as TDMA and FDMA, which localize in either the time- or frequency-domain, respectively. The desired user codes of an orthogonal transmultiplexer for spread spectrum CDMA communications should jointly satisfy the following time-frequency conditions:

(a) The orthogonal user codes cannot be unit sample functions in the time-domain. This condition prevents CDMA from becoming a TDMA communications scheme

(b) The orthogonal user codes should be all-pass like spread spectrum functions with minimized inter- and intracode correlations. This condition ensures that the communications scheme cannot become an FDMA type.

The current spread spectrum CDMA technology uses Walsh functions (Chapter 2) as the user codes for the communication path from the base station to the mobile user terminal. For the path from user terminal to the base station, it utilizes long duration (1024 samples or more) Gold codes (Gold, 1967). In the first case, the multiuser receives the incoming signal synchronously. Therefore, the orthogonality of the user codes is sufficient for this case (e.g., Walsh codes). The inter- and intracode correlations of user codes are critical factors in the performance of the second case (mobile user terminal to base station), which is called an asynchronous communications system. We extend the subband transform theory and optimal basis design methodologies covered in the previous chapters in the following section for spread spectrum CDMA communication applications.

Optimal Design Criteria

The optimal designs of PR-QMFs based on different measures were treated in Section 4.8. Similarly, an optimal design methodology for spread spectrum PR-QMF user codes is presented in this section for the two-band (two-user) case. In addition to the PR-QMF constraints

$$\sum_k h_r(k)h_r(k+2n) = \delta(n), \quad r = 0,1$$

$$\sum_k h_0(k)h_1(k+2n) = 0, \quad \forall n,$$

the following correlation and time-frequency properties of the user codes are included as metrics in the objective function to be optimized (Akansu, Tazebay, and Haddad, 1997; Akansu and Tazebay, 1996):

(a) Minimization of the inter- and intracode correlations

$$R_{00}(k) = \sum h_0(n)h_0(n+k) \quad (k = 1,3,5,...) \tag{7.3}$$

and

$$R_{01}(k) = \sum h_0(n)h_1(n+k) \quad (k = 1,3,5,...), \tag{7.4}$$

where $h_1(n) = (-1)^n h_0(n)$.

(b) Spreading the PR-QMF user codes in both frequency and time domains as evenly as possible. This measure is critical for PR-QMF user codes in spread spectrum CDMA communications. This feature contrasts with the fundamental property of the conventional PR-QMFs which approximate the ideal brick-wall frequency responses in order to overcome the aliasing problem (meeting Nyquist requirements in multirate processing). The frequency selectivity of conventional PR-QMFs (FDMA) is diminished with this consideration and they become orthogonal spread spectrum user codes of the desired CDMA type.

As described in Chapter 5, the time spread of a discrete-time function $\{h_0(n)\}$ is defined as

$$\sigma_n^2 = \frac{1}{E}\sum_n (n-\bar{n})^2 |h_0(n)|^2. \tag{7.5}$$

The energy, E, and the time center,$\bar{n}$, of the function $\{h_0(n)\}$ are

$$E = \sum_n |h_0(n)|^2 \tag{7.6}$$

$$\bar{n} = \frac{1}{E}\sum_n n|h_0(n)|^2. \tag{7.7}$$

Similarly, its frequency spread is defined as

$$\sigma_w^2 = \frac{1}{2\pi E}\int_{-\pi}^{\pi}(w-\bar{w})^2|H_0(e^{jw})|^2 dw, \tag{7.8}$$

where $H_0(e^{jw}) = \sum_n h_0(n)e^{-jwn}$ and

$$\bar{w} = \frac{1}{2\pi E}\int_{-\pi}^{\pi} w|H_0(e^{jw})|^2 dw. \tag{7.9}$$

Therefore, we can now set the objective function for the optimization as

$$J_{max} = \alpha\sigma_n^2 + \beta\sigma_w^2 - \gamma\sum_k |R_{00}(k)| - \eta\sum_k |R_{01}(k)| \tag{7.10}$$

subject to the PR constraint $\sum_n h_0(n)h_0(n+2k) = \delta(k)$, and where $R_{00}(k)$ and $R_{01}(k)$ were defined in Eqs. (7.3) and (7.4), respectively.

Figure 7.13 displays the spectra of a possible 32-length spread spectrum PR-QMF code for the two-user case for $\alpha = \beta = 0$ and $\gamma = \eta = 1$ in Eq. (7.10) along with a 31-length Gold code. This figure demonstrates the significant difference of the spread spectrum PR-QMF codes from the conventional PR-QMF filters. The inter- and intracode correlations of these sample codes are also displayed in Figures 7.14 and 7.15, respectively.

These figures show that the correlation and frequency properties of the spread spectrum PR-QMF code outperforms the comparable duration Gold code case. The parameters $\alpha, \beta, \gamma, \eta$ of Eq. (7.10) can be changed in order to emphasize the corresponding metrics of the objective function.

The bit error rate (BER) performance of a two-user CDMA system for the asynchronous communications scenarios is displayed in Fig. 7.16.

BPSK modulation and antipodal signaling for CDMA are used in these simulations. The channel noise is assumed to be additive white Gaussian (AWGN). The signal to multiuser interference power ratio (SIR) of 0 dB is simulated in Fig. 7.16 (asynchronous case). These performance simulations show that spread spectrum PR-QMF user codes outperform Gold codes under the same test conditions. They imply the theoretical potentials of using PR-QMFs for CDMA communications. Note that the coefficients of these codes are multiple valued while Gold codes have only binary valued coefficients. Therefore, the latter ensures a constant power transmitter in contrast to the first, which naturally requires power variations.

More studies are needed in order to assess the merits of spread spectrum PR-QMF codes in a real-world communications application.

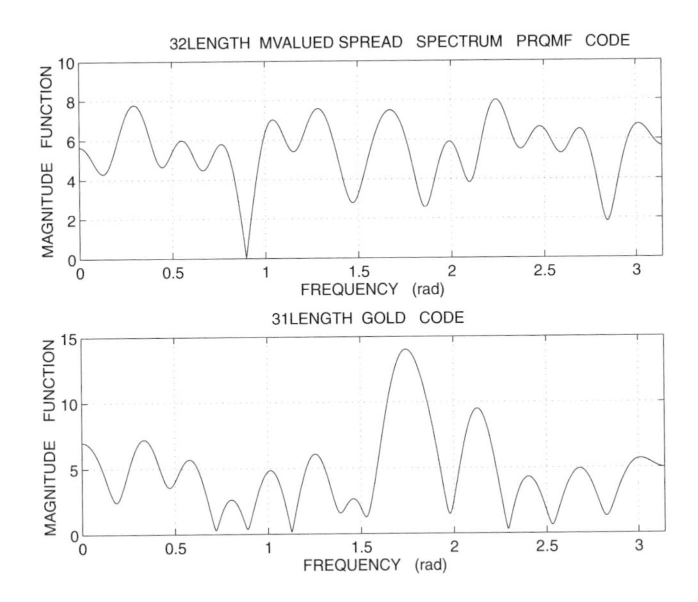

Figure 7.13: Frequency spectra of 32-length M-ary spread spectrum PR-QMF and 31-length Gold codes.

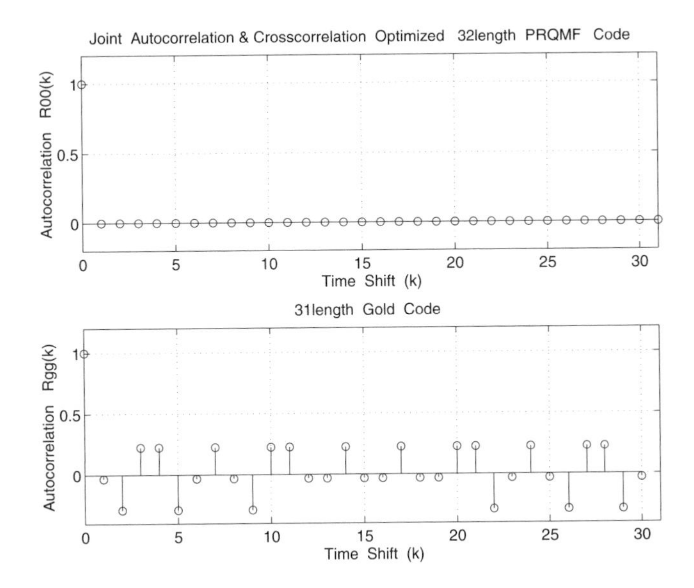

Figure 7.14: Autocorrelation functions of spread spectrum 32-length PR-QMF and 31-length Gold codes.

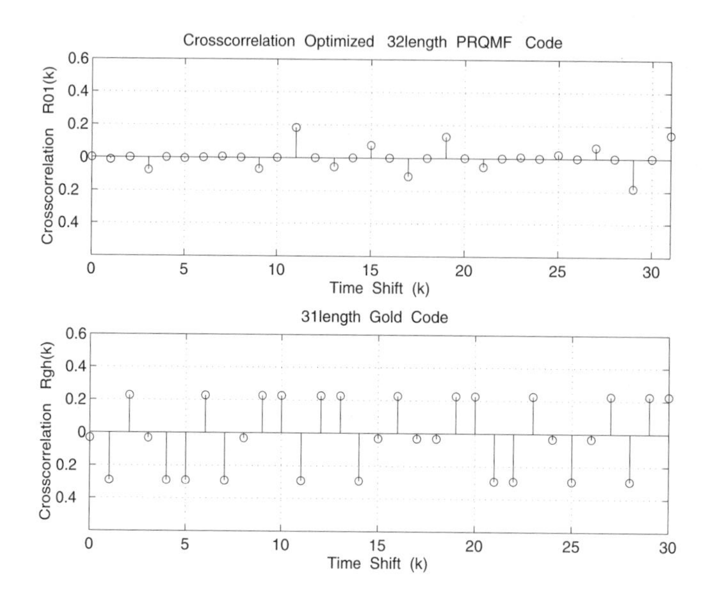

Figure 7.15: Crosscorrelation functions of spread spectrum 32-length PR-QMF and 31-length Gold codes.

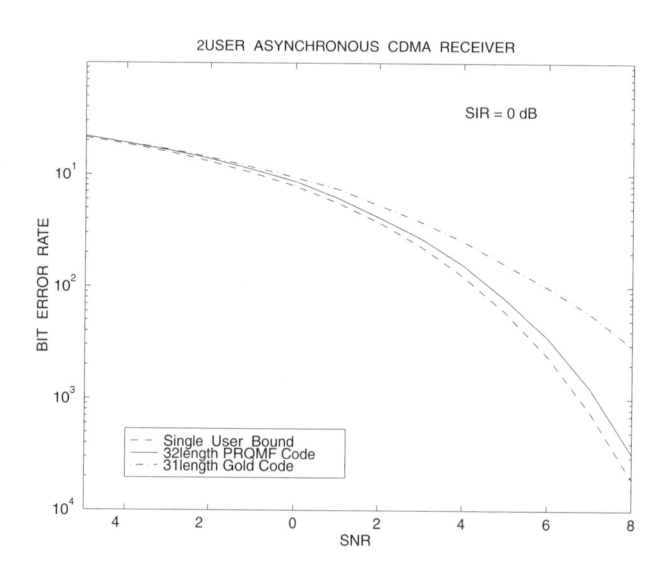

Figure 7.16: BER performance of two-user asynchronous CDMA system for different user code types with SIR = 0 dB.

References

A. N. Akansu and Y. Liu, "On Signal Decomposition Techniques," Optical Engineering Journal, Vol. 30, pp. 912–920, July 1991.

A. N. Akansu and M. J. Medley, Eds., *Wavelet, Subband and Block Transforms in Communications and Multimedia.* Kluwer Academic Publishers, 1999.

A. N. Akansu and M. J. T. Smith, Eds., *Subband and Wavelet Transforms: Design and Applications.* Kluwer Academic Publishers, 1996.

A. N. Akansu and M. V. Tazebay, "Orthogonal Transmultiplexer: A Multiuser Communications Platform from FDMA to CDMA," Proc. European Signal Processing Conference (EUSIPCO), Sept. 1996.

A. N. Akansu, M. V. Tazebay, and R. A. Haddad, "A New Look at Digital Orthogonal Transmultiplexers for CDMA Communications," IEEE Trans. Signal Processing, Vol. 45, No. 1, pp. 263–267, Jan. 1997.

A. N. Akansu, M. V. Tazebay, M. J. Medley, and P. K. Das, "Wavelet and Subband Transforms: Fundamentals and Communication Applications," IEEE Communications Magazine, Dec. 1997.

A. N. Akansu, P. Duhamel, X. Lin, and M. de Courville, "Orthogonal Transmultiplexers in Communication: A Review," IEEE Trans. Signal Processing, Vol. 46, No. 4, pp. 979–995, April 1998.

M. Alard and R. Lasalle, "Principles of Modulation and Channel Coding For Digital Broadcasting for Mobile Receivers," EBU Review, No. 224, pp. 47–69, Aug. 1987.

A. Benyassine, *Theory, Design and Applications of Linear Transforms for Information Transmission.* Ph.D. Thesis. New Jersey Institute of Technology, 1995.

A. Benyassine and A. N. Akansu, "Optimal Subchannel Structuring and Basis Selection for Discrete Multicarrier Modulation," Proc. IEEE Globecom, 1995.

J. A. C. Bingham, "Multicarrier Modulation for Data Transmission: An Idea Whose Time Has Come," IEEE Comm. Magazine, pp. 5–14, May 1990.

J. N. Bradley, C. M. Brislawn, and T. Hopper, "The FBI Wavelet/Scalar Quantization Standard for Gray-Scale Fingerprint Image Compression," Proc. Visual Information Processing II, SPIE, April 1993.

R. W. Chang, "High-speed Multichannel Data Transmission with Bandlimited Orthogonal Signals," Bell Sys. Tech. J., Vol. 45, pp. 1775–1796, Dec. 1966.

W. Chen, *DSL: Simulation Techniques and Standards Development for Digital Subscriber Lines.* Macmillan, 1998.

J. M. Cioffi, "A Multicarrier Primer," Amati Comm. Corp. and Stanford Univ., Tutorial.

R. J. Clarke, *Digital Compression of Still Images and Video.* Academic Press, 1995.

A. Cohen and E. Sere, "Time-Frequency Localization with Non-Stationary Wavelet Packets. in A. N. Akansu and M. J. T. Smith, Eds., *Subband and Wavelet Transforms: Design and Application.* Kluwer, 1996.

R. R. Coifman and M. V. Wickerhauser, "Entropy-Based Algorithms for Best Basis Selection," IEEE Trans. Information Theory, Vol. 38, No. 2, pp. 713–718, March 1992.

S. Davidovici and E. G. Kanterakis, "Narrow-Band Interference Rejection Using Real-Time Fourier Transforms," IEEE Trans. Communications, Vol. 37, No. 7, pp. 713–722, July 1989.

M. de Courville and P. Duhamel, "Orthogonal Frequency Division Multiplexing for Terrestrial Digital Broadcasting," in A. N. Akansu and M. J. Medley, Eds., *Wavelet, Subband and Block Transforms in Communications and Multimedia.* Kluwer, 1999.

M. de Courville, P. Duhamel, P. Madec and J. Palicot, "Blind Equalization of OFDM Systems Based on the minimization of a Quadratic Criterion," Proc. IEEE Int'l Conference on Comm. (ICC), 1996.

Draft American National Standard for Telecommunications, ANSI; Network and Customer Installation Interfaces. Asymmetric Digital Subscriber Line (ADSL) Metallic Interface. T1E1.4 (95–007R2).

B. Girod, F. Hartung and U. Horn, "Subband Image Coding," in A. N. Akansu and M. J. T. Smith, Eds., *Subband and Wavelet Transforms: Design and Applications.* Kluwer, 1996.

R. Gold, "Optimal Binary Sequences for Spread Spectrum Multiplexing," IEEE Trans. Information Theory, pp. 619–621, Oct. 1967.

F. M. Hsu and A. A. Giordano, "Digital Whitening Techniques for Improving Spread Spectrum Communications Performance in the Presence of Narrowband Jamming and Interference," IEEE Trans. Communications, Vol. 26, pp. 209–216, Feb. 1978.

IEEE Trans. Communications. (Special issue on TDM-FDM conversion), Vol. 26, No. 5, May 1978.

IEEE Trans. Communications. (Special issue on Transmultiplexers), Vol. 30, No. 7, July 1982.

IEEE Transactions on Signal Processing special issue on Theory and Application of Filter Banks and Wavelet Transforms, April 1998.

W. W. Jones and K. R. Jones, "Narrowband Interference Suppression Using Filter-Bank Analysis/Synthesis Techniques," Proc. IEEE Military Communications Conference, pp. 898–902, Oct. 1992.

I. Kalet, "Multitone Modulation," in A. N. Akansu and M. J. T. Smith, Eds., *Subband and Wavelet Transforms: Design and Applications.* Kluwer, 1996.

J. W. Ketchum and J. G. Proakis, "Adaptive Algorithms for Estimating and Suppressing Narrow-Band Interference in PN Spread-Spectrum Systems," IEEE Trans. Communications, Vol. 30, pp. 913–924, May 1982.

R. D. Koilpillai, T. Q. Nguyen and P. P. Vaidyanathan, "Some Results in the Theory of Crosstalk-Free Transmultiplexers," IEEE Trans. Signal Processing, Vol. 39, No.10, pp. 2174–2183, Oct. 1991.

X. Lin, *Orthogonal Transmultiplexers: Extensions in Digital Subscriber Line (DSL) Communications.* Ph.D. Thesis. New Jersey Institute of Technology, 1998.

X. Lin and A. N. Akansu, "A Distortion Analysis and Optimal Design of Orthogonal Basis for DMT Transceivers," Proc. IEEE ICASSP, Vol. 3, pp. 1475–1478, 1996.

X. Lin, M. Sorbara, and A. N. Akansu, "Digital Subscriber Line Communications". in A. N. Akansu and M. J. Medley, Eds., *Wavelet, Subband and Block Transforms in Communications and Multimedia.* Kluwer, 1999.

M. J. Medley, *Adaptive Narrow-Band Interference Suppression Using Linear Transforms and Multirate Filter Banks.* Ph.D. Thesis. Rensselaer Polytechnic Institute, 1995.

L. B. Milstein and P. K. Das, "An Analysis of a Real-Time Transform Domain Filtering Digital Communication System-Part I: Narrowband Interference Rejection," IEEE Trans. Communications, Vol. 28, No. 6, pp. 816–824, June 1980.

D. L. Nicholson, *Spread Spectrum Signal Design.* Computer Science Press, 1988.

A. Nosratinia, G. Davis, Z. Xiong and R. Rajagopalan, Subband Image Compression. in A. N. Akansu and M. J. Medley, Eds., *Wavelet, Subband and Block Transforms in Communications and Multimedia.* Kluwer, 1999.

A.Peled and A.Ruiz, "Frequency Domain Data Transmission Using Reduced Computational Complexity Algorithms," Proc. IEEE ICASSP, pp. 964–967, April 1980.

A. Said and W. A. Pearlman, "A New Fast and Efficient Image Codec Based on Set Partitioning in Hierarchical Trees," IEEE Trans. Circuits and Systems for Video Technology, Vol. 6, No. 3, pp. 243–250, June 1996.

B. R. Saltzberg, "Performance of an Efficient Parallel Data Transmission System," IEEE Trans. Comm., Vol. 15, No. 6, pp. 805–811, Dec. 1967.

G. J. Saulnier, M. J. Medley and P. K. Das, "Wavelets and Filter Banks in Spread Spectrum Communication Systems," in A. N. Akansu and M. J. T. Smith, Eds., *Subband and Wavelet Transforms: Design and Applications.* Kluwer Academic Publishers, 1996.

J. M. Shapiro, "Embedded Image Coding Using Zero-trees of Wavelet Coefficients," IEEE Trans. Signal Processing, Vol. 41, No. 12, pp. 3445–3462, Dec. 1993.

T. Starr, J. M. Cioffi and P. Silverman, *Understanding Digital Subscriber Line Technology.* Prentice-Hall, 1999.

M. V. Tazebay and A. N. Akansu, "Progressive Optimality in Hierarchical Filter Banks," Proc. IEEE ICIP, Vol. I, pp. 825–829, Nov. 1994.

M. V. Tazebay and A. N. Akansu, "Adaptive Subband Transforms in Time-Frequency Excisers for DSSS Communications Systems," IEEE Trans. Signal Processing, Vol. 43, No. 11, pp. 2776–2782, Nov. 1995.

M. V. Tazebay, *On Optimal Design and Applications of Linear Transforms.* Ph.D. Thesis. New Jersey Institute of Technology, 1996.

P. Topiwala, *Wavelet Image and Video Compression.* Kluwer, 1998.

M. A. Tzannes et al., "A Multicarrier Transceiver for ADSL Using M-Band Wavelet Transforms," in ANSI T1E1.E4 Committee Contribution, No. 93–67, Miami, FL, March 1993.

M. Vetterli and J. Kovacevic, *Wavelets and Subband Coding.* Prentice-Hall, 1995.

S. B. Weinstein and P. M. Ebert, "Data Transmission by Frequency-Division Multiplexing Using the Discrete Fourier Transform," IEEE Trans. Comm., Vol. 19, No. 5, pp.628–634, Oct. 1971.

J. W. Woods, *Subband Image Coding.* Kluwer, 1991.

R. E. Ziemer and R. L. Peterson, *Digital Communications and Spread Spectrum Systems.* Macmillan Inc., 1985.

Appendix A

Resolution of the Identity and Inversion

Theorem:
Let

$$W_f(a,b) = \int_{-\infty}^{\infty} \psi_{ab}(t) f(t) dt. \tag{A.1}$$

Then
(i)

$$I \triangleq \int_{-\infty}^{\infty} \int_{0}^{\infty} W_f(a,b) W_g^*(a,b) \frac{dadb}{a^2} = C_\psi < f(t), g^*(t) > \tag{A.2}$$

where

$$C_\psi = \int_0^\infty \frac{|\Psi(\omega)|^2}{\omega} d\omega < \infty \tag{A.3}$$

and
(ii)

$$f(t) = \frac{1}{C_\psi} \int_{-\infty}^{\infty} \int_0^\infty \frac{dadb}{a^2} W_f(a,b) \psi_{ab}(t). \tag{A.4}$$

Proof of (i) (Eq. A.2)
From Fourier transform theory, we know that

$$W_f(a,b) = \int \psi_{ab}(t) f(t) dt = \frac{1}{2\pi} \int_{-\infty}^{\infty} F(\Omega) \Psi_{ab}^*(\Omega) d\Omega \tag{A.5}$$

But, from Eq. (5.16), $\Psi_{ab}(\Omega) = \sqrt{a}\Psi(a\Omega)e^{-jb\Omega}$.
Therefore,

$$W_f(a,b) = \frac{1}{2\pi}\int_{-\infty}^{\infty} \sqrt{a}e^{jb\Omega}\Psi^*(a\Omega)F(\Omega)d\Omega. \tag{A.6}$$

Similarly, we can obtain an expression $W_g(a,b)$ of the same form as (A.6) for a function $g(t)$, or

$$W_g^*(a,b) = \frac{1}{2\pi}\int_{-\infty}^{\infty} \sqrt{a}e^{-jb\Omega'}\Psi^*(a\Omega')G^*(a\Omega')d\Omega'. \tag{A.7}$$

Substituting (A.6) and (A.7) into (A.2),

$$I = (\frac{1}{2\pi})^2 \int_{-\infty}^{\infty}\int_0^{\infty} \frac{dadb}{a^2}\int_{-\infty}^{\infty}\int_{-\infty}^{\infty} ad\Omega d\Omega' \underbrace{F(\Omega)G^*(\Omega)\Psi(a\Omega)\Psi^*(a\Omega')e^{jb(\Omega-\Omega')}}_{P(\Omega,\Omega',a)}.$$

Interchanging the order of integration,

$$I = (\frac{1}{2\pi})^2 \int \frac{da}{a} \int\int d\Omega d\Omega' P(\Omega,\Omega',a) \int_{-\infty}^{\infty} e^{jb(\Omega-\Omega')}db \tag{A.8}$$

The integral over b can be shown to be (Papoulis, 1977)

$$\lim_{B\to\infty}\int_{-B}^{B} e^{jb(\Omega-\Omega')}db = \lim_{B\to\infty} 2\pi\frac{sin(\Omega-\Omega')B}{\pi\Omega-\Omega')} = 2\pi\delta(\Omega-\Omega'). \tag{A.9}$$

Substituting (A.9) into (A.8), and integrating over Ω' gives

$$\begin{aligned} I &= \left(\frac{1}{2\pi}\right)\int\frac{da}{a}\int d\Omega \underbrace{\int P(\Omega,\Omega',a)\delta(\Omega-\Omega')}_{P(\Omega,\Omega,a)} \\ &= \frac{1}{2\pi}\int_0^{\infty}\frac{da}{a}\int_{-\infty}^{\infty} F(\Omega)G^*(\Omega)|\Psi(a\Omega)|^2 d\Omega. \end{aligned} \tag{A.10}$$

Again, an interchange in order of integration and a change of variable $x = a\Omega$ gives

$$\begin{aligned} I &= \left(\int_0^{\infty}\frac{|\Psi(x)|^2}{x}\right)\left(\frac{1}{2\pi}\int_{-\infty}^{\infty} F(\Omega)G^*(\Omega)d\Omega\right) \\ &= C_\psi < f(t), g^*(t) > . \end{aligned} \tag{A.11}$$

Proof of (ii), the inversion formula (A.4):
Let $I(t)$ represent the right-hand side of (A.4). Substituting (A.1) into (A.4)

$$\begin{aligned} I(t) &= \frac{1}{C_\psi}\int\int \frac{dadb}{a^2}\psi_{ab}(t)\int \psi_{ab}^*(\tau)f(\tau)d\tau \\ &= \frac{1}{C_\psi}\int f(\tau)d\tau \underbrace{\left[\int\int \psi_{ab}(t)\psi_{ab}^*(\tau)\frac{dadb}{a^2}\right]}_{K(t,\tau)} \\ &= \frac{1}{C_\psi}\int_{-\infty}^{\infty} f(\tau)K(t,\tau)d\tau. \end{aligned} \tag{A.12}$$

The proof is complete if $K(t,\tau) = C_\psi\delta(t-\tau)$.

Using the Fourier transforms of $\psi_{ab}(\cdot)$ in (A.12) gives

$$K(t,\tau) = (\frac{1}{2\pi})^2 \int\int\int\int \left(\frac{dadb}{a}d\Omega d\Omega'\right)\Psi(a\Omega)\Psi^*(a\Omega')e^{-jb(\Omega-\Omega')}e^{j(t\Omega-\tau\Omega')}. \tag{A.13}$$

Following the tactic used previously, we integrate first with respect to b and obtain the impulse $2\pi\delta(\Omega-\Omega')$ as in (A.9). This leaves us with

$$\begin{aligned} K(t,\tau) &= \frac{1}{2\pi}\int\frac{da}{a}\int d\Omega\Psi(a,\Omega)e^{jt\Omega}\int \Psi^*(a\Omega')e^{-j\tau\Omega'}\delta(\Omega-\Omega')d\Omega' \\ &= \frac{1}{2\pi}\int\frac{da}{a}\int d\Omega|\Psi(a\Omega)|^2 e^{j(t-\tau)\Omega}. \end{aligned} \tag{A.14}$$

This separates into

$$\begin{aligned} K(t,\tau) &= \left(\frac{1}{2\pi}\int_{-\infty}^{\infty} e^{j(t-\tau)\Omega}d\Omega\right)\left(\int_0^\infty \frac{|\Psi(a\Omega)|^2}{a}da\right) \\ &= \delta(t-\tau)C_\psi. \end{aligned} \tag{A.15}$$

Appendix B

Orthonormality in Frequency

A set of functions $\phi(., -n)$ form an orthonormal family if and only if their Fourier transforms satisfy

$$\sum_k |\Phi(\Omega + 2\pi k)|^2 = 1. \tag{B.1}$$

Proof: Since $\{\phi(t-n); n \in Z\}$ is an orthonormal family, then they should satisfy

$$||f||^2 = \sum_n |\alpha(n)|^2. \tag{B.2}$$

Then we can expand the $f(t)$ in orthogonal family as

$$f(t) = \sum_n \alpha(n)\phi(t-n) = \phi(t) * \sum_n \alpha(n)\delta(t-n). \tag{B.3}$$

This relation in Fourier domain becomes

$$F(\Omega) = \Phi(\Omega) \sum_n \alpha(n) e^{-j\Omega n}. \tag{B.4}$$

By defining 2π periodic function

$$M(\Omega) = \sum_n \alpha(n) e^{-j\Omega n},$$

this relation becomes

$$F(\Omega) = \Phi(\Omega) M(\Omega). \tag{B.5}$$

Therefore, from the Parseval relation

$$\int_{-\infty}^{\infty} |f(t)|^2 dt \quad = \quad \frac{1}{2\pi} \int_{-\infty}^{\infty} |F(\Omega)|^2 d\Omega$$

$$\begin{aligned}
&= \frac{1}{2\pi}\int_{-\infty}^{\infty} |\Phi(\Omega)|^2 |M(\Omega)|^2 d\Omega \\
&= \frac{1}{2\pi}\sum_{n=-\infty}^{\infty}\int_{2\pi n}^{2\pi(n-1)} |\Phi(\Omega)|^2 |M(\Omega)|^2 d\Omega \\
&= \frac{1}{2\pi}\sum_{n=-\infty}^{\infty}\int_{0}^{2\pi} |M(\Omega)|^2 |\Phi(\Omega+2\pi n)|^2 d\Omega \\
&= \frac{1}{2\pi}\int_{0}^{2\pi} |M(\Omega)|^2 \sum_{n=-\infty}^{\infty} |\Phi(\Omega+2\pi n)|^2 d\Omega. \qquad \text{(B.6)}
\end{aligned}$$

From Eq. (B.2) we have

$$\begin{aligned}
\int_{-\infty}^{\infty} |f(t)|^2 dt &= \sum_n |\alpha(n)|^2 \\
&= \frac{1}{2\pi}\int_0^{2\pi} |M(\Omega)|^2 d\Omega.
\end{aligned}$$

Therefore, $\phi(t)$ must satisfy

$$\sum_n |\Phi(\Omega+2\pi n)|^2 = 1.$$

Appendix C

Problems

Chapter 2

(2.1) Reference Eq. (2.24). Show that $\underline{\epsilon}^T \underline{f} = 0$ is a necessary and sufficient condition for minimizing the least square error J, where

$$\underline{f}^T = [f_0 \cdots f_{N-1}]^T, \quad \hat{\underline{f}}^T = [\hat{f}_0 \cdots \hat{f}_{N-1}]^T, \quad \underline{\epsilon} = \underline{f} - \hat{\underline{f}}$$

$$\hat{f}(k) = \sum_{r=0}^{L-1} \gamma_r \phi_r(k), \qquad L \leq N.$$

(2.2) Show that the Cauchy-Schwarz inequality, Eq. (2.25), becomes an equality if $y(k) = ax(k)$.

(2.3) Derive the extended Parseval theorem, Eq. (2.26), starting with

$$x(k) = \sum_r \alpha_r \phi_r(k), \qquad y(k) = \sum_r \beta_r \phi_r(k).$$

(2.4) Use Eq. (2.16) to show that (infinite support) sequences $\phi_r(n)$ are orthonormal, where

$$\begin{aligned} \phi_r(n) &= a_r \left(\frac{\sin n\pi/M}{n\pi} \right) \left(\cos \frac{r2\pi n}{M} \right), \qquad r = 0, 1, \cdots, M-1 \\ a_r &= \begin{cases} 1, & r = 0 \\ \sqrt{2}, & r = 1, 2, \cdots, M-1 \end{cases}. \end{aligned}$$

(2.5) Derive Eq. (2.39),

$$E\{\underline{\tilde{f}}^T \underline{f}\} = E\{\underline{\tilde{\theta}}^T \underline{\tilde{\theta}}\}.$$

(2.6) Show that

$$\nabla\{\underline{x}^T A \underline{x}\} = 2A\underline{x}$$
$$\frac{d}{d\underline{x}}[\underline{b}^T \underline{x}] = \underline{b}.$$

(2.7) Derive Eq. (2.92). Show that it can also be expressed as

$$\sigma_k^2 2^{-2R_k} = c, \text{a constant}$$

where

$$c = 2^{-2A}, \qquad A = R - \frac{1}{2N}\sum_{k=0}^{N-1} log\sigma_k^2$$

and σ_k^2 is the variance of coefficient θ_k.

(2.8) Derive Eqs. (2.93) and (2.94).

(2.9) Let $\hat{x} = Q(x)$ represent the output of a quantizer Q with input x, and let $\tilde{x} = (x - \hat{x})$ be the resulting quantization error. Show that the mean square error $J = E\{|\tilde{x}|^2\}$ is minimized when Q is chosen such that $E\{\tilde{x}\} = 0$, and $E\{\tilde{x}\hat{x}\} = 0$, i.e., when the mean of $\tilde{x}$ is zero, and the quantized output is orthogonal to the error.

(2.10) Show that the DCT basis functions in Eq. (2.119) are orthonormal.

(2.11) Show that the Binomial sequence $x_r(k)$ of Eq. (2.134) satisfies the two-term recurrence relation of Eq. (2.138).

(2.12) Prove that the discrete Hermite polynomials of Eq. (2.140) are orthogonal with respect to the weight function indicated.

(2.13) Show that Eqs. (2.143) and (2.145) imply $X^2 = 2^N I$ as stated in Eq. (2.141).

(2.14) Derive Eq. (2.150) from Eq. (2.152).

(2.15) (a) Show that $\Phi_r(z)$ in Eq. (2.153) can be written as

$$\Phi_r(z) = (-1)^r(1-\lambda^2)^{1/2}G_r(z)(\frac{z}{z-\lambda})$$

where $G_r(z)$ is the all-pass function

$$G_r(z) = \left(\frac{1-\lambda z}{z-\lambda}\right)^r.$$

(b) By contour integration show that the all-pass sequences $\{g_r(n)\}$ are orthogonal on $[0,\infty)$ with respect to a weight function, i.e.,

$$\sum_{n=0}^{\infty} n g_r(n) g_s(n) = r\delta_{r-s}.$$

(c) With

$$\hat{f}(n) = \sum_{r=0}^{N-1} \alpha_r g_r(n)$$

show that

$$\sum_{n=0}^{\infty} n|f(n) - \hat{f}(n)|^2$$

is minimized by selecting

$$\alpha_r = \begin{cases} \frac{1}{r}\sum_0^\infty n f(n) g_r(n), & r \geq 1 \\ \sum_0^\infty f(n)\lambda^n, & r = 0 \end{cases}.$$

[This last demonstration is not trivial.]

(2.16) Consider the 4×4 Hadamard matrix

$$H_4 = \begin{bmatrix} H_2 & H_2 \\ H_2 & -H_2 \end{bmatrix}, \qquad H_2 = \begin{bmatrix} 1 & 1 \\ 1 & -1 \end{bmatrix}.$$

(a) Show that premultiplication of H_4 by

$$S = \begin{bmatrix} 1 & 0 & 0 & 0 \\ 0 & 0 & 1 & 0 \\ 0 & 0 & 0 & 1 \\ 0 & 1 & 0 & 0 \end{bmatrix}.$$

puts the rows in Walsh sequency order.
(b) Let $h_r(n)$ represent the rth row, or basis sequence of $H \triangleq SH_4$. Sketch $h_r(n)$, $|H_r(e^{j\omega})|$ for $0 \le r \le 3$ and comment on symmetry, mirror-image property; compare with sketches for the 8×8 Walsh transform in Fig. 2.7.

(2.17) Sketch the basis functions for the 4×4 Haar transform. Evaluate and sketch the magnitude of the respective Fourier transforms.

(2.18) Derive Eq. (2.183) from Eqs. (2.182) and (2.173).

(2.19) Calculate the 6×6 KLT block transform basis for an AR(1) source with $\rho = 0.95$. Compare this with the 6×6 DCT.

(2.20) Derive Eq. (2.219).
(2.21) Refer to Eqs. (2.119) and (2.121). Derive the matrix S in Section 2.3.1 and use this to show that the eigenvectors of Q are independent of α.

Chapter 3

(3.1) Given

$$x(n) = \lambda^n u(n).$$

(a) Let $y(n)$ be a down-sampling of $x(n)$ as defined by Eq. (3.2) and Fig. 3.1, for $M = 3$. Evaluate $X(z)$, $Y(z)$ and show the pole-zero plots. Evaluate and sketch $|X(e^{j\omega})|$, and $|Y(e^{j\omega})|$ after normalizing the DC gain to unity.
(b) Let $y(n)$ be an up-sampling of $x(n)$ as defined by Eq. (3.11) and Fig 3.3 with $M = 3$. Repeat (a).
(c) Let $x(n)$ be down-sampled by M and then up-sampled by M to create a signal $y(n)$. Repeat (a) for $M = 3$.
(d) Compare pole-zero patterns and frequency responses for (a), (b), and (c).

(3.2) Repeat problem (3.1) for

$$x(n) = \begin{cases} 1, & 0 \le n \le 9 \\ 0, & \text{otherwise} \end{cases}.$$

(3.3) Let $h(n) = \lambda^n u(n)$. Evaluate the polyphase components $G_k(z)$ defined by Eqs. (3.14) and (3.15) for $M = 3$. Show pole-zero plots. Evaluate and sketch $|G_k(e^{j\omega})|$, $|G_k(e^{j3\omega})|$, for $0 \le k \le 2$.

(3.4) Repeat problem (3.3) for

$$h(n) = \begin{cases} 1, & 0 \le n \le 8 \\ 0, & \text{otherwise} \end{cases}.$$

(3.5) (a) Let $x(n)$ be filtered by $h(n)$ and then downsampled by $M = 2$ to give $y(n)$. Show that

$$y(n) = \sum_k x(k)h(2n - k).$$

(b) Let $x(n)$ be up-sampled by $M = 2$ and then filtered by $g(n)$. Show that the output is

$$y(n) = \sum_k g(n - 2k)x(k)$$

(c) Let the output of (a) be the input to (b). Show that

$$\begin{aligned} y(n) &= \sum_j \phi(n,j)x(j) \\ \phi(n,j) &= \sum_k g(n-2k)h(2k-j). \end{aligned}$$

(d) For parts (a), (b), and (c), evaluate the corresponding Z-transforms. Are these systems time-invariant? Why?
(e) From $Y(z)$ in part (d), show that $y(n)$ can be expressed as

$$y(n) = \frac{1}{2}g(n) * h(n) * x(n) + \frac{1}{2}g(n) * (-1)^n h(n) * (-1)^n x(n).$$

From this, show that

$$\phi(n,j) = \sum_k \left(\frac{1+(-1)^k}{2}\right) g(n-k)h(k-j).$$

(3.6) Let

$$H_0(z) = \frac{A_0(z) + A_1(z)}{2}$$

$$H_1(z) = \frac{A_0(z) - A_1(z)}{2}$$

where $A_0(z)$, $A_1(z)$ are all-pass networks. Show that

$$|H_0(e^{j\omega})|^2 + H_1(e^{j\omega})|^2 = 1$$

$$|H_0(e^{j\omega}) + H_1(e^{j\omega})| = 1.$$

(3.7) Let $h(n)$ be a half-band filter. Prove that

$$h(0) = \sum_{\substack{n=-\infty \\ n\neq 0}}^{\infty} h(n).$$

(3.8) Show the equivalence of the structures given in Fig. 3.7(b).

(3.9) Show that the M-band power complementary property of Eq. (3.32) is satisfied if $H(z)$ is a spectral factor of an Mth band filter and conversely.

(3.10) Prove that the four-band binary tree of Fig. 3.21 is paraunitary if (H_0, H_1), (G_0, G_1) constitute a two-band paraunitary PR structure.

(3.11) Show that there is no linear-phase paraunitary solution for the two-band filter bank.

(3.12) Let

$$h_0(n) = \alpha_0 x_0(n) + \alpha_1 x_1(n),$$

where $x_0(n)$, $x_1(n)$ are the four-tap Binomial sequences of Eq. (2.139). For convenience let $\alpha_0 = 1$. (This denormalizes the filter.) Evaluate α_1 such that $H_0(z)$ is the low-pass paraunitary filter of Eq. (3.50). There are two solutions here, a maximum-phase and a minimum phase. For each solution, calculate and sketch the time and frequency responses of $H_0(z)$, $H_1(z)$, $G_0(z)$, and $G_1(z)$, and compare.

(3.13) Demonstrate that the total delay from input to output in Fig. 3.40(a) is given by Eq. (3.101).

(3.14) Derive Eq. (3.98).

(3.15) (a) Show that an Mth band filter can be constructed by

$$h(n) = \left(\frac{\sin \frac{n\pi}{M}}{\pi n} \right) w(n)$$

$$w(n) = w(-n).$$

(b) Evaluate the M polyphase components of this $H(z)$, and demonstrate power complementarity for the case

$$w(n) = \begin{cases} 1, & -LM \leq n \leq LM \\ 0, & \text{otherwise} \end{cases}.$$

(3.16) (a) Show that

$$\sum_{n=-\infty}^{\infty} \alpha(n) = \sum_{k=0}^{M-1} \sum_{n=-\infty}^{\infty} \alpha(Mn+k).$$

Use this result to derive the 1D polyphase expansion

$$\sum_{-\infty}^{\infty} f(n) z^{-n} = \sum_{k=0}^{M-1} z^{-k} F_k(z^M)$$

where

$$F_k(z) = \sum_{-\infty}^{\infty} f(k+Mn) z^{-n}.$$

(b) Consider a subsampling lattice D and associated coset vectors $\{\underline{k}_l, l = 0, 1, \cdots, M-1\}$. Show that

$$\sum_{n \in \Lambda} \alpha(\underline{n}) = \sum_{l=0}^{M-1} \sum_{n \in \Lambda} \alpha(D\underline{n} + \underline{k}_l).$$

Use this result to fill in the missing steps in the derivation of Eq. (3.281).
(c) Repeat (a) with k replaced by $-k$.
(d) Repeat (b) with $\underline{k}_l$ replaced by $-\underline{k}_l$, and compare with Eq. (3.285).

(3.17) Consider a discrete-time system with r inputs $\{x_i(n), i = 1, 2, ..., r\}$ and p outputs $\{y_j(n), j = 1, 2, ..., p\}$. Let $\underline{Y}(z) = H(z)\underline{X}(z)$, where $\underline{X}(z)$, $\underline{Y}(z)$ are input and output vector transforms, and $H(z)$ is the $p \times r$ transfer function matrix. This system is *lossless* (Vaidyanathan, Aug. 1989) if $E_x = E_y$, where

$$E_x \triangleq \sum_n \underline{x}^T(n)\underline{x}(n), \qquad E_y = \sum_n \underline{y}^T(n)\underline{y}(n).$$

Show that the system is lossless if

$$H^\dagger(e^{j\omega})H(e^{j\omega}) = I_{r\times r}, \quad \text{for} \quad -\pi \le \omega \le \pi$$

where $H^\dagger$ is the conjugate transpose of H. Then give arguments (analytic continuation) demonstrating that losslessness is also satisfied by the paraunitary condition

$$\tilde{\mathcal{H}}(z)\mathcal{H}(z) = I_{r\times r}.$$

(3.18) A causal $M \times M$ FIR matrix transfer function $H(z)$ of degree L is lossless if it can be expressed as

$$H(z) = V_L(z)V_{L-1}(z)\cdots V_1(z)H_0$$

where

$$V_k(z) = [I - (1 - z^{-1})\underline{v}_k\underline{v}_k^T]$$

with

$$\underline{v}_k^T\underline{v}_k = 1.$$

In the text, we proved the sufficiency. Prove the necessity, i.e., the "only if" part.

(3.19) (a) If $H(z)$ is synthesized as in Prob. (3.18), show that any section $V_k(z)$ can be realized with a single scalar delay, and show that $det\{V_k(z)\} = z^{-1}$. (b) Use (a) to show that $det\{H(z)\} = z^{-(N-1)}$, for any causal, lossless, FIR matrix.

(3.20) Show that the two-band linear-phase requirement is

$$z^{-k}\begin{bmatrix} 1 & 0 \\ 0 & -1 \end{bmatrix}\left[\mathcal{H}_p(z^{-1})\right]\begin{bmatrix} 0 & 1 \\ 1 & 0 \end{bmatrix} = \mathcal{H}_p(z)$$

where $\mathcal{H}_p(z)$ is the polyphase matrix.

(3.21) Demonstrate the validity of Eq. (3.219).

(3.22) Starting with $\underline{z}^{\underline{n}}$ and $\underline{z}^{\underline{D}}$ defined by Eq. (3.252), and

$$X(z_1, z_2) \stackrel{\Delta}{=} \sum_{n_1}\sum_{n_2} x(n_1, n_2) z_1^{-n_1} z_2^{-n_2} \stackrel{\Delta}{=} \sum_{\underline{n}} x(\underline{n})\underline{z}^{-\underline{n}},$$

show that (Viscito and Allebach, 1991)
(a) $(\underline{z}^D)^{\underline{n}} = \underline{z}^{D\underline{n}}$
(b) $(-\underline{z})^{\underline{n}} = (-1)^{(n_1+n_2)}\underline{z}^{\underline{n}}$
(c) $(-\underline{z})^{-I} = -\underline{z}^{-I}$
(d) If $y(\underline{n}) = x(-\underline{n})$, then $Y(\underline{z}) = X(\underline{z}^{-I})$.

(3.23) (a) Show that the complex exponentials in Eq. (3.256),

$$\phi(\underline{k}, \underline{n}) = \stackrel{\Delta}{=} e^{j2\pi\underline{k}^T D^{-1}\underline{n}},$$

are periodic in $\underline{k}$, and $\underline{n}$ with periodicity matrices D^T, and D, respectively.
(b) Show that $\{\phi(\underline{k},\underline{n})\}$ are orthogonal over the unit cell I_M specified by D, i.e.,

$$\sum_{\underline{n}\in I_M} \phi(\underline{k},\underline{n})\phi^*(\underline{m},\underline{n}) = M\delta(\underline{k}-\underline{m}).$$

(3.24) Compare the sublattices associated with $D_1 = \begin{bmatrix} 1 & 1 \\ -2 & 2 \end{bmatrix}$ and $D_2 = \begin{bmatrix} 2 & 1 \\ 0 & 2 \end{bmatrix}$. Show the sublattice associated with $D_3 = \begin{bmatrix} 4 & 0 \\ 0 & 1 \end{bmatrix}$ and compare with D_1, D_2.

(3.25) Demonstrate the validity of Eqs. (3.289) and (3.291).

(3.26) Show that a 2D, FIR, M-band filter bank is paraunitary if

$$det\{\mathcal{H}_p(z_1,z_2)\} = z_1^{-m_1} z_2^{-m_2}.$$

(3.27) Show that

$$\mathcal{H}_p(z_1,z_2) = H_0 \prod_{i=0}^{k-1} \{I - (1-z_1^{-1})\underline{u}_i\underline{u}_i^T\}\{I - (1-z_2^{-1})\underline{v}_i\underline{v}_i^T\}$$

is paraunitary where H_0 is an $M \times M$ unitary matrix, and $\underline{u}_i, \underline{v}_i$ are $M \times 1$ column vectors with unit norm.
Hint: This is an extension of the 1D result, Eq. (3.194).

3.28 (a) Show that

$$A(z)B(z^{-1}) + A(-z)B(-z^{-1}) = 1$$

implies that the sequence $a(k)$ is orthogonal to the even translates of $b(k)$, i.e.,

$$\sum_k a(k)b(k-2n) = \delta_n.$$

(b) Define

$$\begin{aligned} \tilde{g}_0(n) &= g_0(-n) \longleftrightarrow \tilde{G}_0(z) = G_0(z^{-1}) \\ \tilde{g}_1(n) &= g_1(-n) \longleftrightarrow \tilde{G}_1(z) = G_1(z^{-1}). \end{aligned}$$

In Eq. (3.59), let $S(z) = \frac{1}{2}S_1(z) - \frac{1}{4}G_1(z)H_0(-z)S_2(z)$, substitute the alias cancellation requirements of Eq. (3.60),

$G_0(z) = -z^{-1}H_1(-z)$, and $G_1(z) = z^{-1}H_0(-z)$, la into S_2 of Eq. (3.58) and show that

$$\sum_k h_1(k)\tilde{g}_0(k-2n) = 0.$$

(c) Substitute Eq. (3.60) into $T(z)$ of (3.61) and derive

$$\begin{aligned}\sum_k h_0(k)\tilde{g}_0(k-2n) &= \delta_n \\ \sum_k h_1(k)\tilde{g}_1(k-2n) &= \delta_n.\end{aligned}$$

(d) Finally, provide symmetry arguments for the last orthogonality

$$\sum_k h_0(k)\tilde{g}_1(k-2n) = 0.$$

(3.29) Reference the IIR lattice filter of Section 3.7.2. Let

$$H_0(z) = a_0(z^2) + z^{-1}a_1(z^2) = N(z)D(z)$$

$$a_0(z) = \frac{a+z^{-1}}{1+az^{-1}}, \qquad a_1(z) = \frac{b+z^{-1}}{1+bz^{-1}}.$$

(a) Evaluate $H_0(z)$ and show that the poles are imaginary located at $z = 0, \pm j\sqrt{a}$, $\pm j\sqrt{b}$.

(b) Determine conditions on a, b such that $N(z)$ is a mirror image polynomial. Indicate the resulting pole-zero pattern for $H_0(z)$.

(c) Evaluate a, b so that the zeros of $H_0(z)$ in part (b) are at $z = -1$, $e^{j(180\pm 30)}$, $e^{j(180\pm 60)}$.

(d) Sketch the resulting magnitude and phase of $H_0(e^{j\omega})$.

(3.30) Show the subsampling lattice, the coset vectors and image subbands generated by (a) $D = \begin{bmatrix} 2 & 0 \\ 0 & 2 \end{bmatrix}$ (b) $D = \begin{bmatrix} 2 & 2 \\ -2 & 2 \end{bmatrix}$.

(3.31) Derivation of Eq. (3.222), Chap. 3.

(a) Show that the partitioned cosine modulation matrix satisfies

$$\begin{aligned} c_0^T c_0 &= 2M[I_m + (-1)^{m-1}J_m] \\ c_1^T c_1 &= 2M[I_m - (-1)^{m-1}J_m] \end{aligned}$$

$$c_0^T c_1 = c_1^T c_0 = \phi. \qquad (1)$$

(See Koilpillai and Vaidyanathan, 1992.)
(b) If $H(z)$ is linear-phase with length $2mM$, the polyphase components satisfy

$$G_k(z) = z^{-(m-1)} \tilde{G}_{2M-1-k}(z). \qquad (2)$$

(c) Calculate $\tilde{H}_p(z)H_p(z)$ from Eq. (3.220). Substitute (1) and (2) into this expression. From this, derive Eq. (3.222).

(3.32) Show that the system in Figure C.1 from $x(n)$ to $y(n)$ is LTI.

(3.33) In Figure C.1 show that $Y(z) = E_0(z)X(z)$, when $E_0(z)$ is the 0th component in the polyphase expansion of $G(z)$.

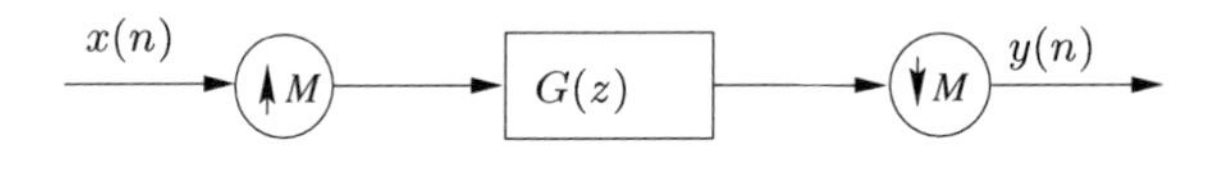

Figure C.1

(3.34) In Figure C.2, show that the transmission from $x(n)$ to

$$y(n) = \begin{cases} 1, & k = 0 \\ 0, & k \neq 0 \end{cases}.$$

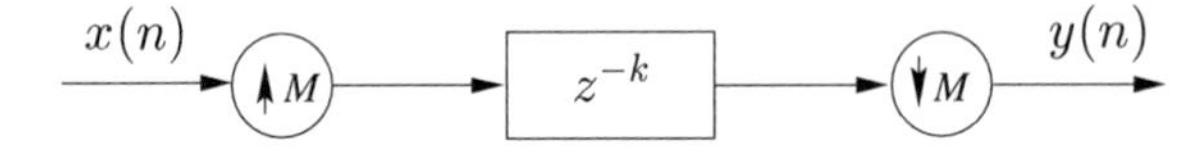

Figure C.2

Chapter 4

(4.1) Find $\{\theta_r\}$ values in the design of the Binomial QMF for $N = 3$. Check your results with Table 4.1. Which one of the possible solutions is minimum-phase? Why? Plot their phase and magnitude responses and comment on them.

(4.2) Show that the Binomial QMF has the maximally flat magnitude square function for $N = 5$. Is this function unique? [See Eq. (4.9).]

(4.3) The input signal to a 2-band, PR-QMF based filter bank is given as

$$x(n) = \sin \omega n T$$

with T=0.01 sec. Employ the 4-tap Binomial PR-QMF in your filter bank. Show that the reconstructed signal $\hat{x}(n)$ is identical to the input signal except a delay. How much this delay differ for the minimum- and nonminimum-phase filters of 4-tap Binomial-QMF?

(4.4) How many zeros at $\omega = \pi$ does the Binomial QMF of $N = 7$ have? What does it mean in wavelet transform context? (See Chapter 5.)

(4.5) Find $V(z)$ in Eq. (4.12) for $N = 5$.

(4.6) Calculate G_{TC} of 4-tap Binomial QMF based 2-band subband decomposition for AR(1) source models with the values of autocorrelation coefficient $\rho = 0.95$ and $\rho = 0.1$. Comment on your results.

(4.7) Assume $\alpha = 0.0348642$ in Eqs. (4.32) and (4.33). Obtain the corresponding 6-tap PR-QMF coefficients. (Check your filter coefficients with the 6-tap most regular wavelet filter.) Compare it with $\alpha = 0$ case (Binomial QMF). Repeat

Problem (4.6) for these two filters and interpret your results.

(4.8) Interpret the significance of wavelet regularity in the PR-QMF context. What is the minimum degree of regularity desired in practice? Why?

(4.9) Plot $f(i)$, $B(f;x)$, and $R(z)$ for 6-tap case with $\alpha = 0$ (similar to Fig. 4.4).

(4.10) Plot the phase and magnitude responses of 8-tap Binomial QMF and Johnston QMF. Comment on the properties of these filters.

(4.11) What is the difference between half-band and half-bandwidth (2-band PR-QMF) filters? Explain.

(4.12) Plot the phase and magnitude functions of LeGall-Tabatabai filters (low- and high-pass). Are these a biorthogonal filter bank? Comment on the properties of these filters.

(4.13) Calculate the aliasing energy component σ_A^2, Eq. (4.50), of the low-pass filter output in Prob. (4.6) for AR(1), $\rho = 0.95$, source.

(4.14) Find the value of E_s, Eq. (4.52), for 8-tap Binomial PR-QMF and Johnston QMF. Comment on this criterion.

(4.15) Does the Smith-Barnwell PR-CQF satisfy the requirements for wavelet filters? Why?

(4.16) Calculate E_p, Eq. (4.55), for 4, 6, and 8-tap Binomial QMF. Does E_p decrease when N increases?

(4.17) Design a 4-tap optimal 2-band PR-QMF based on energy compaction criterion and zero-mean high-pass filter assumption. Assume an AR(1) source with $\rho = 0.95$. Check your result with Table 4.7.

(4.18) Plot the phase and magnitude response of the 8-tap multiplier-free PR-QMF given in Section 4.8.4. Calculate its G_{TC}, for two-band subband tree, and E_s, Eq. (4.52), performance for an AR(1), $\rho = 0.95$ source. Compare them with the performance of 8-tap Binomial QMF.

(4.19) Calculate the band variances σ_L^2 and σ_H^2 of the ideal 2-band PR-QMF bank for an AR(1), $\rho = 0.95$, source, Eq. (4.70). Calculate G_{TC} for this case.

(4.20) Interpret the relations of different performance measures displayed in Fig. 4.7. Show the relation of time vs frequency domain localizations.

(4.21) Find the energy matrix of 8×8 DCT, Eq. (4.80), for an AR(1), $\rho = 0.95$, source.

(4.22) Consider a three level, regular, hierarchical subband tree structure. Find the time and frequency localizations of subband filters (product filters for levels 2 and 3) in 2-, 4-, and 8-band cases (1, 2, 3 level cases respectively). Employ 4-, 6-, and 8-tap Binomial QMFs as the basic decomposition modules for three separate cases. Compare these with the 2×2, 4×4, and 8×8 DCT decomposition. Interpret these results.

(4.23) Derive Eq. (4.83) from Eq. (4.82) and Fig. 4.12.

(4.24) Consider a linear time-invariant system with zero mean input $\underline{x}[n]$, output $\underline{y}[n]$, each an M-vector, and impulse response matrix $W[n]$. Define

$$R_{yx}[k] = E\{\underline{y}[n]\underline{x}^T[n+k]\} \leftrightarrow S_{yx}(z) = \sum_k R_{yx}[k] z^{-k}.$$

Show that

(a) $R_{yx}[k] = R_{xy}^T[-k]$.

(b) $R_{xy}[k] = \sum_j W[-j] R_{xx}[k-j] = W[-k] * R_{xx}[k]$.

(c)

$$\begin{aligned} R_{yy}[m] &= W[-m] * R_{xx}[m] * W^T[m] \\ &= \sum_k \sum_j W[j] R_{xx}[m-k+j] W^T[k] \\ S_{yy}(z) &= \mathcal{W}(z^{-1}) S_{xx}(z) \mathcal{W}(z). \end{aligned}$$

(d) trace $\{R_{yy}[0]\} = \sum_{i=0}^{M-1} \sigma_i^2, \quad \sigma_i^2 = E\{y_i^2[n]\}$.

Chapter 5

(5.1) (a) Derive the uncertainty principle, Eqs. (5.3) and (5.4). Hint: start with Schwarz' inequality,

$$I^2 = \left| \int t f(t) \frac{df}{dt} dt \right|^2 \le \left(\int t^2 |f(t)|^2 dt \right) \left(\int \left| \frac{df}{dt} \right|^2 dt \right),$$

and integrate by parts to show that $I = -\mathcal{E}/2$, where $\mathcal{E} = \int |f(t)|^2 dt$ is the energy in the signal.
(b) Show that the equality holds if $f(t)$ is Gaussian, i.e., a solution of

$$\frac{df}{dt} = -ktf(t), \qquad k > 0.$$

(5.2) Derive the Parseval energy theorem,

$$\int |f(t)|^2 dt = \frac{1}{2\pi} \int \int F(\Omega, \tau) d\Omega d\tau,$$

where $F(\Omega, \tau)$ is the windowed Fourier transform, Eq. (5.2).

(5.3) Show that the signal $f(t)$ can be reconstructed from the windowed FT via

$$f(t) = \frac{1}{2\pi} \int \int F(\Omega, \tau) g(\tau - t) e^{j\Omega t} d\Omega d\tau.$$

(5.4) Consider the discrete windowed Fourier transform of Eqs. (5.9) and (5.11)
(a) Is a Parseval relationship of the form

$$\int |f(t)|^2 dt = k \sum_m \sum_n |F(m, n)|^2$$

valid? Explain, using arguments about linear independence.
(b) Show that when $F(m, n)$ of Eq. (5.9) is substituted into Eq. (5.11) the result is $f(t)$, for the conditions stated.

(5.5) Show that the coefficient vector $\underline{\alpha}_i$ that minimizes the objective function J of Eq. (5.47) is the eigenvector of the matrix E_i^k in Eq. (5.48).

(5.6) Derive Eq. (5.58) from Eq. (5.56) and the relationships for a Fourier transform pair $x(t) \leftrightarrow X(\Omega)$.

(5.7) Prove properties expressed by Eq. (5.59).

(5.8) Derive Eqs. (5.60) and Eq. (5.61).

(5.9) Show that the $P(t, \Omega)$ in Eqs. (5.62) and (5.63) satisfies the marginals.

(5.10) Calculate σ_ω^2 and σ_t^2, frequency and time domain localizations, respectively, of 2×2 DCT and Binomial QMF functions (low- and high-pass)(8-tap). Comment on these properties.

Chapter 6

(6.1) Show that the completeness property of a multiresolution approximation implies that any scaling has a non-zero DC gain, i.e.,

$$\Phi(0) = \int \phi(t)dt \neq 0.$$

(6.2) Show that if $h_0(n)$ is FIR with support on $[0, N-1]$, then the associated scaling function $\phi(t)$ is compactly supported on $[0, (N-1)T_0]$.

(6.3) Start with the 4-tap Binomial QMF-wavelet filter $h_0(n)$ of Section 4.1 and the corresponding frequency response $H_0(e^{j\omega})$. Sketch $H_0(e^{j\omega/2k})$, for $k = 0, 1, 2, 3, 4$, and the partial product $\prod_{k=0}^{4} H_0(e^{j\omega/2k})$. Use a numerical integration routine DSP software to calculate and sketch the resulting approximate scaling function.

(6.4) Provide all the intervening steps in going from Eq. (5.62) to Eq. (5.64).

(6.5) Let $h_0(n)$, $h_1(n)$ be analysis filter, and $g_0(n)$, $g_1(n)$ the synthesis filter in a two-band subband filter bank as in Fig. 3.27.

(a) Trace the signals to the output, and show that

$$\hat{x}(n) = \sum_j \phi(n,j)x(j)$$

$$\phi(n,j) = \sum_k g_0(n-2k)h_0(2k-j) + g_1(n-2k)h_1(2k-j).$$

Show that the condition for perfect reconstruction (causality not imposed) is

$$\phi(n,j) = \delta(n-j)$$

(b) Evaluate $\phi(n,j)$ for

$$g_0(n) = h_0^N(n) = h_0(-n), \quad \textit{and} \quad g_1(n) = h_1^N(n) = h_1(-n),$$

and show that the choice $h_1(n) = (-1)^n h_0(-n+1)$ leads to the orthonomality requirement

$$\sum_k h_0(k)h_0(k+2n) = \delta(n).$$

Show that this last condition can be expressed in the transform domain as

$$H_0(z)H_0(z^{-1}) + H_0(-z)H_0(-z^{-1}) = 1.$$

(c) Evaluate $\phi(n,j)$ for the condition

$$g_0(n) = (-1)^n h_1(n-1)$$

$$g_1(n) = (-1)^{n-1} h_0(n-1)$$

$$\phi(n,j) = \delta(n-j).$$

Show that these lead to the biorthogonality requirement

$$\sum_k h_0(k)\tilde{g}_0(k-2n) = \delta(n), \qquad \tilde{g}_0(n) \triangleq g_0(-n).$$

(6.6) Show that Eq. (5.104) is a solution of Eq. (5.103).

(6.7) Carry out Eq. (5.112) for the simple case of $N = 4$, and obtain explicit equations on $f(\omega)$ to ensure PR.

Index